Handbook of Experimental Pharmacology

Volume 154/I

Editorial Board

G.V.R. Born, London
M. Eichelbaum, Stuttgart
D. Ganten, Berlin
H. Herken, Berlin
F. Hofmann, München
L. Limbird, Nashville, TN
W. Rosenthal, Berlin
G. Rubanyi, Richmond, CA
K. Starke, Freiburg i. Br.

Springer
*Berlin
Heidelberg
New York
Barcelona
Hong Kong
London
Milan
Paris
Tokyo*

Dopamine in the CNS I

Contributors

E. Borrelli, P. Calabresi, A. Carlsson, M.G. Caron, O. Civelli,
J.J. Clifford, J.-A. Girault, P. Greengard, S.N. Haber, B. Hermann,
O. Hornykiewicz, C. Iaccarino, E.J. Nestler, H.B. Niznik,
M. Omori, C. Pifl, J.-C. Schwartz, S.R. Sesack, P. Sokoloff,
K.S. Sugamori, D.J. Surmeier, S. Tan, A. Usiello, J.L. Waddington

Editor:
Gaetano Di Chiara

Springer

Professor
GAETANO DI CHIARA
Department of Toxicology
University of Cagliari
Via Ospedale 72
09125 Cagliari
Italy
e-mail: diptoss@tin.it

With 30 Figures and 8 Tables

ISBN 3-540-42719-8 Springer-Verlag Berlin Heidelberg New York

Library of Congress Cataloging-in-Publication Data

Dopamine in the CNS/editor, Gaetano Di Chiara; contributors, E. Aquas.
    p. cm. – (Handbook of experimental pharmacology; 154)
  Includes bibliographical references and index.
  ISBN 3540427198 (hardcover ; alk. paper) – ISBN 3540427201 (hardcover : alk. paper)
  1. Dopamine. 2. Central nervous system. 3. Dopaminergic mechanisms. 4. Dopaminergic neurons.
I. Di Chiara, Gaetano. II. Aquas, E. III. Series.
  QP905.H3 vol. 154
  [QP364.7]
  615'.1
  [612.8'042 2 21]   2002022665

This work is subject to copyright. All rights are reserved, whether the whole or part of the material is concerned, specifically the rights of translation, reprinting, re-use of illustrations, recitation, broadcasting, reproduction on microfilms or in any other way, and storage in data banks. Duplication of this publication or parts thereof is permitted only under the provisions of the German Copyright Law of September 9, 1965, in its current version, and permission for use must always be obtained from Springer-Verlag. Violations are liable for Prosecution under the German Copyright Law.

Springer-Verlag Berlin Heidelberg New York
a member of BertelsmannSpringer Science+Business Media GmbH

http://www.springer.de

© Springer-Verlag Berlin Heidelberg 2002
Printed in Germany

The use of general descriptive names, registered names, etc. in this publication does not imply, even in the absence of a specific statement, that such names are exempt from the relevant protective laws and regulations and free for general use.

Product liability: The publishers cannot guarantee the accuracy of any information about dosage and application contained in this book. In every individual case the user must check such information by consulting the relevant literature.

Cover design: design & production GmbH, Heidelberg
Typesetting: SNP Best-set Typesetter Ltd., Hong Kong

SPIN: 10705296   27/3020xv – 5 4 3 2 1 0 – printed on acid-free paper

Preface

Dopamine, like Cinderella, has come a long way since its discovery. Initially regarded as a mere precursor of noradrenaline, dopamine has progressively gained its present status of a common target for major drug classes and a substrate for some basic functions and dysfunctions of the central nervous system (CNS). A tangible sign of this status is the fact that dopamine has been the main subject of the studies of the Nobel laureates of 2000, ARVID CARLSSON and PAUL GREENGARD, who also contribute to this book.

The understanding of the function of dopamine was initially marked by the discovery, made in the early 1960s by HORNYKIEWICZ, BIRKMAYER and their associates, that dopamine is lost in the putamen of parkinsonian patients and that the dopamine precursor, L-dopa, reverses their motor impairment. For many years the clinical success of L-dopa therapy was quoted as a unique example of rational therapy directly derived from basic pathophysiology. For the next 10 years, on the wing of this success, dopamine was regarded as the main substrate of basal ganglia functions and was assigned an essentially motor role.

In the early 1970s, studies on the effect of dopamine-receptor antagonists on responding for intracranial self-stimulation and for conventional and drug reinforcers initiated a new era in the understanding of the function of dopamine as related to the acquisition and expression of motivated responding.

This era has merged into the present one, characterized by the notion of dopamine as one of the arousal systems of the brain, modulating the coupling of the biological value of stimuli to patterns of approach behaviour and the acquisition and expression of Pavlovian influences on instrumental responding.

This notion of dopamine has shifted the interest from typically motor areas of the striatum to traditionally limbic ones such as the nucleus accumbens and its afferent areas, the prefrontal cortex, the hippocampal formation and the amygdala. Through these connections, the functional domain of dopamine now extends well into motivational and cognitive functions.

This long development has been marked at each critical step by the contribution of pharmacology: from the association between reserpine akinesia and dopamine depletion and its reversal by L-dopa in the late 1950s, to the blockade of dopamine-sensitive adenylate cyclase by neuroleptics in the early 1970s, to the involvement of the dopamine transporter in the action of cocaine in the 1980s. In no other field of science has pharmacology been as instrumental for the understanding of normal and pathological functions as in the case of dopamine research.

This book intends to provide a rather systematic account of the anatomy, physiology, neurochemistry, molecular biology and behavioural pharmacology of dopamine in the CNS. Nonetheless, the classic extrapyramidal function of dopamine and its role in the action of antiparkinsonian drugs has received relatively little attention here. One reason is that this topic has been the subject of a previous volume of this series. Another reason, however, is that in spite of their systematic layout, even these volumes cannot avoid being a reflection of the times, that is, of the current interests of the research on dopamine.

G. Di Chiara

List of Contributors

BORRELLI, E., Institut de Génétique et de Biologie Moléculaire et Cellulaire, CNRS/INSERM/ULP, BP 163, 67404 Illkirch Cedex, C.U. de Strasbourg, France

CALABRESI, P., Clinica Neurologica – Laboratorio di Neuroscienze, Dipartimento Neuroscienze, Universita' di Tor Vergata, Via di Tor Vergata 135, 00133 Rome, Italy
e-mail: calabre@uniroma2.it

CARLSSON, A., Torild Wulffsgatan. 50, 413 19 Göteborg, Sweden
e-mail: arvid.carlsson@pharm.gu.se

CARON, M.G., Howard Hughes Medical Institute Laboratories, Department of Cellular Biology and Medicine, Duke University Medical Center, Box 3287, Durham, NC 27710, USA
e-mail: m.caron@cellbio.duke.edu

CIVELLI, O., Department of Pharmacology, College of Medicine, University of California, Irvine, 354 Medical Surge II, Irvine, CA 92697-4625, USA
e-mail: ocivelli@uci.edu

CLIFFORD, J.J., Department of Clinical Pharmacology, Royal College of Surgeons in Ireland, St. Stephen's Green, Dublin 2, Ireland

GIRAULT, J.-A., INSERM U 536, Institut du Fer à Moulin, 75005 Paris France
e-mail: girault@ifm.inserm.fr

GREENGARD, P., Laboratory of Molecular and Cellular Neuroscience, The Rockefeller University, 1230 York Avenue, New York, NY 10021, USA
e-mail: greengd@rockvax.rockefeller.edu

HABER, S.N., Department of Neurobiology and Anatomy, University of Rochester School of Medicine, 601 Elmwood Avenue, Rochester, NY 14642, USA
e-mail: suzanne_haber@urmc.rochester.edu

HERMANN, B., Institut de Génétique et de Biologie Moléculaire et Cellulaire, CNRS/INSERM/ULP, BP 163 Illkirch Cedex, C.U. de Strasbourg, France

HORNYKIEWICZ, O., Institute for Brain Research, University of Vienna, Spitalgasse 4, 1090 Vienna, Austria

IACCARINO, C., Institut de Génétique et de Biologie Moléculaire et Cellulaire, CNRS/INSERM/ULP, BP 163 Illkirch Cedex, C.U. de Strasbourg, France

NESTLER, E.J., Department of Psychiatry and Center for Basic Neuroscience, The University of Texas Southwestern Medical Center, 5323 Harry Hines Blvd, Dallas TX 75390-9070
e-mail: eric.nestler@utsouthwestern.edu

NIZNIK, H.B. †, Department of Psychiatry, Laboratory of Molecular Neurobiology, Centre for Addiction and Mental Health, Clark Institute of psychiatry, 250 College Street, Toronto, ON M5T 1R8, Canada

OMORI, M., Institut de Génétique et de Biologie Moléculaire et Cellulaire, CNRS/INSERM/ULP, BP 163 Illkirch Cedex, C.U. de Strasbourg, France

PIFL, C., Institute for Brain Research, University of Vienna, Spitalgasse 4, 1090 Vienna, Austria
christian.pifl@univie.ac.at

SCHWARTZ, J.-C., Unité de Neurobiologie et Pharmacologie de l'INSERM, Centre Paul Broca, 2ter rue d'Alésia, 75014 Paris, France

SESACK, S.R., Department of Neuroscience, University of Pittsburgh, 446 Crawford Hall, Pittsburgh, PA 15260, USA
sesack@bns.pitt.edu

SOKOLOFF, P., Unité de Neurobiologie et Pharmacologie Moléculaire de l'INSERM, Centre Paul Broca, 2ter rue d'Alésia, 75014 Paris, France
sokol@broca.inserm.fr

SUGAMORI, K.S., Laboratory of Molecular Neurobiology, Centre for Addiction and Mental Health, 250 College Street, Toronto, ON M5T 1R8, Canada

SURMEIER, D.J., Department of Physiology, Institute for Neuroscience, Northwestern University Medical School, 320 E. Superior St., 5-447 Searle, Chicago, IL 60611, USA
e-mail: j-surmeier@nwu.edu

TAN, S., Institut de Génétique et de Biologie Moléculaire et Cellulaire, CNRS/INSERM/ULP, BP 163 Illkirch Cedex, C.U. de Strasbourg, France
e-mail: stan@igbmc.u-strasbg.fr; stan@titus.u-strasbg.fr

USIELLO, A., Institut de Génétique et de Biologie Moléculaire et Cellulaire, CNRS/INSERM/ULP, BP 163 Illkirch Cedex, C.U. de Strasbourg, France

WADDINGTON, J.L., Department of Clinical Pharmacology, Royal College of Surgeons in Ireland, St. Stephen's Green, Dublin 2, Ireland
jwadding@rcsi.ie

Contents

CHAPTER 1

Brain Dopamine: A Historical Perspective
O. HORNYKIEWICZ 1

A. Introduction 1
B. The First Half-Century: 1910–1959 2
C. "The Great Brain Serotonin–Catecholamine Debate" 3
D. Striatal Dopamine, Parkinson's Disease and Dopamine
   Replacement 4
E. Dopamine Pathways 5
F. Experimental Models of Parkinson's Disease 6
G. Dopamine Toxins 7
H. Dopamine Uptake 8
I. Physiology and Pathophysiology of Brain Dopamine 9
J. Specific Dopamine Receptors 10
K. Direct-Acting Dopamine Agonists 12
L. Dopamine and Striatal Neurotransmitter Interactions 13
M. Dopamine in Psychiatry 13
N. Concluding Remark 14
Abbreviations 15
References 15

CHAPTER 2

Birth of Dopamine: A Cinderella Saga
A. CARLSSON 23

A. Introduction 23
B. Brodie's Breakthrough Discovery, Focusing on Serotonin 23
C. Catecholamines Entering the Scene 25
D. Discovery of Dopamine 26
E. Facing Rejection by Leaders in the Field 27
F. New Evidence for Monoaminergic Neurotransmission 28
G. A Paradigm Shift: Chemical Transmission in the Brain and
   Emerging Synaptology 29

H. "Awakenings"	30
I. Mode of Action of Antipsychotic Agents	32
J. Dopamine, the Reward System, and Drug Dependence	33
K. Autoreceptors: Discovery and Therapeutic Implications	34
L. Concluding Remarks	37
References	37

CHAPTER 3

The Place of the Dopamine Neurons Within the Organization of the Forebrain
S.N. HABER. With 9 Figures 43

A. Introduction	43
I. Organization of the Midbrain Dopamine Neurons	43
B. The Midbrain and Striatal Circuitry	44
I. Nigrostriatal Pathways	46
II. Striatonigral Pathways	48
III. The Striato-Nigro-Striatal Neuronal Network	49
C. Connections Between the Midbrain and Cortex	52
I. Midbrain Cortical Projections	52
II. Cortical Midbrain Projections	54
D. The Amygdala and Other Forebrain Projections	55
E. Functional Modulation and Integration Through Dopamine Forebrain Pathways	56
References	57

CHAPTER 4

Synaptology of Dopamine Neurons
S.R. SESACK. With 1 Figure 63

A. Introduction	63
B. Methods for Ultrastructural Labeling of DA Axons	64
I. Uptake of DA Analogs or Radiolabeled DA	64
II. Tract-Tracing	65
III. Immunocytochemistry	65
IV. General Morphology and Synaptology	66
C. Regional Observations of DA Synaptology	67
I. Striatum: Dorsal Caudate and Putamen Nuclei	67
1. Ultrastructural Morphology	68
2. Synaptic Incidence, Synapse Types, and General Targets	69
3. Phenotypic Identification of Targets	72
4. Localization of Proteins Involved in DA Transmission	73

II.	Striatum: Nucleus Accumbens	76
	1. Ultrastructural Morphology, Synapses, and General Targets	76
	2. Phenotypic Identification of Targets	77
	3. Localization of Proteins Involved in DA Transmission	78
III.	Globus Pallidus and Basal Forebrain	79
IV.	Lateral Septum and Bed Nucleus of the Stria Terminalis	80
V.	Amygdala	81
VI.	Cortex	82
	1. Ultrastructural Morphology	83
	2. Synaptic Incidence, Synapse Types, and General Targets	84
	3. Phenotypic Identification of Targets	85
	4. Localization of Proteins Involved in DA Transmission	88
VII.	Hypothalamus	89
	1. Tuberoinfundibular DA System	90
	2. Incerto-hypothalamic DA System	90
VIII.	Spinal Cord and Brainstem	91
IX.	Retina	92
	1. DA Neurons and Processes	92
	2. Ultrastructural Morphology, Synapses, and General Targets	92
	3. Localization of Proteins Involved in DA Transmission	94
D. Discussion		96
	I. The Variability of DA Synaptology	96
	II. Parasynaptic DA Transmission	99
	III. Synaptic DA Transmission	101
	IV. Summary and Functional Significance	102
Abbreviations		104
References		105

CHAPTER 5

D_1-Like Dopamine Receptors: Molecular Biology and Pharmacology
H.B. Niznik, K.S. Sugamori, J.J. Clifford, and J.L. Waddington 121

A.	Introduction	121
B.	Molecular Biology of D_1-Like Receptors	122
	I. The D_1/D_{1A} Receptor Gene	122
	II. The D_5/D_{1B} Receptor Gene	122
	III. Primary Structure of D_1-Like Receptors	123
	IV. D_1-Like Receptor Polymorphisms	127

C.	D_1-Like Receptor Pharmacology	129
	I. Selective D_1-Like Agonists	129
	II. Selective D_1-Like Antagonists	130
	III. Selectivity for D_1/D_{1A} Vs D_5/D_{1B} Receptors?	131
D.	Molecular Aspects of Functional Coupling and Signal Transduction for D_1-Like Receptors	132
	I. G-Protein Selectivity	132
	II. Constitutive Activation of D_5/D_{1B} Receptors	133
	III. Desensitization	135
E.	Distribution of D_1-Like Receptors	136
F.	Pharmacology of D_1-Like Receptor-Mediated Function: Behaviour and D_1-Like:D_2-Like Interactions	138
	I. Core Processes at the Level of Behaviour and Their Putative Neuronal Substrates	138
	II. Prototypical D_1-Like Behavioural Phenomena	139
	III. Paradoxes in Relation to the Defining Linkage to Adenylyl Cyclase	140
	IV. D_1-Like Receptors and Anti-parkinsonian Activity	142
	V. Targeted Gene Deletion	143
G.	Are There Additional D_1-Like Receptor Subtypes?	144
	I. D_1-Like Receptors Linked to Phosphoinositide Hydrolysis	144
	II. Expansion of the D_1-Like Receptor Family	145
H.	Conclusions	146
	References	147

CHAPTER 6

Understanding the Function of the Dopamine D_2 Receptor: A Knockout Animal Approach
S. TAN, B. HERMANN, C. IACCARINO, M. OMORI, A. USIELLO, and E. BORRELLI. With 4 Figures 159

A.	Introduction	159
B.	Transcriptional Regulation of the Dopamine D_2 Receptor	160
C.	Dopamine D_2 Receptor Signal Transduction	161
	I. Gi Protein Coupled Pathways and the D_2 Receptor Splice Variants	161
	II. Kinase Pathways Involved in Signaling	163
	III. Receptor Heterodimers	163
D.	D_2 Receptor Function In Vivo	164
	I. Generation of Knockout Mice	164
E.	D_2 Receptor's Role as an Autoreceptor	165
F.	D_2 Receptor Signaling in Physiology	166
	I. Motor Function	166
	II. Drug Abuse	169
G.	Neuronal Protective Pathways via the D_2 Receptor	170

H.	Antiproliferative Role of Dopamine in the Pituitary	171
I.	Genetic Association of the D_2 Receptor with Disease	173
	I. Schizophrenia	173
	II. Alcoholism	174
	III. Parkinson's Disease	174
J.	Distinct Functions of the Dopamine D_2 Receptor Isoforms	175
K.	Conclusion	177
	References	177

CHAPTER 7

The Dopamine D_3 Receptor and Its Implication in Neuropsychiatric Disorders and Their Treatments
P. SOKOLOFF, J.-C. SCHWARTZ. With 6 Figures 185

A.	Introduction	185
B.	Intracellular Signaling of the D_3 Receptor	185
C.	Pre- and Postsynaptic Localizations of the D_3 Receptor in the Brain	187
D.	Coexisting D_1 and D_3 Receptors in Ventral Striatum Mediate Both Synergistic and Opposite Responses	189
E.	D_1/D_3 Receptor Interplay in the Induction and Expression of Behavioral Sensitization: Role of Brain-Derived Neurotrophic Factor	192
F.	D_3 Receptor-Selective Pharmacological Agents	195
G.	The D_3 Receptor and Schizophrenia	198
H.	The D_3 Receptor and Drug Addiction	202
I.	The D_3 Receptor and Depression	205
J.	Conclusions	210
	References	210

CHAPTER 8

Dopamine D_4 Receptors: Molecular Biology and Pharmacology
O. CIVELLI. With 1 Figure 223

A.	The Dopamine D_4 Receptor	223
	I. Discovery	223
	II. Pharmacological Profile	223
	III. Biological Activities	224
	IV. Tissue Localization	224
	V. Physiological Role	225
B.	The Multiple Human D_4 Receptors	225
	I. D_4 Receptors with Variable Third Cytoplasmic Loops	225
	II. D_4 Receptor with Different N-Termini	227
C.	The D_4 Receptor Involvement in Schizophrenia	227
D.	Conclusions	228
	References	229

CHAPTER 9

Signal Transduction by Dopamine D_1 Receptors
J.-A. GIRAULT, and P. GREENGARD. With 2 Figures 235

A. Historical Perspective 235
B. GTP-Binding Proteins Associated with D_1-Family Receptors 236
C. Adenylyl Cyclases and Phosphodiesterases in the Striatum 239
D. cAMP-Dependent Protein Kinase 239
E. cAMP-Activated Phosphorylation Pathways 240
F. The Role of Protein Phosphatase 1 in the Action of
   D_1 Receptors 241
G. DARPP-32 242
H. Other Actions of cAMP 246
I. Concluding Remarks 247
References 247

CHAPTER 10

The Dopamine Transporter: Molecular Biology, Pharmacology and Genetics
C. PIFL, and M.G. CARON 257

A. Introduction 257
B. Molecular Biology 258
   I. Cloning of the Dopamine Transporter 258
   II. Structural Features of the Cloned Dopamine
       Transporter 258
   III. Chimera and Mutagenesis Studies on the Cloned
        Dopamine Transporter 260
   IV. Dopamine Transporter Gene 263
C. Distribution of the Dopamine Transporter 263
   I. Distribution of Dopamine Transporter Binding 264
   II. Distribution of the mRNA 265
   III. Distribution of Dopamine Transporter Immunoreactivity .. 266
D. Pharmacology 268
   I. Uptake by the Dopamine Transporter 268
      1. Ion Dependence of Transport 268
      2. Substrates of the Dopamine Transporter 269
      3. Uptake Blockers 270
   II. Reverse Transport by the Dopamine Transporter 270
   III. The Dopamine Transporter as a Binding Site 271
      1. Ligands for the Dopamine Transporter 271
      2. Imaging Techniques Based on the Dopamine
         Transporter 273
   IV. Electrophysiology of the Dopamine Transporter 273
   V. Regulation of the Dopamine Transporter 274

	VI. Behavioural Studies Related to the Dopamine Transporter	275
	1. Cocaine-Like Substances	275
	2. Amphetamine-Like Substances	276
	3. Dopamine Transporter Knock-out	277
	VII. The Dopamine Transporter as a Gate for Neurotoxins	278
	1. 6-Hydroxydopamine	278
	2. Methamphetamine	279
	3. MPTP	279
	4. Isoquinolines and Carbolines	280
E.	Genetics Related to the Dopamine Transporter	280
	I. Polymorphism	280
	II. Linkage Studies	280
Abbreviations		281
References		282

CHAPTER 11

Cellular Actions of Dopamine
D.J. SURMEIER, and P. CALABRESI. With 3 Figures 299

A.	Introduction	299
B.	DA Receptor Expression in Neostriatal Neurons	299
C.	Dopaminergic Modulation of Intrinsic Properties of Neostriatal Neurons	301
D.	D_1 Receptor Modulation of Synaptic Transmission and Repetitive Activity in Medium Spiny Neurons	304
E.	Modulation of Synaptic Transmission by D_2-Like Receptors	308
F.	Dopaminergic Regulation of Corticostriatal Synaptic Plasticity	309
G.	Summary and Conclusions	312
References		313

CHAPTER 12

Dopamine and Gene Expression
E.J. NESTLER. With 4 Figures 321

A.	Introduction	321
B.	Dopamine Signaling to the Nucleus	321
	I. Overview of Dopamine's Signal Transduction Pathways	321
	II. Regulation of Gene Expression	322
C.	Acute Effects of Dopamine on Gene Expression	324
	I. Regulation of CREB Family Proteins	325
	II. Regulation of Fos Family Proteins	326
	III. Regulation of Jun Family Proteins	329
	IV. Regulation of Egr Family Proteins	330

D. Chronic Effects of Dopamine on Gene Expression 330
   I. Regulation of CREB 331
   II. Regulation of Fos Family Proteins: Induction of ΔFosB 331
References 335

Subject Index 339

Contents of Companion Volume 154/II

Dopamine in the CNS II

CHAPTER 13
Electrophysiological Pharmacology of Mesencephalic Dopaminergic Neurons
M. Diana and J.M. Tepper 1

CHAPTER 14
Presynaptic Regulation of Dopamine Release
J. Glowinski, A. Cheramy, and M.L. Kemel 63

CHAPTER 15
Dopamine – Acetylcholine Interactions
E. Aquas and G. Di Chiara 85

CHAPTER 16
Dopamine – Glutamate Interactions
C. Konrad C. Cepeda, and M.S. Levine 117

CHAPTER 17
Dopamine – Adenosine Interactions
M. Morelli E. Acquas, and E. Ongini 135

CHAPTER 18
Dopamine – GABA Interactions
M.-F. Chesselet 151

CHAPTER 19
Dopamine – Its Role in Behaviour and Cognition in Experimental Animals and Humans
T.W. Robbins and B.J. Everitt 173

CHAPTER 20
Molecular Knockout Approach to the Study of Brain Dopamine Function
G.F. Koob, S.B. Caine, and L.H. Gold 213

CHAPTER 21
Behavioural Pharmacology of Dopamine D_2 und D_3 Receptors: Use of the Knock-out Mice Approach
R. DEPOORTERE, D. BOULAY, G.H. PERRAULT, and D. SANGER 239

CHAPTER 22
Dopamine and Reward
G. DI CHIARA 265

CHAPTER 23
Molecular and Cellular Events Regulating Dopamine Neuron Survival
G.U. CORSINI, R. MAGGIO, and F. VAGLINI 321

CHAPTER 24
Dopamine and Depression
P. WILLNER 387

CHAPTER 25
Dopamine in Schizophrenia
I. WEINER and D. JOEL 417

CHAPTER 26
Atypical Antipsychotics
J.E. LEYSEN 473

CHAPTER 27
Sleep and Wake Cycle
J. BIERBRAUER and L. HILWERLING 491

Subject Index 507

CHAPTER 1
Brain Dopamine: A Historical Perspective

O. Hornykiewicz

A. Introduction

To the student of the history of brain dopamine (DA), the amine offers an excellent example of an endogenous compound that right from the start has presented aspects of both scientific and clinical importance. Although on several points DA shares this characteristic with the other two catecholamines, adrenaline and noradrenaline (NA), what sets DA apart is the tightness of the interdigitation between its basic research and the clinical implications – for brain DA, there has never been a dividing line between the two; each has served as the driving force for the other. To bring out this interconnection has been the primary object of the following "Historical Perspective". The decidedly human relevance of brain DA research also has been the ultimate vindication of the pleasure we take in our work as DA researchers. The writer has tried to convey in this essay some of the excitement of this work.

The reader of this chapter will soon notice that it has been written not by a detached historian, drawing his knowledge from the printed word, but by someone who from the very beginning has been an active party to the events that now, with the passage of many years, have become historically significant. The personal involvement has undoubtedly lent a distinctive colouring to the picture of those historical events, still so freshly remembered. However, the writer has made a special effort to leave intact the period's atmosphere of ideas and to bear witness to those who had been instrumental in translating these ideas into scientific knowledge.

For the most part, the text was written from memory. Therefore, it cannot claim completeness or balance. The limitation on space alone forbade completeness; and in order for the text to be balanced, in depth scrutiny of the literature would have required more time than the writer could allow himself between the pleasures of remembering and the frustrations of forgetting. Thus, for lack of space many important developments had to remain unmentioned; many other events could not, within the given limit on time, be referenced with greater precision. For all these defects, the writer asks his colleagues' indulgence. To the extent that in the last decades the DA literature has taken such dimensions that it would be in fact difficult to include even a fraction of it, the defects of the present article may be excused, if not forgiven.

B. The First Half-Century: 1910–1959

DA is the youngest of the three catecholamines found to occur naturally in the mammalian organism. Although synthesized in the early years of the last century (BARGER and EWINS 1910; MANNICH and JACOBSOHN 1910), DA joined the circle of substances of biological interest proper not before 1951. In that year, GOODALL detected, for the first time, DA in mammalian tissues, specifically sheep heart and adrenal medulla (GOODALL 1951). Also in 1951, DALE (SIR HENRY) coined the name "dopamine" for the, as he felt, confusing term 3-hydroxytyramine, short for the full chemical name β-3,4-dihydroxyphenylethylamine (see footnote in BLASCHKO 1952). Much earlier, in 1910, DALE was also the first to study DA's pharmacological actions, especially on the arterial blood pressure in the cat, declaring the amine to be a weak sympathomimetic substance (BARGER and DALE 1910). In this modest role DA remained for nearly half a century (cf. HORNYKIEWICZ 1986). When GOODALL, in 1951, found DA in mammalian tissues, this discovery was not too surprising. Already earlier, in 1939, BLASCHKO as well as HOLTZ – who, the year before, had discovered the DA-forming enzyme dopa decarboxylase (HOLTZ et al. 1938) – had postulated that DA was the metabolic intermediate, formed from L-dopa, in the biosynthesis of NA and adrenaline (BLASCHKO 1939; HOLTZ 1939). Hence, small amounts of DA and L-dopa could be expected to occur in NA and adrenaline-containing tissues. As for L-dopa, this naturally occurring amino acid precursor of DA had been isolated, in 1913, from *Vicia faba* beans and tested by GUGGENHEIM who found it to be devoid of biological activity (in the rabbit), except for causing nausea and vomiting in a self-experiment (GUGGENHEIM 1913) – unaware that this unpleasant effect was due to the DA formed in his body from the swallowed L-dopa. It is also worth mentioning that FUNK, the inventor of the term "vitamin", in his early search for a "parent substance" of adrenaline, was the first to synthesize the racemic (D,L) form of dopa (FUNK 1911).

Following up his discovery of dopa decarboxylase, HOLTZ soon showed that after administration of L-dopa to animals and humans, DA was excreted in the urine, thereby proving that also in the body was DA the product of L-dopa decarboxylation (HOLTZ and CREDNER 1942). As a matter of historical fact, a year earlier BING had already shown that the in vivo and in vitro artificially perfused ischemic kidney of the cat formed DA from exogenous L-dopa (BING 1941; BING and ZUCKER 1941). HOLTZ, in the course of his studies, also made the intriguing observation that in the rabbit and the guinea-pig, DA had an action on the arterial blood pressure (vasodepression) opposite to the (vasopressor) action of adrenaline (HOLTZ and CREDNER 1942). It is of interest that considerably earlier HASAMA (1930) had observed the corresponding vasodepressor effect of L-dopa in the rabbit, already then wondering about the qualitative difference to the opposite (vasopressor) effect of adrenaline.

The logical idea that should have immediately suggested itself from HOLTZ's observations was that DA might have biological actions of its own, different from those of the other catecholamines. However, HOLTZ gave his discovery a trivial interpretation, suggesting a non-specific vasodepressor action of the DA-derived aldehyde which he assumed to be formed in especially large amounts (through the action of monoamine oxidase) in rabbit and guinea-pig organism. It was not until 15 years later that BLASCHKO, in 1956, for the first time expressed the view that DA may have own physiological actions in the body (BLASCHKO 1957). In the same year, HORNYKIEWICZ, upon BLASCHKO's suggestion, re-examined the vasodepressor action of DA and L-DOPA in the guinea-pig, conclusively showing that it was an intrinsic property of the amine (HORNYKIEWICZ 1958), consistent with BLASCHKO's suggestion of DA having biological activity in its own right.

The study of HORNYKIEWICZ coincided in time with several other crucial observations about DA. First, CARLSSON demonstrated that D,L-dopa (but not D,L-5-hydroxytryptophan) antagonized the reserpine catalepsy in mice and rabbits (CARLSSON et al. 1957); this observation was soon confirmed and enlarged by EVERETT and TOMAN (1959) and BLASCHKO and CHRUSCIEL (1960), and extended to the reserpine "sedation" in humans (DEGKWITZ et al. 1960); second, DA's occurrence in the mammalian brain was discovered (MONTAGU 1957; CARLSSON et al. 1958); third, D,L-dopa was found to restore the brain levels, reduced by reserpine, of DA (and to a much smaller extent of NA) (CARLSSON et al. 1958; EVERETT 1961; EVERETT and WIEGAND 1962). Taken together, these studies laid the groundwork for all future research into the role of DA in normal and abnormal brain function. The final step was taken, again in CARLSSON's laboratory, by BERTLER and ROSENGREN (1959) who found that the bulk of brain DA (in the dog) was concentrated in the corpus striatum, an observation soon confirmed for the human brain (SANO et al. 1959; EHRINGER and HORNYKIEWICZ 1960; BERTLER 1961). Based on their discovery, the Swedish workers suggested that the parkinsonism-like condition produced by reserpine in laboratory animals and humans may be due to lack of DA in the extrapyramidal-motor centres of the corpus striatum (BERTLER and ROSENGREN 1959; CARLSSON 1959).

C. "The Great Brain Serotonin–Catecholamine Debate"

At the time of the brain DA discoveries, a fierce controversy about how best to explain reserpine's central "sedative" actions was raging between two opposing camps. On the one side stood the adherents of the brain serotonin hypothesis of reserpine's central actions; on the other side were the adherents of the brain catecholamine hypothesis. The auspicious arrival, in 1958, of brain DA on the reserpine scene would have been expected to put an end to this dispute. However, the controversy continued unabatedly well into the mid-1960s, forcing the cause of DA into the background. The literature of that time is replete with great debates – which generated much heat but gave little light (see discussion, pp 548–587, in VANE et al. 1960; BRODIE and COSTA 1962;

CARLSSON and LINDQVIST 1962; CARLSSON 1964; general discussion, pp 67–71, in HIMWICH and HIMWICH 1964; CARLSSON 1965). If at all mentioned, DA mostly played the role of a satellite of NA. In the heat of the debates, even the distinction between the specific NA and DA receptors was sometimes forgotten. DA's crucial and so plainly evident role in reserpine parkinsonism and Parkinson's disease (see below) was for some time diminished by claims that the central actions of L-DOPA and the direct (non-natural) NA precursor amino acid 3,4-dihydroxyphenylserine (DOPS) were similar (Carlsson 1964; 1965).

The situation, hardly encouraging for DA, is well illustrated by the fact that at the "Second Symposium on Catecholamines" held in Milan in 1965 (see ACHESON 1966), of a total of 90 contributions, only a single presentation (from CARLSSON's former laboratory), was devoted to brain DA (BERTLER and ROSENGREN 1966). Even in this contribution, the authors – who 7 years earlier had been the first to suggest the connection between striatal DA depletion and (reserpine) parkinsonism – found it safer to say that "relevant to Parkinson's syndrome" was "a decreased catecholamine [!] concentration in the brain". Lamentable as this state of affairs was, three notable exceptions should be mentioned (cf. HORNYKIEWICZ 1992). Two years before the meeting in Milan, at the "Second International Pharmacological Meeting" held in Prague in 1963, a symposium on "Drugs interfering with the extrapyramidal system" included four papers on brain DA (by SOURKES, BERTLER, VAN ROSSUM and HORNYKIEWICZ; cf. TRABUCCHI et al. 1964). Strong experimental support for the cause of brain DA also kept coming from GUY EVERETT, who convincingly defended brain DA against NA, DOPS and serotonin (see EVERETT and WIEGAND 1962; EVERETT 1970). Lastly, GEORGE H. ACHESON, editor of *Pharmacological Reviews* had the right idea at the right time, and, in the summer of 1964, requested "a review of the interesting aspects of dopamine and the brain". The resulting article (HORNYKIEWICZ 1966) finally brought DA to the attention of the scientific community at large, placing the amine permanently on the agenda of modern neuroscience research. However, in the end it was the unprecedented success of the DA replacement therapy with L-dopa that brought the catecholamine–serotonin controversy to an end – simply by making it irrelevant.

D. Striatal Dopamine, Parkinson's Disease and Dopamine Replacement

In retrospect, it appears fortunate that despite the absolute priority claimed by "The Great Brain Serotonin–Catecholamine Debate", research on DA was being carried on, quietly but persistently, by a handful of "DA believers" (cf. HORNYKIEWICZ 1992). Thus, immediately after BERTLER and ROSENGREN had suggested a possible role of striatal DA in reserpine-induced parkinsonism, HORNYKIEWICZ, in Vienna, took up this idea and applied it within a few months to diseases of the human basal ganglia. A study, done in autopsied brains of six patients with Parkinson's disease (PD), two Huntington's disease patients, seventeen controls, and six cases with extrapyramidal symptoms of unknown aetiology, already showed what now is common textbook knowledge: a severe loss of DA in the caudate nucleus and putamen, confined to the patients with PD (EHRINGER and HORNYKIEWICZ 1960). In the same year, Sano (1960) reported on low putamen DA in one case of PD. DA levels in the substantia nigra were also found to be markedly reduced (HORNYKIEWICZ 1963). Independently, SOURKES, MURPHY and BARBEAU in Montreal measured DA in the urine of patients with extrapyramidal disorders and found reduced amounts of the excreted amine in the parkinsonian group (BARBEAU et al. 1961).

In Vienna and Montreal, these studies were immediately followed up independently by the involved researchers who successfully treated PD patients by replacing the missing brain DA with intravenous or oral doses of L-dopa (BIRKMAYER and HORNYKIEWICZ 1961, 1962; BARBEAU et al. 1962; see also GERSTENBRAND et al. 1963). Although for some time there was considerable reluctance, especially among neurologists, to accept L-dopa's anti-Parkinson's effect (cf. BARBEAU 1969; HORNYKIEWICZ 1994), 5 years later COTZIAS definitely established L-dopa by introducing the high-dose oral regimen presently in use (COTZIAS et al. 1967). This was followed, in 1969, by the combination with peripheral (extracerebral) dopa decarboxylase inhibitors (cf. PLETSCHER and DAPRADA 1993). (Interestingly, also Sano tried D,L-DOPA i.v. in (two) PD patients, but dismissed it as of no therapeutic value [SANO 1960; cf. HORNYKIEWICZ 2001].) The high anti-Parkinson's efficacy of L-dopa demonstrated, for the first time, that neurotransmitter replacement was possible even in chronic, progressive, degenerative brain disorders such as PD. Two recent developments have logically grown out of the DA replacement concept, i.e. grafting of DA-producing fetal (nigral) cells into the PD striatum (BJÖRKLUND and STENEVI 1979) and intrastriatal transfer of somatic cells genetically modified (by means of viral vectors) to express DA synthetic enzymes (GAGE et al. 1991).

E. Dopamine Pathways

At the time of DA's discovery in the brain, its place in brain function was unknown. To illustrate the situation, when in 1954 MARTHE VOGT found an uneven regional distribution of brain NA, she suspected, but could not prove, that the amine might be contained in neurons (VOGT 1954). Even 6 years later this uncertainty remained, applying also to DA's (and serotonin's) brain localization. A new approach to the problem of cellular localization soon removed all uncertainty. Studies using the new Falck-Hillarp method for the histochemical visualization of monoamines in brain-tissue slices demonstrated that in the rat, DA was localized to the neuronal cell bodies of the compact zone of the substantia nigra and the nerve terminals in the striatum (ANDÉN et al. 1964a; DAHLSTRÖM and FUXE 1964; FUXE 1965). Thus, in PD the nigral cell loss typical for this disorder (HASSLER 1938), offered, together with the nigral DA loss (HORNYKIEWICZ 1964), a simple explanation for the marked DA loss in the striatum. This conclusion received substantial support from observations in Huntington's disease, where despite the severe loss of the intrinsic striatal neurons, normal nigral cell counts were accompanied by normal levels of striatal DA (EHRINGER and HORNYKIEWICZ 1960; BERNHEIMER and HORNYKIEWICZ 1973).

The great importance of the histofluorescence method for brain research of that time lay in the fact that it extended the knowledge about brain DA (as well as the other monoamines) beyond what could possibly be learnt from other methodologies. Thus, in addition to the nigrostriatal pathway, fluorescence histochemistry soon permitted researchers to define other DA neuron systems in the rat brain. One of these pathways, originating in the midbrain tegmentum, terminated in forebrain limbic areas, notably in nucleus accumbens, olfactory tubercle and areas of the limbic cortex (ANDÉN et al. 1966). These mesolimbic-mesocortical DA pathways were subsequently found to subserve both locomotor, psychomotor/motivational and higher (cognitive, mental) functions (ANDÉN 1972; PIJNENBURG and VAN ROSSUM 1973; BROZOSKI et al. 1979; see also LAVERTY 1974). The knowledge about these extrastriatal forebrain DA systems provided a rational biochemical/ anatomical basis for hypotheses implicating brain DA in psychiatric disorders. A third DA pathway, the tuberoinfundibular neuron system, was found to be involved in the hypothalamic control of hormone secretion of the anterior pituitary (FUXE 1964). Studies of this DA system played an important role in the discovery of ergoline derivatives (e.g. bromocriptine) as direct acting DA receptor agonists.

F. Experimental Models of Parkinson's Disease

For several years, many brain scientists found it difficult to accept the existence of the nigrostriatal DA pathway. Two reasons come to mind for this negative attitude. One argument that was used against a nigrostriatal fibre connection was that after experimental substantia nigra lesions no anterograde fibre degeneration in the striatum could be demonstrated by the available staining techniques. This point gave rise to a heated and protracted controversy between two prominent experimental neuroanatomists: MALCOLM CARPENTER, at Columbia University's Presbyterian Center, who saw no histoanatomical reason for a nigrostriatal fibre connection; and LOUIS POIRIER, at Canada's Laval University, a pioneer of this pathway (see, for example, discussion, pp 190–194, in COSTA et al. 1966). It was only by means of a new silver impregnation technique, introduced by FINK and HEIMER in 1967, that MOORE (1970) furnished irrefutable anatomical evidence for the nigrostriatal fibre connection. This brought the controversy slowly to an end. Another major obstacle for the acceptance of the nigrostriatal (DA) pathway was the rather idiosyncratic attitude towards the substantia nigra of several prominent neuroscientists of that time. To quote only two examples: In the "Index" to DENNY-BROWN's remarkable treatise of 1966 (DENNY-BROWN 1966) on the *Cerebral Control of Movement* (with a chapter on the "Mesencephalic and Pretectal

Organisation") no entry is to be found for "substantia nigra"; and HASSLER, that eminent expert on the functional neuroanatomy of PD, although assiging to substantia nigra an important role, accepted, on some functional considerations, only the caudal and, like Carpenter, tectal and tegmental projections of the nucleus, thinking very little of a rostral nigrostriatal connection.

Of invaluable help in this situation was the development, by POIRIER and SOURKES, of an experimental model of PD in monkeys, by electrolytically lesioning the substantia nigra (cf. SOURKES 2000). In this first valid animal (primate) model of the disease, the essential clinical features of PD, the biochemical changes, and the effectiveness of DA replacement could be reproduced and studied under controlled experimental conditions (POIRIER and SOURKES 1965; SOURKES and POIRIER 1965; GOLDSTEIN et al. 1966). This line of research was later extended by the introduction of the ("rotational") model of PD by unilaterally lesioning the substantia nigra in the rat with locally injected 6-hydroxydopamine (6-OHDA) (UNGERSTEDT 1968; UNGERSTEDT et al. 1973). This model led to much experimental work, especially in the fields of DA receptor and agonist pharmacology and behavioural (psycho)pharmacology (see JONSSON et al. 1975; UNGERSTEDT 1979; ZIGMOND and STRICKER 1989).

G. Dopamine Toxins

6-OHDA was the first experimentally useful, although only locally effective, neurotoxin selective for the DA (and NA) neurons. It is historically of interest that 6-OHDA's discovery goes back to studies, in the late 1950s, on the enzymatic β-hydroxylation of DA to NA. It was then found that a compound, which in more than ten different solvent systems behaved like NA, was formed from DA by auto-oxidation, both in vitro and in vivo (SENOH et al. 1959). The selectivity of 6-OHDA for catecholamine neurons was soon recognized to be due to its active intraneuronal accumulation by way of the specific neuronal (re-)uptake mechanism for DA and NA. The main reason for 6-OHDA's cytotoxicity was later proposed to be its high susceptibility to non-enzymatic oxidation, with simultaneous formation of several very reactive products, including quinones, hydrogen peroxide and several reactive oxygen species (see JONSSON 1980). Although it is unlikely that in the human brain 6-OHDA is formed in large enough quantities to produce selective nigral cell death, this line of research continues to be of considerable interest in connection with the role of auto-oxidative processes and oxyradicals in neurodegeneration as well as the possibility that DA itself might be a potentially cytotoxic agent (see JENNER 1998).

A great advance in this area was made by the discovery of 1-methyl-4-phenyl-1,2,3,6-tetrahydropyridine (MPTP) as a systemically (parenterally) effective, highly selective toxin for nigrostriatal DA neurons. In this discovery, serendipity, as it so often does, played an important part. In a group of drug addicts who after i.v. self-administration of a "synthetic heroin" had developed a permanent PD-like syndrome, MPTP was identified as the toxic contaminant (LANGSTON et al. 1983; BALLARD et al. 1985). Studies on MPTP's mechanisms of action soon revealed the crucial role of (glial) monoamine oxidase B in converting MPTP to the actual DA toxin MPP$^+$; the involvement of the specific plasma membrane DA transporter as the "gate" allowing MPP$^+$ to enter and accumulate specifically in the DA neurons; and the ability of MPP$^+$ to inhibit mitochondrial respiration, i.e. complex I activity, and/or produce oxidative stress (cf. LANGSTON and IRWIN 1986; MARKEY et al. 1986). These possible mechanisms directly triggered numerous studies on the aetiology of PD, focussing especially on environmental and endogenous DA toxins and on genetic defects that might result in defects of detoxifying enzymes or enzymes of the mitochondrial respiratory chain. The recent neuroprotective strategy – especially with neurotrophic factors and antioxidants – aimed at arresting the death of nigral neurons in PD, also owes its origin to MPTP research.

H. Dopamine Uptake

Work on DA uptake, which subsequently proved crucial for the understanding of the 6-OHDA and MPTP experimental parkinsonism models, was a logical extension of the work on the metabolic fate of the synaptically released NA that had led to the discovery of the specific neuronal re-uptake as an important step in terminating the amine's biological actions (cf. IVERSEN 1975).

There appears to be little doubt that the first to have thought of this possibility was the Oxford pharmacologist WILLIAM D.M. PATON. While discussing a paper given by (SIR LINDOR) BROWN at the 1960 Ciba Symposium on "Adrenergic Mechanisms" (see pp 116–124 in VANE et al. 1960), PATON proposed that the presented data could be best explained by assuming that the NA in the nerve endings is "released when the membrane potential is reduced, and sucked back, recovered, returned to store, when the events of excitation are over" (see p 127 in VANE et al. 1960). This perfect definition of the (re-)uptake concept, to which even today nothing substantial needs to be added, was declared – right at the Symposium – as being "heretical" (BROWN) and incompatible with a number of established ideas about synaptic events. Fortunately, at about that time JULIUS AXELROD was busying himself with studies of the metabolic fate of NA (see, e.g. his contribution at the Ciba Symposium, pp 28–39, and numerous discussion remarks in VANE et al. 1960), for which he started using, for the first time, tritium-labelled NA of high specific activity – an approach initiated by SEYMOUR KETY. In the course of these studies, AXELROD, together with HERTTING, proved by experiment what PATON had deduced by sharp logic, namely that the now easily detectable ^{3}H-NA, after being released by adrenergic nerve stimulation, was removed from the surrounding medium by being indeed "sucked back", unmetabolized, into the nerve terminal (HERTTING and AXELROD 1961; HERTTING et al. 1961). For brain DA it was GLOWINSKI who soon afterwards showed that the amine, after intracerebroventricular injection in the rat, was taken up and stored in brain tissue (MILHAUD and GLOWINSKI 1962), specifically in the (striatal) synaptosomal fraction (GLOWINSKI and IVERSEN 1966). Subsequently, the kinetics of the synaptosomal DA uptake and its inhibition by amphetamine was worked out by COYLE and SNYDER (1969a), who also demonstrated that anticholinergic drugs used in treatment of PD were potent (non-competitive) inhibitors of the striatal synaptosomal DA uptake (COYLE and SNYDER 1969b). The recent cloning of the membrane protein responsible for the specific neuronal DA uptake has given an especially potent stimulus for work on the role of the brain DA transporter in normal brain function, addictive drug action and neurodegeneration (MILLER et al. 1999).

I. Physiology and Pathophysiology of Brain Dopamine

The crucial role of DA in the antireserpine and anti-Parkinson's activity of L-dopa quickly attracted the interest of neurophysiologists. The first neurophysiological experiments with DA were performed right after the original DA/L-DOPA studies in PD. McGEER et al. (1961a) and KERKUT and WALKER (1961) tested DA's actions on spontaneously active single neurons of invertebrates, using as models the stretch receptor neuron of the crayfish and molluscan (snail) brain ganglia respectively. In both preparations, DA was found to be highly potent in inhibiting spontaneous neuronal activity and hyperpolarizing the neuronal membrane (KERKUT and WALKER 1962). In mammals, CURTIS and DAVIS (1961) were the first to show that DA, locally applied in vivo to lateral geniculate neurons of the cat, also had predominantly inhibitory effects on either their spontaneous or glutamate-induced discharge rate. Of special importance were observations made on striatal neurons. In the unanaesthetized cat, BLOOM et al. (1965) and McLENNAN and YORK (1967) showed that DA most frequently inhibited caudate neuron activity, be it spontaneous or stimulated by glutamate or acetylcholine. Very important was the demonstration that electrical stimulation of the substantia nigra had an effect on striatal neuron activity (inhibition) identical to the inhibition of the same neurons produced by iontophoretic DA (CONNOR 1970). Since the effects on striatal neuron activity were DA specific, they suggested the existence of DA-sensitive sites (receptors) on the striatal effector cells. Later electrophysiological–behavioural correlations (see EVARTS et al. 1984; DELONG et al. 1983), together with advances related to striatal connectivity and neurotransmitter interactions (GRAYBIEL and RAGSDALE 1979) greatly contributed to our current concepts about the functional organization of the basal ganglia and our understanding of the pathophysiology of PD and other movement disorders (cf. DELONG 1990; PARENT and HAZRATI 1995). The new insights also have influenced the current work on the functional neurosurgery ("the new surgery") for PD, thus reviving the neurosurgical approach originally developed in the early 1950s (cf. NARABAYASHI 1990).

Important for the further development of our ideas about the pathophysiology of the dopaminergically denervated striatum were findings about the functional compensatory mechanisms in the remaining DA neurons. Today, the concept of compensatory overactivity and large adaptive capacity of the nigrostriatal DA system is generally accepted. How did this concept arise? The beginnings of this idea go back to a biochemical study in a patient with unilateral PD (BAROLIN et al. 1964). In this patient, the DA levels in the striatum contralateral to the symptoms were, as expected, severely reduced. However, also the other striatum, corresponding to the side of the body free of symptoms, already had distinctly subnormal DA levels. Could it be that not too severe degrees of DA loss were functionally compensated by the remaining DA neurons? Studies of DA metabolism (by measuring DA's metabolite homovanillic acid) soon gave substance to this possibility. They showed that in the PD striatum the metabolism of DA was indeed greatly increased (BERNHEIMER and HORNYKIEWICZ 1965; BERNHEIMER et al. 1973), thus supporting the possibility that lower degrees of DA loss could indeed be compensated by overactivity of the remaining DA neurons. This observation provided the basis for the concept of the plasticity and high adaptive capacity of the brain DA neurons in the face of partial damage. Subsequent studies measuring various indices of DA neuron activity in the 6-OHDA rotational model of PD fully confirmed the observations made in PD brain (AGID et al. 1973; ZIGMOND et al. 1990), providing an additional stimulus for extensive studies on DA's synthesis, storage, release, metabolism and uptake, as well as a variety of regulatory mechanisms (see GLOWINSKI et al. 1979).

J. Specific Dopamine Receptors

The concept of specific receptors for DA was slow in developing. At that time, the notion of specific neurotransmitter receptors was not, as it is today, a self-evident matter. It is instructive to go a few decades back in the history of this concept. At a meeting in 1943, HENRY DALE expressed his doubts about the adrenergic receptor in the following words: "It is a mere statement of fact to say that the action of adrenaline picks out certain such effector cells and leaves others unaffected; it is a simple deduction that the affected cells have a special affinity of some kind for adrenaline; but I doubt whether the attribution to such cells of 'adrenaline-receptors' does more than re-state the deduction in another form" (DALE 1943). Twelve years later, GOODMAN and GILMAN's pharmacology textbook of 1955 (2nd edn.) still cautions the reader, on pages 412–413, that "it must be realized that the 'receptor' may not be a morphologically demonstrable structure"; also ALQUIST's pioneering classification of adrenergic α and β receptors (ALQUIST 1948) is mentioned (on p 401) as "an interesting classification of these hypothetical [!] receptors". On the other hand, the unexpectedly high efficacy of L-dopa in PD patients (in doses ineffective in normal individuals) could only be understood by assuming that in

PD the striatum developed a denervation supersensitivity to DA. This required the existence of specific DA receptors and their up-regulation after dopaminergic denervation.

For the idea of a specific DA receptor, as for DA itself, the first difficulty was to free itself conceptually from the near-reflex connection, and confusion, with NA. The possibility of DA acting (in the crayfish preparation) on "a type of [inhibitory] receptor for catecholamines" different from the inhibitory receptor for γ-aminobutyric acid (GABA), was first raised in print in 1961 (McGeer et al. 1961a). When Carlsson, in 1963, and Andén (in Carlsson's laboratory), in 1964, showed that neuroleptics (chlorpromazine, haloperidol) increased the metabolism of brain NA and/or DA, respectively, they suggested that this might be due to blockade of brain "monoaminergic" (Carlsson) or "catecholaminergic" (Andén) receptors, with Andén stating that these receptors were "not of the α type" (Carlsson and Lindqvist 1963; Andén et al. 1964b). It seems that the possibility of "specific DA receptors" was for the first time clearly envisaged by Eble (1964) in connection with the DA-induced vasodilation of renal and mesenteric blood vessels in the dog (see Goldberg 1972). In the same year, van Rossum and Hurkmans (1964) already use the term "receptors for DA (as distinct from other catecholamine receptors) in the nuclei of the extrapyramidal system", when discussing their observations on the central actions of psychostimulant drugs (see also van Rossum 1964, 1965). Finally, the DA receptor concept assumed a clearer molecular contour when Ernst, in a series of papers, drew attention to the striking similarity between the striatal effects (compulsive motor behaviour in the rat) of apomorphine and DA, and related these similar effects to the similarity of the molecular structures of the two substances (Ernst 1965, 1967; Ernst and Smelik 1966). Subsequently, the occurrence of specific DA receptors in mammalian, including human, brain was established in a great number of studies, using pharmacological and behavioural models in laboratory animals (e.g. the 6-OHDA rotational rat model); DA's biochemical effects on signal transduction, especially stimulation of adenylyl cyclase (Kebabian et al. 1972); and receptor binding assays, using animal and human brain tissue membrane preparations (Creese et al. 1975; Seeman et al. 1975; for review, see Seeman 1980). By means of the latter method, as well as the more recently developed in vivo brain imaging technique of positron emission tomography, an increased number of DA receptors (i.e. neuroleptic binding sites) was detected in PD striatum (especially putamen; Lee et al. 1978); this, in principle, explained the phenomenon of the high sensitivity of the PD patient to DA substitution.

The finding of a compensatory increase in DA turnover in the (denervated) PD striatum (Bernheimer and Hornykiewicz 1965; Bernheimer et al. 1973) suggested the possibility that the level of DA synthesis and release could be regulated – by way of a feedback mechanism – by the postsynaptic DA receptors located on the striatal effector cells. This possibility was borne out by observations from animal experiments showing that postsynaptic DA receptor blockers (neuroleptics) stimulated brain DA metabolism, and agonists inhibited it (CARLSSON and LINDQVIST 1963; ANDÉN et al. 1967). Later biochemical and electrophysiological studies demonstrated that the presynaptic DA metabolism was under the additional, and very effective, control of DA receptors located directly on DA cell bodies and terminals, the so-called autoreceptors (KEHR et al. 1972; AGHAJANIAN and BUNNEY 1973; cf. also USDIN and BUNNEY 1975).

The understanding of DA receptor function was substantially advanced by work that permitted researchers to distinguish two classes of DA receptor: the D_1 receptor, positively coupled to adenylyl cyclase, and the negatively coupled D_2 receptor (SPANO et al. 1978; KEBABIAN and CALNE 1979), with the latter being localized also presynaptically on DA neurons. Using the recombinant DNA approach, several subtypes of the D_1 and D_2 DA receptor "superfamilies" have been recently cloned and characterized (see SOKOLOFF and SCHWARTZ 1995).

K. Direct-Acting Dopamine Agonists

It was only logical that research on DA receptors should go hand in hand with attempts to develop, for clinical use, DA agonists directly acting on the postsynaptic striatal DA receptor sites. Apomorphine had been tried, in principle successfully, in PD patients as early as 1951 (SCHWAB et al. 1951), without any knowledge about the drug being a direct-acting DA agonist. An even earlier example of an empirical anti-Parkinson's medication involving DA is amphetamine. Found in the late 1930s to be effective in patients with (especially postencephalitic) PD (SOLOMON et al. 1937), amphetamine continued to be recommended for that use until the beginning of the 1970s (cf. GOODMAN and GILMAN's pharmacology textbook, 4th edn., 1970, p 518). Today we know that amphetamine has, in fact, an indirect DA agonistic mode of action via release of synaptic DA and blockade of its re-uptake.

The first compound selected on an experimental basis was bromocriptine. Although originally singled out from among a series of ergot alkaloids and developed for the empirical clinical use as an inhibitor of prolactin secretion (FLÜCKIGER and WAGNER 1968), it soon was shown – in the early 1970s – to act, both in the pituitary and in the striatum, as a direct DA agonist and immediately suggested as a possible anti-Parkinson's drug (FUXE and HÖKFELT 1970; HÖKFELT and FUXE 1972; CORRODI et al. 1973). A year later, the first beneficial results with bromocriptine in patients with PD were reported (CALNE et al. 1974). This was the starting point for the development of many new DA agonists, both ergoline derivatives, like bromocriptine, and non-ergolines (cf. LEVANT et al. 1999). Most of the agonists effective in PD act predominantly at the D_2 receptor sites, but efforts at finding subtype specific compounds with fewer adverse effects are now being actively made. New subtype specific DA agonists might prove especially valuable for a selective control of extrastriatal DA systems, especially in the limbic forebrain and cerebral cortex – brain structures which have been proposed to be involved in affective, motivational and cognitive processes, and which might contain substantial populations of subtype specific D_1 and/or D_2 DA receptor sites (SCHWARTZ et al. 1993).

L. Dopamine and Striatal Neurotransmitter Interactions

Until the introduction of the L-dopa treatment of PD, the only (moderately) effective anti-Parkinson's drugs were the anticholinergics. Immediately after the discovery of DA's pivotal role in PD, the concept of a brain DA-acetylcholine interrelationship was developed (McGEER et al. 1961b; BARBEAU 1962). In PD, the reduction of striatal DA (corrected with L-dopa) was proposed to result in a cholinergic overactivity (responding to anticholinergics). The underlying idea of this concept was that DA tonically inhibited the activity of the cholinergic neurons, which, if released from DA inhibition, aggravated the PD symptoms. Although we now know that the DA-acetylcholine interrelationship is much more complex (cf. DiCHIARA et al. 1994), the original proposal was the starting point for the notion of various neurotransmitter interrelationships as the basis of striatal functioning (cf. LLOYD 1977). This notion was substantially supported by studies in PD brain showing that in addition to the severe DA loss, other brain neurotransmitters were also, although less markedly, affected (cf. HORNYKIEWICZ 1976, 1998). The basal ganglia neurotransmitter interactions concept significantly influenced the current model of the functional anatomy of the neurotransmitter organization of basal ganglia (see WICHMANN and DELONG 1996).

M. Dopamine in Psychiatry

Psychotic reactions to L-dopa were for the first time observed in PD patients after the introduction of the (oral) high-dose regimen of the drug. Before that time, VAN ROSSUM had already hypothesized that L-dopa's behavioural actions (in rats) had certain elements in common with the "psychomotor" effects of psychostimulants (e.g. the amphetamines; VAN ROSSUM 1964). He also discussed, apparently for the first time, the possibility that these effects were due to stimulation of a DA receptor distinct from the NA (or "catecholamine") receptor (VAN ROSSUM and HURKMANS 1964). These ideas were the direct forerunners of the proposals postulating a connection between increased brain DA activity and psychotic, especially schizophrenic, behaviour (cf. HORNYKIEWICZ 1978). Demonstration, by fluorescence histochemistry, of mesolimbic and mesocortical (limbic) DA pathways provided a possible anatomical substrate for this notion. At about the time of van Rossum's studies, CARLSSON and LINDQVIST (1963) showed that chlorpromazine and haloperidol, in addition to their effects on brain NA, also increased brain DA metabolism, suggesting, for the first time, the possibility of blockade of DA (and NA) receptors by neuroleptics. This basic observation was later elaborated in several studies (see VAN ROSSUM 1966; ANDÉN et al. 1970; SEEMAN et al. 1975; CREESE et al. 1976).

In the late 1960s, RANDRUP and MUNKVARD formalized the available behavioural and biochemical evidence on DA and amphetamine and proposed that the amphetamine-induced stereotypy in the rat, mediated by brain DA release and blocked by antipsychotic neuroleptics, was the correlate of abnormal behaviour in schizophrenia (RANDRUP and MUNKVAD 1972; see also SNYDER 1973). Until recently this simple form of the DA hypothesis of schizophrenia was the dominant hypothesis in this field, replacing the earlier serotonin and NA hypotheses. The DA hypothesis proved of tremendous heuristic value, especially in the development of new classes of DA receptor blocking (neuroleptic) drugs with improved therapeutic (antipsychotic) activity and fewer extrapyramidal side-effects. This, on its part, greatly stimulated molecular DA receptor research. Currently, the DA hypothesis is being substantially complemented by new neuroanatomical, pharmacological and neurochemical data, leading to novel concepts and therapeutic possibilities, especially in schizophrenia. In addition to schizophrenia, early clinical observations implicated brain DA also in affective disorders. The main evidence at that time was amphetamine's strong euphoriant potential as well as the neuroleptics' therapeutic effectiveness in mania and their depression-inducing propensity.

VAN ROSSUM's initial studies had already indicated that the central effects of addictive psychostimulants, such as amphetamine and cocaine, were well correlated to an activation of brain DA receptors. Also narcotic drugs, specifically morphine and methadone, were found to affect the metabolism of striatal and limbic DA (CLOUET and RATNER 1970; KUSCHINSKY and HORNYKIEWICZ 1972; SASAME et al. 1972). These and many other studies, including recent studies about the role of DA transporter (GIROS et al. 1996), have suggested a connection between brain DA, especially limbic DA (DICHIARA and IMPERATO 1988; TANDA et al. 1997), and the central actions of drugs of abuse, a notion that today has developed into a specialized field of CNS DA research.

N. Concluding Remark

From the preceding historical account the reader will be right in concluding that the research into brain DA has been an exceptional success story, having no parallel in the history of any of the other brain neurotransmitters. The aspect that most clearly sets DA apart from the rest is the fact that, for brain DA, the greatest possible success came practically the moment it entered the stage of brain science. This gave the whole research field right from the start a forceful impulse that has continued, in undiminished intensity, to this day. As would be expected, in the course of time this success has attracted to the field of DA research some of the best brains in neuroscience.

Although the present essay has, of necessity, been limited in scope, the interested reader may have noticed the unusual number, and the diversity, of fields which DA research has entered over the four decades of its existence.

Today, it would be hard to name a field of neuroscience in which DA does not play a part – from "simple" motor systems to "complex" cognitive and mental processes. This breadth of research is the ultimate reason for the seemingly unending chain of DA's achievements; it also guarantees continuing scientific interest in this most versatile of brain amines whose research momentum shows no signs of slackening. It would appear that we are still far away from the time when it will be possible to write the final chapter on the history of research into brain DA.

Abbreviations

DA	dopamine
dopa	3,4-dihydroxyphenylalanine
DOPS	3,4-dihydroxyphenylserine
GABA	γ-aminobutyric acid
MPP$^+$	1-methyl-4-phenylpyridinium
MPTP	1-methyl-4-phenyl-1,2,3,6-tetrahydropyridine
NA	noradrenaline
6-OHDA	6-hydroxydopamine
PD	Parkinson's disease

References

Acheson GH (ed) (1966) Second symposium on catecholamines, Pharmacol Rev, vol 18. Williams & Wilkins, Baltimore

Aghajanian GK, Bunney BS (1973) Central dopaminergic neurons: neurophysiological identification and responses to drugs. In: Usdin E, Snyder SH (eds) Frontiers in catecholamine research. Pergamon Press, New York Toronto Oxford Sydney Braunschweig, p 643

Agid Y, Javoy F, Glowinski J (1973) Hyperactivity of remaining dopaminergic neurones after partial destruction of the nigro-striatal dopaminergic system in the rat. Nature 245:150–151

Ahlquist RP (1948) A study of the adrenotropic receptors. Am J Physiol 153:586–600

Andén N-E (1972) Dopamine turnover in the corpus striatum and the limbic system after treatment with neuroleptic and anti-acetylcholine drugs. J Pharm Pharmacol 24:905–906

Andén N-E, Carlsson A, Dahlström A, Fuxe K, Hillarp N-A, Larsson K (1964a) Demonstration and mapping out of nigro-striatal dopamine neurons. Life Sci 3:523–530

Andén N-E, Roos B-E, Werdinius B (1964b) Effects of chlorpromazine, haloperidol and reserpine on the levels of phenolic acids in rabbit corpus striatum. Life Sci 3:149–158

Andén N-E, Dahlström A, Fuxe K, Larsson K, Olson L, Ungerstedt U (1966) Ascending monoamine neurons to the telencephalon and diencephalon. Acta Physiol Scand 67:313–326

Andén N-E, Rubenson A, Fuxe K, Hökfelt T (1967) Evidence for dopamine receptor stimulation by apomorphine. J Pharm Pharmacol 19:627–629

Andén N-E, Butcher SG, Corrodi H, Fuxe K, Ungerstedt U (1970) Receptor activity and turnover of dopamine and noradrenaline after neuroleptics. Eur J Pharmacol 11:303–314

Ballard PA, Tetrud JW, Langston JW (1985) Permanent human parkinsonism due to 1-methyl-4-phenyl-1,2,3,6-tetrahydropyridine (MPTP): seven cases. Neurology 35:949–956

Barbeau A (1962) The pathogenesis of Parkinson's disease: a new hypothesis. Can Med Ass J 87:802–807

Barbeau A (1969) L-Dopa therapy in Parkinson's disease: a critical review of nine years' experience. Can Med Ass J 101:791–800

Barbeau A, Murphy GF, Sourkes TL (1961) Excretion of dopamine in diseases of basal ganglia. Science 133:1706–1707

Barbeau A, Sourkes TL, Murphy GF (1962) Les catécholamines dans la maladie de Parkinson. In: de Ajuriaguerra J (ed) Monoamines et système nerveux centrale. Georg, Genève and Masson, Paris, p 247

Barger G, Dale HH (1910) Chemical structure and sympathomimetic action of amines. J Physiol 41:19–59

Barger G, Ewins AJ (1910) Some phenolic derivatives of β-phenylethylamine. J Chem Soc (London) 97:2253–2261

Barolin GS, Bernheimer H, Hornykiewicz O (1964) Seitenverschiedenes Verhalten des Dopamins (3-Hydroxytyramin) im Gehirn eines Falles von Hemi-parkinsonismus. Schweiz Arch Neurol Psychiat 94:241–248

Bernheimer H, Hornykiewicz O (1965) Herabgesetzte Konzentration der Homovanillinsäure im Gehirn von parkinsonkranken Menschen als Ausdruck der Störung des zentralen Dopaminstoffwechsels. Klin Wschr 43:711–715

Bernheimer H, Hornykiewicz O (1973) Brain amines in Huntington's chorea. Adv Neurol 1:525–531

Bernheimer H, Birkmayer W, Hornykiewicz O, Jellinger K, Seitelberger F (1973) Brain dopamine and the syndromes of Parkinson and Huntington. Clinical, morphological and neurochemical correlations. J Neurol Sci 20:415–455

Bertler Å (1961) Occurrence and localization of catecholamines in the human brain. Acta Physiol Scand 51:97–107

Bertler Å, Rosengren E (1959) Occurrence and distribution of dopamine in brain and other tissues. Experientia 15:10–11

Bertler Å, Rosengren E (1966) Possible role of brain dopamine. Pharmacol Rev 18:769–773

Bing RJ (1941) The formation of hydroxytyramine by extracts of renal cortex and by perfused kidneys. Am J Physiol 132:497–503

Bing RJ, Zucker MB (1941) Renal hypertension produced by an amino acid. J Exp Med 74:235–245

Birkmayer W, Hornykiewicz O (1961) Der L-Dioxyphenylalanin (= DOPA)-Effekt bei der Parkinson-Akinese. Wien Klin Wschr 73:787–788

Birkmayer W, Hornykiewicz O (1962) Der L-Dioxyphenylalanin (= DOPA)-Effekt beim Parkinson-Syndrom des Menschen: zur Pathogenese und Behandlung der Parkinson-Akinese. Arch Psychiat Nervenkr 203:560–574

Björklund A, Stenevi U (1979) Reconstruction of the nigrostriatal dopamine pathway by intracerebral nigral transplants. Brain Res 177:555–560

Blaschko H (1939) The specific action of L-dopa decarboxylase. J Physiol 96:50P–51P

Blaschko H (1952) Amine oxidase and amine metabolism. Pharmacol Rev 4:415–458

Blaschko H (1957) Metabolism and storage of biogenic amines. Experientia 13:9–12

Blaschko H, Chrusciel TL (1960) The decarboxylation of amino acids related to tyrosine and their awakening action in reserpine-treated mice. J Physiol 151:272–284

Bloom FE, Costa E, Salmoiraghi GC (1965) Anesthesia and the responsiveness of individual neurons of the caudate nucleus of the cat to acetylcholine, norepinephrine and dopamine administration by microelectrophoresis. J Pharmacol Exp Ther 150:244–252

Brodie BB, Costa E (1962) Some current views on brain monoamines. In: de Ajuriaguerra J (ed) Monoamines et système nerveux central. Georg, Genève and Masson, Paris, p 13

Brozoski TJ, Brown RM, Ptak J, Goldman PS (1979) Dopamine in prefrontal cortex of rhesus monkeys: evidence for a role in cognitive function. In: Usdin E, Kopin IJ, Barchas J (eds) Catecholamines: basic and clinical frontiers, vol 2. Pergamon Press, New York Oxford, p 1681

Calne DB, Teychenne PF, Claveria LE, Eastman R, Greenacre JK, Petrie A (1974) Bromocriptine in parkinsonism. Brit Med J 4:442–444
Carlsson A (1959) The occurrence, distribution and physiological role of catecholamines in the nervous system. Pharmacol Rev 11:490–493
Carlsson A (1964) Functional significance of drug-induced changes in brain monoamine levels. In: Himwich HE, Himwich WA (eds) Progr Brain Res 8: Biogenic amines. Elsevier, Amsterdam, p 9
Carlsson A (1965) Drugs which block the storage of 5-hydroxytryptamine and related amines. In: Eichler O, Farah A (eds) 5-Hydroxytryptamine and related indolealkylamines. Springer, Berlin Heidelberg New York (Handbook of Experimental Pharmacology, vol 19) pp 529–592)
Carlsson A, Lindqvist M, Magnusson T (1957) 3,4-Dihydroxyphenylalanine and 5-hydroxytryptophan as reserpine antagonists. Nature 180:1200
Carlsson A, Lindqvist M, Magnusson T, Waldeck B (1958) On the presence of 3-hydroxytyramine in brain. Science 127:471
Carlsson A, Lindqvist M (1962) DOPA analogues as tools for the study of dopamine and noradrenaline in brain. In: de Ajuriaguerra J (ed) Monoamines et système nerveux central. Georg, Genève and Masson, Paris, p 89
Carlsson A, Lindqvist M (1963) Effect of chlorpromazine or haloperidol on formation of 3-methoxytyramine and normetanephrine in mouse brain. Acta Pharm Tox 20:140–144
Clouet DH, Ratner RL (1970) Catecholamine biosynthesis in brains of rats treated with morphine. Science 168:854–855
Connor JD (1970) Caudate nucleus neurones: correlation of the effects of substantia nigra stimulation with iontophoretic dopamine. J Physiol 208:691–703
Corrodi H, Fuxe K, Hökfelt T, Lidbrink P, Ungerstedt U (1973) Effect of ergot drugs on central catecholamine neurons: evidence for a stimulation of central dopamine neurons. J Pharm Pharmacol 25:409–411
Costa E, Côté LJ, Yahr MD (eds) (1966) Biochemistry and pharmacology of the basal ganglia. Raven Press, Hewlett, New York
Cotzias GC, Van Woert MH, Schiffer IM (1967) Aromatic amino acids and modification of Parkinsonism. New Engl J Med 276:374–379
Coyle JT, Snyder SH (1969a) Catecholamine uptake by synaptosomes in homogenates of rat brain: stereospecificity in different areas. J Pharmacol Exp Ther 170:221–231
Coyle JT, Snyder SH (1969b) Antiparkinsonian drugs: inhibition of dopamine uptake in the corpus striatum as a possible mechanism of action. Science 166:899–901
Creese I, Burt DR, Snyder SH (1975) Dopamine receptor binding: differentiation of agonist and antagonist states with ^{3}H-dopamine and ^{3}H-haloperidol. Life Sci 17:993–1002
Creese I, Burt DR, Snyder SH (1976) Dopamine receptor binding predicts clinical and pharmacological potencies of antischizophrenic drugs. Science 192:481–483
Curtis DR, Davis R (1961) A central action of 5-hydroxytryptamine and noradrenaline. Nature 192:1083–1084
Dahlström A, Fuxe K (1964) Evidence for the existence of monoamine-containing neurons in the central nervous system. I. Demonstration of monoamines in the cell bodies of brain stem neurons. Acta Physiol Scand 62:suppl 232
Dale H (1943) Modes of drug action. General introductory address. Trans Faraday Soc 39:319–322
Degkwitz R, Frowein R, Kulenkampff C, Mohs U (1960) Über die Wirkungen des L-DOPA beim Menschen und deren Beeinflussung durch Reserpin, Chlorpromazin, Iproniazid und Vitamin B6. Klin Wschr 38:120–123
DeLong MR, Georgopoulos AP, Crutcher MD (1983) Cortico-basal ganglia relations and coding of motor performance. Exp Brain Res (Suppl) 7:30–39
DeLong MR (1990) Primate models of movement disorders of basal ganglia origin. Trends Neurosci 13:281–285
Denny-Brown D (1966) The Cerebral Control of Movement (Sherrington lectures for 1963). Liverpool University Press, Liverpool.

DiChiara G, Imperato A (1988) Drugs abused by humans preferentially increase synaptic dopamine concentration in the mesolimbic system of freely moving rats. Proc Natl Acad Sci USA 85:5274–5278

DiChiara G, Morelli M, Consolo S (1994) Modulatory functions of neurotransmitters in the striatum: ACh/dopamine/NMDA interactions. Trends Neurosci 17:228–233

Eble JN (1964) A proposed mechanism for the depressor effect of dopamine in the anesthetized dog. J Pharmacol Exp Ther 145:64–70

Ehringer H, Hornykiewicz O (1960) Verteilung von Noradrenalin und Dopamin (3-Hydroxytyramin) im Gehirn des Menschen und ihr Verhalten bei Erkrankungen des extrapyramidalen Systems. Klin Wschr 38:1236–1239

Ernst AM (1965) Relation between the action of dopamine and apomorphine and their O-methylated derivatives upon the CNS. Psychopharmacologia 7:391–399

Ernst AM (1967) Mode of action of apomorphine and dexamphetamine on gnawing compulsion in rats. Psychopharmacologia 10:316–323

Ernst AM, Smelik PG (1966) Site of action of dopamine and apomorphine on compulsive gnawing behaviour in rats. Experientia 22:837

Evarts EV, Kimura M, Wurtz RH, Hikosaka O (1984) Behavioural correlates of activity in basal ganglia neurons. Trends Neurosci 7:447–453

Everett GM (1961) Some electrophysiological and biochemical correlates of motor activity and aggressive behavior. Neuro-Psychopharmacol 2:479–484

Everett GM (1970) Evidence for dopamine as a central neuromodulator: neurological and behavioral implications. In: Barbeau A, McDowell FH (eds) L-DOPA and Parkinsonism. FA Davis, Philadelphia, p. 364

Everett GM, Toman JEP (1959) Mode of action of Rauwolfia alkaloids and motor activity. Biol Psychiat 2:75–81

Everett GM, Wiegand RG (1962) Central amines and behavioral states: a critique and new data. Proc. 1st Internat Pharmacol Meeting 8:85–92

Flückiger E, Wagner HR (1968) 2-Br-α-Ergokryptin: Beeinflussung von Fertilität und Laktation bei der Ratte. Experientia 24:1130

Funk C (1911) Synthesis of dl-3:4-dihydroxyphenylalanine. J Chem Soc 99:554–557

Fuxe K (1964) Cellular localization of monoamines in the median eminence and the infundibular stem of some mammals. Z Zellforsch 61:710–724

Fuxe K (1965) Evidence for the existence of monoamine neurons in the central nervous system. IV. Distribution of monoamine nerve terminals in the central nervous system. Acta Physiol Scand 64:Suppl 247

Fuxe K, Hökfelt T (1970) Central monoaminergic systems and hypothalamic function. In: Martini L, Motta M, Fraschini F (eds) The hypothalamus. Academic Press, New York, p 123

Gage FH, Kawaja MD, Fisher LJ (1991) Genetically modified cells: applications for intracerebral grafting. Trends Neurosci 14:328–333

Gerstenbrand F, Pateisky K, Prosenz P (1963) Erfahrungen mit L-Dopa in der Therapie des Parkinsonismus. Psychiat Neurol 146:246–261

Giros B, Jaber M, Jones SR, Wightman RM, Caron MG (1996) Hyperlocomotion and indifference to cocaine and amphetamine in mice lacking the dopamine transporter. Nature 379:606–612

Glowinski J, Iversen L (1966) Regional studies of catecholamines in the rat brain – III: subcellular distribution of endogenous and exogenous catecholamines in various brain regions. Biochem Pharmacol 15:977–987

Glowinski J, Cheramy A, Giorguieff MF (1979) In-vivo and in-vitro release of dopamine. In: Horn AS, Korf J, Westerink BHC (eds) The neurobiology of dopamine. Academic Press, London New York San Francisco, p 199

Goldberg LI (1972) Cardiovascular and renal actions of dopamine: potential clinical applications. Pharmacol Rev 24:1–29

Goldstein M, Anagnoste B, Owen WS, Battista AF (1966) The effects of ventromedial tegmental lesions on the biosynthesis of catecholamines in the striatum. Life Sci 5:2171–2176

Goodall McC (1951) Studies of adrenaline and noradrenaline in mammalian hearts and suprarenals. Acta Physiol Scand 24: Suppl 85
Graybiel AM, Ragsdale Jr CW (1979) Fiber connections of the basal ganglia. Progr Brain Res 51:239–283
Guggenheim M (1913) Dioxyphenylalanin, eine neue Aminosäure aus vicia faba. Z Physiol Chem 88:276–284
Hasama B-I (1930) Beiträge zur Erforschung der Bedeutung der chemischen Konfiguration für die pharmakologischen Wirkungen der adrenalinähnlichen Stoffe. Arch Exp Path Pharmakol 153:161–186
Hassler R (1938) Zur Pathologie der Paralysis agitans und des postenzephalitischen Parkinsonismus. J Psychol Neurol 48:387–476
Hertting G, Axelrod J (1961) Fate of tritiated noradrenaline at the sympathetic nerve endings. Nature 192:172–173
Hertting G, Axelrod J, Kopin IJ, Whitby LG (1961) Lack of uptake of catecholamines after chronic denervation of sympathetic nerves. Nature 189:66
Himwich HE, Himwich WA (eds) (1964) Progress Brain Res 8: Biogenic amines. Elsevier, Amsterdam
Hökfelt T, Fuxe K (1972) Effects of prolactin and ergot alkaloids on the tubero-infundibular dopamine (DA) neurons. Neuroendocrinology 9:100–122
Holtz P (1939) Dopadecarboxylase. Naturwissenschaften 27:724–725
Holtz P, Heise R, Lüdtke K (1938) Fermentativer Abbau von l-Dioxyphenylalanin durch die Niere. Arch Exp Path Pharmak 191:87–118
Holtz P, Credner K (1942) Die enzymatische Entstehung von Oxytyramin im Organismus und die physiologische Bedeutung der Dopadecarboxylase. Arch Exp Path Pharmak 200:356–388
Hornykiewicz O (1958) The action of dopamine on the arterial pressure of the guinea pig. Brit J Pharmacol 13:91–94
Hornykiewicz O (1963) Die topische Lokalisation und das Verhalten von Noradrenalin und Dopamin (3-Hydroxytyramin) in der Substantia nigra des normalen und Parkinsonkranken Menschen. Wien Klin Wschr 75:309–312
Hornykiewicz O (1964) Zur Frage des Verlaufs dopaminerger Neurone im Gehirn des Menschen. Wien Klin Wschr 76:834–835
Hornykiewicz O (1966) Dopamine (3-hydroxytyramine) and brain function. Pharmacol Rev 18:925–964
Hornykiewicz O (1976) Neurohumoral interactions and basal ganglia function and dysfunction. In: Yahr MD (ed) The basal ganglia. Raven Press, New York, p 269
Hornykiewicz O (1978) Psychopharmacological implications of dopamine and dopamine antagonists: a critical evaluation of current evidence. Neuroscience 3:773–783
Hornykiewicz O (1986) A quarter century of brain dopamine research. In: Woodruff GN, Poat JA, Roberts PJ (eds) Dopaminergic systems and their regulation. Macmillan, London, p 3
Hornykiewicz O (1992) From dopamine to Parkinson's disease: a personal research record. In: Samson F, Adelman G (eds) The neurosciences: paths of discovery II. Birkhäuser, Boston, p 125
Hornykiewicz O (1994) Levodopa in the 1960s: starting point Vienna. In: Poewe W, Lees AJ (eds) 20 Years of madopar – new avenues. Editiones Roche, Basel, p 11
Hornykiewicz O (1998) Biochemical aspects of Parkinson's disease. Neurology 51: Suppl 2: S2–S9
Hornykiewicz O (2001) How L-DOPA was discovered as a drug for Parkinson's disease 40 years ago. Wien Klin Wschr 113:855–862
Iversen LL (1975) Uptake processes for biogenic amines. In: Iversen LL, Iversen SD, Snyder SH (eds) Handbook of psychopharmacology, vol 3: Biochemistry of biogenic amines. Plenum Press, New York, London, p 381
Jenner P (1998) Oxidative mechanisms in nigral cell death in Parkinson's disease. Movement Disorders 13:24–34
Jonsson G (1980) Chemical neurotoxins as denervation tools in neurobiology. Ann Rev Neurosci 3:169–187

Jonsson G, Malmfors T, Sachs Ch (eds) (1975) Chemical tools in catecholamine research I. 6-Hydroxydopamine as a denervation tool in catecholamine research. North Holland, Amsterdam

Kebabian JW, Petzold GL, Greengard P (1972) Dopamine-sensitive adenylate cyclase in caudate nucleus of rat brain, and its similarity to the "dopamine receptor". Proc Natl Acad Sci 69:2145–2149

Kebabian JW, Calne DB (1979) Multiple receptors for dopamine. Nature 277:93–96

Kehr W, Carlsson A, Lindqvist M, Magnusson T, Attack C (1972) Evidence for a receptor-mediated feedback control of striatal tyrosine hydroxylase activity. J Pharm Pharmacol 24:744–747

Kerkut GA, Walker RJ (1961) The effects of drugs on the neurons of the snail Helix aspersa. Comp Biochem Physiol 3:143–160

Kerkut GA, Walker RJ (1962) The specific chemical sensitivity of Helix nerve cells. Comp Biochem Physiol 7:277–288

Kuschinsky K, Hornykiewicz O (1972) Morphine catalepsy in the rat: relation to striatal dopamine metabolism. Eur J Pharmacol 19:119–122

Langston JW, Ballard PA, Tetrud JW, Irwin I (1983) Chronic parkinsonism in humans due to a product of meperidine-analog synthesis. Science 219:979–980

Langston JW, Irwin I (1986) MPTP: current concepts and controversies. Clin Neuropharmacol 9:485–507

Laverty R (1974) On the roles of dopamine and noradrenaline in animal behaviour. Progr Neurobiol 3:31–70

Lee T, Seeman P, Rajput A, Farley IJ, Hornykiewicz O (1978) Receptor basis for dopaminergic supersensitivity in Parkinson's disease. Nature 273:59–61

Levant B, Ling ZD, Carvey PM (1999) Dopamine D_3 receptors. Relevance for the drug treatment of Parkinson's disease. CNS Drugs 12:391–402

Lloyd KG (1977) Neurotransmitter interactions related to central dopamine neurons. In: Youdim MBH, Lovenberg W, Sharman DF, Lagnado TR (eds) Essays in neurochemistry and neuropharmacology. John Wiley & Sons, Chichester, p 131

Mannich C, Jacobsohn W (1910) Über Oxyphenylalkylamine und Dioxyphenylalkylamine. Ber Deut Chem Ges 43:189–197

Markey SP, Castagnoli Jr N, Trevor AJ, Kopin IJ (eds) (1986) MPTP: a neurotoxin producing a parkinsonian syndrome. Academic Press, Orlando.

McGeer EG, McGeer PL, McLennan H (1961a) The inhibitory action of 3-hydroxytyramine, gamma-aminobutyric acid (GABA) and some other compounds towards the crayfish stretch receptor neuron. J Neurochem 8:36–49

McGeer PL, Boulding JE, Gibson WC, Foulkes RG (1961b) Drug-induced extrapyramidal reactions. JAMA 177:665–670

Milhaud G, Glowinski J (1962) Métabolisme de la dopamine-^{14}C dans le cerveau du Rat. Ètude du mode d'administration. CR Acad Sci (Paris) 255:203–205

Miller GW, Gainetdinov RR, Levey AI, Caron MG (1999) Dopamine transporters and neuronal injury. Trends Pharmacol Sci 20:424–429

McLennan H, York DH (1967) The action of dopamine on neurones of the caudate nucleus. J Physiol 189:393–402

Montagu KA (1957) Catechol compounds in rat tissues and in brains of different animals. Nature 180:244–245

Moore RY (1970) The nigrostriatal pathway: demonstration by anterograde degeneration. In: Barbeau A, McDowell FH (eds) L-DOPA and parkinsonism. FA Davis Company, Philadelphia, p 143

Narabayashi H (1990) Surgical treatment in the levodopa era. In: Stern G (ed) Parkinson's disease. Chapman & Hall, London, p 597

Parent A, Hazrati L-N (1995) Functional anatomy of the basal ganglia I. The cortico-basal ganglia-thalamo-cortical loop. Brain Res Rev 20:91–127

Pijnenburg AJJ, van Rossum JM (1973) Stimulation of locomotor activity following injection of dopamine into the nucleus accumbens. J Pharm Pharmacol 25:1003–1005

Pletscher A, DaPrada M (1993) Pharmacotherapy of Parkinson's disease: research from 1960 to 1991. Acta Neurol Scand 87: Suppl 146:26–31

Poirier LJ, Sourkes TL (1965) Influence of the substantia nigra on the catecholamine content of the striatum. Brain 88:181–192

Randrup A, Munkvad I (1972) Evidence indicating an association between schizophrenia and dopaminergic hyperactivity in the brain. Orthomolec Psychiat 1:2–7

Sano I (1960) Biochemistry of the extrapyramidal system. Shinkei Kennkyu No Shinpo 5:42–48 (First tranlation from the original Japanese in: Parkinsonism Relat Disord (2000) 6:3–6

Sano I, Gamo T, Kakimoto Y, Taniguchi K, Takesada M, Nishinuma K (1959) Distribution of catechol compounds in human brain. Biochim Biophys Acta 32: 586–587

Sasame HA, Perez-Cruet J, DiChiara G, Tagliamonte A, Tagliamonte P, Gessa GL (1972) Evidence that methadone blocks dopamine receptors in the brain. J Neurochem 19:1953–1957

Schwab RS, Amador LV, Lettvin JY (1951) Apomorphine in Parkinson's disease. Trans Amer Neurol Ass 76:251–253

Schwartz J-C, Giros B, Martres M-P, Sokoloff P (1993) Multiple dopamine receptors as molecular targets for antipsychotics. In: Brunello N, Mendlewicz J, Racagni G (eds) New generation of antipsychotic drugs: novel mechanisms of action. Int Acad Biomed Drug Res, vol 4. Karger, Basel, p 1

Seeman P (1980) Brain dopamine receptors. Pharmacol Rev 32:229–313

Seeman P, Chau-Wong M, Tedesco J, Wong K (1975) Brain receptors for antipsychotic drugs and dopamine: direct binding assays. Proc Natl Acad Sci USA 72:4376–4380

Senoh S, Creveling CR, Udenfriend S, Witkop B (1959) Chemical, enzymatic and metabolic studies on the mechanism of oxidation of dopamine. J Am Chem Soc 81: 6236–6240

Snyder SH (1973) Amphetamine psychosis: a model schizophrenia mediated by catecholamines. Am J Psychiat 130:61–67

Sokoloff P, Schwartz J-C (1995) Novel dopamine receptors half a decade later. Trends Pharmacol Sci 16:270–275

Solomon P, Mitchell R, Prinzmetal M (1937) The use of benzedrine sulfate in postencephalitic Parkinson's disease. JAMA 108:1765–1770

Sourkes TL (2000) How dopamine was recognised as a neurotransmitter: a personal view. Parkinsonism Relat Disord 6:63–67

Sourkes TL, Poirier L (1965) Influence of the substantia nigra on the concentration of 5-hydroxytryptamine and dopamine of the striatum. Nature 207:202–203

Spano PF, Govoni S, Trabucchi M (1978) Studies on the pharmacological properties of dopamine receptors in various areas of the central nervous system. Adv Biochem Psychopharmacol 19:155–165

Tanda G, Pontieri FE, DiChiara G (1997) Cannabinoid and heroin activation of mesolimbic dopamine transmission by a common μ_1 opioid receptor mechanism. Science 276:2048–2050

Trabucchi E, Paoletti R, Canal N, Volicer L (eds) (1964) Biochemical and neurophysiological correlation of centrally acting drugs. Pergamon Press, Oxford

Ungerstedt U (1968) 6-Hydroxydopamine induced degeneration of central monoamine neurons. Eur J Pharmacol 5:107–110

Ungerstedt U (1979) Central dopamine mechanisms and unconditioned behaviour. In: Horn AS, Korf J, Westerink BHC (eds) The neurobiology of dopamine. Academic Press, London New York San Francisco, p 577

Ungerstedt U, Avemo A, Avemo E, Ljungberg T, Ranje C (1973) Animal models of parkinsonism. Adv Neurol 3:257–271

Usdin E, Bunney Jr WE (eds) (1975) Pre- and postsynaptic receptors. Marcel Dekker Inc, New York

Vane JR, Wolstenholme GEW, O'Connor M (eds) (1960) Adrenergic mechanisms, Ciba Foundation Symposium. Churchill, London van Rossum JM (1964) Significance of dopamine in psychomotor stimulant action. In: Trabucchi E, Paoletti R, Canal N, Volicer L (eds) Biochemical and neurophysiological correlation of centrally acting drugs. Pergamon Press, Oxford, p 115

van Rossum JM (1965) Different types of sympathomimetic α-receptors. J Pharm Pharmacol 17:202–216

van Rossum JM (1966) The significance of dopamine-receptor blockade for the action of neuroleptic drugs. Excerpta Med Intern Congr Series, no 129:321–329

van Rossum JM, Hurkmans JAThM (1964) Mechanism of action of psychomotor stimulant drugs. Significance of dopamine in locomotor stimulant action. Int J Neuropharmacol 3:227–239

Vogt M (1954) The concentration of sympathin in different parts of the central nervous system under normal condition and after the administration of drugs. J Physiol 123:451–481

Wichmann T, DeLong MR (1996) Functional and pathophysiological models of the basal ganglia. Curr Opin Neurobiol 6:751–758

Zigmond MJ, Stricker EM (1989) Animal models of parkinsonism using selective neurotoxins. Int Rev Neurobiol 31:1–79

Zigmond MJ, Abercrombie ED, Berger TW, Grace AA, Stricker EM (1990) Compensations after lesions of central dopaminergic neurons: some clinical and basic implications. Trends Neurosci 13:290–296

CHAPTER 2
Birth of Dopamine: A Cinderella Saga

A. CARLSSON

A. Introduction

The history of dopamine goes back to the early part of the previous century. This compound was synthesized by WASER and SOMMER (1923). Its physiological significance became evident through the discovery of dopa decarboxylase in mammalian tissue by HOLTZ et al. (1938) and through its identification as a normal urinary constituent (HOLTZ et al. 1942). In 1939 it was proposed to be an intermediate in the biosynthesis of adrenaline (BLASCHKO et al. 1957). However, in certain tissues, including adrenergic nerves, dopamine was found to occur in amounts exceeding those to be expected from a catecholamine precursor (SCHÜMANN 1956; EULER and LISHAJKO 1957). Thus EULER and LISHAJKO (1957) and BLASCHKO (1957) speculated on some additional function of dopamine, besides being a precursor. However, the possible non-precursor function of dopamine in peripheral tissues seemed to be unrelated to neurotransmission, because its occurrence in greater than precursor amounts seemed to be limited to ruminants, and in ruminant tissues it correlated strongly to the occurrence of mast cells (BERTLER et al. 1959). What function dopamine might serve in the mast cells of ruminants remained completely unknown. Speculations about an independent role of dopamine had thus ended up in a blind alley.

A new phase in the history of dopamine, starting in 1958, will be described in the present chapter.

B. Brodie's Breakthrough Discovery, Focusing on Serotonin

Thanks to a series of fortunate events, I had the opportunity to spend a sabbatical half year in the Laboratory of Chemical Pharmacology at the National Institutes of Health, Bethesda, Md., USA, starting in late August, 1955. The head of this laboratory was the renowned Dr. Bernard B. Brodie. He told me that he wanted me to work with him and Dr. Parkhurst A. Shore on the action of reserpine on the storage of serotonin in blood platelets in vitro. I started immediately on this project. Adequate equipment had already been acquired.

We were then able to demonstrate a direct action of reserpine on the storage of serotonin (CARLSSON et al. 1957a).

It can hardly be over-emphasized how lucky I was to get this opportunity to work in Dr. Brodie's laboratory during a very dramatic period, when drug research was undergoing a revolution and psychopharmacology was *in statu nascendi*. This was only 3 years after the discovery of the antipsychotic action of chlorpromazine and 1 or 2 years after the re-discovery of the antipsychotic action of reserpine (reported by Indian psychiatrists three decades earlier). At this stage, Drs. Brodie and Shore introduced me into the most modern methods of biochemical pharmacology as well as into the hottest area of psychopharmacology at that time.

What kind of person was Bernard B. Brodie, the man who has played the most important role in my scientific career? It is difficult to answer this question in a few words. He was obviously richly gifted, with a lot of charisma. His background was in organic chemistry, but he had specialized in drug metabolism, which he had pioneered by developing a multitude of methods for measuring the levels of drugs and their metabolites in tissues and body fluids. At the time of my visit, the prototype of a new instrument had been constructed in Brodie's laboratory by Dr. Robert Bowman in collaboration with Dr. Sidney Udenfriend, i.e., a spectrophotofluorimeter. This instrument was to revolutionize the measurement of drugs as well as endogenous compounds of great physiological interest. It combined a high sensitivity with specificity. For several decades, this instrument played a dominating role in biochemical pharmacology. It has now been surpassed by equipment that is even more powerful.

At the time of my visit and during the following several years, an impressive number of young scientists came to spend some time in Brodie's laboratory. A remarkable number of them then made very successful careers, mainly in pharmacology. It seemed as though Brodie had some magic gift to inspire young scientists, even though he did not show any particular ambition to be a mentor. This remarkable aspect of Brodie's personality has been discussed in a book titled *An Apprentice to Genius* (KANIGEL 1986).

Brodie had a remarkable intuition and intensity. When he sensed that a research area was "hot," he did not hesitate to go into it, even if his knowledge in that area was limited. He liked to call himself a gambler. In his youth, he had tried his luck as a boxer, with some success. When he learned about the antipsychotic actions of chlorpromazine and reserpine and the finding that LSD (lysergic acid diethylamide) seemed to possess affinity for 5HT receptors, he started to do experiments with these drugs in order to find out more about their relation to serotonin. Although some of these experiments were primitive and inconclusive, they culminated in the demonstration of reserpine's dramatic effect on the tissue storage of serotonin (PLETSCHER et al. 1955). This seminal discovery was made only a few months before I joined Brodie's group.

I proposed to Brodie that we should investigate the action of reserpine on some endogenous compounds chemically related to serotonin, such as the catecholamines, but Brodie did not consider it worthwhile. He was convinced that serotonin was the important amine in this context.

C. Catecholamines Entering the Scene

After spending five very fruitful months in Brodie's lab, I returned to my home university in Lund, Sweden, having recently been appointed associate professor of pharmacology. Already before my return, I had contacted Dr. NILS-ÅKE HILLARP and proposed that we collaborate on the action of reserpine on catecholamines. He was working as associate professor of histology at the University of Lund. HILLARP was a highly talented and ingenious scientist with a remarkably broad knowledge in various aspects of neuroscience and endocrinology and had already made a number of seminal discoveries in these areas. HILLARP and I started to work together in early 1956, and this lasted until his untimely death in 1965. In the mid-1950s, HILLARP's interest focused on the adrenal medulla where he had discovered the organelles storing the adrenal medullary hormones and the role of ATP as counter-ions in the storage complex. I had the hunch that reserpine might act on this kind of storage mechanism. With HILLARP, I discovered that reserpine caused depletion of the adrenal medullary hormones (CARLSSON and HILLARP 1956), and soon afterwards, I discovered, together with my students ÅKE BERTLER and EVALD ROSENGREN, that similar depletion took place in other tissues, including brain (BERTLER et al. 1956; CARLSSON et al. 1957b). These findings offered an explanation of the hypotensive action of reserpine, and this was confirmed by experiments where stimulation of sympathetic nerves no longer caused release of the neurotransmitter noradrenaline following reserpine treatment (for review and references, see CARLSSON 1987).

These discoveries made us very excited, but at the same time they placed me in an awkward position in relation to my highly esteemed mentors, Drs. Brodie and Shore. Our results challenged their interpretations in two respects. First, they indicated that the action of reserpine should not necessarily be interpreted as due solely to its effect on serotonin, and second, they argued against the proposal that continuous release of the putative neurotransmitter serotonin onto its receptors is responsible for the action of the drug. Rather, our results suggested that at least the hypotensive action was due to an effect on catecholamines and that this effect was caused by depletion rather than release. Unfortunately, this divergence of opinion was to place my mentors and myself in different "camps" for many years to come and led to a large number of sometimes very intense debates, in writing as well as at various meetings. This was especially unfortunate, because we, despite these divergences, were much more on common ground than a great number of other workers in this field, as will be apparent in the following.

To resolve the issue concerning the mode of action of reserpine, my colleagues and I administered dopa to reserpine-treated rabbits and mice and discovered the central stimulant action of this amino acid, as well as its ability to reverse the akinetic and sedative action of reserpine. Since the serotonin precursor 5-hydroxytryptophan was not capable of reversing the action of reserpine, we suggested that depletion of catecholamines rather than serotonin was responsible for some important behavioral effects of reserpine (CARLSSON et al. 1957c).

D. Discovery of Dopamine

When we analyzed the brains of the animals treated with reserpine and dopa, however, we found them still fully depleted of noradrenaline. Further analysis revealed that the behavioral action of dopa was closely correlated to the accumulation of dopamine in the brain. Moreover, our studies disclosed that dopamine is a normal brain constituent and is released by reserpine, like noradrenaline and serotonin. The data suggested to us that dopamine is not just a precursor to noradrenaline, as was generally believed at that time, but is an endogenous agonist in its own right (CARLSSON et al. 1958). This received further support when BERTLER and ROSENGREN (1959) shortly afterwards discovered the marked difference in regional distribution between dopamine and noradrenaline, the former being largely accumulated in the basal ganglia. We could thus suggest that the parkinsonism induced by reserpine is due to dopamine depletion, which can be restored by L-dopa, and that dopamine is involved in the control of extrapyramidal motor functions. This was further supported by the fact that the motor disturbances in Huntington's chorea can be alleviated by reserpine and similar drugs (CARLSSON 1959).

The discovery of dopamine in the brain (CARLSSON et al. 1958) has to be credited entirely to our research group. However, some authors have challenged this, referring to two papers published in *Nature* the year before (MONTAGU 1957; WEIL-MALHERBE and BONE 1957), even though, to my knowledge, none of these authors have themselves challenged our claims to have discovered dopamine in the brain. It should be clear that they did not present any acceptable evidence for the occurrence of dopamine in the brain, nor proposed any particular function for it. I have commented on these papers in detail in my autobiography (CARLSSON 1998).

Another priority issue deals with the discovery of the regional distribution of dopamine in the brain. Credit for this discovery is sometimes given entirely to BERTLER and ROSENGREN (1959). I think it is fair to mention, however, that BERTLER and ROSENGREN were my graduate students preparing for their theses under my direction. We had agreed that the study of the distribution of dopamine in the brain should be part of their thesis work. Thus, according to publication policies in Sweden at the time, it would be appropriate not to have my name on the first publication on this issue. To maintain the balance, however, the first announcement of dopamine's abundance in the basal ganglia was made by me at the First International Catecholamine Symposium in October 1958 (CARLSSON 1959). Here I also summarized, for the first time, the evidence for a role of dopamine in extrapyramidal functions and disorders and for a therapeutic potential of L-dopa in parkinsonism.

E. Facing Rejection by Leaders in the Field

For the first time, evidence was forthcoming for a role of endogenous agonists, present in brain tissue, in animal behavior. At first serotonin had come into focus, but the subsequent experiments pointed to a role of the catecholamines, and especially dopamine, for the sedative and akinetic actions of reserpine, and the reversal of these actions by L-dopa. We were very excited by these findings and were surprised to meet with considerable resistance to our views by some prominent investigators. The Ciba Symposium on adrenergic mechanisms, held in London in the spring of 1960 (VANE et al. 1960), was an especially strange experience to me. At this meeting, practically all prominent workers and pioneers in the catecholamine field were present. It was rather dominated by the strong group of British pharmacologists, headed by Sir Henry Dale. I was impressed to see the British pharmacologists, as well as many other former Dale associates, behave towards Sir Henry like school children to their teacher, although some of them had indeed reached a mature age. It was also remarkable to find how little disagreement there was among these people, who behaved more or less like a football team. At this meeting, I reported on some of our data indicating a role of the catecholamines in motor functions and alertness. No doubts were expressed about our observations as such. In fact, Drs. Blaschko and Chrusciel presented observations that confirmed our findings on some essential points. Moreover, the anti-reserpine action of dopa had been confirmed in humans (DEGKWITZ et al. 1960).

The discussions recorded in the Symposium volume conveyed a very clear message to me. In his summary of the session on central adrenergic mechanisms, Sir John Gaddum concluded (p 584): "The meeting was in a critical mood, and no one ventured to speculate on the relation between catecholamines and the function of the brain." However, my formal paper at the meeting was entitled "On the biochemistry and possible functions of dopamine and noradrenaline in brain," and a considerable number of remarks that I made during the discussion sessions dealt precisely with this issue. Obviously, in Gaddum's mind I was nobody!

Why did Gaddum and the other British pharmacologists so completely ignore our arguments? At first there was some concern about L-dopa being a "poison." This appeared to be mainly based on the observation by WEIL-MALHERBE, that large doses of L-dopa, given together with a monoamine oxidase inhibitor, could be lethal. This discussion ended by a concluding remark by Sir Henry Dale (p 551) that L-dopa is, in fact, a poison, though he found it remarkable for an amino acid. Then Paton referred to unpublished data by Edith Bülbring, suggesting the presence of catecholamines in glia rather than nerve cells. Responding to a question of Dale, Marthe Vogt concluded (p 551) that there was absolutely no evidence that the catecholamines in the brain act as synaptic transmitters or serve a general hormonal function. The proposal that this may be the case was said to depend on the particular pharmacological agents used. A critical survey of all the available evidence led, according to Marthe Vogt, to the conclusion that any of the theories on a relation between catecholamines or serotonin and behavior is "a construction which some day will be amended" (p 579). She also stated (p 578): "My personal view is that neither of these theories will have a long life."

In order to understand the reluctance of some of the most prominent pioneers in chemical transmission to accept a role of the monoamines in brain function, it may help to recall that at that time brain research was dominated by electrophysiology. A vivid debate had been ongoing between Dale and Eccles about the role of electrical versus chemical transmission in general, and it is possible that Dale, who had a great respect for Eccles, had been impressed by the arguments in favor of electrical transmission, at least concerning the brain. From a classical neurophysiological point of view, it must have seemed hard to accept that a loss of nerve function could be alleviated by administering a chemical such as dopamine (given as its precursor L-dopa). Moreover, pharmacologists who had been unable to detect any significant physiological activity of dopamine, when tested on classical smooth-muscle preparations, were reluctant to accept the idea that dopamine could be an agonist in its own right. With such a perspective, the alternative interpretation of the L-dopa effects as being due to some strange kind of amino acid toxicity would seem less far-fetched.

F. New Evidence for Monoaminergic Neurotransmission

At the meeting in London, HILLARP and I decided to further increase our efforts to convince people about our ideas. I had just been appointed professor and chairman at the Department of Pharmacology, University of Gothenburg. We agreed that HILLARP should join me to work on catecholamines in the new department, provided that he could be set free from his associate professorship in histology in Lund. We applied for the necessary funds at the Swedish Medical Research Council, and our grant was funded. We decided to focus on two problems, (1) to investigate a possible active amine-uptake mechanism by the adrenal medullary granules and its inhibition by reserpine, (2) to try to develop a histochemical fluorescence method to visualize the catecholamines in tissues. Both these projects turned out to be successful. Since detailed accounts of this work have been given elsewhere (CARLSSON et al. 1962a, 1964; ANDÉN et al. 1964b; DAHLSTRÖM and CARLSSON 1986; CARLSSON 1987), they will not be repeated here. I just want to conclude that in my opinion, both discoveries had a considerable impact on the scientific community's acceptance of the concept of chemical transmission in the CNS and on the development of monoaminergic synaptology.

G. A Paradigm Shift: Chemical Transmission in the Brain and Emerging Synaptology

A series of observations made in Sweden by HILLARP, me, and our respective collaborators during the early part of the 1960s, using a combination of histochemical, biochemical and physiological methods and a number of pharmacological tools, provided convincing evidence for a role of biogenic amines as neurotransmitters. This, in turn paved the way for a general acceptance of chemical transmission as an important physiological principle in the brain. That we can speak of a true paradigm shift is evident from a comparison between the proceedings of the meeting in London mentioned above and an international symposium held in Stockholm in February 1965 entitled "Mechanisms of Release of Biogenic Amines" (VON EULER et al. 1966). In his introductory remarks to this symposium, UVNÄS stated that "these amines play an important role as chemical mediators in the peripheral and central system." None of the distinguished participants in this symposium expressed any doubts on this point. This time, there were some controversies regarding the function of various synaptic structures and mechanisms. A few recollections of this debate will be reviewed below.

A major issue dealt with the role of the synaptic vesicles in the transmission mechanism. In the mid-1960s, opinions still differed concerning the subcellular distribution of the monoaminergic transmitters. In the fluorescence microscope the accumulation of monoamines in the so-called varicosities of nerve terminals was obvious. This corresponded to the distribution of synaptic vesicles, as observed in the electron microscope. In fact, HÖKFELT (1968) was able to demonstrate the localization of central as well as peripheral monoamines to synaptic vesicles in the electron microscope. However, there was controversy about the nature and size of the extravesicular (or extragranular) neurotransmitter pool. This is evident from the recorded discussions of the above-mentioned symposium "Mechanisms of Release of Biogenic Amines." For example, Drs. Axelrod and von Euler (p 471) maintained that a considerable part of the transmitter was located outside the granules, mainly in a bound form. This fraction was proposed to be more important than the granular fraction, since it was thought to be more readily available for release. Indeed, the granules were facetiously referred to as "garbage cans." Our group had arrived at a different model of the synapse, based on combined biochemical, histochemical, and pharmacological data (CARLSSON 1966). We were convinced that the granules were essential in transmission, and that the transmitter had to be taken up by them in order to become available for release by the nerve impulse. In favor of this contention was our finding that reserpine's site of action is the amine uptake mechanism of the granules. The failure of adrenergic transmission, as well as the behavioral actions of reserpine, was correlated to the blockade of granular uptake induced by the drug, rather than to the size of the transmitter stores (LUNDBORG 1963). Moreover, extragranular noradrenaline (accumulated in adrenergic nerves by pretreatment with reserpine, followed by an inhibitor of MAO (monoamine oxidase) and systemically administered noradrenaline) was unavailable for release by the nerve impulse, as observed histochemically (MALMFORS 1965). We proposed that under normal conditions, the extragranular fraction of monoaminergic transmitters was very small, owing to the presence of MAO intracellularly, and that the evidence presented to the contrary was an artifact. Subsequent work in numerous laboratories has lent support to these views. Already at the Symposium, Douglas presented evidence suggesting a Ca^{++}-triggered fusion between the granule and the cell membrane, preceding the release. The release is now generally assumed to take place as "exocytosis," even though a fractional rather than complete extrusion of the granule content seems to be the most likely alternative. For a discussion of this issue, see a recent paper by FOLKOW and NILSSON (1997), presented as a tribute to the late Jan Häggendal.

An important issue in the early debate dealt with the site of action of major psychotropic drugs. In their first studies on reserpine, BRODIE and his colleagues had proposed that this agent was capable of releasing serotonin onto receptors, which would suggest the cell membrane to be its site of action. However, our observations, quoted above, demonstrated that reserpine acted on the storage mechanism of the synaptic vesicles. As to the tricyclic antidepressants, BRODIE et al. suggested their site of action to be the synaptic vesicles. In their original studies reported in 1960, AXELROD et al. (see AXELROD 1964) observed that the uptake of circulating catecholamines by adrenergic nerves could be blocked by a variety of drugs, for example, reserpine, chlorpromazine, cocaine, and imipramine. These studies obviously did not distinguish between a number of different pharmacological mechanisms. In our own combined biochemical (CARLSSON et al. 1962b; see also the independent, simultaneous work of KIRSHNER 1962) and histochemical studies (MALMFORS 1965), two different amine-uptake mechanisms could be distinguished, i.e., uptake at the level of the cell membrane, sensitive, e.g., to cocaine and imipramine, and uptake by the storage granules or synaptic vesicles, sensitive, e.g., to reserpine. These two mechanisms have of course to be distinguished because of the different, functional consequences of their inhibition, i.e., enhancement and inhibition, respectively, of monoaminergic neurotransmission.

H. "Awakenings"

Ensuing upon our above-mentioned proposal of a role of dopamine in parkinsonism, some important parallel and apparently independent developments took place in Austria, Canada, and Japan. These will now be briefly commented upon, starting out with Austria.

Later in the same year as the Symposium on Adrenergic Mechanisms, there appeared in Klinische Wochenschrift a paper in German, describing a marked reduction of dopamine in the brains of deceased patients who had suffered from Parkinson's disease and postencephalitic parkinsonism (EHRINGER and HORNYKIEWICZ 1960). This was soon followed by a paper by BIRKMAYER and HORNYKIEWICZ (1961) in which a temporary improvement of akinesia was reported following a single intravenous dose of L-dopa to Parkinson's patients.

As far as I can gather from HORNYKIEWICZ's autobiography (1992), as well as a personal communication from him, the following had happened. I wish to mention this in some detail, because it illustrates how the interaction of different minds can lead to important progress. In 1958, HORNYKIEWICZ was approached by his mentor, Prof. Lindner, or, according to a slightly different version, by his chief, Prof. Brücke, who tried to persuade him to analyze the brain of a Parkinson's patient, which the neurologist WALTER BIRKMAYER wanted analyzed for serotonin. Presumably, BIRKMAYER had been impressed by BRODIE's already-mentioned discovery in 1955 of the depletion of this compound by reserpine, and in contrast to many neurologists at that time, he was aware of its possible implications. Shortly afterwards, in 1959, HORNYKIEWICZ read about our work on dopamine and its role in the Parkinson's syndrome. He then decided to include dopamine and noradrenaline in the study. In fact in the subsequent work, serotonin had to be left out initially because of some technical problems.

HORNYKIEWICZ and his postdoctoral fellow EHRINGER were now facing a challenge because they had no adequate equipment to measure dopamine. But they managed to overcome this problem by using the purification of the brain extracts by ion exchange chromatography that our research group had worked out. The subsequent measurement was performed using the colorimetric method of EULER and HAMBERG (1949). Although this method by itself is highly unspecific, specificity could be obtained by using our purification step together with our finding that dopamine is by far the dominating catecholamine in the basal ganglia, where it occurs in high concentrations. They had to work up several grams of tissue and to concentrate the extracts by evacuation to dryness. Following this heroic procedure, they were richly rewarded because the samples from the parkinsonian brains, in contrast to the controls, turned out to be colorless, as revealed by the naked eye!

The corresponding development of Parkinson's research in Canada is summarized in a paper by BARBEAU et al. (1962), presented at a meeting in Geneva in September the previous year. The main findings of the Canadian workers was a reduction of the urinary excretion of dopamine in Parkinson's patients and an alleviation of the rigidity of such patients following oral treatment with L-dopa.

In Japan some remarkable progress was made, which has not been adequately paid attention to in Western countries (see reviews by NAKAJIMA 1991; FOLEY 2000). In a lecture on 5 August 1959, less than a year after my lecture at the International Catecholamine Symposium mentioned above, the basic concept regarding the role of dopamine in the basal ganglia in Parkinson's disease was presented by SANO (1959). In this lecture, data on the distribution of dopamine in the human brain were presented for the first time. In a lecture in Tokyo on 6 February 1960, SANO reported on reduced amounts of dopamine in the basal ganglia of a Parkinson's patient, analyzed postmortem, and in the same year he published a paper describing alleviation of rigidity in a Parkinson's patient following intravenous administration of DL-dopa (SANO 1960).

Thus, treatment of Parkinson's patients with dopa was initiated simultaneously in three different countries only a few years after the discovery of the anti-reserpine action of this agent and the subsequent formulation of the concept of a role of dopamine in extrapyramidal functions. While this treatment led to results of great scientific interest, it took several years until it could be implemented as routine treatment of Parkinson's patients. The reason was that the treatment regimens used initially were inadequate and led to merely marginal improvement of questionable therapeutic value (HORNYKIEWICZ 1966). It remained for GEORGE COTZIAS (1967) to develop an adequate dose regimen. After that, L-dopa treatment rapidly became the golden standard for the treatment of Parkinson's disease.

When I had seen COTZIAS's impressive film demonstrating the effect of escalating oral doses of L-dopa at a meeting in Canada, I hastened back to Göteborg and initiated studies together with DRS. SVANBORG, STEG, and others, which quickly confirmed COTZIAS's observations (ANDÉN et al. 1970b), as occurred in many other places at the same time. This success story was soon afterwards told to the general public by OLIVER SACKS in *Awakenings* (SACKS 1973), which became a bestseller and was also made into a movie.

I. Mode of Action of Antipsychotic Agents

In the early 1960s, we were puzzled by the fact that the major antipsychotic agents, such as chlorpromazine and haloperidol, have a reserpine-like pharmacological and clinical profile and yet lack the monoamine-depleting properties of the latter drug. We found that chlorpromazine and haloperidol accelerated the formation of the dopamine metabolite 3-methoxytyramine and of the noradrenaline metabolite normetanephrine, while leaving the neurotransmitter levels unchanged. In support of the specificity, neither promethazine, a sedative phenothiazine lacking antipsychotic and neuroleptic properties, nor the adrenergic blocker phenoxybenzamine, caused any change in the turnover of the catecholamines (CARLSSON and LINDQVIST 1963). To us it did not seem farfetched, then, to propose that rather than reducing the availability of monoamines, as does reserpine, the major antipsychotic drugs block the receptors involved in dopamine and noradrenaline neurotransmission. This would explain their reserpine-like pharmacological profile. To account for the enhanced catecholamine turnover, we proposed that neurons can increase their physiological activity in response to receptor blockade. This, I believe, was the first time that a receptor-mediated feedback control of neuronal activity was proposed. These findings and interpretations have been amply confirmed and extended by numerous workers, using a variety of techniques. In the following year, our research group discovered the neuroleptic-induced increase in the concentrations of deaminated dopamine metabolites (ANDÉN et al. 1964a). Later papers by ANDÉN et al. (1970a) from our own laboratory and by NYBÄCK and SEDVALL (1970), emphasized the effect of neuroleptics on dopamine, and the work of AGHAJANIAN and BUNNEY (1974) described the effect of dopaminergic agonists and antagonists on the firing of dopaminergic neurons. Other important, subsequent discoveries were the dopamine-sensitive adenylate cyclase by GREENGARD and his colleagues (KEBABIAN and GREENGARD 1971) and the binding of dopamine to specific cell-membrane sites, from which it could be displaced by neuroleptics (SEEMAN et al. 1976; CREESE et al. 1976).

These observations formed the basis for the "dopamine hypothesis of schizophrenia." It should be noted, however, that the paper by CARLSSON and LINDQVIST did not particularly emphasize dopamine, even though the adrenergic blocker was inactive and the effect of haloperidol was more striking on dopamine than on noradrenaline turnover. In fact, CARLSSON and LINDQVIST did not exclude the possibility that serotonin receptors could also be involved in the antipsychotic action. Even though the subsequent research, referred to above, favored an important role of dopamine-receptor blockade in the antipsychotic action, the data could hardly exclude a contributory role of other monoaminergic receptors. Such a possibility has gained increased interest more recently, thanks to research on clozapine and other atypical antipsychotic agents.

J. Dopamine, the Reward System, and Drug Dependence

That dopamine plays a crucial role in the reward system and in drug dependence is now generally recognized. Our research group became interested in this problem in the mid 1960s, when we and others found that amphetamine releases dopamine and that its stimulating action can be blocked by an inhibitor of catecholamine synthesis, i.e., alpha-methyltyrosine. Somewhat later, GUNNE and his colleagues found that also the stimulant and euphoriant action of amphetamine in humans could be prevented by treatment with alpha-methyltyrosine (JÖNSSON et al. 1971). Our further pursuit of this line of research led to the concept that dopamine is also involved in the psychostimulant and dependence-producing action of some other major drugs of abuse, such as the opiates and ethanol. Thus, in the case of ethanol, we found that its stimulating action in animals is accompanied by an increase in dopamine synthesis (CARLSSON and LINDQVIST 1973) and can be prevented by alpha-methyltyrosine, which can also prevent the stimulating and euphoriant action of ethanol in humans (for review, see ENGEL and CARLSSON 1977).

The fundamental role of dopamine in the reward system has also important implications for the treatment with neuroleptic drugs. Since they are all dopamine-receptor blocking agents, they are likely to impair the reward system, resulting in dysphoria and anhedonia. Such side effects may be at least as serious as the extrapyramidal side effects (EPS). In fact, since the site of action on the reward system is probably in the ventral striatum, which appears to be more sensitive to dopamine-receptor blockade than the dorsal striatum, an impairment of the reward system is likely to show up after lower doses of neuroleptics.

K. Autoreceptors: Discovery and Therapeutic Implications

One area, closely related to the issue of receptor-mediated feedback discussed in the previous section, deals with the autoreceptors. In fact, already in our work published in 1963 we were investigating autoreceptors, even though we did not understand it at that time. The nature of the feedback mechanism that we proposed was obscure to us, apart from its mediation via receptors responding to the neurotransmitter in question. It has sometimes been stated that we proposed a feedback loop, but this is not true. It was not until the early 1970s that we were able to examine the problem further. Meanwhile FARNEBO and HAMERGER (1971) had proposed the existence of presynaptic receptors as one possible explanation for their observation that the release of catecholamines from brain slices following field stimulation could be influenced by receptor agonists and antagonists. After we had developed a method to measure the first, rate-limiting step in the synthesis of catecholamines in vivo, we discovered that the synthesis of dopamine could be inhibited by a dopamine receptor agonist and stimulated by an antagonist even after exclusion of a feedback loop by means of axotomy (KEHR et al. 1972). Thus, we felt convinced that the receptors involved were presynaptic. In order to avoid confusion about the nature of these receptors, which appeared to be located on various parts of the neuron and possessed a special functional significance among the presynaptic receptors, I proposed to call them "autoreceptors" (CARLSSON 1975), and this has later become generally accepted. I also proposed that agents with selective action on autoreceptors may prove useful not only as research tools but also as therapeutic agents. Already at that time, we knew that low doses of a dopaminergic receptor agonist could have a preferential action on autoreceptors and thus cause a paradoxical behavioral inhibition.

Four years later our collaboration with skilful organic chemists led to the discovery of 3-PPP. In our original studies this agent appeared to be highly selective for dopaminergic autoreceptors, but a few years later, when we had the opportunity to study the pure enantiomers of 3-PPP, we discovered that they had different pharmacological profiles. The (+) form turned out to be an agonist rather similar to apomorphine, though with somewhat lower intrinsic activity, whereas the (−) form was found to have agonistic properties, especially on the dopaminergic autoreceptors, though with but moderate intrinsic activity. On postsynaptic dopamine (D_2) receptors the (−) form behaved essentially as an antagonist, although it seemed devoid of cataleptogenic properties (for review, see CLARK et al. 1985).

Subsequent work revealed that the profile of (−)-3-PPP, now also called preclamol, is shared by many other dopaminergic agents, which all appear to have in common the property of being partial receptor agonists. Several such agents, with varying intrinsic activities and specificity, have been or are now being tested in the clinic, mainly as antipsychotic agents. In fact these agents may be said to test two different, though somewhat related hypotheses: (1) that a preferentially or selectively acting dopaminergic autoreceptor agonist may have antipsychotic properties, though with fewer side effects than the classical neuroleptics, and (2) that a partial dopamine receptor agonist may possess a suitable intrinsic activity to avoid extrapyramidal side effects, including tardive dyskinesias, and yet be sufficiently antagonistic on postsynaptic receptors to allow for an antipsychotic action.

In 1986, I received a letter from DR. CAROL TAMMINGA of the Maryland Psychiatric Research Center, in which she enquired about the possibility of trying (−)-3PPP in schizophrenic patients. DR. TAMMINGA had a longstanding interest in the possible usefulness of dopaminergic agonists in the treatment of schizophrenia, starting out from her discovery that apomorphine, given in single doses, can alleviate psychotic symptoms in schizophrenic patients (TAMMINGA et al. 1978). With her letter, a most stimulating and fruitful collaboration began and is still ongoing. I replied that I would be delighted to supply her with the drug and the documents needed for a personal IND (investigational new drug application) and to assist her as much as possible to carry out such a study.

It took some time to obtain the IND from the FDA. In 1989, the first series of patients started to receive single escalating intramuscular doses of (−)-3PPP or of placebo in a double-blind study. The results were encouraging. Psychotic symptoms tended to be reduced, and the drug seemed to be well tolerated (TAMMINGA et al. 1992). Subsequently schizophrenic patients received escalating single oral doses of the drug, aiming to obtain the same plasma levels as in the parenteral study. Again, the results were promising. The next step was to give repeated doses of the drug compared to placebo in a double-blind crossover design. It was found that one weeks' treatment with (−)-3-PPP caused a significant antipsychotic response, but that no therapeutic effect remained after 2 or 3 weeks. Apparently some kind of tolerance had developed (LAHTI et al. 1997). As expected from the preclinical data, no extrapyramidal effects were detectable. In fact, (−)-3PPP is anticataleptic in rats and has been shown to possess mild anti-Parkinson's action in clinical studies (PIRTOSEK et al. 1993). In these studies, the partial dopamine receptor agonism became apparent, in that the drug was able to antagonize dopamine receptor agonists, while at the same time having antiparkinsonian properties.

Further studies are underway to investigate the possibility to develop (−)-3PPP to a therapeutically useful antipsychotic agent. If these efforts are successful, they will no doubt represent a breakthrough by being antipsychotic without manifesting the serious side effects induced by too severe dopamine receptor blockade. These effects are not limited to motor functions but extend to the endocrine system and to severe dysphoria, probably related to interference with the well-established function of dopamine in the reward system.

More recently, the partial dopamine receptor antagonist aripiprazole has been studied fairly extensively in schizophrenic patients and seems to offer considerable promise. It seems to be comparable to haloperidol in terms of efficacy but superior to this agent as regards side effects (KANE et al. 2000).

A group of agents which we call dopaminergic stabilizers, are pure antagonists, again acting on the D_2 family of receptors, and can thus readjust elevated dopamine functions, but in contrast to the currently used antipsychotic agents, they do not cause hypodopaminergia. In fact, they rather antagonize subnormal dopamine function. The reason for this aberrant pharmacological profile seems to be that their action on different subpopulations of dopamine receptors differs from that of the currently used antipsychotic drugs. Thus, whereas they exert a strong action on dopaminergic autoreceptors, they have a weaker effect postsynaptically and seem unable to reach a subpopulation of postsynaptic dopamine receptors (SVENSSON et al. 1986; SONESSON et al. 1994; HANSSON et al. 1995). (In spite of a different mode of action at the molecular level, these agents have a pharmacological profile similar to partial dopamine-receptor agonists; also, the latter agents may be described as "dopaminergic stabilizers.")

In subhuman primates, in which parkinsonism had been induced by 1-methyl-4-phenyl-1,2,3,6-tetrahydropyridine (MPTP), one member of this class of dopaminergic stabilizers, named (−)-OSU6162, given in single doses, could prevent L-dopa-induced dyskinesias without interfering with the therapeutic movement response, and in subsequent trials on parkinsonian patients, the same kind of response was observed (EKESBO et al. 1997; J. Tedroff et al., unpublished data). Subsequent trials on patients with Huntington's disease (TEDROFF et al. 1999) showed a marked reduction of choreatic movements, considerably outlasting the presence of the drug in the blood. These observations support the view that drugs of this class are capable of stabilizing dopaminergic function; that is, they are able to alleviate signs of hyperdopaminergia without inducing any signs of reduced dopaminergic function. If these findings can be extrapolated from neurology to psychiatry, these agents should possess antipsychotic activity without any concomitant signs of hypodopaminergia. Forthcoming trials with such agents in schizophrenia will answer this question. In fact, preliminary studies on a few schizophrenic patients, partly using a double-blind crossover design, have demonstrated an antipsychotic action of (−)-OSU6162 (GEFVERT et al. 2000; L. Lindström, personal communication).

L. Concluding Remarks

Dopamine has rightly been called the Cinderella among the monoaminergic neurotransmitters. After its discovery as a normal brain constituent in 1958, it took a long time for dopamine to be generally recognized as a neurotransmitter, despite the fact that compelling evidence for its role in important brain functions as an agonist in its own right was available within a few years after its discovery. One reason for the slow acceptance was probably dopamine's virtual lack of physiological activity in classical smooth-muscle preparations. I remember the disappointment that the renowned Danish pharmacologist Eric Jacobsen expressed when he, as an external examiner at ÅKE BERTLER's dissertation in 1960, could read that drug-induced behavioral changes correlated better with dopamine than with noradrenaline. No doubt, his reaction was representative for the feelings of pharmacologists in those days. As already mentioned, such data led Sir Henry Dale to conclude that the behavioral action of L-dopa was probably due to an effect of the amino acid itself rather than to dopamine.

To account for this peculiar profile of dopamine, with its almost exclusive role as an agonist in the central nervous system, it seems fruitful to apply an evolutionary perspective. Dopamine is especially dominant in a phylogenetically recent part of the brain, that is the basal ganglia, and particularly so in the most recent part, that is the dorsal striatum. In the ventral striatum, dopamine dominates too, although this part of the striatum contains, in addition, quite significant amounts of noradrenaline and serotonin. It is worth mentioning that in amphibia, the dominating catecholamine is, by far, adrenaline – that is, the most elaborate among the catecholamines. It looks as though the increased need for quick motion in the evolution of vertebrates has necessitated a reduction of the number of steps in the synthesis of the catecholaminergic messenger.

There is emerging evidence that dopamine has actually taken over as a leader in the monoaminergic ensemble of neurotransmitters in the sense that serotonergic and noradrenergic behavioral components cannot be fully expressed in the case of a subnormal level of dopaminergic activity. It is remarkable how the crucial role of dopamine in various aspects of brain function and pathophysiology has become increasingly evident during the several decades that followed its discovery. Thus, there are reasons to suspect that the Cinderella of the present saga has not yet reached her full glory.

References

Aghajanian GK, Bunney BS (1974) Pre- and postsynaptic feedback mechanisms in central dopaminergic neurons. In: Seeman P, Brown GM (eds) Frontiers of Neurology and Neuroscience Research, Toronto: University of Toronto Press 4–11

Andén N-E, Roos B-E, Werdinius B (1964a) Effects of chlorpromazine, haloperidol and reserpine on the levels of phenolic acids in rabbit corpus striatum. Life Sci 3:149–158

Andén N-E, Carlsson A, Dahlström A, Fuxe K, Hillarp N-Å, Larsson K (1964b) Demonstration and mapping out of nigro-neostriatal dopamine neurons. Life Sci 3:523–530

Andén N-E, Butcher SG, Corrodi H, Fuxe K, Ungerstedt U (1970a) Receptor activity and turnover of dopamine and noradrenaline after neuroleptics. Eur J Pharmacol 11:303–314

Andén N-E, Carlsson A, Kerstell J, Magnusson T, Olsson R, Roos B-E, Steen B, Steg G, Svanborg A, Thieme G, Werdinius B (1970b) Oral L-DOPA treatment of Parkinsonism. Acta Med Scand 187:247–255

Axelrod J (1964) The uptake and release of catecholamines and the effect of drugs. Prog Brain Res 8:81–89

Barbeau A, Sourkes TL, Murphy GF (1962) Les catecholamines dans la maladie de Parkinson. In: Monoamines et Système nerveux central. Génève:Georg et Cie S.A. 247–262

Bertler Å, Rosengren E (1959) Occurrence and distribution of dopamine in brain and other tissues. Experientia 15:10

Bertler Å, Carlsson A, Rosengren E (1956) Release by reserpine of catecholamines from rabbits' hearts. *Naturwissenschaften* 22:521 (only)

Bertler Å, Falck B, Hillarp N-Å, Torp A (1959) Dopamine and chromaffin cells. Acta Physiol Scand 47:251–258

Birkmayer W, Hornykiewicz O (1961) Der L-3,4-Dioxyphenylalanin (= L-DOPA)-Effekt bei der Parkinson-Akinese. Wien Klin Wschr 73:787–788

Blascko H. The specific action of L-dopa decarboxylase (1939) J Physiol (Lond) 96:50P–51P

Blaschko H (1957) Metabolism and storage of biogenic amines. Experientia 13:9–12

Carlsson A (1959) The occurrence, distribution and physiological role of catecholamines in the nervous system. Pharmacol Rev 11:490–493

Carlsson A (1966) Physiological and pharmacological release of monoamines in the central nervous system. In: von Euler US, Rosell S, Uvnäs B (eds) Mechanisms of Release of Biogenic Amines. Oxford: Pergamon Press 331–346

Carlsson A (1975) Dopaminergic autoreceptors. In: Almgren O, Carlsson A, Engel J (eds) Chemical Tools in Catecholamine Research, Vol. II. Amsterdam: North-Holland Publishing Company 219–225

Carlsson A (1987) Perspectives on the discovery of central monoaminergic neurotransmission. Ann Rev Neurosci 10:19–40

Carlsson A (1988) The current status of the dopamine hypothesis of schizophrenia. Neuropsychopharmacology 1:179–186

Carlsson A (1998) Autobiography. In: Squire LR (ed) The History of Neuroscience in Autobiography, Volume 2. San Diego, Academic Press 28–66

Carlsson A, Hillarp N-Å (1956) Release of adrenaline from the adrenal medulla of rabbits produced by reserpine. Kgl Fysiogr Sällsk Förhandl 26, no 8

Carlsson A, Lindqvist M (1963) Effect of chlorpromazine or haloperidol on the formation of 3-methoxytyramine and normetanephrine in mouse brain. Acta Pharmacol (Kbh) 20:140–144

Carlsson A, Waldeck B (1958) A fluorimetric method for the determination of dopamine (3-hydroxytyramine). Acta physiol scand 44:293–298

Carlsson A, Shore PA, Brodie BB (1957a) Release of serotonin from blood platelets by reserpine in vitro. J Pharmacol Exp Ther 120:334–339

Carlsson A, Rosengren E, Bertler Å, Nilsson J (1957b) Effect of reserpine on the metabolism of catecholamines. In: Garattini S, Ghetti V (eds) Psychotropic Drugs. Amsterdam: Elsevier 363–372

Carlsson A, Lindqvist M, Magnusson T (1957c) 3,4-Dihydroxyphenylalanine and 5-hydroxytryptophan as reserpine antagonists. Nature (Lond) 180:1200 (only)

Carlsson A, Lindqvist M, Magnusson T, Waldeck B (1958) On the presence of 3-hydroxytyramine in brain. Science 127:471 (only)

Carlsson A, Falck B, Hillarp N-Å (1962a) Cellular localization of brain monoamines. Acta physiol scand 56, Suppl 196:1–27

Carlsson A, Hillarp N-Å, Waldeck B (1962b) A Mg^{++}-ATP dependent storage mechanism in the amine granules of the adrenal medulla. Medicina Experimentalis 6:47–53

Carlsson A, Falck B, Fuxe K, Hillarp N-Å (1964) Cellular localization of monoamines in the spinal cord. Acta Physiol Scand 60:112–119

Clark D, Hjorth S, Carlsson A (1985) Dopamine receptor agonists: Mechanisms underlying autoreceptor selectivity. II. Theoretical Considerations. J Neural Transm 62:171–207

Cotzias GC, Van Woert MH, Schiffer LM (1967) Aromatic amino acids and modification of Parkinsonism. N Engl J Med 276:374–379

Creese I, Burt DR, Snyder SH (1976) Dopamine receptor binding predicts clinical and pharmacological potencies of antischizophrenic drugs. Science 192:481–483

Dahlström A, Carlsson A (1986) Making visible the invisible. (Recollections of the first experiences with the histochemical fluorescence method for visualization of tissue monoamines.) In: Parnham MJ, Bruinvels J (eds) Discoveries in Pharmacology. Vol 3, Amsterdam/New York/Oxford: Elsevier, 97–128

Degkwitz R, Frowein R, Kulenkampff C, Mohs U (1960) Über die Wirkungen des L-dopa beim Menschen und deren Beeinflussung durch Reserpin, Chlorpromazin, Iproniazid und Vitamin B6. Klin Wschr 38:120–123

Ehringer H, Hornykiewicz O (1960) Verteilung von Noradrenalin und Dopamin (3-Hydroxytyramin) im Gehirn des Menschen und ihr Verhalten bei Erkrankungen des extrapyramidalen Systems. Klin Wschr 38:1236–1239

Ekesbo A, Andrén PE, Gunne LM, Tedroff J (1997) (–)-OSU6162 inhibits levodopa-induced dyskinesias in a monkey model of Parkinson's disease. Neuroreport 8:2567–2570

Engel J, Carlsson A (1977) Catecholamines and behavior. Curr Developments in Psychopharmacol 4:1–32

Euler USv, Hamberg U (1949) Colorimetric determination of noradrenaline and adrenaline. Acta Physiol Scand 19:74–84

Euler U S v, Lishajko F (1957) Dopamine in mammalian lung and spleen. Acta Physiol Pharmacol Neerl 6:295–303

Euler USv, Rosell S, Uvnäs B, eds (1966) Mechanisms of Release of Biogenic Amines. Oxford: Pergamon Press 331–346

Farnebo L-O, Hamberger B (1971) Drug-induced changes in the release of 3-H-monoamines from field stimulated rat brain slices. Acta Physiol Scand Suppl 371:35–44

Foley P (2000) The L-DOPA story revisited. Further surprises to be expected? The contribution of Isamo Sano to the investigation of Parkinson's disease. In: Riederer P, Calne DB, Horowski R, Mizuno Y, Olanow CV, Poewe W, Youdim MBH (eds) Advances in Research on Neurodegeneration. Wien: Springer Verlag 8:1–20

Folkow B, Nilsson H (1997) Transmitter release at adrenergic nerve endings: Total exocytosis or fractional release? News Physiol Sci 12:32–36

Gefvert O, Lindström LH, Dahlbäck O, Sonesson C, Waters N, Carlsson A, Tedroff J (2000) (–)-OSU6162 induces a rapid onset of antipsychotic effect after a single dose. A double-blind study. Nordic J Psychiat 54/2:93–94

Hansson LO, Waters N, Holm S, Sonesson C (1995) On the quantitative structure-activity relationships of meta-substituted (S)-phenylpiperidines, a class of preferential dopamine D_2 autoreceptor ligands. Modeling of dopamine synthesis and release in vivo by means of partial least squares regression. J Med Chem 38:3121–3131

Hökfelt T. Thesis (1968) Stockholm: I Häggströms Tryckeri AB

Holtz P, Heise R, Lüdtke K (1938) Fermentativer Abbau von l-Dioxyphenylalanin durch die Niere. Arch Exp Path Pharmakol 191:87

Holtz P, Credner K, Koepp W (1942) Die enzymatische Entstehung von Oxytyramin im Organismus und die physiologische Bedeutung der Dopa decarboxylase. Arch Exp Path Pharmakol 200:356

Hornykiewicz O (1966) Metabolism of brain dopamine in human Parkisonism: Neurochemical and clinical aspects. In: Costa E, Côté LKJ, Yahr MD (eds)

Biochemistry and Pharmacology of the Basal Ganglia. New York: Raven Press 171–186

Hornykiewicz O (1992) From dopamine to Parkinson's disease: A personal research record. In: Samson F, Adelman G (eds) The Neurosciences: Paths of Discovery II. Boston: Birkhäuser 125–148

Jönsson L-E, Änggård E, Gunne L-M (1971) Blockade of intravenous amphetamine euphoria in man. Clin Pharmacol Ther 12:889–896

Kane J, Ingenito G, Ali M (2000) Efficacy of Aripiprazole in psychotic disorders: Comparison with haloperidol and placebo. Poster presented at CINP Meeting

Kanigel R (1986) Apprentice to Genius, The Making of a Scientific Dynasty. New York: Macmillan 1–271

Kebabian JW, Greengard P (1971) Dopamine-sensitive adenylyl cyclase: possible role in synaptic transmission. Science 174:1346–1349

Kehr W, Carlsson A, Lindqvist M, Magnusson T, Atack C (1972) Evidence for a receptor-mediated feedback control of striatal tyrosine hydroxylase activity. J Pharm Pharmacol 24:744–747

Kirshner N (1962) Uptake of catecholamines by a particulate fraction of the adrenal medulla. J Biol Chem 237:2311–2317

Lahti AC, Weiler MA, Corey PK, Lahti RA, Carlsson A, Tamminga CA (1998) Antipsychotic properties of the partial dopamine agonist (−)-3-(3-hydroxyphenyl)-N-n-propylpiperidine (preclamol) in schizophrenia. Biol Psychiat 43:2–11

Lundborg P (1963) Storage function and amine levels of the adrenal medullary granules at various intervals after reserpine treatment. Experientia 19:479

Malmfors T (1965) Studies on adrenergic nerves. Acta Physiol Scand 64, Suppl 248:1–93

Montagu KA (1957) Catechol compounds in rat tissues and in brains of different animals. Nature (Lond) 180:244–245

Nakajima T (1991) Discovery of dopamine deficiency and the possibility of dopa therapy in Parkinsonism. In: Nagatsu T, Narabayashi H, Yoshida M (eds) Parkinson's Disease. From Clinical aspects to Molecular Basis. Wien: Springer Verlag 13–18

Nybäck H, Sedvall G (1970) Further studies on the accumulation and disappearance of catecholamines formed from tyrosine-14-C in mouse brain. Eur J Pharmacol 10:193–205

Pirtosek Z, Merello M, Carlsson A, Stern G (1993) Preclamol and Parkinsonian fluctuations. Clin Neuropharmacol 16:550–554

Pletscher A, Shore PA, Brodie BB (1955) Serotonin release as a possible mechanism of reserpine action. Science 122:374–375

Sacks O (1973) Awakenings. London: Gerald Duckworth 1–408

Sano I (1959) Biochemical studies of aromatic monoamines in the brain. In: Japanese Medicine in 1959. The report on scientific meetings in the 15th General Assembly of the Japan Medical Congress, Vol. V. 607–615

Sano I (2000) Biochemistry of extrapyramidal motor system. Shinkey Kenkyu no Shinpo (Adv Neurol Sci) 1960;5:42–48. English translation in: Parkinsonism and Related Disorders 6:3–6

Schümann HJ (1956) Nachweis von Oxytyramin (Dopamin) in sympathischen Nerven und Ganglien. Arch Exp Path Pharmakol 227:566–573

Seeman P, Lee T, Chau-Wong M, Wong K (1976) Antipsychotic drug doses and neuroleptic/dopamine receptors. Nature 261:717–719

Sonesson C, Lin C-H Hansson, L, Waters N, Svensson K, Carlsson A, Smith MW, Wikström H (1994) Substituted (S)-phenylpiperidines and rigid congeners as preferential dopamine autoreceptor antagonists: Synthesis and structure-activity relationships. J Med Chem 37:2735–2753

Svensson K, Hjorth S, Clark D, Carlsson A, Wikström H, Andersson B, Sanchez D, Johansson AM, Arvidsson L-E, Hacksell U, Nilsson, JLG (1986) (+)-UH 232 and (+)-UH 242: Novel stereoselective DA receptor antagonists with preferential action on autoreceptors. J Neural Transm 65:1–27

Tamminga CA, Schaffer MH, Smith RC, Davis JM (1978) Schizophrenic symptoms improve with apomorphine. Science 200:567–568
Tamminga CA, Cascella NG, Lahti RA, Lindberg M, Carlsson A (1992) Pharmacologic properties of (–)-3PPP (Preclamol) in man. J Neural Transm 88:165–175
Tedroff J, Ekesbo A, Sonesson C, Waters N, Carlsson A (1999) Long-lasting improvement following (–)-OSU6162 in a patient with Huntington's disease. Neurology 53:1605–1606
Vane JR, Wolstenholme GEW, O'Connor M (eds) (1960) Ciba Foundation Symposium on Adrenergic Mechanisms, London: J & A Churchill Ltd 1–632
Waser E, Sommer H (1923) Synthesis of 3,4-dihydroxyphenylethyl amine. Helv Chim Acta 6:54–61
Weil-Malherbe H, Bone AD (1957) Intracellular distribution of catecholamines in the brain. Nature 180:1050–1051

CHAPTER 3
The Place of Dopamine Neurons Within the Organization of the Forebrain

S.N. HABER

A. Introduction

The midbrain dopamine neurons play a unique role in basal ganglia and cortical circuits, modulating a broad range of behaviors from learning and "working memory" to motor control. Dopamine neurons are considered to be key for focusing attention on significant and rewarding stimuli, a requirement for the acquisition of behaviors (SCHULTZ et al. 1997; YAMAGUCHI and KOBAYASHI 1998). This acquisition not only involves limbic, cognitive, and motor pathways, but requires the coordination of information between these pathways. Consistent with its role as a mediator of complex behaviors in response to the environment (LJUNGBERG et al. 1992; SCHULTZ et al. 1995), the dopamine pathways are in a position to provide an interface between the limbic, cognitive, and motor functional domains of the forebrain, through complex forebrain neuronal networks. The differential relationship between dopamine projections to the striatum and cortex further emphasizes the role midbrain dopamine neurons play in the ability to respond appropriately to environmental cues. Subpopulations of dopamine neurons have been associated with different functions such as reward and motivation, cognition and higher cortical processing, and movement and sensorimotor integration. Classically, these functions are related to the mesolimbic, mesocortical system, and striatonigral pathways respectively. This chapter reviews the organization of the midbrain dopamine pathways and how they relate to the integration of limbic, cognitive, and motor functions of the forebrain.

I. Organization of the Midbrain Dopamine Neurons

The midbrain dopamine neurons have been divided in various ways into subgroups based on morphology, position, and pathways. Classically they have been partitioned into the ventral tegmental area (VTA), the substantia nigra, pars compacta (SNc), and the retrorubral cell groups (FUXE et al. 1970; BJORKLUND and LINDVALL 1984; HOKFELT et al. 1984). The VTA consists of several nuclei, including the paranigral nucleus and the parabrachial pigmented nuclei. The prominence of these nuclei varies between different species (HALLIDAY and TORK 1986). The SNc is divided into three groups (OLSZEWSKI and BAXTER 1954; POIRIER et al. 1983; FRANCOIS et al. 1985; HALLIDAY and TORK 1986; HABER et al. 1995b): a dorsal group (the γ group, or pars dorsalis), a densocellular region (the β group), and the cell columns (the α group). The dorsal group is composed of loosely arranged cells, extending dorsolaterally to circumvent the ventral and lateral superior cerebellar peduncle and the red nucleus. These neurons, which form a continuous band with the VTA, are oriented horizontally and do not extend into the ventral parts of the pars compacta or into the pars reticulata. Calbindin (CaBP), a calcium-binding protein, is an important phenotypic marker for both the VTA and the dorsal SNc. The calbindin-positive staining of the adjacent VTA and the dorsal SNc cell groups emphasizes the continuity of these two cell groups (LAVOIE and PARENT 1991; HABER et al. 1995b; McRITCHIE and HALLIDAY 1995). In contrast, the dendritic arbor of the cells in the densocellular region and cell columns are oriented ventrally and can be followed deep into the pars reticulata. The cell columns are finger-like extensions of cell clusters that lie within the pars reticulata and are particularly prominent in primates. The densocellular group and the cell columns are calbindin-negative with high expression levels for the dopamine transporter and for the dopamine 2 receptor (D_2R) mRNAs (HABER et al. 1995b).

Based on these characteristics and specific connections (see below), recently, the midbrain dopamine neurons have been divided into two tiers: a dorsal tier (the dorsal SNc and the contiguous VTA) that is calbindin-positive, has relatively low expression levels for the dopamine transporter and the D_2R mRNAs, and is selectively spared from neurodegeneration; and a ventral tier (the densocellular region and the cell columns) that is calbindin-negative, has relatively high levels of neuromelanin and expression of the dopamine transporter and the D_2R mRNA, and is selectively vulnerable to neurodegeneration (FALLON and MOORE 1978; GERFEN et al. 1985; HIRSCH et al. 1992; PARENT and LAVOIE 1993; HABER et al. 1995b) (Fig. 1). The concept of the dopamine neurons as being divided into two tiers is relatively new. However, it builds on the classical distinctions between cells groups and pathways. It further emphasizes the continuity and similarities between the VTA and the dorsal SNc.

B. The Midbrain and Striatal Circuitry

The striatum projects massively to the midbrain and the main target of the dopaminergic neurons is the striatum. The dorsal and ventral tiers contribute to these pathways creating a striato-nigro-striatal neuronal network. Both the striatonigral and nigrostriatal pathways show a general mediolateral, rostrocaudal topographic organization (SZABO 1967, 1970, 1979, 1980; CARPENTER and PETER 1971; JOHNSON and ROSVOLD 1971; FALLON and MOORE 1978; NAUTA et al. 1978; PARENT et al. 1983, 1984; FRANCOIS et al. 1984; DEUTCH et al. 1986; SMITH and PARENT 1986; HABER et al. 1990; SELEMON and GOLDMAN-RAKIC 1990; HEDREEN and DELONG 1991; LYND-BALTA and HABER 1994a,b; PARENT

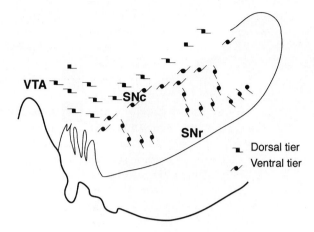

Fig. 1. Schematic of the substantia nigra, illustrating the dorsal and ventral tiers of midbrain dopamine neurons

and HAZRATI 1994). There is also an important inverse dorsoventral topography to this midbrain–forebrain neuronal network. The ventral striatum projects to the dorsal midbrain and the dorsal striatum projects to the ventral midbrain. Likewise, the ventral midbrain projects to the dorsal striatum and dorsal midbrain projects to the ventral striatum. The inverse dorsoventral topography is of particular interest when considered with respect to the functional regions of the striatum. These functional regions are based on the organization of corticostriatal input (GOLDMAN-RAKIC and SELEMON 1986; FLAHERTY and GRAYBIEL 1994; GROENEWEGEN and BERENDSE 1994a; WRIGHT and GROENEWEGEN 1996; HABER and MCFARLAND 1999).

The dorsal and ventral striatum are classically associated with motor and limbic function respectively. Motor and premotor cortex projects to the dorsolateral striatum at rostral levels and to much of the central and caudal putamen (KÜNZLE 1975, 1978). These connections are supported by physiological studies demonstrating dorsolateral striatal involvement in sensorimotor processing and in movement disorders (LILES and UPDYKE 1985; DELONG 1990; KIMURA 1990). Cortical areas are most closely associated with the limbic system and are involved in the development of reward-guided behaviors project to the ventromedial striatum (the nucleus accumbens, and the rostral, ventral caudate nucleus and putamen) (ROLLS et al. 1980; HEIMER et al. 1991; HABER et al. 1995a; GROENEWEGEN et al. 1997; KOOB and NESTLER 1997). The ventromedial striatum contains two subdivisions: the "shell," distinguished by its calbindin-negative staining; and the "core," which is histochemically indistinguishable from the rest of the striatum (ZABORSZKY et al. 1985; ZAHM and BROG 1992; GROENEWEGEN and BERENDSE 1994a,b; MEREDITH et al. 1996; VOORN et al. 1996; HEIMER et al. 1997; HABER and MCFARLAND 1999). The shell is also distinguished by its limited and limbic-specific input from the cortex, midbrain, and thalamus. Between the dorsolateral and ventromedial striatum

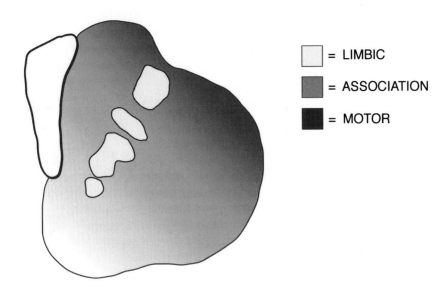

Fig. 2. Schematic of the striatum. Shading represents topographic input from motor (*black*), association (*gray*), and limbic (*light gray*) cortical areas

lies a region (the head of the caudate nucleus and the rostral putamen) that receives input from the dorsolateral prefrontal cortex (SELEMON and GOLDMAN-RAKIC 1985). This prefrontal region is involved in working memory (GOLDMAN-RAKIC 1994). Taken together, the combined corticostriatal projections create a gradient of inputs, from dorsolateral to ventromedial, which correspond to motor, cognitive, and limbic functions (Fig. 2).

I. Nigrostriatal Pathways

The inverse dorsoventral nigrostriatal projection creates a general limbic-to-limbic (VTA-ventral striatum), motor-to-motor (SNc-dorsolateral striatum) dopamine projection gradient. Most tracing studies have not used double-labeling techniques that would demonstrate retrogradely labeled cells to contain tyrosine hydroxylase or other markers indicative of dopamine. However, since the majority of SNc cells are dopaminergic (UNGERSTEDT 1971; PEARSON et al. 1979; FELTEN and SLADEK 1983; HOKFELT et al. 1984), projections from the ventral tier are presumed to be largely dopaminergic. The percentage of cells from the dorsal tier presumed to be dopaminergic may be somewhat smaller, in that there are a greater number of nondopaminergic cells there.

The dorsal tier neurons, (both the VTA and the dorsal SNc), project to the ventromedial striatum; the densocellular neurons of the ventral tier project throughout the striatum, with the exception of the shell; and the cell columns of the ventral tier project to the dorsolateral striatum (Fig. 3). The ventrome-

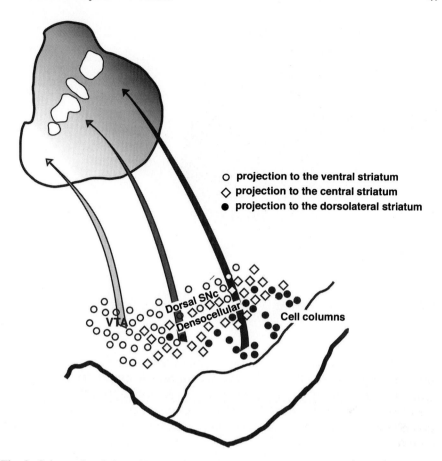

Fig. 3. Schematic of the midbrain illustrating the distribution of cells that project to the ventromedial, central, and dorsolateral striatum from the midbrain. *Open circles*, cells projecting to the ventromedial striatum (limbic region); *diamonds*, cells projecting to the central striatum (association region); *filled circles*, cells projecting to the dorsolateral striatum (motor region)

dial striatum receives the most limited midbrain projection. This projection is derived primarily from the dorsal tier. The shell receives input from the VTA, with only a sparse projection from the dorsal SNc. The core receives input from the entire dorsal tier, including both the VTA and the dorsal SNc. The core also receives some afferent input from the dorsal part of the densocellular region of the ventral tier. The central striatum does not receive input from the dorsal tier, but does from a wide area of the densocellular group. Cells from the dorsal part of the densocellular region that project to the central striatum are intermingled with cells that project to the core of the ventral striatum. The dorsolateral striatum receives the largest midbrain projection. This projection is derived from cells throughout the ventral tier, including the cell

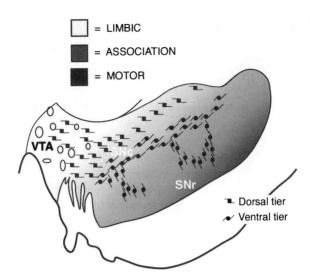

Fig. 4. Schematic of the midbrain illustrating the relationship of nigrostriatal projections to striatal regions defined by cortical inputs. Shading represents motor (*black*), association (*gray*), and limbic (*light gray*) areas

columns. In fact, the cell columns project almost exclusively to the dorsolateral striatum. Neurons in the densocellular region that project to the dorsolateral striatum overlap with those projecting to the central striatum. Thus, the organization of the midbrain dopamine projections onto the striatum, like the corticostriatal pathway, is organized as a functional gradient, but in a dorsomedial (limbic) to a ventrolateral (motor) manner (Fig. 4).

II. Striatonigral Pathways

Projections from the striatum are to the ventral midbrain and terminate in the dorsal and ventral tier dopamine cell groups in addition to the well-known projection to the pars reticulata. Like the nigrostriatal projection, the striatonigral projection has an inverse dorsoventral organization. The ventromedial striatum terminates dorsomedially and the dorsolateral striatum terminates ventrolaterally. This organization links limbic and motor areas from the striatum and midbrain respectively. However, unlike the nigrostriatal pathways, the midbrain receives the largest projection from limbic regions and the most limited input from motor regions.

Projections from the ventromedial striatum, including the shell, have an extensive terminal field. This field is found in the rostromedial midbrain, including the VTA, the medial SNc, and the medial pars reticulata. At central and caudal levels, it extends laterally and includes the dorsal densocellular ventral tier. The ventral striatum, therefore, projects not only throughout the

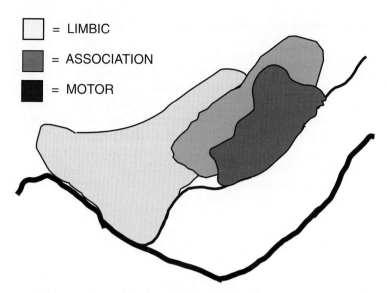

Fig. 5. Schematic of the midbrain illustrating the distribution of terminal fields from the ventromedial, central, and dorsolateral striatum to the midbrain. *Light gray*, terminal field from the ventromedial striatum (limbic region); *gray*, terminal field from the central striatum (association region); *dark gray*, terminal field from the dorsolateral striatum (motor region)

rostrocaudal extent of the substantia nigra, but also to a wide mediolateral area of the SNc. The central striatum projection terminates more ventrally, primarily in the ventral densocellular dopamine cell group, the cell columns, and in the pars reticulata. The dorsolateral striatal projections to the midbrain are the most limited and terminate in the ventrolateral pars reticulata and in the cell columns of the ventral tier. Thus, the ventromedial striatonigral fibers terminate throughout a wide medial and lateral extent of the midbrain and interface with a large region of the dopamine cells in both the dorsal tier and the densocellular region of the ventral tier (HABER et al. 1990; HEDREEN and DELONG 1991). In contrast, efferent projections from the dorsolateral striatum are confined to a relatively small ventrolateral region of the substantia nigra (SMITH and PARENT 1986; LYND-BALTA and HABER 1994b) (Fig. 5). These differences in the proportions of projections from different functional striatal regions have important implications for the type of information flow that can influence dopamine cells.

III. The Striato-Nigro-Striatal Neuronal Network

Taken together, the inverse dorsal–ventral topography both striatonigral and nigrostriatal pathways create complementary systems. The dorsolateral striatum is related to the ventrolateral midbrain and the ventromedial striatum is related to the dorsomedial midbrain. They differ, however, in their relative proportion of nigrostriatal and striatonigral projections. The ventromedial striatum receives a limited midbrain input, but projects to a large region, which includes dorsal and ventral tiers and the dorsal pars reticulata. In contrast, the dorsolateral striatum receives a wide midbrain input, but projects to a limited region. These proportional differences in the network projections suggests that the ventromedial striatum is in a position to influence a wide range of dopamine neurons, but is influenced by a relatively limited group of dopamine cells. On the other hand, the dorsolateral striatum influences a limited midbrain region but is affected by a relatively large midbrain region. The relationship between these striatonigral pathways and the nigrostriatal pathways creates a complex striato-nigro-striatal neuronal network.

For each striatal region there are three midbrain components to the striato-nigro-striatal network – one central reciprocal component and two, adjacent non-reciprocal components. Projections to the midbrain from each striatal region overlap extensively with midbrain cells that project back to that striatal region. This creates a central reciprocal or closed loop. In addition, there are two nonreciprocal connections (or open loops), one dorsal and one ventral to the closed loop. The dorsal open loop component consists of a dorsal group of midbrain cells that project to a specific striatal region, but does not receive a reciprocal efferent connection from that striatal region (Fig. 6, filled arrow heads). The closed loop lies ventrally and comprises a group of cells that project to the same specific striatal region and also lie within its reciprocal connection (Fig. 6, open arrow heads). The ventral, open loop consists of the efferent terminals that do not contain a reciprocally connected group of cells that project back to the specific striatal region (Fig. 6, arrows).

The three components for each region of the striatum occupy a different position within the midbrain. The ventromedial striatal system lies dorsomedially, the dorsolateral striatal system ventrolaterally, and the central striatal system between the two. The ventral open loop of each system overlaps with the dorsal loop of the adjacent system (Fig. 7). Each ventral open loop contains terminals from one striatal region, and each dorsal open loop contains the midbrain cells that project to a specific striatal region. Overlap between the ventral and dorsal open loops from different striatal regions allows dopamine to modulate between different striatal areas. This interaction between the striato-nigro-striatal open loops provides a mechanism for integration of information between different striatal regions.

The nucleus accumbens plays a major role in influencing motor outcome by funneling information from the limbic system to the motor system (the "limbic/motor interface") (NAUTA and DOMESICK 1978; NAUTA et al. 1978; MOGENSON et al. 1980; SOMOGYI et al. 1981; HEIMER et al. 1982; KALIVAS et al. 1993; HABER and FUDGE 1997). This ventromedial striatal modulation of the dorsal striatum via the midbrain dopamine cells has been considered one mechanism by which limbic circuitry can affect motor outcome directly. Since the ventral open loop component of each striato-nigro-striatal system

The Place of Dopamine Neurons 51

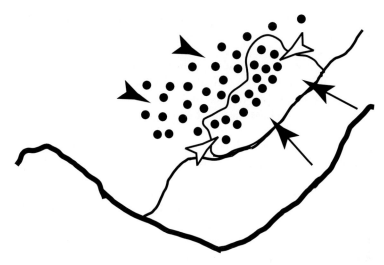

Fig. 6. The striato-nigro-striatal projection system illustrating the three components within the midbrain for each striatal area. Schematic of the midbrain showing the distribution of terminals (*outline*) from the dorsolateral striatum and cells (*filled circles*) that project to the dorsolateral striatum. *Filled (black) arrowheads* indicate cells dorsal to terminals (dorsal open loop), *unfilled (white) arrowheads* indicate the region of cells that project to the dorsolateral striatum within the dorsolateral striatal terminal field (closed loop), and *arrows* point to terminal field, ventral and lateral to cells that project to the dorsolateral striatum (ventral open loop)

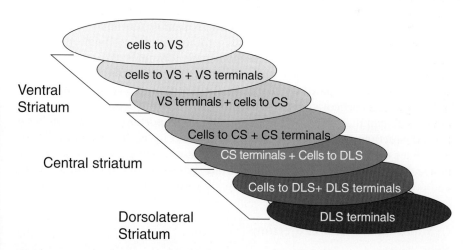

Fig. 7. Diagram of the three striato-nigro-striatal components for each striatal region illustrating an overlapping and interdigitating system in the midbrain. The three midbrain components for each striatal region are represented by *three sequential disks*. The first corresponds to the dorsal, open loop, the second corresponds to the closed loop, and the third corresponds to the ventral open loop. Note that the third midbrain component of a striatal region overlaps the first component of the adjacent, more dorsal striatal region VS = ventral striatum; CS = central striatum; DLS = dorsolateral striatum

overlaps with the dorsal open loop component of the adjacent striatal system (Fig. 7), information from limbic striatal regions reaches motor striatal regions through a series of connections passing through the central or cognitive striatal regions. The shell receives input only from the dorsal tier, but its projection field includes the densocellular area. This area does not project back to the shell, but does project to the core. The core terminates within a wider range of densocellular region, much of which projects to the central striatum. The central striatum is reciprocally connected to the densocellular region and also to the cell columns. The cell columns project to the dorsolateral striatum. Since cortical innervation imposes a functional gradient from limbic to cognitive to motor functions onto the striatum, limbic influence on motor outcome is indirect, passing through cognitive areas of the striatum. Thus, this system of open and closed loops creates an interface between different striatal regions via the midbrain dopamine cells. This interface forms an ascending neuronal network that interconnects functional regions of the striatum from the limbic striatum to reach the motor striatum (Haber et al. 2000) (Fig. 8).

C. Connections Between the Midbrain and Cortex

The presence of dopamine in the cortex is clearly demonstrated by a variety of methods (Levitt et al. 1984, 1987; Gaspar et al. 1989, 1992; Lidow et al. 1991; Verney et al. 1993; Sesack et al. 1998). The dopamine innervation of primate cortex is more extensive than in rats and found not only in granular areas but also in agranular frontal regions, parietal cortex, temporal cortex, and even, albeit sparsely, in occipital cortex. Dopamine terminals in layer I are prevalent throughout cortex and provide a rather general modulation of many cells at the distal apical dendrites. Dopamine terminals are also found in layers V–VI in specific cortical areas. Here they are in a position to provide a more direct modulation of specific cortical efferent projections, including corticostriatal and corticothalamic projections. Through the deep layer projecting neurons, dopamine has an additional potentially important influence on basal ganglia function. Dopamine fibers are also found in the hippocampus, albeit to a lesser extent than found in neocortex. These terminals are densest in the molecular layer and the hilus of the dentate gyrus, and in the molecular layer and polymorphic layer of the subiculum (Samson et al. 1990).

I. Midbrain Cortical Projections

In rats, the majority of midbrain neurons projecting to the cortex arise from the VTA and the medial half of the SNc, throughout its rostrocaudal extent. In primates, the majority of dopamine cortical projections are from the parabrachial nucleus of the VTA and the dorsal SNc (Amaral and Cowan 1980; Porrino and Goldman-Rakic 1982; Fallon and Loughlin 1987; Gaspar et al. 1992). The ventral tier does not project extensively to cortex. The VTA

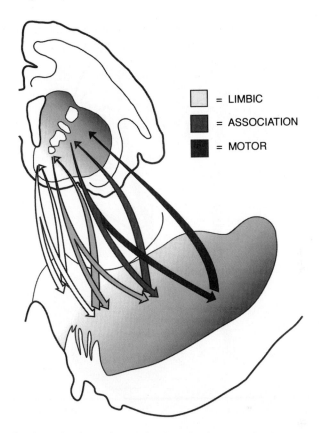

Fig. 8. Organization of striato-nigro-striatal projections. Projections from the shell to the VTA and from the VTA to the shell form a closed striato-nigro-striatal loop. Projections from the shell to the dorsomedial SNc feed-forward to the core form an open loop and the first part of a spiral. The spiral continues through the striato-nigro-striatal projections through the central to the dorsolateral striatum. In this way, ventral striatal regions influence more dorsal striatal regions

projects throughout prefrontal cortex including orbital, medial, and dorsolateral areas, with fewer projections to motor areas. Midbrain projections to the hippocampus are also derived primarily from the VTA, with few cells from the pars compacta proper terminating there. The dorsal SNc primarily innervates dorsolateral prefrontal cortex, motor and premotor regions. The midbrain dopamine cells that project to different regions of motor and premotor cortex are intermingled; double-label studies show many cells sending collateral axons to different cortical regions (GASPAR et al. 1992). The dopamine-cortical projection is a more diffuse and widespread system compared to the nigrostriatal system. Of particular interest is the fact that the dorsal tier neurons project not only to limbic and cognitive cortical areas, but also to motor and premotor regions (Fig. 9). Thus, while divided into the VTA and

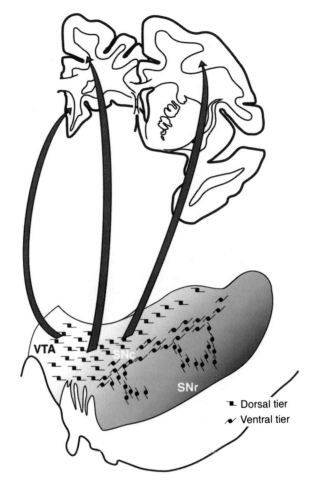

Fig. 9. Schematic of the midbrain illustrating projections primarily from the dorsal tier to the cortex

dorsal SNc respectively, cortical innervation of dopamine is derived primarily from neurons influenced by the limbic system.

II. Cortical Midbrain Projections

There is a general acceptance of descending cortical projections to the substantia nigra, based primarily on studies in rats and cats. Cortical lesions result in reduced glutamate content in the rat substantia nigra as well as fiber degeneration in the corticonigral pathway (AFIFI et al. 1974; CARTER 1980; USUNOFF et al. 1982; KERKERIAN et al. 1983; KORNHUBER 1984). Retrograde tracing studies confirm these projections (BUNNEY and AGHAJANIAN 1976; SESACK et al. 1989). However, these projections have been difficult to definitively demonstrate in primates. (LEICHNETZ and ASTRUC 1976; KÜNZLE 1978) Descending corticonigral fibers have been demonstrated with fiber degeneration and anterograde tracing techniques in primates. However, neither technique clearly showed that fibers actually terminated in the substantia nigra. Therefore, the authors in both studies point out that the results must be interpreted with care.

D. The Amygdala and Other Forebrain Projections

Dopamine neurons project to several other forebrain regions including the bed nucleus of the stria terminalis, the septal nucleus, and the amygdala. These structures are often associated with the limbic system by virtue of their connections with the amygdala and the medial forebrain bundle. Within the amygdala, the basolateral complex and the central amygdala nuclei are most consistently associated with projections to and from the midbrain (FALLON 1978; SADIKOT and PARENT 1990). The densest dopamine innervation is to the central and medial part of the central nucleus (FREEDMAN and CASSELL 1994). Projections arise primarily from the dorsal tier (both the VTA and the dorsal SNc) (AGGLETON et al. 1980; MEHLER 1980; NORITA and KAWAMURA 1980).

The central nucleus provides the main amygdaloid input to the dopamine neurons (HOPKINS 1975; BUNNEY and AGHAJANIAN 1976; KRETTEK and PRICE 1978; PRICE and AMARAL 1981; GONZALES and CHESSELET 1990). This projection is to the dorsal tier and to some extent the densocellular region, but not to the cell columns. The central nucleus is further subdivided into regions characterized by histochemical and morphologic features as well as by differential inputs (CASSELL et al. 1999). These subregions project differentially to the midbrain. The medial central nucleus projects to a wide mediolateral extent of the dopamine neurons, including the dorsal tier, the rostrocentral medial SN and lateral SN, the pars lateralis (CASSELL et al. 1986; GONZALES and CHESSELET 1990). The lateral capsular area of the central nucleus and the amygdalo-striatal area project primarily to the dorsal tier. The lateral central nucleus is unique in its restricted projection to the lateral SN, the pars lateralis. In addition to the amygdala proper, the bed nucleus of the stria terminalis and the sublenticular substantia innominata project to the VTA and the dorsomedial SNc. These fibers travel with those from the amygdala. Together these structures form the "extended amygdala" (ALHEID and HEIMER 1988; HEIMER et al. 1991).

The amygdala is generally involved in determining the emotional significance of complex sensory inputs (AGGLETON 1993; ADOLPHS et al. 1994) thereby providing an important limbic input to the dopamine neurons. The central amygdala nucleus is particularly critical in increasing attention during unexpected shifts in predictive relationships (NISHIJO et al. 1988; WILSON and ROLLS 1993; GALLAGHER and HOLLAND 1994). The projection from specific subregions of the central nucleus to dorsal and ventral tier cells indicates that these subnuclear regions differentially modulate dopamine neurons. Thus, a direct limbic input influences the firing of certain dopamine cells that in turn project to striatum and cortex. In this way, dopamine is in a pivotal position to coordinate the initiation of appropriate responses which are carried out through basal ganglia and cortical circuits.

E. Functional Modulation and Integration Through Dopamine Forebrain Pathways

The dorsal tier neurons are tightly linked with the limbic system. Afferent projections to the dorsal tier are from the limbic-related striatum and from the central amygdala nucleus. In turn, the dorsal tier innervates the shell and core of the striatum, the amygdala, the extended amygdala, and limbic cortex. The fact that the dorsal tier neurons do not extend their processes into the ventral tier and receive little direct information from dorsal striatonigral pathways indicates that this cell group is selectively and reciprocally connected with limbic structures. However, while the dorsal tier projection to the striatum is limited, it projects throughout cortex, placing it in a position to influence more than the limbic system through this widespread cortical innervation.

The ventral tier is divided into two groups. The densocellular region is in a unique position of receiving information from both the limbic and association striatum, and as well as some input from the central nucleus of the amygdala. Unlike the dorsal tier, the densocellular cell group projects widely throughout the striatum and thus can have a profound influence on all striatal output, including the motor system. However, these cells do not send a significant projection to cortex. The cell columns are most tightly linked to the motor system and are the least influenced by descending fibers from limbic and association areas. Furthermore, they do not project to widespread regions of striatum or cortex.

Investigation of the basal ganglia link between motivation and motor outcomes has focused primarily on dopamine pathways of the nucleus accumbens (Mogenson et al. 1980, 1993; Groenewegen et al. 1996). Behavioral studies of dopamine pathways have lead to the association of the mesolimbic and nigrostriatal pathways with reward and motor activity, respectively. While the role of dopamine and reward is well established (Wise and Rompre 1989), its primary function is to direct attention to important stimuli likely to bring about a desired outcome (Ljungberg et al. 1992; Schultz et al. 1997). This requires processing a complex chain of events beginning with motivation and proceeding through cognitive processing that shapes final motor outcomes. This sequence is reflected in the complex organization of dopamine connections to the forebrain. Information is channeled through the limbic and association regions of cortex and striatum to mediate motor outcome through motor regions of striatum and cortex. These cells are influenced by other forebrain structure, most notably the amygdala. Motor decision-making processes are thus influenced by motivation and cognitive inputs, allowing the animal to respond based on information from both internal and external stimuli.

References

Adolphs R, Tranel D, Damasio H, Damasio A (1994) Impaired recognition of emotion in facial expressions following bilateral damage to the human amygdala. Nature 372:669–672

Aggleton JP (1993) The contribution of the amygdala to normal and abnormal emotional states. Trends Neurosci 16:328–333

Afifi AK, Bahuth NB, Kaelber WW, Mikhael E, Nassar S (1974) The cortico-nigral fibre tract. An experimental Fink-Heimer study in cats. J Anat 118(3):469–476

Aggleton JP, Burton MJ, Passingham RE (1980) Cortical and subcortical afferents to the amygdala of the rhesus monkey (Macaca mulatta). Brain Res 190:347–368

Alheid GF, Heimer L (1988) New perspectives in basal forebrain organization of special relevance for neuropsychiatric disorders: the striatopallidal, amygdaloid, and corticopetal components of substantia innominata. Neuroscience 27:1–39

Amaral DG, Cowan WM (1980) Subcortical afferents to the hippocampal formation in the monkey. J Comp Neurol 189:573–591

Bjorklund A, Lindvall O (1984) Dopamine-containing systems in the CNS. In: Bjorklund, Hokfelt (eds) Handbook of Chemical Neuroanatomy, Vol. II: Classical Transmitters in the CNS, Part I, pp 55–122. Amsterdam: Elsevier

Bunney BS, Aghajanian GK (1976) The precise localization of nigral afferents in the rat as determined by a retrograde tracing technique. Brain Res 117:423–435

Carpenter MB, Peter P (1971) Nigrostriatal and nigrothalamic fibers in the rhesus monkey. J Comp Neurol 144:93–116

Carter CJ (1980) Glutamatergic pathways from the medial pre-frontal cortex to the anterior striatum, nucleus accumbens and substantia nigra. Brti J Phermacol 70: 50–51

Cassell MD, Freedman LJ, Shi C (1999) The Intrinsic Organization of the Central Extended Amygdala. Annals of the New York Academy of Sciences 877:217–241

Cassell MD, Gray TS, Kiss JZ (1986) Neuronal architecture in the rat central nucleus of the amygdala: a cytological, hodological, and immunocytochemical study. Journal of Comparative Neurology 246(4):478–99

DeLong MR (1990) Primate models of movement disorders of basal ganglia origin. Trends Neurosci 13:281–285

Deutch AY, Goldstein M, Roth RH (1986) The ascending projections of the dopaminergic neurons of the substantia nigra, zona reticulata: a combined retrograde tracer-immunohistochemical study. Neuroscience Letters 71:257–263

Fallon JH (1978) Catecholamine Innervation of the Basal Forebrain. II. Amygdala, suprarhinal cortex and entorhinal cortex. J Comp Neurol 180(3):509–532

Fallon JH, Loughlin SE (1987) Monoamine innervation of cerebral cortex and a theory of the role of monoamines in cerebral cortex and basal ganglia. In: Jones EG, Peters A (eds) Cerebral Cortex, pp 41–109. Plenum Press

Fallon JH, Moore RY (1978) Catecholamine innervation of the basal forebrain. IV. Topography of the dopamine projections to the basal forebrain and neostriatum. J Comp Neurol 180(3):545–580

Felten DL, Sladek JR, Jr. (1983) Monoamine distribution in primate brain. V. Monoaminergic nuclei: anatomy, pathways and local organization. Brain Res Bull 10:171–284

Flaherty AW, Graybiel AM (1994) Input-output organization of the sensorimotor striatum in the squirrel monkey. J Neurosci 14:599–610

Francois C, Percheron G, Yelnik J (1984) Localization of nigrostriatal, nigrothalamic and nigrotectal neurons in ventricular coordinates in macaques. Neuroscience 13, No. 1:61–76

Francois C, Percheron G, Yelnik J, Heyner S (1985) A histological atlas of the macaque (*Macaca mulatta*) substantia nigra in ventricular coordinates. Brain Res Bull 14:349–367

Freedman LJ, Cassell MD (1994) Distribution of dopaminergic fibers in the central division of the extended amygdala of the rat. Brain Research 633(1–2):243–52

Fuxe K, Hokfelt T, Ungerstedt U (1970) Morphological and Functional Aspects of Central Monoamine Neurons. Int Review of Neurobiology 13:93–126

Gallagher M, Holland PC (1994) The amygdala complex: multiple roles in associative learning and attention. Proceedings of the National Academy of Sciences of the United States of America 91(25):11771–6

Gaspar P, Berger B, Febvret A, Vigny A, Henry JP (1989) Catecholamine innervation of the human cerebral cortex as revealed by comparative immunohistochemistry of tyrosine hydroxylase and dopamine-beta-hydroxylase. J Comp Neurol 279:249–271

Gaspar P, Stepneiwska I, Kaas JH (1992) Topography and collateralization of the dopaminergic projections to motor and lateral prefrontal cortex in owl monkeys. J Comp Neurol 325:1–21

Gerfen CR, Baimbridge KG, Miller JJ (1985) The neostriatal mosaic: Compartmental distribution of calcium-binding protein and parvalbumin in the basal ganglia of the rat and monkey. Proc Natl Acad Sci USA 82:8780–8784

Goldman-Rakic PS (1994) Working memory dysfunction in schizophrenia. J Neuropsychiatry Clin Neurosci 6:348–357

Goldman-Rakic PS, Selemon LD (1986) Topography of corticostriatal projections in nonhuman primates and implications for functional parcellation of the neostriatum. In: Jones EG, Peters A (eds) Cerebral Cortex Vol. 5, pp 447–466. New York: Plenum Publishing Corporation

Gonzales C, Chesselet M-F (1990) Amygdalonigral Pathway: An anterograde study in the rat with phaseolus vulgaris leucoagglutinin. J Comp Neurol 297:182–200

Groenewegen HJ, Berendse HW (1994a) Anatomical relationships between the prefrontal cortex and the basal ganglia in the rat In: A.-M. T (ed) Motor and Cognitive Functions of the Prefrontal Cortex, pp 51–77. Berlin Heidelberg: Springer-Verlag

Groenewegen HJ, Berendse HW (1994b) The specificity of the "nonspecific" midline and intralaminar thalamic nuclei. Trends Neurosci 17:50

Groenewegen HJ, Wright CI, Beijer AVJ (1996) The nucleus accumbens: gateway for limbic structures to reach the motor system? In: Holstege G, Bandler R, Saper CP (eds) Progress in Brain Research, pp 485–511: Elsevier Science

Groenewegen HJ, Wright CI, Uylings HB (1997) The anatomical relationships of the prefrontal cortex with limbic structures and the basal ganglia. Journal of Psychopharmacology 11(2):99–106

Haber SN, Fudge JL (1997) The primate substantia nigra and VTA: Integrative circuitry and function. Crit Rev Neurobiol 11(4):323–342

Haber SN, Fudge JL, McFarland N (2000) Striatonigrostriatal pathways in primates form an ascending spiral from the shell to the dorsolateral striatum. J Neurosci 20(6):2369–2382

Haber SN, Kunishio K, Mizobuchi M, Lynd-Balta E (1995a) The orbital and medial prefrontal circuit through the primate basal ganglia. J Neurosci 15:4851–4867

Haber SN, Lynd E, Klein C, Groenewegen HJ (1990) Topographic organization of the ventral striatal efferent projections in the rhesus monkey: An anterograde tracing study. J Comp Neurol 293:282–298

Haber SN, McFarland NR (1999) The concept of the ventral striatum in nonhuman primates. In: McGinty JF (ed) Advancing from the ventral striatum to the extended amygdala, pp 33–48. New York: The New York Academy of Sciences

Haber SN, Ryoo H, Cox C, Lu W (1995b) Subsets of midbrain dopaminergic neurons in monkeys are distinguished by different levels of mRNA for the dopamine transporter: Comparison with the mRNA for the D2 receptor, tyrosine hydroxylase and calbindin immunoreactivity. J Comp Neurol 362:400–410

Halliday GM, Tork I (1986) Comparative anatomy of the ventromedial mesencephalic tegmentum in the rat, cat, monkey and human. J Comp Neurol 252:423–445

Hedreen JC, DeLong MR (1991) Organization of striatopallidal, striatonigral, and nigrostriatal projections in the Macaque. J Comp Neurol 304:569–595

Heimer L, Alheid GF, de Olmos JS, Groenewegen HJ, Haber SN, Harlan RE, Zahm DS (1997) The Accumbens: Beyond the core-shell dichotomy. J Neuropsychiatry Clin Neurosci 9 (3):354–381

Heimer L, de Olmos J, Alheid GF, Zaborszky L (1991) "Perestroika" in the basal forebrain: Opening the border between neurology and psychiatry. In: Holstege G (ed) Progress in Brain Research, Vol. 87, pp 109–165: Elsevier Science Publishers

Heimer L, Switzer RD, Van Hoesen GW (1982) Ventral striatum and ventral pallidum. Components of the motor system? Trends Neurosci Vol. 5:83–87

Hirsch EC, Mouatt A, Thomasset M, Javoy-Agid F, Agid Y, Graybiel AM (1992) Expression of calbindin D_{28K}-like immunoreactivity in catecholaminergic cell groups of the human midbrain: normal distribution and distribution in Parkinson's disease. Neurodegeneration 1:83–93

Hokfelt T, Martensson R, Bjorklund A, Kleinau S, Goldstein M (1984) Distributional maps of tyrosine-hydroxylase immunoreactive neurons in the rat brain. In: Bjorklund A, Hokfelt T (eds) Handbook of Chemical Neuroanatomy, Vol. II: Classical Neurotransmitters in the CNS, Part I, pp 277–379. Amsterdam: Elsevier

Hopkins DA (1975) Amygdalotegmental projections in the rat, cat, and rhesus monkey. Neuroscience Letters 1:263–270

Johnson TN, Rosvold HE (1971) Topographic projections on the globus pallidus and the substantia nigra of selectively placed lesions in the precommissural caudate nucleus and putamen in the monkey. Exp Neurol 33:584–596

Kalivas PW, Churchill L, Klitenick MA (1993) The Circuitry Mediating the Translation of Motivational Stimuli into Adaptive Motor Responses. In: Kalivas PW, Barnes CD (eds) Limbic Motor Circuits and Neuropsychiatry, pp 237–275. Boca Raton: CRC Press, Inc

Kerkerian L, Nieoullon A, Dusticier N (1983) Topographic changes in high-affinity glutamate uptake in the cat red nucleus, substantia nigra, thalamus, and caudate nucleus after lesions of sensorimotor cortical areas. Experimental Neurology 81(3):598–612

Kimura M (1990) Behaviorally contingent property of movement-related activity of the primate putamen. J Neurophysiol 63:1277–1296

Koob GF, Nestler EJ (1997) The Neurobiology of Drug Addiction. The Journal of Neuropsychiatry and Clinical Neurosciences 9:482–497

Kornhuber J (1984) The cortico-nigral projection: reduced glutamate content in the substantia nigra following frontal cortex ablation in the rat. Brain Res 322:124–126

Krettek JE, Price JL (1978) Amygdaloid projections to subcortical structures within the basal forebrain and brainstem in the rat and cat. J Comp Neurol 178:225–254

Künzle H (1975) Bilateral projections from precentral motor cortex to the putamen and other parts of the basal ganglia. An autoradiographic study in *Macaca fascicularis*. Brain Res 88:195–209

Künzle H (1978) An autoradiographic analysis of the efferent connections from premotor and adjacent prefrontal regions (areas 6 and 9) in *Macaca fascicularis*. Brain Behav Evol 15:185–234

Lavoie B, Parent A (1991) Dopaminergic neurons expressing calbindin in normal and parkinsonian monkeys. Neuroreport 2, No. 10:601–604

Leichnetz GR, Astruc J (1976) The efferent projections of the medial prefrontal cortex in the squirrel monkey (*Saimiri sciureus*). Brain Research 109(3):455–472

Levitt P, Rakic P, Goldman-Rakic P (1984) Region-specific distribution of cat cholamine afferents in primate cerebral cortex: A fluorescence histochemical analysis. J Comp Neurol 227:23–36

Lewis DA, Campbell MJ, Foote SL, Goldstein M, Morrison JH (1987) The distribution of tyrosine hydroxylase-immunoreactive fibers in primate neocortex is widespread but regionally specific. J Neurosci 7(1):279–290

Lidow MS, Goldman-Rakic PS, Gallager DW, Rakic P (1991) Distribution of dopaminergic receptors in the primate cerebral cortex: quantitative autoradiographic analysis using [3H] raclopride, [3H] spiperone and [3H]sch23390. Neuroscience 40, No. 3:657–671

Liles SL, Updyke BV (1985) Projection of the digit and wrist area of precentral gyrus to the putamen: relation between topography and physiological properties of neurons in the putamen. Brain Res 339:245–255

Ljungberg T, Apicella P, Schultz W (1992) Responses of monkey dopamine neurons during learning of behavioral reactions. J Neurophysiol 67(1):145–163

Lynd-Balta E, Haber SN (1994a) The organization of midbrain projections to the striatum in the primate: Sensorimotor-related striatum versus ventral striatum. Neuroscience 59:625–640

Lynd-Balta E, Haber SN (1994b) Primate striatonigral projections: A comparison of the sensorimotor-related striatum and the ventral striatum. J Comp Neurol 343:1–17

McRitchie DA, Halliday GM (1995) Calbindin D28K-containing neurons are restricted to the medial substantia nigra in humans. Neuroscience 65:87–91

Mehler WR (1980) Subcortical afferent connections of the amygdala in the monkey. J Comp Neurol 190:733–762

Meredith GE, Pattiselanno A, Groenewegen HJ, Haber SN (1996) Shell and core in monkey and human nucleus accumbens identified with antibodies to calbindin-D28 k. J Comp Neurol 365:628–639

Mogenson GJ, Brudzynski SM, Wu M, Yang CR, Yim CCY (1993) From motivation to action: A review of dopaminergic regulation of limbic-nucleus accumbens-pedunculopontine nucleus circuitries involved in limbic-motor integration. In: Kalivas PW, Barnes CD (eds) Limbic Motor Circuits and Neuropsychiatry, pp 193–236. Boca Raton: CRC Press

Mogenson GJ, Jones DL, Yim CY (1980) From motivation to action: Functional interface between the limbic system and the motor system. Prog Neurobiol 14:69–97

Nauta WJH, Domesick VB (1978) Crossroads of limbic and striatal circuitry: hypothalamic-nigral connections. In: Livingston KE, Hornykiewicz O (eds) Limbic Mechanisms, pp 75–93. New York: Plenum Publishing Corp

Nauta WJH, Smith GP, Faull RLM, Domesick VB (1978) Efferent connections and nigral afferents of the nucleus accumbens septi in the rat. Neuroscience 3:385–401

Nishijo H, Ono T, Nishino H (1988) Single neuron responses in amygdala of alert monkey during complex sensory stimulation with affective significance. J Neurosci 8:3570–3583

Norita M, Kawamura K (1980) Subcortical afferents to the monkey amygdala: an HRP study. Brain Res 190:225–230

Olszewski J, Baxter D (1954) Cytoarchitecture of the Human Brain Stem. Basil: S. Karger

Parent A, Bouchard C, Smith Y (1984) The striatopallidal and striatonigral projections: two distinct fiber systems in primate. Brain Res 303:385–390

Parent A, Hazrati L-N (1994) Multiple striatal representation in primate substantia nigra. J Comp Neurol 344:305–320

Parent A, Lavoie B (1993) The heterogeneity of the mesostriatal dopaminergic system as revealed in normal and Parkinsonian monkeys. Adv Neurol 60:25–20

Parent A, Mackey A, De Bellefeuille L (1983) The subcortical afferents to caudate nucleus and putamen in primate: a fluorescence retrograde double labeling study. Neuroscience 10(4):1137–1150

Pearson J, Goldstein M, Brandeis L (1979) Tyrosine hydroxylase immunohistochemistry in human brain. Brain Res 165:333–337

Poirier LJ, Giguere M, Marchand R (1983) Comparative morphology of the substantia nigra and ventral tegmental area in the monkey, cat and rat. Brain Res Bull 11:371–397
Porrino LJ, Goldman-Rakic PS (1982) Brainstem innervation of prefrontal and anterior cingulate cortex in the rhesus monkey revealed by retrograde transport of HRP. J Comp Neurol 205:63–76
Price JL, Amaral DG (1981) An autoradiographic study of the projections of the central nucleus of the monkey amygdala. J Neurosci 1:1242–1259
Rolls ET, Burton MJ, Mora F (1980) Neurophysiological analysis of brain-stimulation reward in the monkey. Brain Research 194:339–357
Sadikot AF, Parent A (1990) The monoaminergic innervation of the amygdala in the squirrel monkey: an immunohistochemical study. Neuroscience 36:431–447
Samson Y, Wu JJ, Friedman AH, Davis JN (1990) Catecholaminergic innervation of the hippocampus in the cynomolgus monkey. J Comp Neurol 298:250–263
Schultz W, Dayan P, Montague PR (1997) A neural substrate of prediction and reward. [Review] [37 refs]. Science 275:1593–1599
Schultz W, Romo R, Ljungberg T, Mirenowicz J, Hollerman JR, Dickinson A (1995) Reward-related signals carried by dopamine neurons. In: Houk JC, Davis JL, Beiser DG (eds) Models of Information Processing in the Basal Ganglia, pp 233–248: Cambridge: MIT Press
Selemon LD, Goldman-Rakic PS (1985) Longitudinal topography and interdigitation of corticostriatal projections in the rhesus monkey. J Neurosci 5:776–794
Selemon LD, Goldman-Rakic PS (1990) Topographic intermingling of striatonigral and striatopallidal neurons in the rhesus monkey. J Comp Neurol 297:359–376
Sesack SR, Deutch AY, Roth RH, Bunney BS (1989) Topographical organization of the efferent projections of the medial prefrontal cortex in the rat: an anterograde tract-tracing study with Phaseolus vulgaris leucoagglutinin. Journal Of Comparative Neurology 290:213–242
Sesack SR, Hawrylak VA, Melchitzky DS, Lewis DA (1998) Dopamine innervation of a subclass of local circuit neurons in monkey prefrontal cortex: ultrastructural analysis of tyrosine hydroxylase and parvalbumin immunoreactive structures. Cerebral Cortex 8(7):614–22
Smith Y, Parent A (1986) Differential connections of the caudate nucleus and putamen in the squirrel monkey (*Saimiri sciureus*). Neuroscience 18(2):347–371
Somogyi P, Bolam JP, Totterdell S, Smith AD (1981) Monosynaptic input from the nucleus accumbens-ventral striatum region to retrogradely labelled nigrostriatal neurones. Brain Res 217:245–263
Szabo J (1967) The efferent projections of the putamen in the monkey. Exp Neurol 19:463–476
Szabo J (1970) Projections from the body of the caudate nucleus in the rhesus monkey. Exp Neurol 27:1–15
Szabo J (1979) Cellular and synaptic organization of basal ganglia. Appl Neurophysiol 42:9–12
Szabo J (1980) Organization of the ascending striatal afferents in monkeys. J Comp Neurol 189:307–321
Ungerstedt U (1971) Stereotaxic mapping of the monoamine pathways in the rat brain. Acta Physiol Scand 367:153–174
Usunoff KG, Romansky KV, Malinov GB, Ivanov DP, Blagov ZA, Galabov GP (1982) Electron microscopic evidence for the existence of a corticonigral tract in the cat. Journal fur Hirnforschung 23(1):23–9
Verney C, Milosevic A, Alvarex C, Berger B (1993) Immunocytochemical evidence of well-developed dopaminergic and noradernergic innervations in the frontal cerebral cortex of human fetuses at midgestation. J Comp Neurol 336:331–344
Voorn P, Brady LS, Berendse HW, Richfield EK (1996) Densitometrical analysis of opioid receptor ligand binding in the human striatum –I. Distribution of µ opioid receptor defines shell and core of the ventral striatum. Neuroscience 75:777–792

Wilson FA, Rolls ET (1993) The effects of stimulus novelty and familiarity on neuronal activity in the amygdala of monkeys performing recognition memory tasks. Experimental Brain Research 93(3):367–82

Wise RA, Rompre PP (1989) Brain dopamine and reward. Annual Review of Psychology 40:191–225

Wright CI, Groenewegen HJ (1996) Patterns of overlap and segregation between insular cortical, intermediodorsal thalamic and basal amygdaliod afferents in the nucleus accumbens of the rat. Neuroscience 73:359–373

Yamaguchi S, Kobayashi S (1998) Contributions of the dopaminergic system to voluntary and automatic orienting of visuospatial attention. Journal Of Neuroscience 18:1869–1878

Zaborszky L, Alheid GF, Beinfeld MC, Eiden LE, Heimer L, Palkovits M (1985) Cholecystokinin innervation of the ventral striatum: A morphological and radioimmunological study. Neuroscience 14, No. 2:427–453

Zahm DS, Brog JS (1992) On the significance of subterritories in the "accumbens" part of the rat ventral striatum. Neuroscience 50, No. 4:751–767

CHAPTER 4
Synaptology of Dopamine Neurons

S.R. SESACK

A. Introduction

Central dopamine (DA) neurons comprise several cell groups whose long range and local projections modulate neuronal systems mediating voluntary motor control, autonomic function, reproductive behavior, light/dark adaptation, and cognitive processes such as reward learning and working memory. The synaptology of DA cells reflects the richness of these diverse functions by exhibiting a continuum of morphological specializations that ranges from complete absence of synaptic connections to highly specific synapses onto selective dendritic compartments of target neurons. In addition to this diversity, certain common themes emerge from an examination of DA synaptology in various brain regions. For example, one of the most common synapses reported for DA nerve fibers is onto distal targets, such as spines or small-caliber dendrites in the immediate vicinity of convergent synaptic inputs from non-DA terminals. Such convergence suggests an anatomical substrate for DA modulation of afferent drive, as documented in electrophysiological studies of DA systems (MANTZ et al. 1988; CEPEDA and LEVINE 1998). Another common observation is the absence of identifiable DA synapses at particular sites where physiological studies suggest that DA mediates important regulatory control, or even at sites known to express DA receptors. Such findings have argued for the mediation of DA actions through non-synaptic mechanisms, sometimes referred to as volume transmission (GRACE 1991; GONON 1997; ZOLI et al. 1998).

This chapter extends previous reviews on the ultrastructure of DA neurons in the rodent (PICKEL and SESACK 1999) and primate (LEWIS and SESACK 1997) by focusing on the architectural variability expressed by DA axons both within and across regions. The review is intended for a varied audience, including those (1) seeking a critical evaluation of the methods used to study DA synaptology, (2) desiring a current synopsis of DA synaptic connections within various brain regions, and (3) interested in the functional implications of DA synaptology, including the extent to which this transmitter may communicate via non-conventional means. The chapter thus begins with a description of the methodology used for identification of synapses formed by DA neurons and the general morphological features exhibited by these processes and synapses. Then, current knowledge regarding DA synaptology within various brain regions is catalogued. All of the DA projection cells are reviewed, and retinal DA neurons are included as a specific example of DA cells that are local circuit neurons. Following the review of regional synaptology, the evidence for a continuum of DA synaptic architecture is integrated with findings regarding the localization of DA transporter and receptor proteins. It is then argued that the available anatomical and physiological data indicate that DA transmission occurs via a synaptic mode as well as a *parasynaptic* mode (parasynaptic terminology according to SCHMITT 1984; see Sect. D.II, this chapter). The specific structural components that may support one or both of these modes are discussed, along with the assertion that structural variability is dictated by the specific functions required.

B. Methods for Ultrastructural Labeling of DA Axons

I. Uptake of DA Analogs or Radiolabeled DA

Autoradiographic detection of radiolabeled DA uptake was among the first methods used to localize DA axons and varicosities (DESCARRIES et al. 1980; DOUCET et al. 1986, 1988). In brain regions lacking other monoamines, this can be an effective marker of DA axons. However, norepinephrine (NE) fibers also avidly take up DA, and uptake has also been reported in serotonin (5-HT) fibers. To some extent, these limitations can be resolved by the use of selective uptake inhibitors (DOUCET et al. 1986, 1988). However, an uptake barrier has also been described in areas like the striatum, where the tight packing of DA axons can limit the diffusion of radiolabeled DA (DOUCET et al. 1986). Finally, there are issues regarding the resolution of the autoradiographic signal, which is not always localized precisely to the site of emission, although accumulation of multiple silver grains reasonably guarantees the specific labeling of underlying processes. Thus, this method is still utilized for quantitative studies (DESCARRIES et al. 1996).

The false transmitter, 5-hydroxydopamine is applied to brain slices or injected into the lateral ventricle and subsequently taken up into monoamine neurons. It is visualized by precipitation with glutaraldehyde and oxidation by osmium tetroxide (GROVES 1980). There is some question as to the specificity of this method for labeling monoamine axons. For example, in initial studies of the striatum, most of the labeled profiles were reported to form synapses that are now known to be atypical of DA processes (GROVES 1980; see Sect. C.I.2, this chapter). However, the utility of this technique has been improved by recent observations that presumed monoamine axons label heavily with reaction product, while sparsely labeled profiles are likely to represent non-monoamine fibers (GROVES et al. 1994). Nevertheless, 5-hydroxydopamine uptake cannot distinguish between the various monoamine axons. Furthermore, early degenerative changes characteristic of toxicity have been noted in some 5-hydroxydopamine-labeled profiles (GROVES et al. 1994).

II. Tract-Tracing

Anterograde tract-tracing from cells of origin can also be used to identify the synaptic connections of DA axons. However, the main problem with this approach is the need to identify the DA phenotype of the axons labeled with tracer; often such dual labeling is not performed. Moreover, the characteristics of the specific tracers used may cause more or less labeling of non-DA axons. For example, use of radiolabeled amino acids for anterograde transport is complicated by uncertainty in defining the injection-site boundaries. Amino acids can also be transported transneuronally, making it difficult to define the source of labeled axons observed in the electron microscope (EDWARDS and HENDRICKSON 1981). Use of biotinylated dextran amine is limited by the fact that it can also be transported retrogradely, with subsequent labeling of axon collaterals of retrogradely labeled neurons (PARÉ and SMITH 1996).

Use of anterograde degeneration to trace DA projections is perhaps the least-specific method of identification. Physical destruction of DA cell bodies cannot be performed without damage to non-DA neurons and processes, including fibers of passage. This problem is alleviated to some extent by the use of selective neurotoxins. However, loss of DA and other monoamines can cause significant synaptic reorganization within target structures that includes degeneration of non-DA elements. For example, use of 6-hydroxydopamine to lesion the striatal DA innervation causes reorganization of spiny dendrites and their synaptic inputs that is not confined to DA axons (INGHAM et al. 1998). Nevertheless, some laboratories have made use of subtle, early morphological changes after neurotoxin lesion (for a review, see ZÁBORSKY et al. 1979) or have combined lesion-induced degeneration with immunocytochemistry (HORVATH et al. 1993) to improve the identification of probable DA axons.

III. Immunocytochemistry

The bulk of evidence for the synaptology of DA neurons comes from immunocytochemical studies. The use of highly selective antibodies allows identification of DA itself, its synthetic enzyme, tyrosine hydroxylase (TH), and its receptors and transporters. Immunocytochemistry also presents certain limitations (for a review, see LERANTH and PICKEL 1989), the most serious of which are the possibility of crossreaction with undesired antigens and non-specific binding of antibodies. The careful selection and characterization of antibodies and the use of appropriate blocking methods can usually minimize these problems. However, many antigens are sensitive to the conditions of tissue fixation that are essential for electron microscopic visualization of brain tissue. Moreover, the large size of antibodies limits their penetration. The methods used to enhance penetration (cycles of freezing and thawing or addition of detergent) can damage ultrastructure if excessive. Thus, immunoelectron microscopic studies are restricted to examinations of the outer surface of tissue, and even here, false-negative results can be expected (DESCARRIES et al. 1996).

Nevertheless, immunocytochemical methods are advantageous because of their flexibility and excellent specificity for particular antigens. In this chapter, we will emphasize the results obtained with immunocytochemical methods. However, the findings from other methods are included where appropriate.

While immunoreactivity for DA ought to be the preferred method for studying DA synaptology, most available DA antibodies work only in tissue fixed with high glutaraldehyde concentrations, a condition incompatible with the localization of many other antigens. Thus, many single-label, but few double-label studies utilize these antibodies. Conversely, TH antibodies offer considerable advantage for labeling of DA axons because they perform optimally in many different fixation conditions. By direct comparison with immunolabeling for dopamine-β-hydroxylase (DBH), DA, or the dopamine transporter (DAT), TH immunoreactivity has often been found to distinctly label DA as opposed to NE axons, particularly in the cortex (GASPAR et al. 1989; NOACK and LEWIS 1989; LEWIS and SESACK 1997), but also in other brain regions (LINDVALL and BJÖRKLUND 1978; VERNEY et al. 1987; ASAN 1993, 1998). The evidence to support this selective immunolabeling of DA axons by TH antisera has been extensively reviewed in the past (ASAN 1993, 1998; LEWIS and SESACK 1997) and will not be elaborated here. In addition, it should be noted that several studies directly comparing DA- and TH-immunoreactive profiles, have reported no significant differences in synaptology (GOLDMAN-RAKIC et al. 1989; TOTTERDELL and SMITH 1989; VERNEY et al. 1990; SESACK et al. 1995c; KARLE et al. 1996). Nevertheless, it is important to note that TH immunoreactivity cannot always be assumed to represent DA axons (GASPAR et al. 1989), particularly in regions where the DA innervation overlaps with dense NE or adrenergic afferents (LERANTH et al. 1988; ASAN 1998).

Immunostaining for the DAT is also an appropriate marker for DA axons in the striatum and nucleus accumbens, where its distribution closely matches that seen with DA or TH immunoreactivity (NIRENBERG et al. 1996b; HERSCH et al. 1997; NIRENBERG et al. 1997b; SESACK et al. 1998b). However, many DA axons in the rat cortex, particularly the prelimbic prefrontal cortex, appear to lack appreciable DAT protein (SESACK et al. 1998b). Thus, this marker is not useful for studying the synaptology of DA axons in some cortical regions and perhaps in other brain areas such as the hypothalamus and retina, where DAT levels are reportedly low (CERRUTI et al. 1993; LORANG et al. 1994).

IV. General Morphology and Synaptology

Using any of the above methods, DA axons have typically been found to be thin (0.1–0.3 µm; ARLUISON et al. 1984; SMILEY and GOLDMAN-RAKIC 1993b; GROVES et al. 1994; ASAN 1997b) and usually, though not always, unmyelinated. They give rise to varicosities whose small size (0.3–1.5 µm; PICKEL et al. 1981; ONTENIENTE et al. 1984; SEGUÉLA et al. 1988; SMILEY and GOLDMAN-RAKIC 1993b; DESCARRIES et al. 1996; ASAN 1997b) makes it sometimes difficult to define the limits of the varicose portion of DA axons (PICKEL et al. 1981;

FREUND et al. 1984; SMILEY and GOLDMAN-RAKIC 1993a; GROVES et al. 1994; HANLEY and BOLAM 1997; KUNG et al. 1998). In other instances, DA fibers possess large, well-defined varicosities (ARLUISON et al. 1984; ONTENIENTE et al. 1984; GROVES et al. 1994; DESCARRIES et al. 1996; ASAN 1997b). In the retina, DA fibers have features of both axons and dendrites (KOLB et al. 1990) and differ substantially from the axons of midbrain and diencephalic DA neurons. The vesicles contained in DA fibers are primarily small with clear centers; dense-cored vesicles are sometimes also observed, particularly within limbic and cortical regions (ONTENIENTE et al. 1984; VAN EDEN et al. 1987; SÉGUÉLA et al. 1988; PHELIX et al. 1992; SMILEY et al. 1992; SMILEY and GOLDMAN-RAKIC 1993a; SMILEY and GOLDMAN-RAKIC 1993b; ASAN 1997b). Peptides localized to these dense-cored vesicles typically include cholecystokinin or neurotensin (IBATA et al. 1983; LOOPUIJT and VAN DER KOOY 1985; BAYER et al. 1991a,b).

In most cases, DA axons form *en passant* synapses that are small, punctate junctions with parallel apposed membranes, a cleft that is only slightly widened, and intracleft dense material (PICKEL et al. 1981; BUIJS et al. 1984; HOKOC and MARIANI 1987; LERANTH et al. 1988; SÉGUÉLA et al. 1988; GROVES et al. 1994; DESCARRIES et al. 1996; ANTONOPOULOS et al. 1997; ASAN 1997b; RODRIGO et al. 1998). In keeping with their symmetric (Gray's type II; GRAY 1959) morphology, a postsynaptic thickening is often absent, which makes these specializations difficult to recognize when morphology is compromised or presynaptic immunolabeling is dense. Recognition of DA synapses is also limited by their small size and low probability of being detected in single sections (GROVES et al. 1994; DESCARRIES and UMBRIACO 1995; DESCARRIES et al. 1996), as well as the fact that synapses can be formed by intervaricose as well as varicose portions of DA axons (PICKEL et al. 1981; FREUND et al. 1984; PICKEL et al. 1988b; GROVES et al. 1994; SMITH et al. 1994; HANLEY and BOLAM 1997; KUNG et al. 1998). Occasionally, DA varicosities make asymmetric synapses with prominent postsynaptic densities (Gray's type I; GRAY 1959), although the frequency with which this occurs depends on the region and species examined (see below).

C. Regional Observations of DA Synaptology
I. Striatum: Dorsal Caudate and Putamen Nuclei

The DA nigrostriatal pathway provides a crucial modulatory influence on basal ganglia function that facilitates adaptive movement control. Numerous studies have described the DA neurons in the substantia nigra (SN) and ventral tegmental area (VTA) that project to the striatum and other forebrain targets in rodent and primate, as well as the distribution of DA fibers within the striatal complex. The reader is referred to primary and review articles on these subjects (FALLON and MOORE 1978; LINDVALL and BJÖRKLUND 1978; SWANSON 1982; BJÖRKLUND and LINDVALL 1984; OADES and HALLIDAY 1987; LAVOIE et al. 1989; FALLON and LOUGHLIN 1995; LEWIS and SESACK 1997). For clarity, the ventral portion of the striatum (nucleus accumbens) will be presented separately, as the morphology and synaptology of DA axons in this region differs somewhat from the more dorsal caudate and putamen nuclei.

1. Ultrastructural Morphology

The initial studies of presumed DA axons in the striatum were performed in the rat using uptake of radiolabeled DA or 5-hydroxydopamine (TENNYSON et al. 1974; ARLUISON et al. 1978; DESCARRIES et al. 1980; GROVES 1980; DOUCET et al. 1986). Transport into NE fibers should be a low probability event in this region, although uptake by 5-HT axons is a possibility, particularly in the ventral striatum (VAN BOCKSTAELE and PICKEL 1993). Moreover, these early studies suffered from problems of specificity and difficulty in identifying synapses. A more recent comparison of radiolabeled DA uptake with immunocytochemistry for DA is described below (DESCARRIES et al. 1996), as is a detailed, corrected analysis of 5-hydroxydopamine uptake (GROVES et al. 1994).

Surprisingly few studies have examined the nigrostriatal projection by anterograde tract-tracing. In one investigation, transport of radiolabeled amino acids and degeneration induced by 6-hydroxydopamine were used to examine this pathway (HATTORI et al. 1991). In both cases, terminals forming asymmetric axospinous synapses were prevalent, although symmetric synapses were also seen. These results are at variance with the findings from studies using other labeling methods. Furthermore, as mentioned previously, neither of the techniques used is selective for labeling DA axons, suggesting that the striatal collaterals of non-DA neurons contributed to the results. The argument put forth that DA axons branch and give rise to collaterals with separate morphology and transmitter phenotype (HATTORI et al. 1991) is not compelling, because evidence that the asymmetric morphology originates from DA neurons is lacking.

In a more recent investigation (HANLEY and BOLAM 1997), the DA innervation to the patch versus matrix compartments (GERFEN 1992) of the rat striatum was examined by both anterograde transport and TH immunocytochemistry. This elegant study reported no differences in any measure of size, synapse type, or target structure for profiles labeled by either method. Thus, even though the DA projection to these two compartments originates from different nigral neurons, and the compartments differ in their afferents and neurochemistry, the observed synaptology of the resident DA axons is virtually identical. This investigation also provided evidence for a non-DA nigrostriatal projection with γ-aminobutyric acid (GABA) phenotype that is directed solely to the matrix compartment (HANLEY and BOLAM 1997; see also RODRFGUEZ and GONZÁLEZ-HERNÁNDEZ 1999).

The first full immunocytochemical study of the striatal DA innervation was published in 1981 (PICKEL et al. 1981), and since that time, numerous immunolabeling studies have described the morphology and synaptology of

DA axons in the rat (ARLUISON et al. 1984; BOUYER et al. 1984b; FREUND et al. 1984; DECAVEL et al. 1987b; ZAHM 1992; DESCARRIES et al. 1996; HANLEY and BOLAM 1997), mouse (TRIARHOU et al. 1988), bird (KARLE et al. 1996), lizard (HENSELMANS and WOUTERLOOD 1994), monkey (SMILEY and GOLDMAN-RAKIC 1993b; SMITH et al. 1994), and, most recently, human (KUNG et al. 1998). Most of these studies have utilized antibodies against TH, since the striatum receives little NE input. Data from studies using antibodies directed against DA (HENSELMANS and WOUTERLOOD 1994; DESCARRIES et al. 1996; KARLE et al. 1996) or the DAT (NIRENBERG et al. 1996b; HERSCH et al. 1997) largely confirm the results with TH immunoreactivity, and in general, there is considerable agreement among the various studies regarding the morphology of DA axons and synapses. Striatal DA axons are small (0.1–0.3µm) and unmyelinated and give rise to varicosities (0.15–1.5µm) containing almost exclusively small clear vesicles. Dense-cored vesicles have been described only rarely in this region. In the human, striatal DA axons are slightly larger and occasionally myelinated (KUNG et al. 1998). DA varicosities have been estimated to comprise 15%–21% of the axon terminals in the rat striatum (PICKEL et al. 1981; ARLUISON et al. 1984; KARLE et al. 1996) and 16% of the synapses in the human striatum (KUNG et al. 1998). Significant intrastriatal regional differences have not been observed (PICKEL et al. 1981; ARLUISON et al. 1984; KARLE et al. 1996; HANLEY and BOLAM 1997), even between the human caudate and putamen nuclei (KUNG et al. 1998).

2. Synaptic Incidence, Synapse Types, and General Targets

The incidence with which DA striatal axons form synapses remains controversial. In studies reporting synaptic frequency from analysis of varicosities in single sections, the results vary considerably: 12%–30% in rats (ARLUISON et al. 1984; DESCARRIES et al. 1996), 24% in mice (TRIARHOU et al. 1988), 49% in birds (KARLE et al. 1996), 18% in lizards (HENSELMANS and WOUTERLOOD 1994), and 11% in humans (KUNG et al. 1998). The small length (i.e., diameter) of DA synapses is estimated at 0.21–0.27µm (mean synaptic area of 0.039µm^2) in rodent and human (GROVES et al. 1994; DESCARRIES et al. 1996; KUNG et al. 1998), suggesting that many DA synapses will be out of the plane of single sections. However, by serial section analysis, the reported frequency of DA synapses also varies. In the primate, the estimated synaptic frequency is nearly 100%, although only a small number of vesicle-containing profiles were examined in that study (SMITH et al. 1994). In an extensive evaluation of this question in the rat, DESCARRIES has estimated that DA varicosities identified either by radiolabeled DA uptake or DA immunocytochemistry form synapses 35%–40% of the time, whether calculated by extrapolation from single sections or full serial reconstruction (DESCARRIES et al. 1996).

An important contribution to this debate was provided by Groves, who re-examined material labeled by 5-hydroxydopamine uptake to identify probable DA axons and to reconstruct long portions of these fibers from serial sections (GROVES et al. 1994). While some of these axons are alternatingly thin and varicose, many DA axons maintain a nearly constant, small diameter throughout. Moreover, these axons clearly form synapses along their intervaricose segments, with no obvious preference for synapse formation at the varicosities. Similar observations had been suggested previously based on TH-immunocytochemistry (PICKEL et al. 1981; FREUND et al. 1984). The significance of these observations is that they call into question the custom of defining synaptic incidence per varicosity (see Sect. D.II, this chapter). Thus, there appears to be consensus that the varicose portions of DA axons do not always form synapses (PICKEL et al. 1981; FREUND et al. 1984; GROVES et al. 1994; DESCARRIES et al. 1996). However, regarding synapse formation along intervaricose segments of DA axons, DESCARRIES reportedly observed few such occurrences, while PICKEL, FREUND, and GROVES describe them as frequent events. The reasons for this discrepancy are unclear, although differences in the methods used for labeling DA axons and bias in the selection of profiles for measurement are possible contributors. Others have verified in human and non-human primates that DA axons form synapses along intervaricose segments (SMITH et al. 1994; KUNG et al. 1998), suggesting that a focus on varicosities as preferential sites of synapse formation may be unfounded (see Sect. D.II, this chapter), at least in the striatum.

By all recent accounts, striatal DA synapses are almost exclusively symmetric, regardless of the species or region of striatum examined. The rare occurrence of asymmetric DA synapses in the striatum varies from 0%–9% in different studies (Table 1). The vast majority of DA synapses target the spines and shafts of striatal dendrites, with only a few synapses formed on perikarya. Other DA axons are closely apposed to somal membranes without synapsing. The estimated frequency of spine versus shaft synapses varies considerably between studies (Table 1) for reasons that are unclear, beyond the obvious issue of sampling bias. The diameter of the DA axon and the area of the DA synapse are both smaller when the target structure is a spine compared to a dendrite (GROVES et al. 1994), suggesting that axospinous synapses will be observed less frequently in single-section analyses. However, this has not always been the case. Moreover, axospinous synapses are also observed less often in serial section analyses (SMITH et al. 1994; DESCARRIES et al. 1996). In the human striatum, an unbiased stereological estimate indicates that DA profiles form 55% of all the symmetric axospinous and axodendritic synapses in the striatum. However, unbiased stereology was not used to determine the relative proportion of DA synapses that are axospinous versus axodendritic. It is important to note that individual DA axons can make synapses onto both spines and dendrites (FREUND et al. 1984; GROVES et al. 1994). Moreover, the same dendrite can receive synaptic input from multiple DA axons.

In early striatal studies, axo-axonic contacts that were potentially synaptic were often described. However, most investigators now agree that true axo-axonic synapses are rare in this region. Nevertheless, there is one report of a few DA synapses on axon initial segments (FREUND et al. 1984), and

Synaptology of Dopamine Neurons

Table 1. Synaptic morphology and targets of dopamine axons in the striatal complex of different species

Investigation	Species	Labeling Method	No. of synapse-forming profiles examined	Symmetric synapses[a]	Synaptic target		
					Spine	Dendrite	Soma
Dorsal Striatum							
Arluison et al. 1984	Rat	TH immunocytochemistry	323	98%	51%	45%	4%
Freund et al. 1984	Rat	TH immunocytochemistry	411 total	99%	57%	36%	6%
			280 onto striatonigral cells	100%	59%	35%	6%
Triarhou et al. 1988	Mouse	TH immunocytochemistry	250	96%	47%	50%	3%
Zahm 1992	Rat	TH immunocytochemistry	422	[b]	66%	32%	2%
Groves et al. 1994	Rat	5-OHDA uptake	50	100%	56%	42%	2%
Smith et al. 1994	Monkey	TH immunocytochemistry	204[c]	100%	23%	72%	5%
Henselmans and Wouterlood 1994	Lizard	DA immunocytochemistry	48	[b]	8%	54%	38%
Descarries et al. 1996	Rat	^{3}H-DA uptake or DA immuno	43 uptake; 80 immuno	98%	30%	67%	3%
Karle et al. 1996	Pigeon	TH or DA immunocytochemistry	266 TH; 165 DA	100%	34%	60%	6%
Hanley and Bolam 1997	Rat	TH immuno or anterograde transport	254 immuno; 144 transport	98%	51%	46%	3%
Kung et al. 1998	Human	TH immunocytochemistry	125[d]	91%	60%	40%	0%
Nucleus Accumbens							
Arluison et al. 1984	Rat	TH immunocytochemistry	49	88%	39%	61%	0%
Voorn et al. 1986	Rat	DA immunocytochemistry	220 NAc	97%	38%	54%	8%
			171 OT	100%	37%	60%	4%
Totterdell and Smith 1989	Rat	TH immunocytochemistry	116	100%	34%	65%	<1%
Zahm 1992	Rat	TH immunocytochemistry	332 core	[b]	56%	39%	5%
			418 shell	[b]	30%	65%	5%
Henselmans and Wouterlood 1994	Lizard	DA immunocytochemistry	46 NAc	[b]	13%	78%	9%
			12 OT	[b]	17%	83%	0%
Ikemoto et al. 1996	Monkey	DA immunocytochemistry	217; NAc shell	67%	51%	44%	5%

[a] The remainder were asymmetric.
[b] "Nearly all" were symmetric.
[c] Most common profile type.
[d] Combination of simple profile counts and unbiased stereological analysis.

occasional axo-axonic synapses involving DA profiles are reported in the lizard striatal complex (HENSELMANS and WOUTERLOOD 1994). Moreover, DA axons are commonly in direct apposition to other axons and varicosities, including other DA profiles and the class of terminals forming asymmetric axospinous synapses (PICKEL et al. 1981; BOUYER et al. 1984b; DESCARRIES et al. 1996; KUNG et al. 1998).

3. Phenotypic Identification of Targets

The fact that DA axons often synapse on spines or on the shafts of spiny dendrites indicates that the GABA-containing medium spiny projection neuron is the preferred target of these afferents. This suggestion has been directly confirmed by demonstration of DA synapses on the dendrites and even soma of neurons retrogradely labeled from the substantia nigra (FREUND et al. 1984). Furthermore, DA axons synapse on dendrites immunoreactive for enkephalin or substance P, the peptides expressed in striatopallidal and striatonigral projection neurons, respectively (KARLE et al. 1992; PICKEL et al. 1992; KARLE et al. 1994). Somal contacts on these neuron types have also been observed (KUBOTA et al. 1986a; KUBOTA et al. 1986b; KUBOTA et al. 1987a), although the "mirror technique" used in these studies can only identify contacts on large profiles seen in two adjacent sections, only one of which is labeled. Since few DA axons synapse on soma, most of these identified contacts are likely to be appositions rather than synapses (PICKEL et al. 1992).

When contacting spines, DA axons synapse on the neck or on the side of the spine head, with the head itself always receiving asymmetric synaptic input (BOUYER et al. 1984b; FREUND et al. 1984; TRIARHOU et al. 1988; GROVES et al. 1994; HERSCH et al. 1997; KUNG et al. 1998). This architecture, which has been identified throughout the striatum in all species examined to date, provides the geometry necessary for DA to modulate specific excitatory afferents without appreciably altering the electrical and/or metabolic activity of the parent dendrite (FREUND et al. 1984; KOCH and ZADOR 1993; YUSTE and DENK 1995). Interestingly, it has been estimated that 47% of the spines of striatonigral projection neurons receive convergent asymmetric and symmetric synapses, and of the latter synapses, 83% are from DA afferents (FREUND et al. 1984). Admittedly, this analysis involved a sampling bias, and stereological estimates in the human striatum suggest that DA axons make up only 55% of symmetric axospinous synapses (KUNG et al. 1998). Nevertheless, DA appears to be one of the principal modulators of excitatory afferents to spines in this region. Many of the putative glutamate terminals forming convergent asymmetric synapses originate from the cortex (BOUYER et al. 1984b; FREUND et al. 1984; SMITH et al. 1994), a phenomenon that has been observed for multiple cortical areas and in multiple species. Thus, DA afferents are in a position to strategically modulate cortical transmission throughout the striatal complex (SMITH and BOLAM 1990). The thalamus is another possible source of excitatory afferents that could converge synaptically with DA axons. However, a recent study identified no such associations in the monkey striatum (SMITH et al. 1994). DA may also modulate convergent inputs to common dendritic shafts. For example, DA axons reportedly converge with enkephalin-immunoreactive local axon collaterals of medium spiny neurons (PICKEL et al. 1992). However, compared to spines, convergence on dendrites provides a lower degree of spatial selectivity for particular afferents.

The determination of whether DA axons also synapse on aspiny local circuit neurons in the striatum is complicated by the low density of these cells (less than 10% of striatal neurons) (KAWAGUCHI et al. 1995). However, many have widely spreading processes, suggesting that some of the dendritic shafts postsynaptic to DA axons may belong to interneurons. Historically, there is reason to predict that cholinergic aspiny cells would be synaptically innervated by DA (LEHMANN and LANGER 1983). Nevertheless, clear evidence for such a synaptic input is lacking (PICKEL and CHAN 1990). Some studies have described close contacts between DA axons and the soma or proximal dendrites of cells immunolabeled for choline acetyltransferase (KUBOTA et al. 1987b; CHANG 1988; DIMOVA et al. 1993). However, obvious synaptic specializations are typically absent at these sites (PICKEL and CHAN 1990). Interestingly, similar observations have been made in the striatal complex of reptiles (HENSELMANS and WOUTERLOOD 1994). DA and cholinergic terminals rarely synapse on common targets (CHANG 1988), but they are frequently in close apposition to each other (CHANG 1988; PICKEL and CHAN 1990). The latter observation, in addition to the known expression of DA receptors by cholinergic striatal neurons (see Sect. C.1.4, this chapter), is consistent with parasynaptic DA modulation of cholinergic transmission.

In other studies of striatal interneurons, few if any synaptic contacts occur between DA axons and dendrites labeled for neuropeptide Y (AOKI and PICKEL 1988; KUBOTA et al. 1988; VUILLET et al. 1989) or nicotinamide adenine dinucleotide phosphate, reduced (NADPH)-diaphorase (FUJIYAMA and MASUKO 1996). Somal appositions do occur in some cases. Furthermore, DA axons and terminals immunoreactive for neuropeptide Y sometimes converge on common postsynaptic elements or are directly apposed to each other (AOKI and PICKEL 1988). Together, these findings suggest that DA afferents to the striatum primarily target spiny projection neurons and rarely synapse on local circuit neurons.

4. Localization of Proteins Involved in DA Transmission

It is important to determine the distribution of DA receptors and transporters in relation to DA axons in order to gain information about the probable sphere of DA's parasynaptic influence. In the striatum, numerous studies have described the cellular localization of DA receptors by light microscopic immunocytochemistry or in situ hybridization. However, only a few studies have conducted ultrastructural examinations of the subcellular distribution of these receptors alone or in relation to DA axons. With regard to light microscopic studies, all five DA receptor subtypes have been localized to the striatum of rats, monkeys, and humans (BERGSON et al. 1995; LE MOINE and BLOCH 1995; JABER et al. 1996; MEADOR-WOODRUFF et al. 1996; LIDOW et al. 1998). The predominant D_1 and D_2 receptor subtypes have been localized to largely segregated populations of medium spiny projection neurons: D_1 on striatonigral (substance P containing) neurons and D_2 on striatopallidal (enkephalin containing) cells (LE MOINE and BLOCH 1995). It is generally appreciated that this anatomical segregation is at odds with electrophysiological observations of D_1 and D_2 receptor coexpression. However, a thorough discussion of this controversy is beyond the scope of this chapter, and the reader is referred to more pertinent references on the subject (HERSCH et al. 1995; LE MOINE and BLOCH 1995; SURMEIER et al. 1996). DA receptors are also strongly expressed by cholinergic interneurons in the striatum (LE MOINE et al. 1990; JONGEN-RELO et al. 1995), consistent with a prominent parasynaptic effect of DA on this cell class.

Ultrastructural studies largely confirm that D_1, D_2, and D_5 receptors are expressed predominantly in medium spiny neurons, by virtue of their being localized mainly to spines and distal dendrites (LEVEY et al. 1993; SESACK et al. 1994; BERGSON et al. 1995; HERSCH et al. 1995; YUNG et al. 1995; CAILLÉ et al. 1996), as well as by verification using morphological criteria (HERSCH et al. 1995; CAILLÉ et al. 1996), Golgi impregnation studies (FISHER et al. 1994), and dual immunocytochemistry for D_2 receptors and GABA (DELLE DONNE et al. 1997). Moreover, the electron microscopic study of HERSCH and colleagues supports the lack of D_1 and D_2 receptor co-localization within spines and distal dendrites. The possible distribution of DA receptors, particularly the D_2 subtype, to local circuit neurons has been noted (FISHER et al. 1994; HERSCH et al. 1995; YUNG et al. 1995) but little explored at the ultrastructural level.

Presynaptic localization of D_2 receptors is commonly reported in the striatum (LEVEY et al. 1993; FISHER et al. 1994; SESACK et al. 1994; HERSCH et al. 1995; SESACK et al. 1995b; YUNG et al. 1995; DELLE DONNE et al. 1997; HERSCH et al. 1997), with occasional reports as well of presynaptic D_1 receptors (LEVEY et al. 1993; BERGSON et al. 1995; YUNG et al. 1995). The presence of D_2 autoreceptors has been verified by localization of receptor immunoreactivity to axons also labeled for TH (SESACK et al. 1994; AN et al. 1998) or DAT (HERSCH et al. 1997). The localization of D_2 receptor immunolabeling to axon terminals with glutamate morphology has been reported in some studies (FISHER et al. 1994; SESACK et al. 1994; YUNG et al. 1995; DELLE DONNE et al. 1997). Other axon varicosities immunoreactive for the D_2 receptor and forming symmetric synapses (FISHER et al. 1994; SESACK et al. 1994; HERSCH et al. 1995) include GABA nerve terminals (DELLE DONNE et al. 1997). The localization of such heteroreceptors in the striatum is consistent with parasynaptic effects of DA to regulate release of other neurotransmitters, in particular GABA and glutamate.

Regarding the subcellular localization of DA receptors, studies utilizing immunoperoxidase methods have described a dense distribution of D_1 and D_2

receptor labeling to extrasynaptic plasma membranes, postsynaptic densities, and intracellular membranous structures such as the smooth endoplasmic reticulum (LEVEY et al. 1993; SESACK et al. 1994; HERSCH et al. 1995; YUNG et al. 1995; CAILLÉ et al. 1996; AN et al. 1998). However, some of this distribution may reflect diffusion of the peroxidase reaction product from the site of generation, particularly when the reaction is excessive (NOVIKOFF et al. 1972; COURTOY et al. 1983). This is supported by the less-extensive association of D_1 receptor immunoreactivity with synaptic membranes when peroxidase staining is less intense (YUNG et al. 1995; AN et al. 1998).

Immunogold labeling methods provide a more stable marker for subcellular localization, although these techniques are typically less sensitive than peroxidase immunocytochemistry (CHAN et al. 1990). Using a pre-embedding immunogold method, the principal localization of both D_1 and D_2 receptors to extrasynaptic plasmalemmal surfaces has been confirmed in rat striatal sections (YUNG et al. 1995; CAILLÉ et al. 1996; DUMARTIN et al. 1998). The likelihood that this immunoreactivity represents functional receptors is suggested by the internalization of D_1-immunogold labeling following systemic administration of D_1 agonists (DUMARTIN et al. 1998). Immunogold particles are occasionally observed at or near asymmetric or symmetric synapses, the latter indicating the potential localization of receptors postsynaptic to DA axons (YUNG et al. 1995). However, with dual immunogold labeling for D_1 receptor and immunoperoxidase staining for TH, the synapses formed by TH-immunoreactive axons rarely exhibit immunogold labeling for the D_1 receptor, although D_1 labeling sometimes appears close to these synaptic sites or along plasmalemmal surfaces closely apposed to TH-labeled axons (CAILLÉ et al. 1996). While these observations argue for a predominant extrasynaptic localization of D_1 receptors, it should be noted that the relative lack of synaptic receptors is likely to reflect both the low sensitivity of the immunogold method and the possible exclusion of gold-tagged antibodies from the synaptic protein complex in thick sections (BAUDE et al. 1995; NUSSER et al. 1995).

In a preliminary study of the monkey dorsal caudate nucleus, we have performed dual immunolabeling of TH and D_1 and D_2 receptors by reversing the markers and using the more sensitive immunoperoxidase method for localization of the DA receptors (AN et al. 1998). Despite the concern regarding false-positive labeling with peroxidase, we have found that the majority of synapses formed by TH-immunoreactive axons on D_1-positive dendrites (62%) do not exhibit synaptic D_1 receptor labeling. Nevertheless, a synaptic localization of D_1 immunoreactivity is seen in 38% of the TH synapses on these structures. Conversely, the majority of synapses formed by TH-immunoreactive axons on D_2-positive dendrites (73%) exhibit synaptic D_2 receptor labeling, with the remainder exhibiting only extrasynaptic D_2 immunoreactivity. These results support the predominant extrasynaptic localization of DA receptors, but suggest that some D_1 and especially D_2 receptors may be localized in the immediate vicinity of DA synapses.

The function of the DAT is crucial for determining the extent and duration of DA diffusion in the extracellular space and the degree to which DA can access its receptors at pharmacologically relevant concentrations (GARRIS and WIGHTMAN 1994; GIROS et al. 1996). In the striatum, this protein has been localized to DA axon varicosities and intervaricose segments (NIRENBERG et al. 1996b; HERSCH et al. 1997; SESACK et al. 1998b), and investigations utilizing immunogold methods (NIRENBERG et al. 1996b; HERSCH et al. 1997) have identified the subcellular distribution of the DAT as being along the plasma membrane throughout DA axons and in the perisynaptic region immediately at the edges of synaptic contacts. However, evidence of synaptic DAT localization is lacking. Whether this reflects the true distribution of the DAT or an inability of the pre-embedding methods used to detect proteins complexed in synaptic densities (BAUDE et al. 1995; NUSSER et al. 1995) has not yet been determined. Nevertheless, the extensive perisynaptic and extrasynaptic localization of this crucial protein suggests that the extracellular diffusion of DA in the striatum is tightly controlled, a conclusion supported by neurochemical studies (GARRIS and WIGHTMAN 1994; GIROS et al. 1996).

II. Striatum: Nucleus Accumbens

1. Ultrastructural Morphology, Synapses, and General Targets

The synaptology of DA axons in the ventral striatum, or nucleus accumbens (NAc) core and shell, has been examined by immunocytochemistry for TH or DA in rats and monkeys (ARLUISON et al. 1984; BOUYER et al. 1984a; VOORN et al. 1986; PICKEL et al. 1988b; TOTTERDELL and SMITH 1989; ZAHM 1992; IKEMOTO et al. 1996). This region receives a minimal NE input, suggesting that TH immunolabeling is an adequate marker for DA axons (TOTTERDELL and SMITH 1989). NAc DA fibers resemble those in the striatum in their small size (0.1–0.2μm), absence of myelination, presence of varicose portions (0.3–1.5μm), content of primarily small clear vesicles, and formation of mainly symmetric synapses. DA varicosities may comprise as many as 33% of the boutons in the NAc (BOUYER et al. 1984a). However, compared to their striatal counterparts, they more frequently: exhibit dense-cored vesicles (1–2 or more per varicosity), contain flattened clear vesicles, form synapses that are shorter in length (0.1–0.15μm), and synapse on dendrites with larger diameter than comparable postsynaptic elements in the striatum. These observations are not unexpected, based on the different midbrain source of afferents to the ventral versus dorsal striatum. Similar observations have been made on the basis of anterograde tract-tracing in rats (ZAHM 1992), although as much as 20% of the mesoaccumbens projection is non-DA (SWANSON 1982; VAN BOCKSTAELE and PICKEL 1995). Estimates of synaptic incidence in single sections range from 15%–52% in the rat and monkey (BOUYER et al. 1984a; VOORN et al. 1986; ZAHM 1989; IKEMOTO et al. 1996), and in one limited serial section analysis, a synaptic incidence of 72% was reported in the monkey NAc (SMITH et al. 1999). DA axons have been reported to form synapses along their intervaricose segments (VOORN et al. 1986; PICKEL et al. 1988b), and nonsynaptic axo-axonic appositions involving DA fibers are common (BOUYER 1984a; et al. PICKEL et al. 1988b; ZAHM 1989; IKEMOTO et al. 1996).

Early studies of DA axons in the NAc were conducted prior to the parcellation of this region into core and shell districts (ZAHM and BROG 1992). Using DA immunocytochemistry, VOORN et al. did not report subregional differences within the NAc (VOORN et al. 1986). However, sampling in this study (described as ventromedial, dorsomedial, and septal pole) was probably confined to the NAc shell. In a subsequent direct comparison of DA axons in the dorsal striatum and NAc core and shell in the rat (ZAHM 1992), it was determined that DA axon varicosities in the NAc shell are smaller in median area (0.18 versus $0.26\,\mu m^2$ for striatum and core) and more frequently synapse on dendrites than spines. The studies by VOORN and ZAHM also represent the only ultrastructural examinations of DA fibers in the rat olfactory tubercle. Within this region, the synaptic incidence and formation of symmetric synapses on dendrites, spines, and soma is similar to that seen in the NAc, with the exception that asymmetric synapses have not been reported.

The observation of smaller size of DA profiles in the NAc shell of the rat (ZAHM 1992) is surprising in light of the larger size reported for a homologous region in the monkey (IKEMOTO et al. 1996): i.e., $0.33\,\mu m^2$ compared to $0.14\,\mu m^2$ in the striatum (SMITH et al. 1994). Moreover, in the primate NAc shell, DA axons form asymmetric synapses on dendrites, spines, and soma with much greater frequency (33%) than in the rat. In our own preliminary comparison of the monkey striatum and NAc using TH antibodies (AN et al. 1998, 1999), we have confirmed the morphological observations of IKEMOTO. Thus, the reported differences in synaptology appear to be related to species rather than labeling methods. The functional significance of this marked species difference in the mesoaccumbens DA innervation has yet to be determined. In the rat, hypothalamic afferents to the NAc shell have been described (SIM and JOSEPH 1991; BROG et al. 1993). However, it remains to be determined whether these (1) include projections from DA cell groups, (2) form asymmetric synapses, and (3) constitute a greater proportion of DA afferents to the primate NAc shell.

2. Phenotypic Identification of Targets

Compared to the striatum, the finding that DA axons in the NAc synapse more often on dendrites than spines, and more frequently on larger dendrites, suggests either that DA's effects in this region are mediated by more proximal contacts on medium spiny neurons or that DA axons more frequently target aspiny cells (BOUYER et al. 1984a). However, few if any DA synapses have been reported on dendrites or soma of neurons that are cholinergic (PICKEL and CHAN 1990), neuropeptide Y-containing (AOKI and PICKEL 1988), or GABA-labeled aspiny (PICKEL et al. 1988b). Thus, mesoaccumbens DA axons appear to exhibit no greater innervation of interneurons than observed for nigrostriatal fibers and therefore may target more proximal portions of medium spiny neurons. The latter conclusion is consistent with observations of DA synapses on dendrites immunoreactive for GABA (PICKEL et al. 1988b).

The spines postsynaptic to DA axons also receive asymmetric synaptic input on the head (VOORN et al. 1986; IKEMOTO et al. 1996), as observed in the striatum. The sources of this convergent excitatory input have been well examined in the NAc, and include all of the known cortical afferents: hippocampal (TOTTERDELL and SMITH 1989; SESACK and PICKEL 1990), prefrontal (SESACK and PICKEL 1992), and amygdaloid (JOHNSON et al. 1994). In addition, some of these excitatory, asymmetric terminals in the rat derive from the thalamic paraventricular nucleus (PINTO and SESACK 1998), an observation in contrast to reports in the monkey striatum, where thalamic afferents appear not to converge with DA axons (SMITH et al. 1994). In the NAc, the dendrites postsynaptic to DA axons also receive convergent symmetric or asymmetric synapses, some of which are immunoreactive for GABA, substance P, or 5-HT (PICKEL et al. 1988a,b; VAN BOCKSTAELE and PICKEL 1993). DA axons are also directly apposed to cortical terminals (SESACK and PICKEL 1990, 1992) and axons labeled for GABA (PICKEL et al. 1988a), 5-HT (VAN BOCKSTAELE and PICKEL 1993), or substance P (PICKEL et al. 1988a).

3. Localization of Proteins Involved in DA Transmission

Most DA receptors have been localized to neurons in the NAc, with a predominance of D_1, D_2, and D_3 subtypes (LANDWEHRMEYER et al. 1993; LE MOINE and BLOCH 1995; JABER et al. 1996). At the ultrastructural level, D_2 receptors have been localized to dendrites, spines, soma, and axons in the NAc core and shell of rats and monkeys (SESACK et al. 1995b; DELLE DONNE et al. 1996, 1997; AN et al. 1999). In the rat NAc shell, some of the soma, dendrites, and spines expressing D_2 receptors also contain GABA and probably derive from medium spiny neurons (DELLE DONNE et al. 1997). Other D_2-immunoreactive somatodendritic structures contain neurotensin and/or exhibit morphology consistent with aspiny local circuit neurons (DELLE DONNE et al. 1996, 1997). In the monkey, many of the axons expressing D_2 immunoreactivity are also labeled for TH, indicating the presence of autoreceptors, or form asymmetric axospinous synapses, consistent with their being glutamatergic (AN et al. 1999). In the rat, other D_2-immunoreactive axons are GABA labeled and occur in greater density than observed in the dorsal striatum (DELLE DONNE et al. 1997). Fewer ultrastructural studies have examined D_1 receptors in the NAc. In a preliminary analysis in the monkey, D_1 immunoreactivity has been localized to spines and dendrites, but not axons in the NAc core and shell (AN et al. 1999). Moreover, the proportion of D_1- or D_2-immunolabeled targets postsynaptic to TH-immunoreactive axons that exhibit evidence for synaptic receptor localization are similar to those reported for the striatum (AN et al. 1999).

The cellular and subcellular localization of the DAT in the rat NAc is similar to that observed in the striatum, including the distribution to varicose and intervaricose DA axon segments and predominance along extrasynaptic portions of the plasma membrane (NIRENBERG et al. 1997b). However, in the NAc shell, the density per axon of immunogold labeling for the DAT protein is lower than observed in the core (and likely the striatum as well). This observation is consistent with the reportedly reduced efficiency of DA uptake in the NAc shell compared to the core (JONES et al. 1996). While reduced uptake suggests a higher degree of parasynaptic actions in the NAc, it is interesting that this is accompanied by a lower density of DA D_2 receptors (MENGOD et al. 1989; MANSOUR et al. 1990; SESACK et al. 1995b; DELLE DONNE et al. 1997).

III. Globus Pallidus and Basal Forebrain

Few studies have examined the ultrastructure of DA axons within the globus pallidus proper, although the ventromedial pallidum corresponding to the nucleus basalis of Meynert has been investigated (see below). In one study of the rat globus pallidus, the majority of TH-immunoreactive profiles were axons of passage, and labeled varicosities forming synapses were rarely observed (ARLUISON et al. 1984). These observations are consistent with the finding that immunoreactivity for D_1 and D_2 receptors is sparse in this region, with D_1 receptors being mainly presynaptic on presumed striatal axons (YUNG et al. 1995).

By light microscopic examination of the rat ventral forebrain, the comparative distribution and morphology of axons labeled for TH versus DBH suggests that many TH-immunoreactive fibers represent DA axons (GAYKEMA and ZÁBORSZKY 1996; ZÁBORSZKY and CULLINAN 1996; RODRIGO et al. 1998). Nevertheless, the possible contribution of NE fibers to the synaptic connections observed for TH-labeled axons has not been ruled out (MILNER 1991a,b). Fibers labeled by TH-immunocytochemistry are frequently apposed to cholinergic perikarya in the diagonal band nuclei, substantia innominata, bed nucleus of the stria terminalis, and ventral pallidum (GAYKEMA and ZÁBORSZKY 1996; ZÁBORSZKY and CULLINAN 1996; RODRIGO et al. 1998). Within the ventral pallidum and diagonal band nuclei, synaptic contacts between TH-labeled fibers and cholinergic, as well as non-cholinergic structures, have been verified by electron microscopy (MILNER 1991a; MILNER 1991b; ZÁBORSZKY and CULLINAN 1996; RODRIGO et al. 1998). However, in both regions, the principal synaptic targets are distal dendrites and spines, with synapses on soma being uncommon. Most TH-immunolabeled profiles in the pallidum form symmetric synapses (RODRIGO et al. 1998), while those in the diagonal band are primarily asymmetric (MILNER 1991b). TH-immunoreactive axons in both areas converge synaptically on common targets with unlabeled or cholinergic terminals (MILNER 1991a; MILNER 1991b; RODRIGO et al. 1998).

The observations that TH-labeled axons rarely synapse on soma in the pallidum and diagonal band (MILNER 1991a,b; RODRIGO et al. 1998) and that cholinergic neurons in the diagonal band are heavily invested with astrocytic processes (MILNER 1991a) suggests that few of the light microscopic contacts described in these regions (GAYKEMA and ZÁBORSZKY 1996; ZÁBORSZKY and CULLINAN 1996; RODRIGO et al. 1998) represent synaptic input from DA axons to cholinergic perikarya. Nevertheless, some synaptic DA input to cholinergic neurons in the ventral pallidum and substantia innominata is supported by anterograde tract-tracing from the SN and VTA (GAYKEMA and ZÁBORSZKY 1996). In addition, fibers anterogradely labeled from midbrain DA cell regions synapse on parvalbumin-immunoreactive, presumed GABA-containing dendrites and soma in the basal forebrain (GAYKEMA and ZÁBORSZKY 1997). Together, these observations suggest that DA midbrain neurons exert complex modulatory actions on a number of forebrain and cortical areas via direct projections to these regions and via synapses on the cholinergic and GABA neurons that maintain widespread connections to the same targets.

IV. Lateral Septum and Bed Nucleus of the Stria Terminalis

DA projections to the rat septal complex and bed nucleus of the stria terminalis contribute to the regulation of autonomic, endocrine, and behavioral responses to stressful environmental circumstances. Within the lateral septum, DA fibers identified by DA or TH immunoreactivity are conspicuous in the rostral basal portion and more caudally and dorsally in a band curving from the lateral ventricles to the medial septum (ONTENIENTE et al. 1984; DECAVEL et al. 1987b; JAKAB and LERANTH 1990). DA varicosities display similar morphological features in both areas, although those in the caudal dorsal septum tend to be larger (ONTENIENTE et al. 1984). In both regions, DA axons synapse on dendrites and perikarya (ONTENIENTE et al. 1984; VERNEY et al. 1987; JAKAB and LERANTH 1990; ANTONOPOULOS et al. 1997), with the latter being more frequently detected in the caudal septum where DA fibers form pericellular basket-like arrays (ONTENIENTE et al. 1984; JAKAB and LERANTH 1990). The DA nature of these pericellular baskets has been established by their absence of immunoreactivity for DBH (VERNEY et al. 1987). Occasional axospinous synapses are seen, some consisting of somatic spines, and DA profiles are sometimes apposed to unlabeled terminals. The frequency of synapse formation per DA profile has been estimated at 45% by extrapolation from single sections (ANTONOPOULOS et al. 1997).

With regard to identified cellular targets, presumed DA axons labeled by TH-immunoreactivity synapse on somatospiny neurons in the mediolateral lateral septum, some of which are GABA-immunoreactive (JAKAB and LERANTH 1990). This finding confirms a prior observation using a more indirect approach (ONTENIENTE et al. 1987). On the same population of somatospiny cells, TH-labeled axons converge with hippocampal, presumably glutamate afferents synapsing on common soma and dendrites (JAKAB and LERANTH 1990). Whether these afferents also converge synaptically on common distal dendrites of these same cells has not been determined but is suggested by the frequent observation of TH-labeled and hippocampal terminals synapsing on common dendrites in the neuropil (JAKAB and LERANTH 1990).

There exists some controversy regarding the morphology of DA synapses in the septum, with some investigators reporting primarily asymmetric type (ONTENIENTE et al. 1987; VERNEY et al. 1987) and others describing mainly symmetric synapses (JAKAB and LERANTH 1990; ANTONOPOULOS et al. 1997). Interestingly, the density of perineuronal baskets (VERNEY et al. 1987) and both asymmetric and axosomatic synapses (ANTONOPOULOS et al. 1997) increase with postnatal age. This suggests the possibility that the septum is innervated by two different DA fiber systems, one developing early and forming primarily symmetric axodendritic synapses and the second developing towards the end of the second week postnatal and forming primarily asymmetric axosomatic synapses (ANTONOPOULOS et al. 1997). It is not yet known whether these inputs correspond to the differential innervation of the septum by VTA and hypothalamic DA cell groups (JAKAB and LERANTH 1993). Furthermore, some of the discrepancy may derive from different groups examining different regions of the lateral septum and focusing on proximal versus distal synaptic inputs. Finally, the functional consequences of a dual DA innervation have yet to be delineated, although the electrophysiological actions of this transmitter in the septum are certainly complex.

Within the rat bed nucleus of the stria terminalis, presumed DA axons labeled by TH immunocytochemistry are most pronounced in the dorsolateral region. Based on comparison to the distribution of DBH-immunoreactive fibers, the majority of TH-labeled profiles in this dorsolateral area appear to represent DA axons (PHELIX et al. 1992). By electron microscopic examination, approximately 30% of TH-labeled profiles in the dorsolateral bed nucleus form synapses in single sections. Axodendritic synapses are the most common type of contact observed, with many being asymmetric in character. Symmetric axospinous and symmetric or asymmetric axosomatic synapses are also found, and axo-axonic contacts lacking synaptic specializations have been described (PHELIX et al. 1992). At least one of the cellular targets of presumed DA axons in the bed nucleus involves neurons expressing corticotropin-releasing factor (PHELIX et al. 1994).

V. Amygdala

Through projections to the amygdala, DA modulates the learning of autonomic and operant responses to aversive and appetitive conditions. Within the rat amygdala, DA axons labeled by TH immunocytochemistry are densely distributed within the intercalated cell groups, central nucleus (medial and rostral intermediate divisions and the medial portion of the centrolateral region), and basal nuclei (ASAN 1997b, 1998). The overlap with fibers labeled for DBH or phenylethanolamine N-methyltransferase (PNMT) is greatest in the medial central nucleus and least in the intercalated cell groups. Based on extensive comparative studies, TH-immunoreactivity appears to be absent from many fibers labeled by DBH but to faithfully represent all PNMT-stained axons (ASAN 1998). Thus, the extent to which TH-immunolabeling represents DA axons varies depending on which amygdala subdivision is examined, being most representative in the intercalated nuclei and the medial portion of the centrolateral central nucleus. Moreover, the low sensitivity of the methods used for immunoelectron microscopy enhance the selectivity of TH-immunoreactivity for DA axons (ASAN 1998). It should be noted that the distribution of TH immunolabeling in the amygdala matches well the light microscopic expression of DA receptors (LEVEY et al. 1993) and DAT (REVAY et al. 1996).

The majority of synapses formed by TH-immunoreactive axons are symmetric, with a synaptic incidence estimated at 7%–20% in single sections (ASAN 1997b, 1998). Larger diameter TH-immunolabeled varicosities occasionally form asymmetric synapses with an incidence ranging from 11%–22% in the central nucleus, 4% in the basal nuclei, and none in the intercalated cell groups (ASAN 1997b). Such large profiles closely resemble those labeled by PNMT antibodies, suggesting that many derive from adrenergic fibers (ASAN 1998). Within the central and basal nuclei, TH-labeled axons synapse primarily on dendrites (approximately 64%–76%), frequently on spines (23%–28%), and rarely on soma (<1%–3%). Although few soma are contacted by TH-positive fibers, those that do receive input exhibit multiple points of contact. Within the intercalated amygdala cell groups, probable DA profiles are exclusively symmetric and occur more frequently on soma (as high as 20%) and proximal dendrites; otherwise, the synaptology of TH-immunolabeled axons is similar to that of the central and basal nuclei.

The fact that projection neurons in amygdaloid nuclei are often spiny cells (ASAN 1998) suggests that many of the DA afferents target output neurons. Furthermore, DA axons synapse on neurotensin-containing (BAYER et al. 1991b) and somatostatin-expressing (ASAN 1997a) cells of the centrolateral subdivision of the central nucleus that are believed to project to brainstem and forebrain targets. Unlabeled dendrites postsynaptic to DA afferents may belong to projection or local circuit neurons. The spines and dendrites innervated by DA varicosities also receive synaptic input from unlabeled terminals. The triadic arrangement of a DA profile contacting a spine that also receives asymmetric input from a presumed glutamate terminal is a common observation in the amygdala (ASAN 1998), suggesting that DA modulates glutamate transmission in this region in a manner similar to the striatum and other target areas. Candidate sources of this excitatory drive include sensory and limbic cortices, the thalamus, and intra-amygdaloid projections. Appositions between DA varicosities and unlabeled terminals, some of which exhibit the morphology of glutamate afferents, are also frequently observed in amygdaloid nuclei (ASAN 1997b, 1998).

VI. Cortex

The mesocortical DA projection plays an important role in facilitating cognitive, motor, and to a lesser extent sensory functions. In the rodent, DA fibers are most dense in limbic regions: the medial prefrontal cortex (prelimbic and infralimbic cortices), suprarhinal area, other rhinal cortices, and the cingulate cortex (DESCARRIES et al. 1987; VAN EDEN et al. 1987; BERGER et al. 1991). Within prefrontal and rhinal cortices, DA axons are expressed primarily in layers 5 and 6, while the cingulate cortex receives a more substantial input to layers 1–3. In motor and sensory areas, sparse DA axons are observed mainly in layer 6 (DESCARRIES et al. 1987; PAPADOPOULOS et al. 1989; BERGER et al. 1991). Deep layer DA fibers originate primarily from cells in the VTA, while superficial layer afferents derive mainly from SN neurons (DESCARRIES et al. 1987; VAN EDEN et al. 1987; BERGER et al. 1991). The primate mesocortical DA innervation is more extensive, particularly within motor and association areas, where DA fibers are most dense in superficial layers, with a more moderate density in deep layers (BERGER et al. 1991; LEWIS and SESACK 1997). Compared to the rat, the midbrain neurons giving rise to the cortical DA innervation in the monkey are more broadly and loosely distributed within the SN, VTA, and retrorubral fields (WILLIAMS and GOLDMAN-RAKIC 1998).

1. Ultrastructural Morphology

Following the discovery of a cortical DA innervation separate from NE fibers (THIERRY et al. 1973), ultrastructural studies examining these inputs were first performed using uptake of radiolabeled DA. Selective uptake blockers can be used to prevent spurious labeling of NE axons, and the nearly complete loss of labeling following destruction of midbrain DA neurons argues for the specificity of the findings (DESCARRIES et al. 1987; DOUCET et al. 1988). Varicosities labeled in this manner range in diameter from 0.5 to 0.9μm, contain mostly small clear vesicles, and arise from unmyelinated axons (DESCARRIES et al. 1987). However, the synaptic connections of these profiles are often difficult to discern with this methodology.

Ultrastructural investigation of mesocortical projections by use of tract-tracing has been attempted in only one study to date (KURODA et al. 1996). Anterograde transport of wheat germ agglutinin-conjugated horseradish peroxidase (WGA-HRP) from the VTA to the prefrontal cortex primarily labels large varicosities (0.7–2.8μm) that contact dendritic shafts at prominent symmetric or asymmetric synapses. Such morphological features are at odds with the characteristics of DA axons labeled by immunocytochemistry (see below) and are more consistent with synapses formed by GABA terminals (for symmetric synapses) or the collaterals of retrogradely labeled cortical neurons (for asymmetric synapses). We have recently confirmed that GABA neurons contribute substantially to the mesocortical projection in the rat (SESACK and CARR 1998), of which only 30% is dopaminergic (SWANSON 1982).

Thus, a precise view of the synaptology of cortical DA axons generally requires application of immunocytochemical methods. Studies directly comparing antibodies against TH or DA find that these markers produce remarkably similar results (GOLDMAN-RAKIC et al. 1989; VERNEY et al. 1990; SESACK et al. 1995c), in keeping with evidence that TH antibodies primarily label DA axons in the cortex (LEWIS and SESACK 1997). However, most of these investigations have focused on frontal cortices that receive a substantial DA input in rodents and primates. In sensory cortical areas, where the density of NE axons exceeds that of DA afferents, TH-immunolabeling is present in a higher percentage of fibers that are also immunoreactive for DBH (GASPAR et al. 1989; NOACK and LEWIS 1989).

Within the deep layers of the rat medial prefrontal cortex, DA axons labeled by TH or DA immunocytochemistry are generally unmyelinated and give rise to varicosities filled with small clear, usually pleomorphic vesicles, as well as occasional dense-cored vesicles (VAN EDEN et al. 1987; SÉGUÉLA et al. 1988; VERNEY et al. 1990; SESACK et al. 1995c). Similar morphological features have been described in the superficial layers of the medial prefrontal, suprarhinal, cingulate, and primary visual cortices (VAN EDEN et al. 1987; SÉGUÉLA et al. 1988; PAPADOPOULOS et al. 1989; VERNEY et al. 1990). In the monkey, most of the ultrastructural studies examining DA axons have also focused on frontal cortices: prefrontal, cingulate, and motor (GOLDMAN-RAKIC et al. 1989; SMILEY and GOLDMAN-RAKIC 1993a,b; SESACK et al. 1995a,c, 1998a). The morphological features of these axons are overall similar to those described in the rodent and do not exhibit laminar heterogeneity, with the exception of a population of large diameter fibers in layer 1 that represent parent axons (SMILEY and GOLDMAN-RAKIC 1993a). Similar morphological observations have also been made in the human temporal cortex from surgically excised tissue (SMILEY et al. 1992).

2. Synaptic Incidence, Synapse Types, and General Targets

The frequency with which DA axons form synapses in cortical regions has been the subject of some controversy. Monoamine cortical systems were initially reported to form synapses infrequently and instead, to mediate diffuse, non-specific actions primarily via parasynaptic communication (BEAUDET and DESCARRIES 1978; DESCARRIES and UMBRIACO 1995). However, subsequent analysis of DA varicosities in the rat medial prefrontal cortex has revealed a high incidence of synaptic contacts (93%) when viewed in complete serial sections (SÉGUÉLA et al. 1988). Interestingly, DA varicosities in the suprarhinal area exhibit a lower synaptic incidence of 56% in serial sections. In a more limited sample of DA-immunoreactive varicosities in the primary visual cortex, an overall synaptic incidence of 90% per varicosity has been reported (PAPADOPOULOS et al. 1989). Ensuing analyses have also demonstrated a high incidence of synapse formation by NE and 5-HT cortical axons (PARNAVELAS and PAPADOPOULOS 1989, 1991; BLOOM 1991), although there may be regional differences and not all studies agree (DESCARRIES and UMBRIACO 1995). These observations underscore the need to assess synaptic incidence by full serial reconstruction of axons, including an examination of the intervaricose segments. Nevertheless, the findings suggest that the mesocortical DA system may be characterized primarily by synaptic actions on its targets, although parasynaptic effects cannot be ruled out and may be regionally specific. This is further suggested by observations in the primate prefrontal cortex, in which the overall synaptic incidence is lower than observed in the rodent (39%, with laminar variation of 27%–45%; SMILEY and GOLDMAN-RAKIC 1993a).

The majority of DA profiles in the prefrontal cortex form symmetric synapses (VAN EDEN et al. 1987; SÉGUÉLA et al. 1988; SESACK et al. 1995c), although asymmetric junctions are sometimes observed: 16%–22% in the rat medial prefrontal and suprarhinal areas (SÉGUÉLA et al. 1988), 3% in the monkey prefrontal cortex (SMILEY and GOLDMAN-RAKIC 1993a), and 13% in human temporal cortex (SMILEY et al. 1992). In the rat primary visual cortex, a greater number of asymmetric compared to symmetric synapses are formed by DA-immunoreactive varicosities (PAPADOPOULOS et al. 1989), suggesting that the functions of this transmitter may differ in primary sensory areas. The length of the DA synaptic contact has been estimated at 0.21μm (SÉGUÉLA et al. 1988).

The synaptic targets of cortical DA axons include both dendritic spines and small- to medium-diameter dendrites (VAN EDEN et al. 1987; SÉGUÉLA et al. 1988; PAPADOPOULOS et al. 1989; SESACK et al. 1995c). Dendritic shafts comprise 62% and 89% of the postsynaptic targets in the rat medial prefrontal and suprarhinal cortices, respectively (SÉGUÉLA et al. 1988) and 64% in the monkey prefrontal cortex (SMILEY and GOLDMAN-RAKIC 1993a). Most of the remaining synapses are axospinous. However, in the human temporal cortex, the synaptic input to spines appears to be more extensive than to shafts (57% versus 43%; SMILEY et al. 1992). This observation may relate to the fact that most of the synapses were detected in deep layers, as laminar comparisons in the monkey reveal that dendritic shafts are the principle targets in superficial layers (79%), while DA axons are equally likely to synapse on spines as dendrites (50%) in deep layers (SMILEY and GOLDMAN-RAKIC 1993a). Axosomatic synapses are observed only rarely in any species, although DA axons sometimes course along the plasmalemma of soma and proximal dendrites. Furthermore, DA axons are frequently in direct apposition to other axonal structures, and while membrane thickenings are occasionally seen at these sites, true axo-axonic synapses have not been identified. Finally, DA axons have recently been reported to lie in close apposition to the basal lamina of small cerebral blood vessels in the superficial layers of frontal cortical areas, suggesting a role in blood flow regulation (KRIMER et al. 1999).

3. Phenotypic Identification of Targets

The fact that cortical DA profiles synapse on dendritic spines suggests that some of the targets are spiny, glutamate-containing pyramidal neurons, a hypothesis confirmed by Golgi staining in the monkey frontal cortex (GOLDMAN-RAKIC et al. 1989). A similar relationship is thus likely in the rodent (see especially VERNEY et al. 1990). Whether DA axons differentially target subpopulations of pyramidal neurons is difficult to ascertain because of the relative neurochemical homogeneity of pyramidal cells and the failure of most retrograde tracers to label the distal dendrites where DA fibers synapse. We have recently applied retrograde transport of pseudo-rabies virus (CARD et al. 1993) to address this issue, because transsynaptic passage of viral particles requires transport into the distal dendrites where most synapses are located (DeFELIPE and FARIÑAS 1992). By using this approach in the rat prefrontal cortex, we have shown that DA axons synapse on the spines and dendrites of pyramidal neurons that project to the NAc (CARR et al. 1997). We are currently examining whether mesocortical DA axons synaptically innervate other classes of pyramidal neurons. Another approach to this question is to combine immunocytochemistry for DA or TH with intracellular filling of individual pyramidal cells (KRIMER et al. 1997). While this method has been applied primarily at the light microscopic level, some of the points where TH-labeled fibers contact pyramidal cell dendrites may be synaptic, as demonstrated by electron microscopy.

The dendrites postsynaptic to DA axons could belong either to pyramidal cells or to GABA-containing local circuit neurons that are sparsely spiny to aspiny (KUBOTA et al. 1994; GABBOTT and BACON 1996a). The unique morphological features of local circuit neuron dendrites (varicose shape, multiple axodendritic synapses and lack of spines) can be used to estimate the extent to which DA axons synapse on these cells. By serial section analysis in the primate (SMILEY and GOLDMAN-RAKIC 1993a), DA axons are estimated to synapse on local circuit neuron dendrites with a frequency of 30%–50% in superficial layers and 12.5% in deep layers. These results suggest that DA axons synapse on both local circuit and pyramidal neurons in superficial layers but preferentially target pyramidal cells in deep layers. However, the data derive from only a few observations in the deep layers of a single animal.

Dual immunocytochemical analyses have also been performed to verify that cortical DA axons in fact synapse on GABA dendrites. While the first such study reported close contacts but few synapses between these cell types (VERNEY et al. 1990), the overall synaptic incidence in this investigation was low. Subsequent analyses using pre-embedding immunogold labeling for GABA have demonstrated that DA axons do indeed synapse on the dendrites of GABA local circuit neurons in the deep layers of the rat prefrontal cortex and the superficial layers of the monkey prefrontal, motor, and entorhinal cortices (SESACK et al. 1995c; ERICKSON et al. 1999). The overall incidence of these contacts is roughly 37%, similar to the estimates made using local circuit neuron morphology (SMILEY and GOLDMAN-RAKIC 1993a). Thus, DA axons in all cortical areas examined to date synapse on both pyramidal and non-pyramidal neurons, although there may be regional and laminar differences in the extent of these inputs. The finding of DA synaptic input to local circuit neurons in the cortex but not the striatum (see Sect. C.I.3, this chapter) indicates likely differences in DA function in these target areas. It should be noted that DA axons in the rat prefrontal cortex have been described in light microscopic experiments as closely contacting the soma of GABA local circuit neurons, as well as pyramidal cells (BENES et al. 1993). However, the failure of most ultrastructural studies to identify synapses along somatic membranes suggests that the majority of these contacts represent close appositions or near juxtapositions separated by glia or neuropil.

The fact that cortical interneurons belong to different morphological classes that correlate with their content of various calcium-binding proteins (CONDÉ et al. 1994; KUBOTA et al. 1994; GABBOTT and BACON 1996a,b) provides the opportunity to test whether DA axons selectively target different interneuron subtypes in these regions. In a series of studies in the superficial layers of the monkey prefrontal cortex, we have determined that DA axons preferentially synapse on the dendrites of local circuit neurons that contain parvalbumin (SESACK et al. 1998a) but not those that label for calretinin (SESACK et al. 1995a). Moreover, the synapses onto parvalbumin dendrites are observed preferentially in layers 3b–4, as opposed to layers 1–3a. Thus, mesocortical DA axons exhibit both phenotypic and laminar specificity in their synapses onto local circuit neurons and selectively innervate cells whose axons target proximal as opposed to distal sites on pyramidal neurons (SESACK et al. 1998a). Regarding interneurons that contain calbindin, this protein is also expressed by a population of pyramidal cells in superficial layers (CONDÉ et al. 1994), which complicates the interpretation of immunocytochemical data. Nevertheless, our preliminary observations suggest that DA axons are closely apposed to and may synapse on both spiny and aspiny calbindin-labeled dendrites, although contacts on spiny structures are more common (LEWIS et al. 1997).

As observed in the striatum, DA axons in the cortex often converge synaptically with unlabeled terminals on common targets, including convergence on spines with terminals forming asymmetric synapses (GOLDMAN-RAKIC et al. 1989; SMILEY et al. 1992; SMILEY and GOLDMAN-RAKIC 1993a). Such arrangements offer anatomical substrates through which DA selectively modulates excitatory cortical drive. Similar convergence onto dendritic shafts of DA profiles and processes forming asymmetric (and sometimes symmetric) synapses is also observed (VERNEY et al. 1990; SMILEY et al. 1992; SESACK et al. 1995c, 1998a). However, such arrangements are less spatially restricted than those involving spines. Our laboratory has sought to identify the sources of glutamate terminals that converge synaptically with DA axons. In this regard, only 5%–30% of spines in the cortex receive both asymmetric and symmetric synapses (JONES and POWELL 1969; KOCH and POGGIO 1983; BEAULIEU and COLONNIER 1985). Thus, DA axons are likely to modulate excitatory glutamate inputs from highly specific sources. To date, neither hippocampal nor callosal prefrontal afferents have been found to converge synaptically with DA axons (CARR and SESACK 1996, 1998). In preliminary studies, we are continuing to examine whether thalamic or amygdaloid afferents converge onto common targets with DA axons (PINTO and SESACK 1998, 1999). Furthermore, we have been exploring the possibility that DA fibers converge synaptically with the intrinsic collaterals of pyramidal neurons (SESACK and MINER 1997). To date, we have observed instances where DA profiles synapse on dendritic shafts in close proximity to asymmetric synapses from intrinsic pyramidal collaterals. Further studies are needed to determine whether such convergence also occurs onto dendritic spines and to identify the sources of convergent glutamate terminals in the primate.

4. Localization of Proteins Involved in DA Transmission

In the rat and monkey cortex, DA D_1 and D_5 receptors are expressed primarily by pyramidal neurons (SMILEY et al. 1994; BERGSON et al. 1995; GASPAR et al. 1995; LIDOW et al. 1998; MULY et al. 1998). At the ultrastructural level, D_1 receptor expression is highest in spines, lower in dendritic shafts, and difficult to detect in proximal dendrites or soma. Conversely, D_5 receptor labeling is more abundant in dendritic shafts than spines (BERGSON et al. 1995). Presynaptic D_1 or D_5 receptors are occasionally observed in axons forming symmetric or asymmetric synapses. D_5 labeling is also sometimes seen in pyramidal cell axon initial segments. When D_1 receptor immunoreactivity is examined in relationship to TH-containing axons, labeling is never found postsynaptic to these profiles; instead, it often appears near the side of asymmetric synapses (SMILEY et al. 1994). As discussed for the striatum (see Sect. C.I.4, this chapter), such negative data may partly reflect problems with signal detection. This is especially true in the cortex where the density of DA receptors is low.

It has been reported that the D_1 receptor in the monkey cortex is extensively localized to parvalbumin- and calbindin-containing local circuit neurons and less frequently to calretinin cells (MULY et al. 1998). Unfortunately, the D_1 antibody used in this case cross-reacts with an unknown protein in the Golgi apparatus of many primate neurons (SMILEY et al. 1994); the same is not true for rodents (S.R. Sesack, unpublished observations). Thus, this antibody is not suitable for quantitative light microscopic studies of cellular expression. Nevertheless, it can be used to examine D_1 receptor distribution in distal dendrites where the Golgi complex is absent. In this regard, D_1 receptor labeling is found in distal parvalbumin-containing dendrites (MULY et al. 1998), consistent with the DA synaptic input to these cells (SESACK et al. 1998a). D_1 labeling can also be found occasionally in parvalbumin-positive axon terminals (MULY et al. 1998). Moreover, mRNA for both D_1 and D_2 receptors has been localized primarily to parvalbumin- as opposed to calbindin-containing local circuit neurons in the rodent cortex (LE MOINE and GASPAR 1998).

Fewer ultrastructural studies have been conducted to localize D_2 receptors in the cortex. By in situ hybridization, D_2, D_3, and D_4 receptors have been observed in several classes of pyramidal projection neurons (GASPAR et al. 1995; LIDOW et al. 1998) and local circuit neurons (LE MOINE and GASPAR 1998). This is consistent with preliminary ultrastructural observations of D_2 immunolabeling in spines and dendritic shafts in rat and monkey prefrontal cortex (SESACK et al. 1995b). In addition, D_2 receptor immunoreactivity is prominently seen in presynaptic structures, many of which form asymmetric synapses. This is similar to the presynaptic localization of D_2 receptors in presumed glutamate terminals in the striatum (see Sect. C.I.4, this chapter). The DA D_4 receptor has been localized to the primate cortex, where it is expressed in both pyramidal and non-pyramidal neurons (MRZLJAK et al. 1996). Regarding the latter cells, D_4 labeling is especially evident in the parvalbumin class of local circuit neurons, consistent with their synaptic DA innervation (SESACK et al. 1998a). At the ultrastructural level, D_4 immunoreactivity is seen along the plasmalemmal surface of soma and dendrites.

In studies localizing the DAT within the rodent prelimbic prefrontal cortex, it is obvious that the level of DAT expression under represents the known DA innervation to this region, as observed with DA or TH antibodies (CILIAX et al. 1995; SESACK et al. 1998b). At the ultrastructural level, DAT is expressed primarily in intervaricose axon segments and rarely in axon varicosities (SESACK et al. 1998b). Similar observations have been made in the primate prefrontal cortex, although the overall level of DAT expression is higher than seen in the rat (D. Melchitzky, S.R. Sesack, and D. Lewis, in preparation). While not examined systematically, the DAT-immunoreactive intervaricose axon segments rarely appear to form synapses. These findings differ markedly from observations in the striatum, where DAT is a faithful marker of DA axons and varicosities and is localized near synaptic junctions (CILIAX et al. 1995; NIRENBERG et al. 1996b; HERSCH et al. 1997; SESACK et al. 1998b). The findings are also inconsistent with observations in the cingulate cortex, where DAT is expressed more robustly (CILIAX et al. 1995) in both axons and their varicose portions (SESACK et al. 1998b). Given that synapses occur frequently at varicosities, at least in the rat prefrontal cortex (SÉGUÉLA et al. 1988), DAT appears to be localized at a distance from sites of DA release and may even be absent from some DA axons. While this appears to argue for greater parasynaptic actions of DA in this region, it must be considered in the context of DA receptor distribution, which may be sufficiently low as to allow diffusion to terminate DA's cellular actions (SESACK et al. 1998b).

VII. Hypothalamus

In their hypothalamic functions, compared to midbrain or retinal DA systems, diencephalic DA neurons are more likely to mediate actions that are truly endocrine in nature. To a large extent, this is to be expected, given their role in the regulation of pituitary and autonomic functions (VAN DEN POL et al. 1984). Hypothalamic DA cells generally belong to either the tuberoinfundibular system or the incerto-hypothalamic system (HÖKFELT et al. 1977; BJ ÖRKLUND and LINDVALL 1984; VAN DEN POL et al. 1984). Tuberoinfundibular DA neurons in the arcuate nucleus and ventral periventricular nucleus send their axons into the median eminence and, to some extent, into the anterior and intermediate pituitary regions. Incerto-hypothalamic DA neurons consist of cells in the caudal thalamus, caudal dorsal hypothalamus, medial zona incerta, and anteroventral periventricular nucleus. They give rise to extensive intra-hypothalamic projections, as well as having both rostrally and caudally directed axons that extend from the septal complex to the spinal cord (HÖKFELT et al. 1977; BJÖRKLUND and LINDVALL 1984).

1. Tuberoinfundibular DA System

By most ultrastructural investigations in the rat, DA axons in the median eminence and pituitary do not form synapses on neuronal targets (AJIKA and HÖKFELT 1973; CUELLO and IVERSEN 1973; CHETVERUKHIN et al. 1979; VAN DEN POL et al. 1984). Many of these axons are varicose (1–3µm diameter) and contain numerous small clear and occasional dense-cored vesicles; others contain abundant dense-cored vesicles. The colocalization of GABA has been reported in some of these fibers (SCHIMCHOWITSCH et al. 1991). Within the medial zone of the median eminence, DA axons frequently abut the portal pericapillary space; in the lateral zone, they are often closely ensheathed by tanycytes lining this space (CHETVERUKHIN et al. 1979; MEISTER et al. 1988). The localization of DARPP-32 immunoreactivity to some of these tanycytes suggests that DA may have modulatory actions on glial cells within this region (MEISTER et al. 1988). Infrequently, DA axons form "synaptoid contacts" on these tanycytes (CHETVERUKHIN et al. 1979). DA axons are frequently apposed to fibers containing luteinizing hormone-releasing hormone (LH-RH) (AJIKA 1979); in the sheep, occasional synapses have been reported at these sites of close contact (KULJIS and ADVIS 1989). The LH-RH fibers are putative axons, suggesting that DA modulates LH-RH release via an axo-axonic influence that may be partly synaptic. If verified, this connection would represent a singular example of a DA axo-axonic synapse in the nervous system. The localization of DAT protein to arcuate nucleus neurons and fibers in the median eminence is interesting (REVAY et al. 1996) because it suggests that DA diffusion is limited to some extent by reuptake, despite the fact that DA's actions in this region are, by design, endocrine. Nevertheless, the levels of DAT message and protein are noticeably low in these cells, as they are in other hypothalamic DA neurons (CERRUTI et al. 1993; LORANG et al. 1994; REVAY et al. 1996).

2. Incerto-hypothalamic DA System

In various hypothalamic areas, TH-labeled axons synapse on somatodendritic targets (VAN DEN POL et al. 1984; MITCHELL et al. 1988). In the dorsal nucleus, these are primarily symmetric synapses, which occur with a frequency of 25% in single sections, and some of which contact enkephalin-labeled soma (MITCHELL et al. 1988). Within the arcuate nucleus, TH-immunoreactive varicosities synapse on the soma and dendrites of neurons labeled for neurotensin (MARCOS et al. 1996), adrenocorticotropic hormone (ACTH) (KOZASA and NAKAI 1987), or opioid peptides (HORVATH et al. 1992). However, in each case, the dopaminergic nature of the labeled axons has not been established, and both DA and NE afferents are abundant in these regions. In the supraoptic and paraventricular nuclei, the DA innervation has been examined by DA antibodies. In this case, labeled axons give rise to varicosities that contain primarily clear vesicles and frequently synapse on the soma and dendrites of magnocellular neurons (BUIJS et al. 1984; DECAVEL et al. 1987a). Many of these synapses have symmetric thickenings, although asymmetric synapses are also observed. In the paraventricular nucleus, some of the neurons receiving synaptic input from TH-immunoreactive axons may synthesize oxytocin (HORIE et al. 1993) or thyrotropin-releasing hormone (SHIODA et al. 1986). However, at least some of the TH-labeled contacts may derive from NE afferents.

In the medial preoptic area, presumed DA axons labeled by TH-immunocytochemistry or uptake of radiolabeled DA or 5-hydroxydopamine synapse on the soma and dendrites of LH-RH-positive neurons and GABA cells (WATANABE and NAKAI 1987; LERANTH et al. 1988; HORVATH et al. 1993). Both symmetric and asymmetric junctions have been described at these contacts, although the majority appears to be symmetric (LERANTH et al. 1988). The DA nature of some of the TH profiles synapsing on GABA cells, and perhaps all of the TH synapses on LH-RH neurons, has been determined by tract-tracing, 6-hydroxydopamine lesions, and transection of the ascending NE bundles (LERANTH et al. 1988; HORVATH et al. 1993). The DA inputs to the medial preoptic area originate from the anteroventral periventricular nucleus and not from the medial zona incerta; the latter DA neurons synapse in the dorsomedial and paraventricular hypothalamic nuclei (HORVATH et al. 1993).

VIII. Spinal Cord and Brainstem

The functions of the DA innervation to the spinal cord are not clearly understood. However, DA-immunoreactive fibers from the caudal dorsal hypothalamus have been described in the dorsal horn (primarily layers III and IV), intermediolateral cell column, and ventral horn of the rat and lamprey (SHIROUZU et al. 1990; RIDET et al. 1992; SCHOTLAND et al. 1996). At the ultrastructural level, DA varicosities in the spinal cord are larger than those typically described in the brain (mean diameter $0.75–0.91\,\mu m$). Within the intermediolateral cell column and ventral horn, a majority of DA profiles form classical synapses (31%–53% observed in single sections, extrapolated to nearly 100%). However, in the dorsal horn, DA axon varicosities vary from low synaptic incidence in cervical spinal segments (13% observed, 34% extrapolated) to more frequent synapse formation in thoracic and lumbar segments (27% observed, 76% extrapolated) (RIDET et al. 1992). Both symmetric and asymmetric DA synapses occur in the spinal cord, with the principal targets being small, medium, and large sized dendrites (more than 85% of synapses). Synapses on soma or other axons are rare.

DA neurons provide little input to the brainstem, although clusters of axons originating from the SN, VTA, and medial hypothalamus have been described in the mesencephalic trigeminal nucleus, with extension to the parabrachial nucleus and locus ceruleus (COPRAY et al. 1990; MAEDA et al. 1994). Within the mesencephalic trigeminal nucleus, DA axons appear to make symmetric synapses on soma, dendrites, and axons. The latter contacts, which were reported but not illustrated in this study, may derive from the hypothalamus, as midbrain DA neurons almost exclusively target somatodendritic structures. A hypothalamic origin of some of the DA projections to this region is further suggested by the reportedly large size of the labeled boutons (mean diameter, 0.98 μm) and the formation of primarily asymmetric synapses (MAEDA et al. 1994).

IX. Retina

1. DA Neurons and Processes

Unlike midbrain and diencephalic neurons, retinal DA cells are local circuit neurons. Depending on species, they belong to one or both of two classes in the inner nuclear layer: amacrine cells that extend processes to the inner plexiform layer and specialized interplexiform cells, that project to both the inner and outer plexiform layers (DJAMGOZ and WAGNER 1992; NGUYEN-LEGROS et al. 1997b). A few of the amacrine cells may be displaced into the ganglion cell layer, particularly in primates. It is difficult to classify the processes of amacrine cells as true axons, so that many of DA's retinal actions are exerted via release from beaded, axon-like processes (NGUYEN-LEGROS et al. 1997b) that may also have features of dendrites (KOLB et al. 1990). In most studies, DA retinal cells have been identified by TH immunocytochemistry, which is an adequate marker given the general absence of other catecholamines in retinal neurons (DJAMGOZ and WAGNER 1992; NGUYEN-LEGROS et al. 1997b). In some species, retinal DA cells may colocalize GABA (DJAMGOZ and WAGNER 1992; NGUYEN-LEGROS et al. 1997b). DA's physiological actions in the retina typically involve modulation of light–dark adaptation in a manner that attenuates rod transmission in favor of the cone pathway. These functions are accomplished mainly by actions on amacrine cells in the rod pathway, but also in some cases via modulation of bipolar, horizontal, and photoreceptor cells (DJAMGOZ and WAGNER 1992; NGUYEN-LEGROS et al. 1997b).

2. Ultrastructural Morphology, Synapses, and General Targets

In the monkey retina, the first identified population of DA amacrine cells (type I) reside in the innermost inner nuclear layer, with synaptic processes extending into the outermost portion of the inner plexiform layer. They contain mostly small, clear vesicles and form synapses primarily onto non-DA amacrine cell bodies and dendrite-like processes and rarely onto bipolar cell terminals in the form of reciprocal synapses (DOWLING et al. 1980; HOKOC and MARIANI 1987). The synapses are often slight, and although not strictly conforming, are commonly considered symmetric (DOWLING et al. 1980). In 80% of the synapses, the DA profile is the presynaptic element; in the remaining 20% of contacts, the DA structure is the postsynaptic element (HOKOC and MARIANI 1987). A second, more numerous class of DA amacrine cells (type 2) resides in the inner nuclear, inner plexiform, and ganglion cell layers (MARIANI 1991). Processes of these neurons extend into the center of the inner plexiform layer. The synaptic output of this cell class is sparse, being identified in only 5% of processes, with the remaining 95% involving the DA structure as the postsynaptic element. The few presynaptic DA profiles form exclusively symmetric synapses, directed primarily to non-DA amacrine cells (MARIANI 1991). These observations are consistent either with a convergent funneling of information from widespread areas of the retina to a small number of amacrine cells or with a predominant parasynaptic function for type 2 retinal DA cells (MARIANI 1991). Little evidence supports the presence of DA interplexiform cells in the rhesus monkey, although such cells have been described in New World monkeys (DJAMGOZ and WAGNER 1992; NGUYEN-LEGROS et al. 1997b), where they synapse on horizontal and bipolar cells and processes in the outer plexiform layer (DOWLING et al. 1980).

In the cat, rat, and rabbit retina (DOWLING et al. 1980; POURCHO 1982; VOIGT and WÄSSLE 1987; KOLB et al. 1990; KOLB et al. 1991), the morphology and synaptology of DA amacrine and interplexiform cells is similar to that observed in the monkey. The synapses are formed at enlarged bead-like varicosities (up to 2μm in diameter) and not along interbead segments (0.15–0.2μm). Some of the structures postsynaptic to DA profiles contain glycine or GABA (KOLB et al. 1991). DA amacrine cells in the cat, like type I cells in the primate, form more output than input synapses (KOLB et al. 1990). In the outer plexiform layer, DA processes are less numerous and appear to synapse on the dendrites of non-DA interplexiform cells (KOLB et al. 1990). Some of the postsynaptic targets are GABA-containing and may include horizontal cell processes (KOLB et al. 1991). The synapses in the outer plexiform layer are remarkably small and punctate, which may explain why they are not always identified in ultrastructural studies. Other nonclassical junctions have also been identified in the inner and outer plexiform layers along the non-beaded portions of DA profiles (KOLB et al. 1990). Whether these represent synapse variants or sites of non-synaptic release is not known. In species where cone transmission dominates, such as the ground squirrel and tree shrew, the synaptology of DA amacrine and interplexiform neurons is similar to species whose vision includes a greater rod component, although the number of DA neurons in the ground squirrel is noticeably low (MÜLLER and PEICHL 1991; LUGO-GARCÍA and BLANCO 1993). Interestingly, DA processes in the cat, rat, and monkey retina also exhibit close appositions to blood vessels and glial cell end feet, suggesting a possible regulation of blood flow (FAVARD et al. 1990; KOLB et al. 1990).

In non-mammalian vertebrates, the synaptology of retinal DA neurons depends in large part on whether DA is present mainly in interplexiform cells or primarily in amacrine cells (DJAMGOZ and WAGNER 1992). Nevertheless, the ultrastructure of retinal DA cells appears to be remarkably conserved throughout phylogeny. In the inner plexiform layer of the fish, the processes of DA interplexiform cells form symmetric synapses (DOWLING and EHRINGER

1975; Yazulla and Zucker 1988) that have been characterized as "rather meager," i.e., exhibiting few clustered vesicles and no prominent membrane densification (Yazulla and Zucker 1988). Such marginal synapses are also formed by amacrine cells in the amphibian retina (Gábriel et al. 1992; Watt and Glazebrook 1993), but are not restricted to non-mammalian vertebrates, having also been described in monkey (Dowling et al. 1980). Approximately 40% of DA profiles exhibit synapses in single sections and contact primarily non-DA amacrine cells, some of which accumulate radiolabeled GABA. In most of the inner plexiform layer synapses, the DA profile is the presynaptic element (Yazulla and Zucker 1988; Gábriel et al. 1992; Watt and Glazebrook 1993). Other contacts formed by DA processes exhibit no synaptic specializations and have been termed "junctional appositions" (Yazulla and Zucker 1988). They target amacrine cell processes, bipolar cell terminal endings, soma within the ganglion cell layer (i.e., possibly displaced amacrine cells), and the processes of other DA cells.

In the outer plexiform layer, the processes of DA interplexiform cells in the lamprey closely appose horizontal cells; membrane densifications, but not classical synaptic specializations, are sometimes observed at these sites (Dalil-Thiney et al. 1996). In the fish retina, similar connections involving horizontal cells have been described, with a widened and parallel intercellular cleft as their main morphological features (Van Haesendonck et al. 1993; Wagner and Behrens 1993). Whether such junctions represent specialized functional contacts is not known but has been suggested (Van Haesendonck et al. 1993). Interestingly, DA profiles with this morphology are sometimes closely apposed to the horizontal cells that engage in gap junction complexes, known to be modulated by DA (Van Haesendonck et al. 1993) and to rod and cone pedicles (Wagner and Wulle 1990; Wagner and Wulle 1992; Van Haesendonck et al. 1993). Nearly identical contacts have been described in the toad retina, which possesses DA interplexiform cells with extensive processes in the outer plexiform layer (Gábriel et al. 1991). In other cases in the fish, classical symmetric synapses have been observed from DA interplexiform neurons onto horizontal cell soma and processes and occasionally onto bipolar cell dendrites (Dowling and Ehringer 1975; Wagner and Wulle 1992; Van Haesendonck et al. 1993; Wagner and Behrens 1993).

3. Localization of Proteins Involved in DA Transmission

Retinal D_1 receptors in rodents and primates have been localized by immunocytochemistry to a dense band in the outer plexiform layer, cells in the inner nuclear layer, and diffuse punctate processes in the inner plexiform layer (Bjelke et al. 1996; Verucki and Wáussle 1996; Nguyen-Legros et al. 1997a). Receptor-expressing profiles most likely derive from horizontal, bipolar, and amacrine cells, with the dense localization in the outer plexiform layer being consistent with D_1 receptor regulation of horizontal cell electrical coupling (Djamgoz and Wagner 1992). Amacrine cells expressing the D_1 receptor are not immunoreactive for TH (i.e., retinal D_1 receptors are not autoreceptors) (NGUYEN-LEGROS et al. 1997a). In addition, D_1 labeling of photoreceptor outer segments has been reported in one study (BJELKE et al. 1996).

D_2 receptor immunoreactivity has been localized in several species, including rat, rabbit, cow, chick, turtle, frog, and fish (WAGNER and BEHRENS 1993; WAGNER et al. 1993; ROHRER and STELL 1995; BJELKE et al. 1996; DEROUICHE and ASAN 1999). The densest expression of D_2 receptors occurs in the inner and outer plexiform layers, within which considerable immunolabeling reflects autoreceptors on amacrine and interplexiform cells and their processes (ROHRER and STELL 1995; BJELKE et al. 1996; DEROUICHE and ASAN 1999). Other D_2 labeling is on target elements, particularly non-DA amacrine cells and possibly bipolar cells. D_4 receptor mRNA has also been identified in retinal neurons, but the D_3 receptor appears to be absent (ROHRER and STELL 1995; DEROUICHE and ASAN 1999). Controversy surrounds the localization of D_2 receptors to photoreceptors or their processes, with some investigators reporting their presence in inner and even outer segments (WAGNER and BEHRENS 1993; WAGNER et al. 1993; ROHRER and STELL 1995; BJELKE et al. 1996) and other studies failing to observe D_2 receptors at these sites (DEROUICHE and ASAN 1999). The reasons for this discrepancy are unclear, although the different species examined and different antibodies used in each case may have produced varying results. Furthermore, photoreceptor outer segments may have a tendency for non-specific immunostaining (DEROUICHE and ASAN 1999). In the most recent study in the rat, photoreceptor cells reportedly express D_4 mRNA but no D_2 receptor immunoreactivity or mRNA (DEROUICHE and ASAN 1999). Conversely, photoreceptors in the chick retina express both D_2 message and protein (ROHRER and STELL 1995). It should be noted that the presence of D_2 receptors in photoreceptors is not unexpected, given the known modulation of retinomotor movement by this receptor (DJAMGOZ and WAGNER 1992).

While the ultrastructural localization of D_1 receptors has not been examined, the reported extrasynaptic localization of D_2 receptors (DEROUICHE and ASAN 1999), and the observed spatial distance between TH-immunolabeled profiles and neuronal structures expressing either D_1 or D_2 receptors, argues for a neuromodulatory influence of DA that may be largely parasynaptic. For example, if photoreceptor segments express DA receptors, they are likely to respond to transmitter that is released more than 10μm distant and that travels across the outer limiting membrane (WAGNER et al. 1993). The reportedly extrasynaptic localization of D_2 receptors in the inner plexiform layer (DEROUICHE and ASAN 1999) is perhaps more puzzling, as DA processes do synapse on targets in this region. However, the pre-embedding immunogold method used is known to have limited sensitivity and to be restricted from synaptic densities in many instances (BAUDE et al. 1995; NUSSER et al. 1995) (see Sect. C.I.4, this chapter). The mRNA for the DAT is localized to cells in the inner nuclear layer of the rat retina, although at levels that are only 5%–10% of those expressed by SN cells (CERRUTI et al. 1993). The low observed mRNA may relate to a different translational efficiency, smaller terminal fields, or lower expression of DAT protein.

D. Discussion
I. The Variability of DA Synaptology

DA neurons project to their target structures via axons or axon-like profiles that share the common morphological features of small size, general lack of myelination, and content of primarily small clear vesicles. However, the transmitting elements of DA axons exhibit a wide range of structural features (see Fig. 1), including the complete absence of synapses, the presence of subtle, synapse-like contacts, and the formation of bona fide symmetric or even asymmetric synapses. This diversity of structure argues for the specificity of these systems. Moreover, it is the contention of this review that the morphology expressed by individual DA axons is that which best fulfills their specific functions, at specific points of connection, and, perhaps, for specific periods of time. Even though the anatomical discipline usually ordains that structure forms the basis for function, it is here proposed that form follows function for DA axon connections.

Fig. 1. Schematic diagrams illustrating common features of the synaptology of DA axons in the striatal complex. Panel **B** shows an enlargement of the area outlined in panel **A**. The extracellular space has been exaggerated for effect. *1*, DA axons release high concentrations of transmitter (*shown in red*) at synapses formed by varicosities. Synaptic DA acts on receptors (*green curved rectangles in panel B*) immediately within or closely adjacent to the synapse. *2*, Some of the synaptically released DA diffuses away from the synaptic cleft, being reduced in concentration (*shown in strong blue*) and exerting parasynaptic actions on receptors located at a distance from the synapse, including receptors in neighboring synapses. *3*, At some distance, the concentration of DA falls to levels that are ineffective for activating extrasynaptic DA receptors (*shown in weak blue*). *4*, DA axons also release transmitter at synapses formed by intervaricose segments. *5*, It remains to be established whether DA axons contribute to parasynaptic actions by release of transmitter from varicosities that fail to form synapses. At present, there is no clear evidence to prove or disprove this mode of transmission. However, if such release occurs, it would produce locally high concentrations of DA. *6*, DA axons frequently synapse in close proximity to asymmetric synapses, providing an anatomical substrate for DA modulation of glutamate transmission. *7*, DA axons also converge synaptically with terminals forming symmetric synapses, consistent with DA modulation of inhibition mediated by GABA and other transmitters. *8*, Localization of the DAT (*purple bars in panel B*) to the immediate perisynaptic region limits the diffusion of synaptically released DA. *9*, DAT distribution to the extrasynaptic membrane of DA axons further reduces DA diffusion in the extracellular space and ultimately terminates its actions. Although DA diffusion is shown as spherical, the actual dimensions depend on tortuosity factors as well as the local density of the DAT and DA receptors. For simplicity, only D_2 receptors are illustrated, including postsynaptic receptors and auto- and hetero-presynaptic receptors. *Structures in the darkest gray* represent glial processes

Synaptology of Dopamine Neurons   97

Some of the variability in DA synaptology may reflect the nervous system's capacity to utilize the same substance for widely different functions. For example, opioid peptides in the brainstem and spinal cord modulate nociception, while they serve completely disparate functions in the basal ganglia or enteric nervous system. Likewise, the DA projection from the arcuate nucleus to the median eminence serves a clear endocrine role in

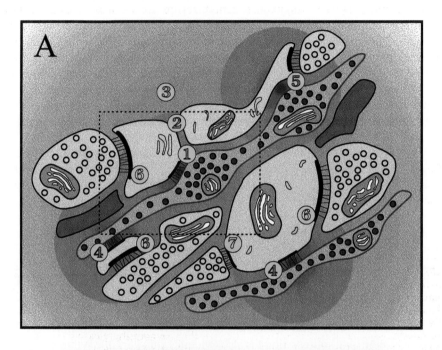

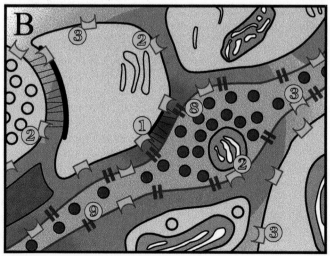

altering the release of pituitary hormones, a function reflected in the near absence of synapses at these nerve endings (VAN DEN POL et al. 1984). However, aside from such obvious exceptions, DA neurons often exert similar functions in different target areas, most typically the modulation of transmission evoked by other afferents. Thus, the occurrence of varying morphology, asynaptic to asymmetric, is more difficult to understand when it occurs across regions with similar function or even within a given brain area.

Variable structure might also reflect different neuronal origins of DA afferents within a region. For example, the nearly unvarying synaptology observed in the dorsal striatum would be expected from its rather homogeneous innervation by the SN (SMITH et al. 1994; HANLEY and BOLAM 1997). Conversely, DA axons in cortical, limbic, and autonomic forebrain areas exhibit a larger number of dense-cored vesicles and a greater tendency to form asymmetric synapses. Such features may indicate a more substantial contribution to these projections from DA neurons in the VTA or even the hypothalamus (SWANSON 1982). However, the physiological significance of asymmetric versus symmetric DA synapses remains to be determined. It has been hypothesized that midbrain DA neurons colocalize and release glutamate at asymmetric synapses in the forebrain (HATTORI et al. 1991; SULZER et al. 1998). While a thorough critique of this argument is beyond the scope of this chapter, it should be mentioned that the anatomical basis for this supposition is based on tract-tracing studies with serious technical limitations. More recent experiments using more appropriate tracers are not consistent with DA axons mediating glutamate excitation in the striatum (HANLEY and BOLAM 1997; see Sect. C.I.1, this chapter). Nevertheless, it cannot be assumed that the morphology of a synapse always indicates its physiological action, as this correlation is only relative and not absolute.

While variability in DA axon morphology may reflect mixed afferents from different neuronal sources, there is also evidence that variable synaptology can be exhibited by different processes emanating from the same neuronal population. For example, hypothalamic DA projections to the spinal cord are primarily synaptic in the ventral horn but more frequently asynaptic in the dorsal horn, suggesting that modulation of motor and autonomic functions requires the temporal and spatial fidelity associated with synaptic transmission, while modulation of pain and other sensory functions may be served by parasynaptic control (RIDET et al. 1992; see Sect. C.VII, this chapter). Other evidence suggests that variable synaptology may even be observed for different processes emanating from the same DA neurons. For example, in the toad retina, the processes of DA interplexiform cells form conventional synapses in the inner plexiform layer but no synapses in the outer plexiform layer (GÁBRIEL et al. 1991, 1992). As cells of this class do not arborize solely in the outer plexiform layer, both the synaptic and the asynaptic profiles must derive from the same neurons. DA processes in the outer plexiform layer do not function as dendrites, as they are never postsynaptic to other structures (GÁBRIEL et al. 1991). Moreover, DA release is likely to occur from these asynaptic structures, by virtue of the known physiological regulation of horizontal cell electrical coupling and photoreceptor movement during light adaptation (DJAMGOZ and WAGNER 1992; NGUYEN-LEGROS et al. 1997b). If these functions could be served by DA diffusion from the inner plexiform layer, then the extension of processes into the outer plexiform layer would seem to serve no purpose. Thus, the morphology and synaptology of DA profiles appears to be matched to the function exerted in a particular region regardless of the neuronal source.

II. Parasynaptic DA Transmission

Observations such as those in the retina and median eminence suggest that if the function of a particular DA system can be achieved without synaptic connections, then such specialized contacts may not be formed. There are numerous other examples of parasynaptic DA transmission in the brain. For example, in the striatum, DA regulates gene expression via mechanisms that appear not to require functional synapses (BERKE et al. 1998). In addition, striatal cholinergic neurons express high levels of DA receptors (LE MOINE et al. 1990; JONGEN-RELO et al. 1995) despite the rare occurrence of DA synapses on these cells (PICKEL and CHAN 1990). Likewise, the physiological actions of DA on the apical dendrites of cortical pyramidal cells (YANG and SEAMANS 1996; GULLEDGE and JAFFE 1998) match the distribution of DA receptors (SMILEY et al. 1994; BERGSON et al. 1995) but not the relatively few DA synapses onto proximal dendrites (SÉGUÉLA et al. 1988; GOLDMAN-RAKIC et al. 1989; SMILEY and GOLDMAN-RAKIC 1993a; KRIMER et al. 1997). Moreover, the many reported examples of presynaptic actions of DA (VIZI 1991; LANGER 1997) are consistent with the axonal localization of DA receptors (see Fig. 1) but not the virtual absence of axo-axonic synapses in most target areas.

Many of these examples of DA's parasynaptic actions can be explained by diffusion of transmitter from synaptic sites of release (see Fig. 1). However, historically, there has been an additional argument that DA can diffuse from non-synaptic release sites located at axon varicosities that do not form conventional junctions (BEAUDET and DESCARRIES 1978; DESCARRIES and UMBRIACO 1995). However, reports that not all DA varicosities form synapses must be considered in light of several caveats. First, determining that a DA varicosity fails to make a synapse requires complete serial section analysis in order to verify that these small junctions are not missed. Second, some DA synapses have been described as "small and subtle" (SMILEY and GOLDMAN-RAKIC 1993b; SMILEY and GOLDMAN-RAKIC 1993a), "delicate" (GROVES et al. 1994), or "rather meager" (YAZULLA and ZUCKER 1988), suggesting that they may be rejected by some investigators because they fail to meet criteria that could be debated. Third, with few exceptions (DESCARRIES et al. 1996), the intervaricose segments of DA axons have been reported to make synapses (FREUND et al. 1984; GROVES et al. 1994; SMITH et al. 1994; HANLEY and BOLAM 1997; KUNG et al. 1998), and evidence that the varicosities are the preferential sites of synapse formation is lacking (see Fig. 1). In this regard, it should be noted that DA and other monoamine axons clearly differ from non-monoamine fibers, which typically synapse at the varicose portions (DESCARRIES and UMBRIACO 1995).

These observations call into question the function of monoamine varicosities, if not to provide repositories of vesicles available for release at morphologically defined junctions. Indeed, it has been suggested that the varicose portions of DA axons are associated more with the presence of mitochondria than vesicle accumulation (FREUND et al. 1984; GROVES et al. 1994; KUNG et al. 1998). As such, varicosities might represent structurally unstable elements whose position relative to synaptic junctions varies over time. Alternatively, or in addition, monoamine varicosities may function primarily as sites of transmitter uptake and storage (BLOOM 1991). Since the varicose portions of DA axons cannot be considered a priori to be synonymous with release sites, the practice of calculating the incidence with which these structures form synapses should probably be abandoned. Rather, in individual brain regions, it should be determined whether DA axons form conventional synapses, and attention should then be shifted to identification of the targets of these synapses and the localization of DA receptors relative to these junctions.

This position does not imply that DA varicosities are incapable of transmitter release. While small clear vesicles are thought to be released only at morphologically defined junctions, there is as yet no concrete proof of this, and some evidence to the contrary does exist. For example, DA axons in the median eminence contain a predominance of small clear vesicles and make few, if any, synapses (VAN DEN POL et al. 1984). Nevertheless, DA is released from these profiles into the hypophyseal blood supply. Another example of such potential asynaptic release has been described above for the fibers of DA interplexiform neurons in the toad retina. Thus, there must be some mechanism for release of small clear vesicles in the absence of identifiable synapses, just as there is for exocytosis of dense-cored vesicles (THURESON KLEIN and KLEIN 1990). Moreover, DA is known to be released from the dendrites of midbrain neurons (CHERAMY et al. 1981; HEERINGA and ABERCROMBIE 1995), despite the infrequent occurrence of conventional synaptic vesicles or junctional specializations at these sites (GROVES and LINDER 1983; HALLIDAY and TORK 1984; BAYER and PICKEL 1990; NIRENBERG et al. 1996a). As dendritic DA release is dependent on calcium and on compartmental sequestration by the vesicular monoamine transporter (CHERAMY et al. 1981; HEERINGA and ABERCROMBIE 1995), it is interesting to note that the vesicular transporter is extensively expressed along large tubulovesicular structures throughout DA dendrites (NIRENBERG et al. 1996a). While release from this compartment remains to be demonstrated, it should be noted that such tubulovesicles are also seen in striatal DA axons (NIRENBERG et al. 1997a).

Thus, DA appears to be capable of parasynaptic actions (see Fig. 1), regardless of whether extracellular DA originates as overflow from conventional synapses, non-junctional fusion of synaptic vesicles, exocytosis from novel compartments, or even reversal of the DAT (BANNON et al. 1995). The contribution that anatomical investigations can make to the resolution of these issues, as exemplified by the study of NIRENBERG and colleagues (NIRENBERG et al. 1996a), is to identify the molecular machinery necessary for non-synaptic release (for example, in the median eminence) and to provide evidence for the existence of this machinery at sites reputed to release DA. It is also essential that the distribution of functional DA receptors be delineated, as neurons are capable of selectively trafficking receptors and hence, determining which compartments are receptive to DA signals (see Fig. 1). While most available studies provide little evidence to support the presence of DA receptors within or near DA synapses (SMILEY et al. 1994; CAILLÉ et al. 1996; DEROUICHE and ASAN 1999; but, see AN et al. 1998, 1999), most of the methods utilized lack the sensitivity necessary to demonstrate the presence of low abundance proteins embedded in a synaptic complex (BAUDE et al. 1995; NUSSER et al. 1995). Thus, the final resolution of this issue awaits application of postembedding immunocytochemical approaches for localization of DA receptors and/or additional technological advancements. Application of these methods to the localization of other G-protein coupled receptors, such as metabotropic glutamate receptors, has shown that they are distributed predominantly to perisynaptic sites (SOMOGYI et al. 1998).

The concept of parasynaptic DA transmission is certainly not novel and has been reviewed previously, most recently under the nomenclature of "volume transmission" (ZOLI et al. 1998). While defined as "diffusion of signals in the extracellular fluid... for distances larger than the synaptic cleft," the term itself seems to imply communication over considerable distances. Thus, this review has used the alternative term, "parasynaptic," defined by SCHMITT in 1984 as "in 'parallel with' or 'alongside' (not instead of or in competition with) neuronal circuitry" (SCHMITT 1984). Parasynaptic transmission simply characterizes signaling that is not purely synaptic and so, implies little with regard to distance. Thus, it would encompass short-range signaling initiated either by synaptic release and subsequent overflow or non-synaptic release events. Understanding the parasynaptic mode of transmission places emphasis on defining the microenvironment of DA releasing elements, so that in addition to synaptic targets, the cellular structures in close proximity to DA axons also need to be characterized, particularly those that express DA receptors (see Fig. 1).

III. Synaptic DA Transmission

A consideration of DA's substantial parasynaptic actions raises the question of why DA axons make synapses at all. One might speculate that by virtue of their sometimes meager morphology, DA specializations are not true synapses but some form of attachment point, such that DA axons are "tacked" onto their preferred targets during development, facilitating release in relatively close proximity to the desired areas of function. As suggested by DESCARRIES (DESCARRIES et al. 1996), such junctions "could be considered as structural devices to stabilize, in time as well as in space, the relationships and, hence, the interactions between these varicosities and certain cellular targets in their immediate vicinity." However, it should be noted that DA junctions do not meet the morphological criteria for *puncta adherentia*, which exhibit electron densities that are extensively and symmetrically distributed on both sides of the membrane and are not associated with accumulated vesicles (PETERS et al. 1991). Furthermore, if DA axons are designed to be adhered close to their sites of action, then it is unclear why, in the cortex, they rarely contact apical dendrites (KRIMER et al. 1997), despite exerting significant actions on voltage gated channels at these sites (YANG and SEAMANS 1996), or why, in the retinal outer plexiform layer of the fish, they are simply in close apposition to horizontal cell gap junction complexes, without forming obvious specializations (VAN HAESENDONCK et al. 1993). Clearly, DA axons can be located close to their functional targets without being physically attached to them. Moreover, many DA junctions actually do meet all the criteria for symmetric or asymmetric synapses (PETERS et al. 1991), including the accumulation of vesicles at the presynaptic specialization.

Thus, it must be considered that DA axons form true synapses at sites where the spatial and temporal fidelity of synaptic transmission is required and for which parasynaptic transmission is functionally insufficient. This may be true regardless of whether these specific functions have yet been identified in electrophysiological studies. In this regard, it is interesting that DA axons in the rat cortex appear to have a greater synaptic incidence than those in the striatum (SÉGUÉLA et al. 1988; PAPADOPOULOS et al. 1989; DESCARRIES et al. 1996). In the striatum, DA axons primarily synapse on projection neurons, converge with axospinous synapses from the cortex, and exert parasynaptic actions on interneurons (see Sect. C.I.3, this chapter). Conversely, cortical DA fibers synapse on both pyramidal projection cells and selective subclasses of local circuit neurons, and the sources of excitatory afferents with which they converge synaptically appear to be highly selective (see Sect. C.VI.3, this chapter). Thus, a greater synaptic incidence may reflect a greater need for target specificity in cortical versus striatal regions. To some extent, this is contradicted by the lower synaptic incidence in primate cortex compared to rodent (SMILEY and GOLDMAN-RAKIC 1993a). While this might imply a heavier reliance on parasynaptic actions in the course of phylogeny, evolutionary studies suggest that synaptic actions represent the more recently evolved form of transmission (SCHMITT 1984). Moreover, the true synaptic incidence in the primate cortex has not yet been ascertained, because only the varicose portions of DA axons have been examined. Furthermore, a lower synaptic incidence does not translate a priori into lower specificity, as even in the striatum, the non-uniform distribution of DA synapses (GROVES et al. 1994) suggests that these inputs are directed only to specific dendritic processes.

IV. Summary and Functional Significance

The synaptology of DA neurons argues for anatomical specificity and supports multiple modes of neurotransmission lying along a continuum that includes synaptic, parasynaptic, and endocrine. The parasynaptic mode either originates as overflow from synapses or non-synaptic mechanisms of release that remain to be defined in their spatial and molecular characteristics. Its limits are dictated by the distribution of DA receptors in relation to release sites, as well as the efficiency of DA reuptake and/or enzymatic breakdown (GONON 1997; see also Fig. 1). In this regard, the extent of parasynaptic DA transmission is likely to exhibit regional variability, given differences in the subcellular localization and density of the DAT protein (SESACK et al. 1998b) and differences in DA receptor density. Parasynaptic actions of DA are likely to communicate the tonic level of DA neuronal activity in behaving animals when not responding to environmental challenges (SCHULTZ 1998). Significant changes in extracellular DA levels may communicate gross alterations in systems function in a broadcast manner.

The synaptic mode of DA transmission is mediated by release from junctional specializations and binding to receptors that are immediately within or in close proximity to the synapse (see Fig. 1), as observed for other G-protein coupled receptors (SOMOGYI et al. 1998). Synaptic transmission is best suited to communicate phasic alterations in DA cell activity that are temporally linked to behaviorally relevant events, such as those observed in response to stimuli that are rewarding or predictive of reward, and that are believed to constitute "learning" or "switching" signals (SCHULTZ 1998; REDGRAVE et al. 1999). Such phasic and tonic modes of DA transmission have been described previously (GRACE 1991) but only recently linked to synaptic and parasynaptic sites of action (MOORE et al. 1999). An alternative or additional function of DA synapses may be to ensure the reliability of transmission through signal amplification and/or reduction of variability (BARBOUR and HÄUSSER 1997). This suggestion is consistent with the observation that individual striatal DA axons form multiple synaptic contacts in close proximity on the same dendritic target (GROVES et al. 1994). Synaptic DA connections may also function chronically with regard to plasticity (LOVINGER and TYLER 1996; CALABRESI et al. 1997; CEPEDA and LEVINE 1998; OTANI et al. 1998), such that activation of DA synapses in some temporal relationship to one or more afferents strengthens or weakens these synapses in a spatially specific manner. DA-mediated plasticity may regulate not only long-term synaptic strength, but also anatomical stability. This is suggested by studies of the DA denervated striatum, in which a subpopulation of terminals forming asymmetric synapses, and their spine targets, are selectively lost, while perforated asymmetric synapses are increased in number (INGHAM et al. 1998).

Finally, it should be considered that synaptic and parasynaptic DA signals do not always serve independent functions but rather interact to facilitate DA modulation of target structures. Thus, a tonic level of parasynaptic DA activity might enable subsequent phasic alterations in modulatory drive when stimuli of relevance to the animal are encountered. This idea is reminiscent of the phenomenon of stochastic resonance, whereby background noise of a particular level boosts rather than masks weak inputs (MOSS and WIESENFELD 1995; MAYER-KRESS 1998). In this manner, extracellular DA may act on receptors localized to extrasynaptic membrane or neighboring DA synapses (BARBOUR and HÄUSSER 1997) to raise the probability of response to ensuing synaptic DA release. Given the likely importance of communication via multiple modes of transmission, it is perhaps not surprising that raising or lowering cortical DA levels degrades cognitive function (ARNSTEN 1997), particularly when these adjustments are applied without temporal reference to behaviorally significant events. Such functional degradation might be understood within the context of stochastic resonance, which if operative, would require an optimal level of parasynaptic DA to enhance the effect of behaviorally linked, albeit weak synaptic DA signals (Moss and WIESENFELD 1995). Likewise, with pathological loss of DA axons and their attendant synapses, such as occurs in Parkinson's disease, it could be predicted that restoration of extracellular DA levels would only partially alleviate symptoms and might even induce alternate forms of motor dysfunction (CHASE 1998; CHASE et al. 1998).

In summary, the range of DA ultrastructure parallels the range of DA functions mediated on target structures. Each mode of transmission is specialized to convey a particular type of behaviorally relevant signal, and, in addition, the parasynaptic and synaptic modes may act synergistically according to the rules of stochastic resonance. It is argued that DA axons do not simply express a synaptology to which target structures must then conform. Rather, the ultrastructure expressed by DA axons is that which has developed in concert with target cells in order to best serve their specific modulatory needs. Many examples have been presented of morphological substrates for delivery of DA along a continuum from maximum spatial specificity to more distributed, but still spatially limited exposure, and finally to widespread extracellular diffusion. In addition to morphology, the localization and efficacy of DA receptors and transporter molecules imposes constraints on the boundaries of these modes of transmission that are still being defined. The divergent modes of DA synaptology and transmission described in this review have been presented in the context of normal communication. However, variability in DA ultrastructure may also occur as a result of plasticity in the functions of the target neurons or during the course of pathological processes. In this regard, the ultrastructure of DA axons described in any particular study may represent only a single snapshot from a portfolio of variable DA synaptology.

Abbreviations

5-HT	serotonin
ACTH	adrenocorticotropic hormone
DA	dopamine
DAT	dopamine transporter
DBH	dopamine-β-hydroxylase
GABA	γ-aminobutyric acid
LH-RH	luteinizing hormone-releasing hormone
NAc	nucleus accumbens

NADPH   nicotinamide adenine dinucleotide phosphate, reduced
NE   norepinephrine
PNMT   phenylethanolamine N-methyltransferase
SN   substantia nigra
TH   tyrosine hydroxylase
VTA   ventral tegmental area

References

Ajika K (1979) Simultaneous localization of LHRH and catecholamines in rat hypothalamus. Journal of Anatomy 128:331–347

Ajika K, Hökfelt T (1973) Ultrastructural identification of catecholamine neurones in the hypothalamic periventricular-arcuate nucleus-median eminence complex with special reference to quantitative aspects. Brain Research 57:97–117

An X, Lewis DA, Sesack SR (1998) Ultrastructural relationship of dopamine D1 or D2 receptor immunoreactivity to synapses formed by tyrosine hydroxylase-positive terminals in the monkey striatum. Society for Neuroscience Abstracts 24:857

An X, Lewis DA, Sesack SR (1999) Tyrosine hydroxylase-immunoreactive terminals in the monkey accumbens core and shell: synaptic relationship to dopamine D1 and D2 receptors. Society for Neuroscience Abstracts 25:in press

Antonopoulos J, Dinopoulos A, Dori I, Parnavelas JG (1997) Distribution and synaptology of dopaminergic fibers in the mature and developing lateral septum of the rat. Developmental Brain Research 102:135–141

Aoki C, Pickel VM (1988) Neuropeptide Y-containing neurons in the rat striatum: ultrastructure and cellular relations with tyrosine hydroxylase-containing terminals and with astrocytes. Brain Research 459:205–225

Arluison M, Agid Y, Javoy F (1978) Dopaminergic nerve endings in the neostriatum of the rat. I. Identification by intracerebral injections of 5-hydroxydopamine. Neuroscience 3:657–673

Arluison M, Dietl M, Thibault J (1984) Ultrastructural morphology of dopaminergic nerve terminals and synapses in the striatum of the rat using tyrosine hydroxylase immunocytochemistry: a topographical study. Brain Research Bulletin 13:269–285

Arnsten AFT (1997) Catecholamine regulation of the prefrontal cortex. Journal of Psychopharmacology 11:151–162

Asan E (1993) Comparative single and double immunolabelling with antisera against catecholamine biosynthetic enzymes: criteria for the identification of dopaminergic, noradrenergic and adrenergic structures in selected rat brain areas. Histochemistry 99:427–442

Asan E (1997a) Interrelationships between tyrosine hydroxylase-immunoreactive dopaminergic afferents and somatostatinergic neurons in the rat central amygdaloid nucleus. Histochemistry and Cell Biology 107:65–79

Asan E (1997b) Ultrastructural features of tyrosine-hydroxlyase-immunoreactive afferents and their targets in the rat amygdala. Cell & Tissue Research 288:449–469

Asan E (1998) The catecholaminergic innervation of the rat amygdala. Advances in Anatomy, Embryology, & Cell Biology 142:1–118

Bannon MJ, Granneman JG, Kapatos G (1995) The dopamine transporter: potential involvement in neuropsychiatric disorders. In: Bloom FE, Kupfer DJ (eds) Psychopharmacology: The Fourth Generation of Progress, pp 179–187. New York: Raven Press, Ltd

Barbour B, Häusser M (1997) Intersynaptic diffusion of neurotransmitter. Trends in Neuroscience 20:377–384

Baude A, Nusser Z, Molnár E, McIlhinney RAJ, Somogyi P (1995) High-resolution immunogold localization of AMPA type glutamate receptor subunits at synaptic and non-synaptic sites in rat hippocampus. Neuroscience 69:1031–1055

Bayer VE, Pickel VM (1990) Ultrastructural localization of tyrosine hydroxylase in the rat ventral tegmental area: relationship between immunolabeling density and neuronal associations. Journal of Neuroscience 10:2996–3013

Bayer VE, Towle AC, Pickel VM (1991a) Ultrastructural localization of neurotensin-like immunoreactivity within dense core vesicles in perikarya, but not terminals, colocalizing tyrosine hydroxylase in the rat ventral tegmental area. Journal of Comparative Neurology 311:179–196

Bayer VE, Towle AC, Pickel VM (1991b) Vesicular and cytoplasmic localization of neurotensin-like immunoreactivity (NTLI) in neurons postsynaptic to terminals containing NTLI and/or tyrosine hydroxylase in the rat central nucleus of the amygdala. Journal of Neuroscience Research 30:398–413

Beaudet A, Descarries L (1978) The monoamine innervation of rat cerebral cortex: synaptic and nonsynaptic axon terminals. Neuroscience 3:851–860

Beaulieu C, Colonnier M (1985) A laminar analysis of the number of round asymmetrical and flat-symmetrical synapses on spines, dendritic trunks, and cell bodies in area 17 of the cat. Journal of Comparative Neurology 231:180–189

Benes FM, Vincent SL, Molloy R (1993) Dopamine-immunoreactive axon varicosities form nonrandom contacts with GABA-immunoreactive neurons of rat medial prefrontal cortex. Synapse 15:285–295

Berger B, Gaspar P, Verney C (1991) Dopaminergic innervation of the cerebral cortex: unexpected differences between rodents and primates. Trends in Neuroscience 14:21–27

Bergson C, Mrzljak L, Smiley JF, Pappy M, Levenson R, Goldman-Rakic PS (1995) Regional, cellular, and subcellular variations in the distribution of D_1 and D_5 receptors in primate brain. Journal of Neuroscience 15:7821–7836

Berke JD, Paletzki RF, Aronson GJ, Hyman SE, Gerfen CR (1998) A complex program of striatal gene expression induced by dopaminergic stimulation. Journal of Neuroscience 18:5301–5310

Bjelke B, Goldstein M, Tinner B, Andersson C, Sesack SR, Steinbusch HWM, Lew JY, He X, Watson S, Tengroth B, Fuxe K (1996) Dopaminergic transmission in the rat retina: evidence for volume transmission. Journal of Chemical Neuroanatomy 12:37–50

Björklund A, Lindvall O (1984) Dopamine-containing systems in the CNS. In: Björklund A, Hökfelt T (eds) Handbook of Chemical Neuroanatomy, Vol. 2: Classical Neurotransmitters in the CNS, Part 1, pp 55–122. Amsterdam: Elsevier Science Publishers

Bloom FE (1991) An integrative view of information handling in the CNS. In: Fuxe K, Agnati LF (eds) Volume Transmission in the Brain: Novel Mechanisms for Neural Transmission, pp 11–23. New York: Raven Press, Ltd

Bouyer JJ, Joh TH, Pickel VM (1984a) Ultrastructural localization of tyrosine hydroxylase in rat nucleus accumbens. Journal of Comparative Neurology 227:92–103

Bouyer JJ, Park DH, Joh TH, Pickel VM (1984b) Chemical and structural analysis of the relation between cortical inputs and tyrosine hydroxylase-containing terminals in rat neostriatum. Brain Research 302:267–275

Brog JS, Salyapongse A, Deutch AY, Zahm DS (1993) The patterns of afferent innervation of the core and shell in the "accumbens" part of the rat ventral striatum: immunohistochemical detection of retrogradely transported fluoro-gold. Journal of Comparative Neurology 338:255–278

Buijs RM, Geffard M, Pool CW, Hoorneman EMD (1984) The dopaminergic innervation of the supraoptic and paraventricular nucleus. A light and electron microscopic study. Brain Research 323:65–72

Caillé I, Dumartin B, Bloch B (1996) Ultrastructural localization of D1 dopamine receptor immunoreactivity in rat striatonigral neurons and its relation with dopaminergic innervation. Brain Research 730:17–31

Calabresi P, Pisani A, Centonze D, Bernardi G (1997) Synaptic plasticity and physiological interactions between dopamine and glutamate in the striatum. Neuroscience & Biobehavioral Reviews 21:519–523

Card JP, Rinaman L, Lynn RB, Lee B-H, Meade RP, Miselis RR, Enquist LW (1993) Pseudorabies virus infection of the rat central nervous system: ultrastructural characterization of viral replication, transport, and pathogenesis. Journal of Neuroscience 13:2515–2539

Carr DB, O'Donnell P, Card JP, Sesack SR (1997) Dopamine terminals in the rat prefrontal cortex synapse on pyramidal cells that project to the nucleus accumbens. Society for Neuroscience Abstracts 23:in press

Carr DB, Sesack SR (1996) Hippocampal afferents to the rat prefrontal cortex: synaptic targets and relation to dopamine terminals. Journal of Comparative Neurology 369:1–15

Carr DB, Sesack SR (1998) Callosal terminals in the rat prefrontal cortex: synaptic targets and association with GABA-immunoreactive structures. Synapse 29:193–205

Cepeda C, Levine MS (1998) Dopamine and N-methyl-D-aspartate receptor interactions in the neostriatum. Developmental Neuroscience 20:1–18

Cerruti C, Walther DM, Kuhar MJ, Uhl GR (1993) Dopamine transporter mRNA expression is intense in rat midbrain neurons and modest outside midbrain. Molecular Brain Research 18:181–186

Chan J, Aoki C, Pickel VM (1990) Optimization of differential immunogold-silver and peroxidase labeling with maintenance of ultrastructure in brain sections before plastic embedding. Journal of Neuroscience Methods 33:113–127

Chang HT (1988) Dopamine-acetylcholine interaction in the rat striatum: a dual-labeling immunocytochemical study. Brain Research Bulletin 21:295–304

Chase TN (1998) Levodopa therapy: consequences of the nonphysiologic replacement of dopamine. Neurology 50:S17–25

Chase TN, Oh JD, Blanchet PJ (1998) Neostriatal mechanisms in Parkinson's disease. Neurology 51:S30–35

Cheramy A, Leviel V, Glowinski J (1981) Dendritic release of dopamine in the substantia nigra. Nature 289:537–542

Chetverukhin VK, Belenky MA, Polenov AL (1979) Quantitative radioautographic light and electron microscopic analysis of the localization of monoamines in the median eminence of the rat. I. Catecholamines. Cell and Tissue Research 203:469–485

Ciliax BJ, Heilman C, Demchyshyn LL, Pristupa ZB, Ince E, Hersch SM, Niznik HB, Levey AI (1995) The dopamine transporter: immunocytochemical characterization and localization in brain. Journal of Neuroscience 15:1714–1723

Condé F, Lund JS, Jacobowitz DM, Baimbridge KG, Lewis DA (1994) Local circuit neurons immunoreactive for calretinin, calbindin D-28 k or parvalbumin in monkey prefrontal cortex: distribution and morphology. Journal of Comparative Neurology 341:95–116

Copray JCVM, Liem RSB, Ter Horst GJ, van Willigen JD (1990) Dopaminergic afferents to the mesencephalic trigeminal nucleus of the rat: a light and electron microscopic immunocytochemistry study. Brain Research 514:343–348

Courtoy PJ, Picton DH, Farquhar MG (1983) Resolution and limitations of the immunoperoxidase procedure in the localization of extracellular matrix antigens. Journal of Histochemistry and Cytochemistry 31:945–951

Cuello AC, Iversen LL (1973) Localization of tritiated dopamine in the median eminence of the rat hypothalamus by electron microscopic autoradiography. Brain Research 63:474–478

Dalil-Thiney N, Versaux-Botteri C, Nguyen-Legros J (1996) Electron microscopic demonstration of tyrosine hydroxylase-immunoreactive interplexiform cells in the lamprey retina. Neuroscience Letters 207:159–162

Decavel C, Geffard M, Calas A (1987a) Comparative study of dopamine- and noradrenaline-immunoreactive terminals in the paraventricular and supraoptic nuclei of the rat. Neuroscience Letters 77:149–154

Decavel C, Lescaudron L, Mons N, Calas A (1987b) First visualization of dopaminergic neurons with a monoclonal antibody to dopamine: a light and electron microscopic study. Journal of Histochemistry and Cytochemistry 35:1245–1251

DeFelipe J, Fariñas I (1992) The pyramidal neuron of the cerebral cortex: morphological and chemical characteristics of the synaptic inputs. Progress in Neurobiology 39:563–607

Delle Donne KT, Sesack SR, Pickel VM (1996) Ultrastructural immunocytochemical localization of neurotensin and the dopamine D_2 receptor in the rat nucleus accumbens. Journal of Comparative Neurology 371:552–566

Delle Donne KT, Sesack SR, Pickel VM (1997) Ultrastructural immunocytochemical localization of the dopamine D_2 receptor within GABAergic neurons of the rat striatum. Brain Research 746:239–255

Derouiche A, Asan E (1999) The dopamine D_2 receptor subfamily in rat retina: ultrastructural immunogold and in situ hybridization studies. European Journal of Neuroscience 11:1391–1402

Descarries L, Bosler O, Berthelet F, Des Rosiers MH (1980) Dopaminergic nerve endings visualized by high-resolution autoradiography in adult rat neostriatum. Nature 284:620–622

Descarries L, Lemay B, Doucet G, Berger B (1987) Regional and laminar density of the dopamine innervation in adult rat cerebral cortex. Neuroscience 21:807–824

Descarries L, Umbriaco D (1995) Ultrastructural basis of monoamine and acetylcholine function in CNS. Seminars in the Neurosciences 7:309–318

Descarries L, Watkins KC, Garcia S, Bosler O, Doucet G (1996) Dual character, asynaptic and synaptic, of the dopamine innervation in adult rat neostriatum: a quantitative autoradiographic and immunocytochemical analysis. Journal of Comparative Neurology 375:167–186

Dimova R, Vuillet J, Nieoullon A, Kerkerian-Le Goff L (1993) Ultrastructural features of the choline acetyltransferase-containing neurons and relationships with nigral dopaminergic and cortical afferent pathways in the rat striatum. Neuroscience 53:1059–1071

Djamgoz MBA, Wagner H-J (1992) Localization and function of dopamine in the adult vertebrate retina. Neurochemistry International 20:139–191

Doucet G, Descarries L, Audet MA, Garcia S, Berger B (1988) Radioautographic method for quantifying regional monoamine innervations in the rat brain. Application to the cerebral cortex. Brain Research 441:233–259

Doucet G, Descarries L, Garcia S (1986) Quantification of the dopamine innervation in adult rat neostriatum. Neuroscience 19:427–445

Dowling JE, Ehringer B (1975) Synaptic organization of the amine-containing interplexiform cells of the goldfish and cebus monkey retina. Science 188:270–273

Dowling JE, Ehringer B, Florén I (1980) Fluorescence and electron microscopical observations of amine-accumulating neurons of the Cebus monkey retina. Journal of Comparative Neurology 192:665–685

Dumartin B, Caillé I, Gonon F, Bloch B (1998) Internalization of D1 dopamine receptor in striatal neurons *in vivo* as evidence of activation by dopamine agonists. Journal of Neuroscience 18:1650–1661

Edwards SB, Hendrickson A (1981) The autoradiographic tracing of axonal connections in the central nervous system. In: Heimer L, Robards MJ (eds) Neuroanatomical Tract-Tracing Methods, pp 171–205. New York: Plenum Press

Erickson SL, Sesack SR, Lewis DA (1999) The dopamine innervation of monkey entorhinal cortex: postsynaptic targets of tyrosine hydroxylase terminals. Synapse: in press

Fallon JH, Loughlin SE (1995) Substantia nigra. In: Paxinos G (ed) The Rat Nervous System, Second Edition, pp 215–237. San Diego: Academic Press

Fallon JH, Moore RY (1978) Catecholamine innervation of the basal forebrain: IV. Topography of the dopamine projection to the basal forebrain and neostriatum. Journal of Comparative Neurology 180:545–580

Favard C, Simon A, Vigny A, Nguyen-Legros J (1990) Ultrastructural evidence for a close relationship between dopamine cell processes and blood capillary walls in Macaca monkey and rat retina. Brain Research 523:127–133

Fisher RS, Levine MS, Sibley DR, Ariano MA (1994) D_2 dopamine receptor protein localization: golgi impregnation-gold toned and ultrastructural analysis of the rat neostriatum. Journal of Neuroscience Research 38:551–564

Freund TF, Powell JF, Smith AD (1984) Tyrosine hydroxylase-immunoreactive boutons in synaptic contact with identified striatonigral neurons, with particular reference to dendritic spines. Neuroscience 13:1189–1215

Fujiyama F, Masuko S (1996) Association of dopaminergic terminals and neurons releasing nitric oxide in the rat striatum: an electron microscopic study using NADPH-diaphorase histochemistry and tyrosine hydroxylase immunohistochemistry. Brain Research Bulletin 40:121–127

Gabbott PLA, Bacon SJ (1996a) Local circuit neurons in the medial prefrontal cortex (areas 24a,b,c, 25 and 32) in the monkey: I. Cell morphology and morphometrics. Journal of Comparative Neurology 364:567–608

Gabbott PLA, Bacon SJ (1996b) Local circuit neurons in the medial prefrontal cortex (areas 24a,b,c, 25 and 32) in the monkey: II. Quantitative areal and laminar distributions. Journal of Comparative Neurology 364:609–636

Gábriel R, Zhu B, Straznicky C (1991) Tyrosine hydroxylase-immunoreactive elements in the distal retina of *Bufo marinus*: a light and electron microscopic study. Brain Research 559:225–232

Gábriel R, Zhu B, Straznicky C (1992) Synaptic contacts of tyrosine hydroxylase-immunoreactive elements in the inner plexiform layer of the retina of *Bufo marinus*. Cell and Tissue Research 267:

Garris PA, Wightman RM (1994) Different kinetics govern dopaminergic transmission in the amygdala, prefrontal cortex, and striatum: an *in vivo* voltammetric study. Journal of Neuroscience 14:442–450

Gaspar P, Berger B, Febvret A, Vigny A, Henry JP (1989) Catecholamine innervation of the human cerebral cortex as revealed by comparative immunohistochemistry of tyrosine hydroxylase and dopamine-beta-hydroxylase. Journal of Comparative Neurology 279:249–271

Gaspar P, Bloch B, Le Moine C (1995) D1 and D2 receptor gene expression in the rat frontal cortex: cellular localization in different classes of efferent neurons. European Journal of Neuroscience 7:1050–1063

Gaykema RP, Záborszky L (1996) Direct catecholaminergic-cholinergic interactions in the basal forebrain. II. Substantia nigra-ventral tegmental area projections to cholinergic neurons. Journal of Comparative Neurology 374:555–577

Gaykema RPA, Záborszky L (1997) Parvalbumin-containing neurons in the basal forebrain receive direct input from the substantia nigra-ventral tegmental area. Brain Research 747:173–179

Gerfen CR (1992) The neostriatal mosaic: multiple levels of compartmental organization. Trends in Neuroscience 15:133–139

Giros B, Jaber M, Jones SR, Wightman RM, Caron MG (1996) Hyperlocomotion and indifference to cocaine and amphetamine in mice lacking the dopamine transporter. Nature 379:606–612

Goldman-Rakic PS, Leranth C, Williams SM, Mons N, Geffard M (1989) Dopamine synaptic complex with pyramidal neurons in primate cerebral cortex. Proceedings of the National Academy of Sciences 86:9015–9019

Gonon F (1997) Prolonged and extrasynaptic excitatory action of dopamine mediated by D1 receptors in the rat striatum *in vivo*. Journal of Neuroscience 17:5972–5978

Grace AA (1991) Phasic versus tonic dopamine release and the modulation of dopamine system responsivity: a hypothesis for the etiology of schizophrenia. Neuroscience 41:1–24

Gray EG (1959) Axo-somatic and axo-dendritic synapses of the cerebral cortex: an electron microscope study. Journal of Anatomy 93:420–433

Groves PM (1980) Synaptic endings and their postsynaptic targets in neostriatum: synaptic specializations revealed from an analysis of serial sections. Proceedings of the National Academy of Sciences 77:6926–6929

Groves PM, Linder JC (1983) Dendro-dendritic synapses in substantia nigra: descriptions based on analysis of serial sections. Experimental Brain Research 49:209–217

Groves PM, Linder JC, Young SJ (1994) 5-hydroxydopamine-labeled dopaminergic axons: three-dimensional reconstructions of axons, synapses and postsynaptic targets in rat neostriatum. Neuroscience 58:593–604

Gulledge AT, Jaffe DB (1998) Dopamine decreases the excitability of layer V pyramidal cells in the rat prefrontal cortex. Journal of Neuroscience 18:9139–9151

Halliday GM, Tork I (1984) Electron microscopic analysis of the mesencephalic ventromedial tegmentum in the cat. Journal of Comparative Neurology 230:393–412

Hanley JJ, Bolam JP (1997) Synaptology of the nigrostriatal projection in relation to the compartmental organization of the neostriatum in the rat. Neuroscience 81:353–370

Hattori T, Takada M, Moriizumi T, van der Kooy D (1991) Single dopaminergic nigrostriatal neurons form two chemically distinct synaptic types: possible transmitter segregation within neurons. Journal of Comparative Neurology 309:391–401

Heeringa MJ, Abercrombie ED (1995) Biochemistry of somatodendritic dopamine release in substantia nigra: an in vivo comparison with striatal dopamine release. Journal of Neurochemistry 65:192–200

Henselmans JML, Wouterlood FG (1994) Light and electron microscopic characterization of cholinergic and dopaminergic structures in the striatal complex and the dorsal ventricular ridge of the lizard *Gekko gecko*. Journal of Comparative Neurology 345:69–83

Hersch SM, Ciliax BJ, Gutekunst CA, Rees HD, Heilman CJ, Yung KKL, Bolam JP, Ince E, Yi H, Levey AI (1995) Electron microscopic analysis of D1 and D2 dopamine receptor proteins in the dorsal striatum and their synaptic relationships with motor corticostriatal afferents. Journal of Neuroscience 15:5222–5237

Hersch SM, Yi H, Heilman CJ, Edwards RH, Levey AI (1997) Subcellular localization and molecular topology of the dopamine transporter in the striatum and substantia nigra. Journal of Comparative Neurology 388:211–227

Hökfelt T, Johansson O, Fuxe K, Elde R, Goldstein M, Park D, Efendic S, Luft R, Fraser H, Jeffcoate S (1977) Hypothalamic dopamine neurons and hypothalamic peptides. Advances in Biochemical Psychopharmacology 16:99–108

Hokoc JN, Mariani AP (1987) Tyrosine hydroxylase immunoreactivity in the rhesus monkey retina reveals synapses from bipolar cells to dopaminergic amacrine cells. Journal of Neuroscience 7:2785–2793

Horie S, Shioda S, Nakai Y (1993) Catecholaminergic innervation of oxytocin neurons in the paraventricular nucleus of the rat hypothalamus as revealed by double-labeling immunoelectron microscopy. Acta Anatomica 147:184–192

Horvath TL, Naftolin F, Leranth C (1992) GABAergic and catecholaminergic innervation of mediobasal hypothalamic β-endorphin cells projecting to the medial preoptic area. Neuroscience 51:391–399

Horvath TL, Naftolin F, Leranth C (1993) Luteinizing hormone-releasing hormone and gamma-aminobutyric acid neurons in the medial preoptic area are synaptic targets of dopamine axons originating in anterior periventricular areas. Journal of Neuroendocrinology 5:71–79

Ibata Y, Fukui K, Okamura H, Tanaka M, Obata HL, Tsuto T, Terubayashi H, Yanaihara C, Yanaihara N (1983) Coexistence of dopamine and neutotensin in hypothalamic arcuate and periventricular neurons. Brain Research 269:177–179

Ikemoto K, Satoh K, Kitahama K, Geffard M, Maeda T (1996) Electron-microscopic study of dopaminergic structures in the medial subdivision of the monkey nucleus accumbens. Experimental Brain Research 111:41–50

Ingham CA, Hood SH, Taggart P, Arbuthnott GW (1998) Plasticity of synapses in the rat neostriatum after unilateral lesion of the nigrostriatal dopaminergic pathway. Journal of Neuroscience 18:4732–4743

Jaber M, Robinson SW, Missale C, Caron MG (1996) Dopamine receptors and brain function. Neuropharmacology 35:1503–1519

Jakab RL, Leranth C (1990) Catecholaminergic, GABAergic, and hippocamposeptal innervation of GABAergic "somatospiny" neurons in the rat lateral septal area. Journal of Comparative Neurology 302:305–321

Jakab RL, Leranth C (1993) Presence of somatostatin or neurotensin in lateral septal dopaminergic axon terminals of distinct hypothalamic and midbrain origins: convergence on the somatospiny neurons. Experimental Brain Research 92:420–430

Johnson LR, Aylward RLM, Hussain Z, Totterdell S (1994) Input from the amygdala to the rat nucleus accumbens: its relationship with tyrosine hydroxylase immunoreactivity and identified neurons. Neuroscience 61:851–865

Jones EG, Powell TPS (1969) Morphological variations in the dendritic spines of the neocortex. Journal of Cell Science 5:509–529

Jones SR, O'Dell SJ, Marshall JF, Wightman RM (1996) Functional and anatomical evidence for different dopamine dynamics in the core and shell of the nucleus accumbens in slices of rat brain. Synapse 23:224–231

Jongen-Relo AL, Docter GJ, Jonker AJ, Voorn P (1995) Differential localization of mRNAs encoding dopamine D1 or D2 receptors in cholinergic neurons in the core and shell of the rat nucleus accumbens. Molecular Brain Research 28:169–174

Karle EJ, Anderson KD, Medina L, Reiner A (1996) Light and electron microscopic immunohistochemical study of dopaminergic terminals in the striatal portion of the pigeon basal ganglia using antisera against tyrosine hydroxylase and dopamine. Journal of Comparative Neurology 369:109–124

Karle EJ, Anderson KD, Reiner A (1992) Ultrastructural double-labeling demonstrates synaptic contacts between dopaminergic terminals and substance P-containing striatal neurons in pigeons. Brain Research 572:303–309

Karle EJ, Anderson KD, Reiner A (1994) Dopaminergic terminals form synaptic contacts with enkephalinergic striatal neurons in pigeons: an electron microscopic study. Brain Research 646:149–156

Kawaguchi Y, Wilson CJ, Augood SJ, Emson PC (1995) Striatal interneurones: chemical, physiological and morphological characterization. Trends in Neurosciences 18:527–535

Koch C, Poggio T (1983) Electrical properties of dendritic spines. Trends in Neurosciences 6:80–83

Koch C, Zador A (1993) The function of dendritic spines: devices subserving biochemical rather than electrical compartmentalization. Journal of Neuroscience 13:413–422

Kolb H, Cuenca N, Dekorver L (1991) Postembedding immunocytochemistry for GABA and glycine reveals the synaptic relationships of the dopaminergic amacrine cell of the cat retina. Journal of Comparative Neurology 310:267–284

Kolb H, Cuenca N, Wang H-H, Dekorver L (1990) The synaptic organization of the dopaminergic amacrine cell in the cat retina. Journal of Neurocytology 19:343–366

Kozasa K, Nakai Y (1987) Electron-microscopic cytochemistry of the catecholaminergic innervation of ACTH-containing neurons in the rat hypothalamic arcuate nucleus. Acta Anatomica 128:243–249

Krimer LS, Jakab RL, Goldman-Rakic PS (1997) Quantitative three-dimensional analysis of the catecholaminergic innervation of identified neurons in the macaque prefrontal cortex. Journal of Neuroscience 17:7450–7461

Krimer LS, Muly EC, Williams GV, Goldman-Rakic PS (1999) Dopaminergic regulation of cerebral cortical microcirculation. Nature Neuroscience 1:286–289

Kubota Y, Hattori R, Yui Y (1994) Three distinct subpopulations of GABAergic neurons in rat frontal agranular cortex. Brain Research 649:159–173

Kubota Y, Inagaki S, Kito S (1986a) Innervation of substance P neurons by catecholaminergic terminals in the neostriatum. Brain Research 375:163–167

Kubota Y, Inagaki S, Kito S, Shimada S, Okayama T, Hatanaka H, Pelletier G, Takagi H, Tohyama M (1988) Neuropeptide Y-immunoreactive neurons receive synaptic inputs from dopaminergic axon terminals in the rat neostriatum. Brain Research 458

Kubota Y, Inagaki S, Kito S, Takagi H, Smith AD (1986b) Ultrastructural evidence of dopaminergic input to enkephalinergic neurons in rat neostriatum. Brain Research 367:374–378

Kubota Y, Inagaki S, Kito S, Wu JY (1987a) Dopaminergic axons directly make synapses with GABAergic neurons in the rat neostriatum. Brain Research 406:147–156

Kubota Y, Inagaki S, Shimada S, Kito S, Eckenstein F, Tohyama M (1987b) Neostriatal cholinergic neurons receive direct synaptic inputs from dopaminergic axons. Brain Research 413:179–184

Kuljis RO, Advis JP (1989) Immunocytochemical and physiological evidence of a synapse between dopamine and luteinizing hormone releasing hormone-containing neurons in the ewe median eminence. Endocrinology 124:1579–1581

Kung L, Force M, Chute DJ, Roberts RC (1998) Immunocytochemical localization of tyrosine hydroxylase in the human striatum: a postmortem ultrastructural study. Journal of Comparative Neurology 390:52–62

Kuroda M, Murakami K, Igarashi H, Okada A (1996) The convergence of axon terminals from the mediodorsal thalamic nucleus and ventral tegmental area on pyramidal cells in layer V of the rat prelimbic cortex. European Journal of Neuroscience 8:1340–1349

Landwehrmeyer B, Mengod G, Palacios JM (1993) Differential visualization of dopamine D2 and D3 receptor sites in rat brain. A comparative study using *in situ* hybridization histochemistry and ligand binding autoradiography. European Journal of Neuroscience 5:145–153

Langer SZ (1997) 25 years since the discovery of presynaptic receptors: present knowledge and future perspectives. Trends in Pharmacological Sciences 18:95–99

Lavoie B, Smith Y, Parent A (1989) Dopaminergic innervation of the basal ganglia in the squirrel monkey as revealed by tyrosine hydroxylase immunohistochemistry. Journal of Comparative Neurology 289:36–52

Le Moine C, Bloch B (1995) D1 and D2 dopamine receptor gene expression in the rat striatum: sensitive cRNA probes demonstrate prominent segregation of D1 and D2 mRNAs in distinct neuronal populations of the dorsal and ventral striatum. Journal of Comparative Neurology 355:418–426

Le Moine C, Gaspar P (1998) Subpopulations of cortical GABAergic interneurons differ by their expression of D1 and D2 dopamine receptor subtypes. Molecular Brain Research 58:231–236

Le Moine C, Tison F, Bloch B (1990) D2 dopamine receptor gene expression by cholinergic neurons in the rat striatum. Neuroscience Letters 117:248–252

Lehmann J, Langer SZ (1983) The striatal cholinergic interneuron: synaptic target of dopaminergic terminals? Neuroscience 10:1105–1120

Leranth C, MacLusky NJ, Shanabrough M, Naftolin F (1988) Catecholaminergic innervation of luteinizing hormone-releasing hormone and glutamic acid decarboxylase immunopositive neurons in the rat medial preoptic area. Neuroendocrinology 48:591–602

Leranth C, Pickel VM (1989) Electron microscopic pre-embedding double immunostaining methods. In: Heimer L, Zaborsky L (eds) Neuroanatomical Tract Tracing 2, 2nd Edition, pp 129–172. New York: Plenum Publishing

Levey A, Hersch S, Rye D, Sunahara R, Niznik H, Kitt C, Price D, Maggio R, Brann M, Ciliax B (1993) Localization of D_1 and D_2 dopamine receptors in brain with subtype-specific antibodies. Proceedings of the National Academy of Sciences 90:8861–8865

Lewis DA, Hawrylak VA, Sesack SR (1997) Calbindin-immunoreactive neurons in the monkey prefrontal cortex: ultrastructural associations with dopamine terminals. Society for Neuroscience Abstracts 23:1213

Lewis DA, Sesack SR (1997) Dopamine systems in the primate brain. In: Bloom FE, Björklund A, Hökfelt T (eds) Handbook of Chemical Neuroanatomy, The Primate Nervous System, Part I, pp 261–373. New York: Elsevier Science Publishers

Lidow MS, Wang F, Cao Y, Goldman-Rakic PS (1998) Layer V neurons bear the majority of mRNAs encoding the five distinct dopamine receptor subtypes in the primate prefrontal cortex. Synapse 28:10–20

Lindvall O, Björklund A (1978) Anatomy of the dopaminergic neuron systems in the rat brain. In: Roberts PJ (ed) Advances in Biochemical Psychopharmacology, pp 1–23. New York: Raven Press

Loopuijt LD, VAN DER Kooy D (1985) Simultaneous ultrastructural localization of cholecystokinin- and tyrosine hydroxylase-like immunoreactivity in nerve fibers of the rat nucleus accumbens. Neuroscience Letters 56:329–334

Lorang D, Amara SG, Simerly RB (1994) Cell-type specific expression of catecholamine transporters in the rat brain. Journal of Neuroscience 14:4903–4914

Lovinger DM, Tyler E (1996) Synaptic transmission and modulation in the neostriatum. International Review of Neurobiology 39:77–111

Lugo-García N, Blanco RE (1993) Dopaminergic neurons in the cone-dominated ground squirrel retina: a light and electron microscopic study. Journal für Hirnforschung 35:561–569

Maeda T, Kitahama K, Geffard M (1994) Dopaminergic innervation of rat locus coeruleus: a light and electron microscopic immunohistochemical study. Microscopy Research and Technique 29:211–218

Mansour A, Meador-Woodruff JH, Bunzow JR, Civelli O, Akil H, Watson SJ (1990) Localization of dopamine D_2 receptor mRNA and D_1 and D_2 receptor binding in the rat brain and pituitary: an *in situ* hybridization-receptor autoradiographic analysis. Journal of Neuroscience 10:2587–2600

Mantz J, Millla C, Glowinski J, Thierry AM (1988) Differential effects of ascending neurons containing dopamine and noradrenaline in control of spontaneous activity and of evoked responses in the rat prefrontal cortex. Neuroscience 27:517–526

Marcos P, Corio M, Dubourg P, Tramu G (1996) Reciprocal synaptic connections between neurotensin- and tyrosine hydroxylase-immunoreactive neurons in the mediobasal hypothalamus of the guinea pig. Brain Research 715:63–70

Mariani AP (1991) Synaptic organization of type 2 catecholamine amacrine cells in the rhesus monkey retina. Journal of Neurocytology 20:332–342

Mayer-Kress G (1998) Non-linear mechanisms in the brain. Zeitschrift für Naturforschung 53:677–685

Meador-Woodruff JH, Damask SP, Wang JC, Haroutunian V, Davis KL, Watson SJ (1996) Dopamine receptor mRNA expression in human striatum and neocortex. Neuropsychopharmacology 15:17–29

Meister B, Hökfelt T, Tsuruo Y, Hemmings H, Ouimet C, Greengard P, Goldstein M (1988) DARPP-32, a dopamine and cyclic AMP-regulated phosphoprotein in tanycytes of the mediobasal hypothalamus: distribution and relation to dopamine and luteinizing hormone-releasing hormone neurons and other glial elements. Neuroscience 27:607–622

Mengod G, Martinez-Mir MI, Vilaró MT, Palacios JM (1989) Localization of the mRNA for the dopamine D_2 receptor in the rat brain by *in situ* hybridization histochemistry. Proceedings of the National Academy of Sciences 86:8560–8564

Milner TA (1991a) Cholinergic neurons in the rat septal complex: ultrastructural characterization and synaptic relations with catecholaminergic terminals. Journal of Comparative Neurology 314:37–54

Milner TA (1991b) Ultrastructural localization of tyrosine hydroxylase immunoreactivity in the rat diagonal band of Broca. Journal of Neuroscience Research 30:498–511

Mitchell V, Beauvillain J-C, Poulain P, Mazzuca M (1988) Catecholamine innervation of enkephalinergic neurons in guinea pig hypothalamus: demonstration by an in vitro autoradiographic technique combined with a post-embedding immunogold method. Journal of Histochemistry and Cytochemistry 36:533–542

Moore H, West AR, Grace AA (1999) The regulation of forebrain dopamine transmission: relevance to the pathophysiology and psychopathology of schizophrenia. Biological Psychiatry 46:40–55

Moss F, Wiesenfeld K (1995) The benefits of background noise. Scientific American August:66–69

Mrzljak L, Bergson C, Pappy M, Huff R, Levenson R, Goldman-Rakic PS (1996) Localization of dopamine D4 receptors in GABAergic neurons of the primate brain. Nature 381:245–248

Müller B, Peichl L (1991) Morphology and distribution of catecholaminergic amacrine cells in the cone-dominated tree shrew retina. Journal of Comparative Neurology 308:91–102

Muly ECI, Szigeti K, Goldman-Rakic PS (1998) D_1 receptor in interneurons of macaque prefrontal cortex: distribution and subcellular localization. Journal of Neuroscience 18:10553–10565

Nguyen-Legros J, Simon A, Caillé I, Bloch B (1997a) Immunocytochemical localization of dopamine D_1 receptors in the retina of mammals. Visual Neuroscience 14:545–551

Nguyen-Legros J, Versaux-Botteri C, Savy C (1997b) Dopaminergic and GABAergic retinal cell populations in mammals. Microscopy Research and Technique 36:26–42

Nirenberg MJ, Chan J, Liu Y, Edwards RH, Pickel VM (1996a) Ultrastructural localization of the vesicular monoamine transporter-2 in midbrain dopaminergic neurons: potential sites for somatodendritic storage and release of dopamine. Journal of Neuroscience 16:4135–4145

Nirenberg MJ, Chan J, Liu Y, Edwards RH, Pickel VM (1997a) Vesicular monoamine transporter-2: immunogold localization in striatal axons and terminals. Synapse 26:194–198

Nirenberg MJ, Chan J, Pohorille A, Vaughan RA, Uhl GR, Kuhar MJ, Pickel VM (1997b) The Dopamine transporter: comparative ultrastructure of Dopaminergic axons in limbic and motor compartments of the nucleus accumbens. Journal of Neuroscience 17:6899–6907

Nirenberg MJ, Vaughan RA, Uhl GR, Kuhar MJ, Pickel VM (1996b) The dopamine transporter is localized to dendritic and axonal plasma membranes of nigrostriatal dopaminergic neurons. Journal of Neuroscience 16:436–447

Noack HJ, Lewis DA (1989) Antibodies directed against tyrosine hydroxylase differentially recognize noradrenergic axons in monkey neocortex. Brain Research 500:313–324

Novikoff A, Novikoff P, Quintana N, Davis C (1972) Diffusion artifacts in 3,3′-diaminobenzidine cytochemistry. Journal of Histochemistry and Cytochemistry 20:745–749

Nusser Z, Roberts JD, Baude A, Richards JG, Sieghart W, Somogyi P (1995) Immunocytochemical localization of the α1 and β2/3 subunits of the $GABA_A$ receptor in relation to specific GABAergic synapses in the dentate gyrus. European Journal of Neuroscience 7:630–646

Oades RD, Halliday GM (1987) Ventral tegmental (A10) system: neurobiology. 1. Anatomy and connectivity. Brain Research Reviews 12:117–165

Onteniente B, Geffard M, Calas A (1984) Ultrastructural immunocytochemical study of the dopaminergic innervation of the rat lateral septum with anti-dopamine antibodies. Neuroscience 13:385–393

Onteniente B, Simon H, Taghzouti K, Geffard M, Le Moal M, Calas A (1987) Dopamine-GABA interactions in the nucleus accumbens and lateral septum of the rat. Brain Research 421:391–396

Otani S, Blond O, Desce JM, Crepel F (1998) Dopamine facilitates long-term depression of glutamatergic transmission in rat prefrontal cortex. Neuroscience 85:669–676

Papadopoulos GC, Parnavelas JG (1991) Monoamine systems in the cerebral cortex: evidence for anatomical specificity. Progress in Neurobiology 36:195–200

Papadopoulos GC, Parnavelas JG, Bujis RM (1989) Light and electron microscopic immunocytochemical analysis of the dopamine innervation of the rat visual cortex. Journal of Neurocytology 18:303–310

Paré D, Smith Y (1996) Thalamic collaterals of corticostriatal axons: their termination field and synaptic targets in cats. Journal of Comparative Neurology 372:551–567

Parnavelas JG, Papadopoulos GC (1989) The monoaminergic innervation of the cerebral cortex is not diffuse and nonspecific. Trends in Neurosciences 12:315–319

Peters A, Palay SL, Webster Hd (1991) The Fine Structure of the Nervous System. Neurons and Their Supporting Cells, 3rd Edition. New York: Oxford

Phelix CF, Liposits Z, Paull WK (1992) Monoamine innervation of the bed nucleus of the stria terminalis: an electron microscopic investigation. Brain Research Bulletin 28:949–965

Phelix CF, Liposits Z, Paull WK (1994) Catecholamine-CRF synaptic interaction in a septal bed nucleus: afferents of neurons in the bed nucleus of the stria terminalis. Brain Research Bulletin 33:109–119

Pickel VM, Beckley SC, Joh TH, Reis DJ (1981) Ultrastructural immunocytochemical localization of tyrosine hydroxylase in the neostriatum. Brain Research 225:373–385

Pickel VM, Chan J (1990) Spiny neurons lacking choline acetyltransferase immunoreactivity are major targets of cholinergic and catecholaminergic terminals in rat striatum. Journal of Neuroscience Research 25:263–280

Pickel VM, Chan J, Sesack SR (1992) Cellular basis for interactions between catecholaminergic afferents and neurons containing leu-enkephalin-like immunoreactivity in rat caudate-putamen nuclei. Journal of Neuroscience Research 31:212–230

Pickel VM, Joh TH, Chan J (1988a) Substance P in the rat nucleus accumbens: ultrastructural localization in axon terminals and their relation to dopaminergic afferents. Brain Research 444:247–264

Pickel VM, Sesack SR (1999) Electron microscopy of central dopamine systems. In: Bloom FE, Kupfer DJ (eds) Psychopharmacology: The Fourth Generation of Progress, CD-ROM Version. New York: Raven

Pickel VM, Towle A, Joh TH, Chan J (1988b) Gamma-aminobutyric acid in the medial rat nucleus accumbens: ultrastructural localization in neurons receiving monosynaptic input from catecholaminergic afferents. Journal of Comparative Neurology 272

Pinto A, Sesack SR (1998) Paraventricular thalamic afferents to the rat prefrontal cortex and nucleus accumbens shell: synaptic targets and relation to dopamine afferents. Society for Neuroscience Abstracts 24:1595

Pinto A, Sesack SR (1999) Basolateral amygdala afferents to the rat prefrontal cortex: ultrastructure and relation to dopamine afferents. Society for Neuroscience Abstracts 25: in press

Pourcho RG (1982) Dopaminergic amacrine cells in the cat retina. Brain Research 252:101–109

Redgrave P, Prescott TJ, Gurney K (1999) Is the short-latency dopamine response too short to signal reward error? Trends in Neuroscience 22:146–151

Revay R, Vaughan R, Grant S, Kuhar MJ (1996) Dopamine transporter immunohistochemistry in median eminence, amygdala, and other areas of the rat brain. Synapse 22:93–99

Ridet J-L, Sandillon F, Rajaofetra N, Geffard M, Privat A (1992) Spinal dopaminergic system of the rat: light and electron microscopic study using an antiserum against dopamine, with particular emphasis on synaptic incidence. Brain Research 598:233–241

Rodrigo J, Fernández P, Bentura ML, Martínez de Velasco J, Serrano J, Uttenthal O, Martínez-Murillo R (1998) Distribution of catecholaminergic afferent fibres in the rat globus pallidus and their relations with cholinergic neurons. Journal of Chemical Neuroanatomy 15:1–20

Rodríguez A, González-Hernández T (1999) Electrophysiological and morphological evidence for a GABAergic nigrostriatal pathway. Journal of Neuroscience 19: 4682–4694

Rohrer B, Stell WK (1995) Localization of putative dopamine D_2-like receptors in the chick retina, using in situ hybridization and immunocytochemistry. Brain Research 695:110–116

Schimchowitsch S, Vuillez P, Tappaz ML, Klein MJ, Stoeckel ME (1991) Systematic presence of GABA-immunoreactivity in the tubero-infundibular and tubero-hypophyseal dopaminergic axonal systems: an ultrastructural immunogold study on several mammals. Experimental Brain Research 83:575–586

Schmitt FO (1984) Molecular regulators of brain function: a new view. Neuroscience 13:991–1001

Schotland JL, Shupliakov O, Grillner S, Brodin L (1996) Synaptic and nonsynaptic monoaminergic neuron systems in the lamprey spinal cord. Journal of Comparative Neurology 372:229–244

Schultz W (1998) Predictive reward signal of dopamine neurons. Journal of Neurophysiology 80:1–27

Séguéla P, Watkins KC, Descarries L (1988) Ultrastructural features of dopamine axon terminals in the anteromedial and the suprarhinal cortex of adult rat. Brain Research 442:11–22

Sesack S, Hawrylak V, Melchitzky D, Lewis D (1998a) Dopamine innervation of a subclass of local circuit neurons in monkey prefrontal cortex: ultrastructural analysis of tyrosine hydroxylase and parvalbumin immunoreactive structures. Cerebral Cortex 8:614–622

Sesack SR, Aoki C, Pickel VM (1994) Ultrastructural localization of D2 receptor-like immunoreactivity in midbrain dopamine neurons and their striatal targets. Journal of Neuroscience 14:88–106

Sesack SR, Bressler CN, Lewis DA (1995a) Ultrastructural associations between dopamine terminals and local circuit neurons in the monkey prefrontal cortex: a study of calretinin-immunoreactive cells. Neuroscience Letters 200:9–12

Sesack SR, Carr DB (1998) Projections from the rat prefrontal cortex to the ventral tegmental area. II. Association with GABA- or TH-immunoreactive mesocortical neurons. Society for Neuroscience Abstracts 24: 1597

Sesack SR, Hawrylak VA, Guido MA, Levey AI (1998b) Dopamine axon varicosities in the rat prefrontal cortex exhibit sparse immunoreactivity for the dopamine transporter. Journal of Neuroscience 18:2697–2708

Sesack SR, King SW, Bressler CN, Watson SJ, Lewis DA (1995b) Electron microscopic visualization of dopamine D2 receptors in the forebrain: cellular, regional, and species comparisons. Society for Neuroscience Abstracts 21:365

Sesack SR, Miner LAH (1997) In the rat prefrontal cortex, dopamine terminals converge with local axon collaterals onto common dendritic processes. Society for Neuroscience Abstracts 23:1213

Sesack SR, Pickel VM (1990) In the rat medial nucleus accumbens, hippocampal and catecholaminergic terminals converge on spiny neurons and are in apposition to each other. Brain Research 527:266–279

Sesack SR, Pickel VM (1992) Prefrontal cortical efferents in the rat synapse on unlabeled neuronal targets of catecholamine terminals in the nucleus accumbens septi and on dopamine neurons in the ventral tegmental area. Journal of Comparative Neurology 320:145–160

Sesack SR, Snyder CL, Lewis DA (1995c) Axon terminals immunolabeled for dopamine or tyrosine hydroxylase synapse on GABA-immunoreactive dendrites in rat and monkey cortex. Journal of Comparative Neurology 363:264–280

Shioda S, Nakai Y, Sunayama S, Shimoda Y (1986) Electron-microscopic cytochemistry of the catecholaminergic innervation of TRH neurons in the rat hypothalamus. Cell and Tissue Research 245:247–252

Shirouzu M, Anraku T, Iwashita Y, Yoshida M (1990) A new dopaminergic terminal plexus in the ventral horn of the rat spinal cord. Immunohistochemical studies at the light and electron microscopic levels. Experientia 46:201–204

Sim LJ, Joseph SA (1991) Arcuate nucleus projections to brainstem regions which modulate nociception. Journal of Chemical Neuroanatomy 4:97–109

Smiley JF, Goldman-Rakic PS (1993a) Heterogeneous targets of dopamine synapses in monkey prefrontal cortex demonstrated by serial section electron microscopy: a laminar analysis using the silver-enhanced diaminobenzidine sulfide (SEDS) immunolabeling technique. Cerebral Cortex 3:223–238

Smiley JF, Goldman-Rakic PS (1993b) Silver-enhanced diaminobenzidine-sulfide (SEDS): a technique for high-resolution immunoelectron microscopy demonstrated with monoamine immunoreactivity in monkey cerebral cortex and caudate. Journal of Histochemistry and Cytochemistry 41:1393–1404

Smiley JF, Levey AI, Ciliax BJ, Goldman-Rakic PS (1994) D_1 dopamine receptor immunoreactivity in human and monkey cerebral cortex: predominant and extrasynaptic localization in dendritic spines. Proceedings of the National Academy of Sciences 91:5720–5724

Smiley JF, Williams SM, Szigeti K, Goldman-Rakic PS (1992) Light and electron microscopic characterization of dopamine-immunoreactive axons in human cerebral cortex. Journal of Comparative Neurology 321:325–335

Smith AD, Bolam JP (1990) The neural network of the basal ganglia as revealed by the study of synaptic connections of identified neurones. Trends in Neurosciences 13:259–265

Smith Y, Bennett BD, Bolam JP, Parent A, Sadikot AF (1994) Synaptic relationship between dopaminergic afferents and cortical or thalamic input in the sensorimotor territory of the striatum in monkey. Journal of Comparative Neurology 344:1–19

Smith Y, Kieval J, Couceyro P, Kuhar MJ (1999) CART peptide-immunoreactive neurons in the nucleus accumbens in monkeys: ultrastructural analysis, colocalization studies, and synaptic interactions with dopaminergic afferents. Journal of Comparative Neurology 407:491–511

Somogyi P, Tamás G, Lujan R, Buhl EH (1998) Salient features of synaptic organisation in the cerebral cortex. Brain Research Reviews 26:113–135

Sulzer D, Joyce MP, Lin L, Geldwert D, Haber SN, Hattori T, Rayport S (1998) Dopamine neurons make glutamatergic synapses *in vitro*. Journal of Neuroscience 18:4588–4602

Surmeier DJ, Song W-J, Yan Z (1996) Coordinated expression of dopamine receptors in neostriatal medium spiny neurons. Journal of Neuroscience 16:6579–6591

Swanson LW (1982) The projections of the ventral tegmental area and adjacent regions: a combined fluorescent retrograde tracer and immunofluorescence study in the rat. Brain Research Bulletin 9:321–353

Tennyson VM, Heikkila R, Mytilineou C, Côté L, Cohen G (1974) 5-hydroxydopamine 'tagged' neuronal boutons in rabbit striatum: interrelationship between vesicles and axonal membrane. Brain Research 82:341–348

Thierry AM, Blanc G, Sobel A, Stinus L, Glowinski J (1973) Dopminergic terminals in the rat cortex. Science 182:499–501

Thureson-Klein AK, Klein RL (1990) Exocytosis from neuronal large dense-cored vesicles. International Review of Cytology 121:67–126

Totterdell S, Smith AD (1989) Convergence of hippocampal and dopaminergic input onto identified neurons in the nucleus accumbens of the rat. Journal of Chemical Neuroanatomy 2:285–298

Triarhou LC, Norton J, Ghetti B (1988) Synaptic connectivity of tyrosine hydroxylase immunoreactive nerve terminals in the striatum of normal, heterozygous and homozygous weaver mutant mice. Journal of Neurocytology 17:221–232

Van Bockstaele EJ, Pickel VM (1993) Ultrastructure of serotonin-immunoreactive terminals in the core and shell of the rat nucleus accumbens: Cellular substrates for interactions with catecholamine afferents. Journal of Comparative Neurology 334:603–617

Van Bockstaele EJ, Pickel VM (1995) GABA-containing neurons in the ventral tegmental area project to the nucleus accumbens in rat brain. Brain Research 682: 215–221

Van den Pol AN, Herbst PS, Powell JF (1984) Tyrosine hydroxylase-immunoreactive neurons of the hypothalamus: a light and electron microscopic study. Neuroscience 13:1117–1156

Van Eden CG, Hoorneman EMD, Buijs RM, Matthijssen MAH, Geffard M, Uylings HBM (1987) Immunocytochemical localization of dopamine in the prefrontal cortex of the rat at the light and electron microscopic level. Neuroscience 22: 849–862

Van Haesendonck E, Marc RE, Missotten L (1993) New aspects of dopaminergic interplexiform cell organization in the goldfish retina. Journal of Comparative Neurology 333:503–518

Verney C, Alvarez C, Geffard M, Berger B (1990) Ultrastructural double-labelling study of dopamine terminals and GABA-containing neurons in rat anteromedial cerebral cortex. European Journal of Neuroscience 2:960–972

Verney C, Gaspar P, Alvarez C, Berger B (1987) Postnatal sequential development of dopaminergic and enkephalinergic perineuronal formations in the lateral septal nucleus of the rat correlated with local neuronal maturation. Anatomy and Embryology 176:463–475

Verucki ML, Wässle H (1996) Immunohistochemical localization of dopamine D_1 receptors in rat retina. European Journal of Neuroscience 8:2286–2297

Vizi ES (1991) Nonsynaptic inhibitory signal transmission between axon terminals: physiological and pharmacological evidence. In: Fuxe K, Agnati LF (eds) Volume Transmission in the Brain: Novel Mechanisms for Neural Transmission, pp 89–96. New York: Raven Press, Ltd

Voigt T, Wässle H (1987) Dopaminergic innervation of A II amacrine cells in mammalian retina. Journal of Neuroscience 7:4115–4128

Voorn P, Jorritsma-Byham B, Van Dijk C, Buijs R (1986) The dopaminergic innervation of the ventral striatum in the rat: a light- and electron-microscopical study with antibodies against dopamine. Journal of Comparative Neurology 251:84–99

Vuillet J, Kerkerian L, Kachidian P, Bosler P, Nieoullon A (1989) Ultrastructural correlates of functional relationships between nigral dopaminergic or cortical afferent fibers and neuropeptide Y-containing neurons in the rat striatum. Neuroscience Letters 100:99–104

Wagner H-J, Behrens UD (1993) Microanatomy of the dopaminergic system in the rainbow trout retina. Vision Research 33:1345–1358

Wagner H-J, Luo B-G, Ariano MA, Sibley DR, Stell WK (1993) Localization of D_2 dopamine receptors in vertebrate retinae with anti-peptide antibodies. Journal of Comparative Neurology 331:469–481

Wagner H-J, Wulle I (1990) Dopaminergic interplexiform cells contact photoreceptor terminals in catfish retina. Cell and Tissue Research 261:359–365

Wagner H-J, Wulle I (1992) Contacts of dopaminergic interplexiform cells in the outer retina of the blue acara. Visual Neuroscience 9:325–333

Watanabe T, Nakai Y (1987) Electron microscopic cytochemistry of catecholaminergic innervation of LHRH neurons in the medial preoptic area of the rat. Archivum Histologicum Japonicum 50:103–112

Watt CB, Glazebrook PA (1993) Synaptic organization of dopaminergic amacrine cells in the larval tiger salamander retina. Neuroscience 53:527–536

Williams SM, Goldman-Rakic PS (1998) Widespread origin of the primate mesofrontal dopamine system. Cerebral Cortex 8:321–345

Yang CR, Seamans JK (1996) Dopamine D1 receptor actions in layers V-VI rat prefrontal cortex neurons *in vitro*: modulation of dendritic-somatic signal integration. Journal of Neuroscience 16:1922–1935

Yazulla S, Zucker CL (1988) Synaptic organization of dopaminergic interplexiform cells in the goldfish retina. Visual Neuroscience 1:13–29

Yung KK, Bolam JP, Smith AD, Hersch SM, Ciliax BJ, Levey AI (1995) Immunocytochemical localization of D1 and D2 dopamine receptors in the basal ganglia of the rat: light and electron microscopy. Neuroscience 65:709–730

Yuste R, Denk W (1995) Dendritic spines as basic functional units of neuronal integration. Nature 375:682–684

Záborsky L, Léránth C, Palkovits M (1979) Light and electron microscopic identification of monoaminergic terminals in the central nervous system. Brain Research Bulletin 4:99–117

Záborszky L, Cullinan WE (1996) Direct catecholaminergic-cholinergic interactions in the basal forebrain. I. Dopamine-β-hydroxylase and tyrosine hydroxylase input to cholinergic neurons. Journal of Comparative Neurology 374:535–554

Zahm DS (1989) Evidence for a morphologically distinct subpopulation of striatipetal axons following injections of WGA-HRP into the ventral tegmental area in the rat. Brain Research 482:145–154

Zahm DS (1992) An electron microscopic morphometric comparison of tyrosine hydroxylase immunoreactive innervation in the neostriatum and the nucleus accumbens core and shell. Brain Research 575:341–346

Zahm DS, Brog JS (1992) On the significance of subterritories in the "accumbens" part of the rat ventral striatum. Neuroscience 50:751–767

Zoli M, Torri C, Ferrari R, Jansson A, Zini I, Fuxe K, Agnati LF (1998) The emergence of the volume transmission concept. Brain Research Reviews 26:136–147

CHAPTER 5
D_1-Like Dopamine Receptors: Molecular Biology and Pharmacology

H.B. NIZNIK, K.S. SUGAMORI, J.J. CLIFFORD, and J.L. WADDINGTON

A. Introduction

Initial classification of dopamine (DA) receptors into D_1 and D_2 subtypes on the basis of stimulatory and no/inhibitory linkage to adenylyl cyclase (AC), respectively (SPANO et al. 1978; KEBABIAN and CALNE 1979), endured in substance for approximately a decade until the molecular cloning of D_1 and D_2 receptors provided additional criteria for distinguishing these two receptors in terms of genomic structure/localization, primary structure and mRNA tissue distribution profile. Thereafter, primarily during the early 1990s, further molecular cloning studies revealed the mammalian DA receptor family to be yet more heterogeneous (see MISSALE et al. 1998; NEVE and NEVE 1997; NIZNIK 1994): in particular, cloning both of primate D_1 (rodent homologue D_{1A}) and of primate D_5 (rodent homologue D_{1B}) receptors indicated the original designation of D_1 to encompass a family of D_1-like receptors whose properties are the focus of this chapter, in juxtaposition with a family of D_2-like receptors ($D_{2L/S}$, D_3, D_4) whose properties are the focus of subsequent chapters.

Historically, interest in DA receptors at functional and clinical levels focused on the D_2-like receptor family, due in large part to the substantial correlations observed between the functional/clinical potencies of DA receptor antagonists/antipsychotic drugs and their in vitro affinities for blocking the D_2 receptor in the initial absence of any selective D_1 antagonist (SEEMAN 1992). In contrast, the D_1 receptor was described in 1983 to be a site "in search of a function" (LADURON 1983); indeed, some scepticism for the existence of a distinct D_1 receptor entity endured until evidence for the physical resolution of D_1 and D_2 receptor protein moieties was obtained (DUMBRILLE-ROSS et al. 1985), and the time-lag between molecular advances and identification of pharmacological tools for exploring the functionality of the entities cloned has been a significant generic impediment to progress in this regard.

This chapter will describe some of the salient molecular, pharmacological and functional properties encoded by members of the D_1-like receptor family. To the extent possible, recent advances with regard to identifying subtype-specific functional roles for D_1/D_{1A} vs D_5/D_{1B} receptors will be summarized also.

B. Molecular Biology of D_1-Like Receptors

I. The D_1/D_{1A} Receptor Gene

The gene for the D_1 receptor was first cloned based on the high degree of sequence homology between members of the superfamily of G protein-coupled receptors (GPCRs; DEARRY et al. 1990; MONSMA et al. 1990; SUNAHARA et al. 1990; ZHOU et al. 1990). Unlike the D_2 receptor gene, the coding region of the D_1 receptor gene was found to be unimpeded by intronic sequences and encoded by a seven transmembrane (TM) topology protein with a short third cytoplasmic loop (3rd CL) and large carboxyl terminus, typical of GPCRs that stimulate AC activity. The D_1 receptor shares a higher degree of amino-acid identity to β-adrenergic receptors (AR; ~30%) than to the D_2 receptor (24%). Using fluorescent in situ hybridization to metaphase chromosomes, the gene for the D_1 receptor was mapped to the long arm of human chromosome 5 at 5q35.1. In addition, two restriction fragment length polymorphisms (RFLPs) were reported including a Taq I and an Eco RI.

Cloning of the 5' upstream regulatory region of the human D_1 receptor gene revealed the presence of an intron of 116 bp at positions −599 to −484 upstream from the first putative translation initiation site (MOURADIAN et al. 1994). The primary structure of the promoter region for the D_1 receptor was found to lack both TATA and CAAT box sequences, but contained a high G+C content and putative sequences for transcription factors such as Sp1 (MINOWA et al. 1993), all of which are features typical of "housekeeping" type genes. The transcriptional activity of the 5' upstream region was shown to be cell-specific, as only cells that have the appropriate machinery to interact with this promoter could display transcriptional activity. A second functional TATA- and CAAT-less promoter (LEE et al. 1996) has since been discovered in the intron mentioned above, and appears to be transactivated by the POU transcription factor, Brn-4, which co-localizes with D_1 mRNA in the striatum (OKAZAWA et al. 1996). Similarly, the 5' non-coding sequence of the rat D_{1A} receptor gene has a promoter region characterized by a high G+C content, putative sites for the transcription factors Sp1, Ap1 and Ap2, absence of TATA or CAAT sequences, and an intron of 115 bp (ZHOU et al. 1992).

II. The D_5/D_{1B} Receptor Gene

An additional member of the D_1-like receptor family was cloned and termed the D_5 receptor from humans, with its rodent orthologue termed the D_{1B} receptor (GRANDY et al. 1991; SUNAHARA et al. 1991; TIBERI et al. 1991; WEINSHANK et al. 1991). The D_5 receptor shares approximately 53% overall and 80% transmembrane amino acid identities with the D_1 receptor, lacks introns in the coding region, and possesses the same putative seven TM topology with a relatively small third CL and large carboxyl terminus. The D_5/D_{1B} receptor can be distinguished from the D_1 receptor on the basis of its chromosomal localization to human chromosome 4p15.1–15.3 (EUBANKS et al. 1992; GRANDY et al. 1992; SHERRINGTON et al. 1993).

The transcription initiation site for the D_5 receptor has been defined to 2125 bp upstream from the translational initiation site and, similar to the D_1 promoter region, the region 5′ to the transcription initiation site lacks both TATA and CAAT canonical sequences but contains putative sites for Sp1 and Ap1 (BEISCHLAG et al. 1995). An intron of either 179 or 155 bp is found in the 5′ untranslated region of the D_5 receptor gene. In addition, two pseudogenes, $D_{5\varphi1}$ and $D_{5\varphi2}$, share high sequence identity (>94% nucleotide identity) with the D_5 receptor and localize to human chromosomes 2p11.1-p11.2 and 1q21.1. These possess mutations/deletions resulting in early termination of the gene products and have been identified exclusively in higher primates (GRANDY et al. 1992; MARCHESE et al. 1995). These two pseudogenes share significant sequence homology with the D_5 receptor in the 5′ untranslated region, only showing divergence beyond nucleotide −1916. It is postulated that these two pseudogenes may have arisen from two gene duplication events mediated by Alu sequences (MARCHESE et al. 1995).

III. Primary Structure of D_1-Like Receptors

Since D_1-like receptors resemble structurally the seven α-helical membrane-spanning topology of GPCRs coupled to the activation of AC, much of what can be predicted about D_1-like receptor structure–function relationships can be derived from the $β_2$-AR (DOHLMAN et al. 1991). Conserved residues in the D_1 receptor include: an aspartic acid residue in TM2 (Asp^{70}) that appears to interact allosterically with agonists (TOMIC et al. 1993); an aspartic acid residue in TM3 (Asp^{103}) that is believed to bind the amine group of catecholamine ligands; two cysteine residues (Cys^{96}, Cys^{186}) that may form a receptor stabilizing disulphide bridge between the first and second extracellular loops (NODA et al. 1994) and a phenylalanine residue in TM6 (Phe^{289}) which is postulated to stabilize the catechol ring. Also conserved in the D_1 receptor is the AspArgTyr (DRY) sequence in the 2nd CL which, based on receptor modelling and site-directed mutagenesis, is believed to be constrained in a hydrophilic pocket formed by conserved polar residues in TM1, TM2 and TM7 (SCHEER et al. 1996). Most characteristic and distinctive of catecholamine receptors, though, are the three conserved serine residues in TM5, two of which for the $β_2$-AR are believed to form hydrogen bonds with the hydroxyl groups of the catechol moiety. Mutation of these conserved serine residues in TM5 (Ser^{198}, Ser^{199}, Ser^{202}) of the D_1 receptor (Table 1) resulted in severe alterations in agonist and antagonist binding and in AC activity (POLLOCK et al. 1992).

Within the carboxyl terminus of the D_1 receptor is a conserved cysteine residue (Cys^{347}); the corresponding residue in the $β_2$-AR is palmitoylated to form a putative fourth intracellular loop (IL). Mutation of this cysteine residue or truncation of the carboxyl tail at this residue results in the inability of the D_1 receptor to desensitize, in addition to displaying a decreased affinity for

Table 1. Effect of mutations on D_1 receptor function

Mutation/Type	Region	Effect	Reference
Point mutations			
Ser56 ⇒ Leu56	TM1	No change in receptor function	Jin 1998
Ser127 ⇒ Ala127	IL2	Altered desensitization time-course	Kozell 1997
Ser198 ⇒ Ala198	TM5	No change in down-regulation No [³H]SCH-23390 binding ⇓ DA potency (50-fold)	Pollock 1992
Ser199 ⇒ Ala199	TM5	⇓SKF-38393 potency (14-fold), ⇓V_{max} ⇓ [³H]SCH-23390 affinity (26-fold) ⇓ DA, α-flupentixol, SKF-38393 affinity ⇓ DA potency (23-fold), no change in V_{max} ⇓ SKF-38393 potency and V_{max}	Pollock 1992
Ser202 ⇒ Ala202	TM5	No change in benzazepine binding ⇓ DA affinity (48-fold), potency (48-fold) ⇓SKF-38393 V_{max}	Pollock 1992
Ile205 ⇒ Ala205	TM5	No change in receptor function	Cho 1996
Ile205 ⇒ Tyr205	TM5	⇓ [³H]SCH-23390 affinity ⇓ [³H]SCH-23390 Bmax ⇓ DA, SKF-38393, SKF-82958 affinity ⇓ DA potency, V_{max}	Cho 1996
Phe264 ⇒ Ile264	IL3	⇑ constitutive (basal) activity ⇑ DA affinity and potency	Charpentier 1996
Ile288 ⇒ Phe288	IL3 (D1B)	⇓ constitutive (basal) activity ⇓ DA affinity and potency	Charpentier 1996
Arg266 ⇒ Lys266	IL3	No change in receptor function	Charpentier 1996
Ser263 ⇒ Ala263	IL3	No change in desensitization	Kozell 1997
Leu286 ⇒ Ala286	TM6	No change in down-regulation ⇑ constitutive (basal) activity ⇑ DA potency (2.7-fold) ⇑ SCH-23390 partial agonism	Cho 1996
Leu286 ⇒ Tyr286	TM6	⇓ [³H]SCH-23390 Bmax ⇓ DA, SKF-38393, SKF-82958 affinity ⇓ DA potency, V_{max}	Cho 1996

Mutation	Region	Effect	Reference
Point mutations			
Cys347 $\Rightarrow$ Gly347	COOH tail	No change [^{3}H]SCH-23390 K_d $\Downarrow$ [^{3}H]SCH-23390 B_{max} $\Downarrow$ DA V_{max}	Jensen 1995
Cys351 $\Rightarrow$ Gly351	COOH tail	No desensitization response No change [^{3}H]SCH-23390 K_d Desensitization response ($\Downarrow$ DA V_{max})	Jensen 1995
Cys347 + Cys351 $\Rightarrow$ Gly347 + Gly351	COOH tail	No change [^{3}H]SCH-23390 K_d $\Downarrow$ [^{3}H]SCH-23390 B_{max} $\Downarrow$ DA V_{max} No desensitization response	Jensen 1995
Cys347 $\Rightarrow$ Ala347	COOH tail	No change [^{3}H]SCH-23390 K_d No change in DA V_{max}	Jin 1997
Cys351 $\Rightarrow$ Ala351	COOH tail	No change [^{3}H]SCH-23390 K_d $\Uparrow$ constitutive (basal) activity	Jin 1997
Cys347 + Cys351 $\Rightarrow$ Ala347 + Ala351	COOH tail	No change [^{3}H]SCH-23390 K_d	Jin 1997
Ser380 $\Rightarrow$ Ala380	COOH tail	No change in desensitization	Kozell 1997
Thr135 + Ser229 +Thr268 + Ser380 $\Rightarrow$ Val135 + Ala229 + Val 268 + Ala380	IL2, IL3 COOH tail	No change in down-regulation $\Downarrow$ onset of desensitization No other changes in function	Jiang 1997
Truncations			
Cys347 $\Rightarrow$ stop (Truncation)	COOH tail	No [^{3}H]SCH-23390 binding $\Downarrow$ DA V_{max} $\Downarrow$ butaclamol, α-flupentixol affinity No desensitization response	Jensen 1995
Tyr348 $\Rightarrow$ stop (Truncation)	COOH tail	No change [^{3}H]SCH-23390 K_d No desensitization response	Jensen 1995
Cys351 $\Rightarrow$ stop (Truncation)	COOH tail	No change [^{3}H]SCH-23390 K_d No desensitization response	Jensen 1995
Pro352 $\Rightarrow$ stop (Truncation)	COOH tail	No change [^{3}H]SCH-23390 K_d No desensitization response	Jensen 1995

Table 1. *Continued*

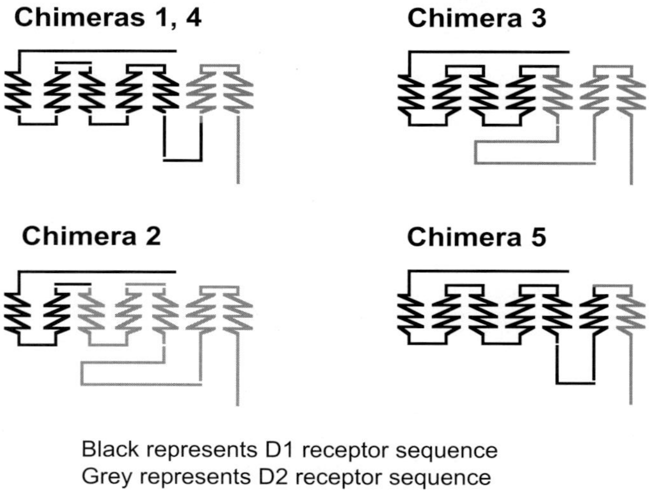

Mutation/Type	Region	Effect	Reference
Receptor chimeras			
1. D1(Thr272)/D2	TM6-COOH	⇓ [^{3}H]SCH-23390 affinity (20-fold) ⇓ DA affinity, potency, V_{max} ⇓ SKF-38393 affinity and potency	MacKenzie 1993
2. D1(Cys96)/D2 3. D1(Arg191)/D2	TM3-COOH TM5-COOH	[^{3}H]SCH-23390 not quantifiable ⇓ [^{3}H]SCH-23390 affinity ⇓ butaclamol, α-flupentixol affinity	Kozell 1995 Kozell 1995
4. D1(Lys271)/D2	TM6-COOH	⇓ [^{3}H]SCH-23390 affinity ⇓ DA affinity, potency, V_{max} ⇓ butaclamol, α-flupentixol affinity	Kozell 1995
5. D1(Asp309)/D2	TM7-COOH	⇓ [^{3}H]SCH-23390 affinity ⇓ DA potency, V_{max} ⇓ butaclamol, α-flupentixol affinity	Kozell 1995

Chimeras 1, 4

Chimera 3

Chimera 2

Chimera 5

Black represents D1 receptor sequence
Grey represents D2 receptor sequence

some antagonists, and a decreased efficacy for some agonists (JENSEN et al. 1995; MOFFETT et al. 1993). These findings were not, however, replicated in an independent study (JIN et al. 1997). The above residues are also conserved in the D_5 receptor and include Asp^{87} in TM2, Asp^{120} in TM3, Ser^{229}, Ser^{230} and Ser^{233} in TM5, Phe^{313} in TM6, Cys^{113} and Cys^{217} in the first and second extracellular loops, and Cys 375 in the carboxyl terminus. Within the amino terminus, two potential N-linked glycosylation sites are present with an additional site found in the second extracellular loop for both D_1 and D_5 receptors. The precise role for glycosylation is unknown but may be necessary for the proper insertion/trafficking of the D_5 receptor (BERGSON et al. 1996).

IV. D_1-Like Receptor Polymorphisms

The possibility that single amino acid mutations can severely effect the ligand binding and functional characteristics of these receptors has led to the search for their possible involvement in the maintenance and expression of neuropsychiatric disorders, such as schizophrenia (OKUBO et al. 1997). While much intensive effort has focused on the DA system, studies on either D_1 or D_5 receptors and schizophrenia, Gilles de la Tourette syndrome, or bipolar disorder have revealed no linkage of either subtype with these diseases (BRETT et al. 1995; see BALDESSARINI 1997; BARR et al. 1997).

Genetic association studies have also yielded mainly negative results. These studies in which DNA sequence variations (VAPSE) in the D_1 or D_5 receptor genes were scanned in several cases of schizophrenia or bipolar disorder and compared with healthy controls revealed several sequence mutations (Table 2) that resulted mainly in silent mutations occurring at the same frequency between patients and controls (CICHON et al. 1994, 1996; DOLLFUS et al. 1996; FENG et al. 1998; LIU et al. 1995; O'HARA et al. 1993; SAVOYE et al. 1998; SOBELL et al. 1995; THOMPSON et al. 1998). The D_5 receptor, in particular, displays a number of sequence changes (SOBELL et al. 1995), five of which result in either nonsense ($C^{335}X$) or missense changes ($N^{351}D$; $A^{269}V$; $S^{453}C$; $P^{330}Q$). The nonsense change, $C^{355}X$, which causes premature truncation of the protein between TM domains 6 and 7 seemed particularly interesting; however, this mutation failed to show a co-segregation with schizophrenia, and the allele frequency of this mutation was not statistically different from controls. Neuropsychological testing of heterozygotes for the $C^{355}X$ allele did reveal a trend toward poor performance in tests sensitive to frontal lobe impairment. The four missense changes leading to amino acid substitutions also did not reveal an association with schizophrenia or any other neuropsychiatric disease, nor did other D_5 receptor sequence polymorphisms (ASHERSON et al. 1998). A polymorphic dinucleotide repeat (TC)13 has been localized to the D_5 receptor promoter transactivation domain with additional allelic forms (TC)12 found in one patient with schizophrenia and one with Huntington's disease, and (TC)14 found in another case of schizophrenia. However, neither of these alleles altered D_5-promoter-mediated luciferase

Table 2. D_1 and D_5 receptor gene polymorphisms

Mutation/Type	Region	Comments	Reference
Dopamine D_1 receptor			
T-2218 $\Rightarrow$ C Polymorphism	5'-UTR	Same allelic frequency[a]	Cichon 1996
C-2102 $\Rightarrow$ A Polymorphism	5'-UTR	Same allelic frequency[a]	Cichon 1996
T-2030 $\Rightarrow$ C Polymorphism	5'-UTR	Same allelic frequency[a]	Cichon 1996
G-1992 $\Rightarrow$ A Polymorphism	5'-UTR	Same allelic frequency[a]	Cichon 1996
G-1251 $\Rightarrow$ C Polymorphism	5'-UTR	Same allelic frequency[a]	Cichon 1996
T-800 $\Rightarrow$ C Polymorphism	5'-UTR	Same allelic frequency[a]	Cichon 1996
G-94 $\Rightarrow$ A Polymorphism	5'-UTR	Same allelic frequency[a]	Cichon 1994a
A-48 $\Rightarrow$ G Polymorphism	5'-UTR	Same allelic frequency[a]	O'Hara 1993; Cichon 1994a; Liu 1995
Leu $\Rightarrow$ Leu30 Silent (A90G)	TM1		O'Hara 1993
Ile49 $\Rightarrow$ Ile Silent (C147T)	TM1		Thompson 1998
Leu66 $\Rightarrow$ Leu Silent (G198A)	TM2	Asian-specific variant	Liu 1995
Ala 74 $\Rightarrow$ Ala Silent (A222C)	TM2		O'Hara 1993
Ser421 $\Rightarrow$ Ser Silent (G1263A)	COOH	Same allelic frequency[a]	Cichon 1994a; Liu 1995
T1403 $\Rightarrow$ C Silent	3'-UTR	Same allelic frequency[a]	Cichon 1994a
Dopamine D_5 receptor			
Ala22 $\Rightarrow$ Ala Silent	NH_2		SoBell 1995
Leu88 $\Rightarrow$ Phe Missense	TM2	Found in one autistic patient	Feng 1998
Ala269 $\Rightarrow$ Val Missense	IL3	No association with schizophrenia	SoBell 1995
Pro326 $\Rightarrow$ Pro Silent (T978C)	IL3	No association with schizophrenia	SoBell 1995
Pro330 $\Rightarrow$ Pro Silent	EL3		
Pro330 $\Rightarrow$ Gln Missense	EL3	No association with schizophrenia	SoBell 1995
Cys335 $\Rightarrow$ Stop Nonsense	EL3	No association with schizophrenia	SoBell 1995; Feng 1998

Table 2. *Continued*

Mutation/Type	Region	Comments	Reference
Asn351 ⇒ Asp Missense	TM7	No association with schizophrenia	SoBell 1995
Ser453 ⇒ Cys	COOH	No association with schizophrenia	SoBell 1995
C1481 ⇒ T Polymorphism	3'-UTR		SoBell 1995

3'-UTR, 3' untranslated; 5'-UTR, 5' untranslated; COOH, carboxyl terminus; EL, extracellular loop; IL, intracellular loop; NH_2, amino terminus.
[a] Between schizophrenic and bipolar affective disorder patients and controls.

activity in SK-N-SH cells and thus were not functional (Beischlag et al. 1996). In addition, an association study of two RFLPs of the D_1 receptor, a Taq I and an Eco RI polymorphism, failed to show an association with bipolar disorder (Savoye et al. 1998). Some recent interest, however, has focused on chromosome 4 (for review see Kennedy and Macciardi 1998) and linkage disequilibrium of markers on chromosome 4p are suggestive of a site relevant to schizophrenia in close proximity to the D_5 receptor gene. There is some preferential transmission of D_5 receptor polymorphic alleles (Daly et al. 1999) in families with attention deficit disorder, another putative "hyperdopaminergic" disease state (Pliszka et al. 1996).

C. D_1-Like Receptor Pharmacology

There has been a steady increase in the identification of compounds exhibiting selectivity for D_1-like over D_2-like receptors, as pharmaceutical companies and academic laboratories have directed medicinal chemistry resources in accordance with evolving pre-clinical research and evidence for therapeutic potential; however, identification of compounds exhibiting selectivity within the D_1-like family, e.g. for D_{1A} over D_{1B} receptors, has yet to match these advances.

I. Selective D_1-Like Agonists

Identification in the late 1970s of the 1-phenyl-3-benzazepine SKF 38393 as a partial agonist in terms of stimulation of DA-sensitive AC preceded even initial D_1/D_2 receptor classification but was soon recognized as the first selective D_1 (D_1-like) receptor (partial) agonist, and its status as a reference functional tool endures; however, its limited selectivity, low in vivo potency and only partial agonist activity, together with increasing appreciation of the biological role and therapeutic potential of the D_1 receptor, engendered a search for superior agents, the primary objective being identification of a highly selective and potent full efficacy agonist (WADDINGTON and O'BOYLE 1989).

Within this extensive benzazepine series, a large number of homologues have been examined to reveal D_1 (D_1-like) agonists having from very low (SK&F 75670), through intermediate (SK&F 77434), to very high (SK&F 81297, SK&F 83189) efficacies to stimulate AC (O'BOYLE et al. 1989; ANDERSEN and JANSEN 1990). It should be emphasized that while one of these, SK&F 82958, has been marketed and used widely as a full efficacy agonist, this status is not straightforward. This compound demonstrates a complex and anomalous interaction with D_1 (D_1-like) receptors in terms of supramaximal stimulation of AC (O'BOYLE et al. 1989) or stimulation with a very shallow dose-response relationship, that indicates an unconventional and/or indirect mechanism at D_1-like receptors (MOTTOLA et al. 1992), with anomalous neurophysiological effects (NISENBAUM et al. 1998; RUSKIN et al. 1998); these properties necessitate considerable caution in its use as a putative full efficacy D_1 (D_1-like) agonist (WADDINGTON et al. 1995, 1998a).

Among non-benzazepine compounds, the thienopyridines SK&F 89145 and 89641 were identified as partial and their congeners SK&F 89615 and 89626 as full efficacy selective D_1 (D_1-like) agonists, but with little penetration through the blood-brain barrier (WADDINGTON and O'BOYLE 1989). Similar to many of the benzazepines, the indolophenanthridine CY 208–243 shows only partial D_1-like agonist activity and only limited selectivity over D_2-like and non-dopaminergic receptors (WADDINGTON and O'BOYLE 1989; MURRAY and WADDINGTON 1990); while the related benzophenanthridine dihydrexidine (MOTTOLA et al. 1992) demonstrates full efficacy to stimulate AC, it too shows only limited selectivity over D_2-like receptors. Thus, it is the isochroman A 68930 (DENINNO et al. 1991) which constitutes the first high-affinity, full-efficacy selective D_1-like agonist; its congener A 77636 (KEBABIAN et al. 1992) demonstrates lower selectivity and reduced biological potency, but was preferred as a (unsuccessful) candidate for clinical development. Thereafter, the thienophenanthrene A 86929 was identified also as a high-affinity, full-efficacy selective D_1-like agonist, with its diacetyl prodrug offering improved pharmacokinetic characteristics (SHIOSAKI et al. 1998). Conformational studies on novel D_1-like agonists (and antagonists) complement molecular biological studies in resolving the pharmacophore for D_1-like receptor recognition and transduction. Additionally, it must be noted that while D_1-like pharmacological interactions with AC have been studied primarily in rodent tissue, it has been argued that there exist species differences therein, particularly vis-à-vis (non-human and human) primates; these issues are considered further below.

II. Selective D_1-Like Antagonists

Identification in the early 1980s of the 7-halogenated 1-phenyl-3-benzazepine SCH 23390 as the first agent having all the defining characteristics of a selective D_1 antagonist, in terms of high affinity for inhibition of DA-sensitive AC and selectivity for D_1 (D_1-like) over D_2 (D_2-like) receptor binding sites, had a fundamental impact on DA receptor pharmacology, and its status as a reference functional tool endures; however, while its material affinity for 5-HT$_2$ receptors presented some practical limitations, at a conceptual level its unexpectedly profound pharmacological effects, in an era where prepotency of the D_2 receptor was assumed, indicated potential danger in fundamentally revising parcellation of function between DA receptor subtypes on the basis of a single agent from one chemical class (WADDINGTON and O'BOYLE 1989). Thus, a series of modifications to the basic benzazepine moiety were explored.

The 6-Br homologue of SCH 23390 was identified as an enantiomeric D_1 (D_1-like) antagonist pair (R->>S-SK&F 83566); more importantly, the benzonaphthazepine SCH 39166 was subsequently introduced as an important high-affinity selective D_1 (D_1-like) receptor antagonist with reduced affinity for 5-HT$_2$ receptors (WADDINGTON and O'BOYLE 1989). Further exploration of variants to the benzazepine moiety resulted in identification of the benzofuranylbenzazepines NNC 112, NNC687 and NNC 756 (ANDERSEN et al. 1992), the aminoalkyl/cinnamylbenzazepines (SHAH et al. 1996), and the thienobenzazepine LY 270411 (DEVENEY and WADDINGTON 1996) as further selective D_1 (D_1-like) antagonists. Perhaps yet more important, however, has been the identification of the isoquinolines A 69024 (KERKMAN et al. 1989) and BW 737C (RIDDALL 1992), the dibenzoquinolizines (MINOR et al. 1994), and the benzoquinoxoline SDZ PSD 958 (MARKSTEIN et al. 1996) as selective D_1 (D_1-like) antagonists chemically distinct from the benzazepines or their congeners.

III. Selectivity for D_1/D_{1A} Vs D_5/D_{1B} Receptors?

It has been argued that drug-based discrimination of the various subtypes of D_1-like receptor is best described in general by a "profile" of binding affinities for several drugs that can readily be compared in different species rather than by the particular properties of a single "specific" ligand; indeed, a quandary in receptor classification may occur when minor sequence differences between orthologous receptors (species homologues) are recognized by specific ligands or, conversely, when paralogous receptors (true subtypes) are not distinguished by different ligands (VERNIER et al. 1995). Assignment of distinct functional roles between D_1/D_{1A} and D_5/D_{1B} receptors has been impeded by the relatively high pharmacological similarity that exists between them. At present, only a few ligands have been identified that can differentiate these two cloned receptor subtypes, at least in recombinant cell lines, in terms either of ligand binding assays or of AC activation. These include DA itself and the agonist 6,7-ADTN which for the D_5 receptor display a tenfold higher affinity than for D_1 receptor (SUNAHARA et al. 1991); the magnitude of this increased affinity, however, was not as great for the rat D_{1B} receptor, for which only a threefold increase in affinity was observed (TIBERI et al. 1991).

In contrast, some antagonists such as *cis*(Z)-flupentixol and (+)-butaclamol exhibit a much lower (seven- to tenfold) affinity for D_5/D_{1B}

than for D_1/D_{1A} receptors. These differences in the pharmacological profiles of DA/ADTN and butaclamol/flupenthixol for cloned D_1 vs D_5 receptors appear to define the pharmacological signature of D_1 and D_5 receptors. Recent evidence suggests that part of the molecular substrate which may confer some of the inherent agonist pharmacological characteristics defining D_1 vs D_5 receptors appears to reside within amino acid motifs of the carboxyl terminal tail. Thus, the characteristic tenfold higher affinity exhibited for DA and 6,7-ADTN by the D_5 receptor can be completely converted to one displaying D_1 receptor characteristics by D_5 receptor mutants in which the carboxyl terminal domains have been replaced by sequences encoded by the D_1 receptor tail (DEMCHYSHYN et al. 1997). These changes are specific for these agonists, while discriminating antagonists such as butaclamol and cis(Z)-fluphentixiol, and virtually all other non-selective D_1/D_5 agonists and antagonists, retain their "normal" pharmacological profiles.

However, while these ligands can be used to differentiate D_1/D_{1A} and D_5/D_{1B} receptors using radioligand binding and functional assay systems, none of these ligands can be used meaningfully in vivo to differentiate between subtype-specific responses. Ongoing efforts are continuing to identify specific D_5/D_{1B} receptor ligands, and to study structure-activity relationships that may govern receptor specificity.

D. Molecular Aspects of Functional Coupling and Signal Transduction for D_1-Like Receptors

I. G-Protein Selectivity

Binding of an agonist to a receptor is believed to cause a conformational change of individual transmembrane domains that is propagated to the intracellular receptor surface whereby coupling to specific classes of G proteins is mediated; the agonist catalyzes the exchange of GDP for GTP on the α-subunit with the resultant dissociation of the αβγ-heterotrimer. The activated α-subunit and the free βγ-subunit can then act on specific effector systems (AC, ion channels, phospholipases) to mediate specific biological responses. The highly variable 3rd CL or IL of heptahelical receptors is believed to be one of the main structural determinants forming part of the molecular interface conferring either selectivity or efficiency of coupling (see WESS 1997). This has been confirmed for DA receptors as revealed by the generation of D_1/D_2 receptor chimeras in which replacement of the D_1 3rd CL with D_2 receptor sequence or addition of the D_1 minigene encoding the 3rd CL abolished the ability of the receptor to stimulate cyclic adenosine monophosphate (cAMP) production (KOZELL et al. 1994; LUTTRELL et al. 1993). Conversely, a chimeric D_2 receptor with D_1 receptor sequence from TM5 to TM6, encompassing the 3rd CL, coupled to G_s (KOZELL and NEVE 1997). The regions within the 3rd CL that mediate selectivity or activation of G protein has been further refined to the amino and/or carboxyl terminal motifs at the junctions of TM5 and 6

(DOHLMAN et al. 1991; GUAN et al. 1995; OKAMOTO et al. 1991) and form the contact site to both α and β subunits of G proteins (TAYLOR et al. 1994, 1996).

In as much as the amino and/or carboxyl regions of the 3rd CL appear to affect coupling and in some cases selectivity of coupling, these regions cannot solely account for receptor–G-protein interactions. Studies have also implicated the carboxyl terminus (OKAMOTO et al. 1991; NAMBA et al. 1993; SANO et al. 1997) and other intracellular domains such as the 1st and 2nd CL in receptor–G-protein-coupling interactions (MORO et al. 1993; SCHREIBER et al. 1994; VARRAULT et al. 1994; VERRALL et al. 1997). For the D_{1A} receptor, synthetic peptides directed to the 2nd and 3rd IL and the N-terminal part of the carboxyl terminus prevented interaction with G_s (KONIG and GRATZEL 1994). Thus, it appears that depending on the receptor, several cytoplasmic domains may act in concert to determine either selectivity or coupling efficiency by forming a three-dimensional binding site. Alterations in these cytoplasmic domains may result in functionally promiscuous receptors (WONG and ROSS 1994) that can couple to multiple G proteins, or in a constitutively active receptor state (KJELSBERG et al. 1992; KOZELL and NEVE 1997).

Functionally, D_1 and D_5 receptors both stimulate the activity of AC when expressed in a number of eukaryotic cell lines and are believed to couple to either $G\alpha_{olf}$ or $G\alpha_s$, (HERVE et al. 1993). Promiscuity of coupling to multiple G proteins, however, may provide a means by which D_1 and D_5 receptors may possibly differentiate their functional status. This has been demonstrated in cell lines either endogenously expressing the D_1 receptor, or transfected with either the D_1 or the D_5 receptor, and immunoprecipitation with antisera directed against specific G proteins (KIMURA et al. 1995a,b; SIDHU et al. 1998). D_1 receptor-mediated events in stably transfected rat pituitary GH4C1 cells are responsive to both cholera and pertussis toxin treatment and are co-immunoprecipitated with antisera to $G\alpha_o$ and $G\alpha_s$ and not $G\alpha_i$ (KIMURA et al. 1995a). In the neuroblastoma cell line SK-N-MC, which endogenously expresses the D_1 receptor (SIDHU 1997) but lacks $G\alpha_o$, only cholera toxin treatment abolished high affinity D_1 receptor interactions. In contrast, the human D_5 receptor, when stably expressed in GH4C1 cells, appeared to associate with $G\alpha_s$ but not $G\alpha_o$, $G\alpha_{i1}$, $G\alpha_{i2}$, $G\alpha_{i3}$ or $G\alpha_q$ and a pertussis toxin-insensitive G protein, recently identified as $G\alpha_z$ (SIDHU et al. 1998). It may be speculated that since $G\alpha_z$ inhibits AC type I and type V (KOZASA and GILMAN 1995), the D_5 receptor in some cells may have a negative effect on AC activity. For instance, a dopaminergic D_5 inhibition of catecholamine secretion observed in chromaffin cells is associated with an unidentified pertussis toxin-insensitive G protein (DAHMER and SENOGLES 1996a,b).

II. Constitutive Activation of D_5/D_{1B} Receptors

One of the unique characteristics distinguishing between D_1 and D_5 receptors is that the D_5/D_{1B} receptor is known to be constitutively activated (TIBERI and CARON 1994), defined as an increase in agonist-independent basal activity. A number of GPCRs have been shown to be constitutively activated in the absence of agonist, or can be modified to become constitutively activated by specific mutations. Based on studies with AR, a modified allosteric ternary complex model has been proposed to explain the endogenous activation of receptors in the absence of agonist (LEFKOWITZ et al. 1993). In agreement with the peptide studies on G-protein activation, the carboxyl region of the 3rd CL adjacent to TM6 and the endofacial region of TM6 appear to be involved in the active vs inactive conformation of the receptor (see PAUWELS and WURCH 1998; JAVITCH et al. 1997).

Similar to the AR, the carboxyl terminal region of the 3rd CL has been implicated in the constitutive activity displayed by D_{1B} receptors (TIBERI and CARON 1994). Site-directed mutagenesis of an amino acid (Ile288) in the carboxyl region of the 3rd CL of the rat D_{1B} receptor by the counterpart Phe residue present at the corresponding site of the D_{1A} receptor attenuated the constitutive activity of the mutant D_{1B} receptor (CHARPENTIER et al. 1996). Conversely, replacement of the Phe264 residue by an Ile residue in the D_{1A} receptor resulted in a constitutive activity for the D_{1A} receptor, thus implicating the importance of this residue in mediating the constitutive activity phenotype of the D_{1B} receptor. Another study in which a conserved Leu residue adjacent to a proline residue in TM6 was mutated to an Ala residue (Leu286Ala) also displayed a constitutive activity phenotype for the D_1 receptor, indicating that this mutation may lead to a release of constraint in the adjacent carboxyl terminal (CT) region of the 3rd CL or alteration in the configuration of TM6 (CHO et al. 1996). Consistent with the observation of one critical conserved Ile residue at the CT end of the 3rd CL (CHARPENTIER et al. 1996), the Xenopus (SUGAMORI et al. 1998) and eel D_{1B} (CARDINAUD et al. 1997) receptors also display an enhanced basal or constitutive activity. Thus, constitutive activity is absolutely conserved throughout the evolutionary course of the D_5/D_{1B} receptor subfamily.

While a majority of the amino acid domains involved in the constitutive activation of either mutated GPCRs and those found in pathology appear restricted to sites within the 3rd IL and TM6 (PAUWELS and WURCH 1998), it is clear that additional sequence motifs, particularly those within the CT-tail, regulate the expression of D_5 receptor-constitutive activity. It appears, from work with D_5-successive deletion/truncation mutants (DEMCHYSHYN et al. 1997) that D_5 receptor-mediated constitutive activity, but not the receptor's ability to stimulate AC in response to agonists, is dependent upon sequences encoded by the amino acids of the D_5 CT-tail. These data would argue for involvement of the CT-tail of the receptor, either directly or indirectly, with the Ile residue defined in the 3rd CL to regulate this receptor's ability to interact with subtype-specific G proteins. In any event, constitutive activation of AC by the D_5/D_{1B} receptor system appears to be fundamental to this subtype, and thus may be functionally relevant to the physiology of DA in the vertebrate nervous system.

III. Desensitization

Desensitization is the phenomenon of attenuation or waning of response upon continued activation by an agonist. With GPCRs two forms of desensitization may occur: (1) homologous desensitization, which is mediated by specific G protein-receptor kinases (GRKs), such as β-ARKs or rhodopsin kinase, which phosphorylate the agonist-occupied receptor and promote the binding of arrestin proteins; and (2) heterologous desensitization, in which activation of the receptor turns on second messenger-dependent kinases such as protein kinase A or protein kinase C which, in turn, can phosphorylate that receptor and others at specific Ser/Thr residues (see ; PREMONT et al. 1995; PITCHER et al. 1998; KRUPNICK and BENOVIC 1998; LEFKOWITZ 1998).

Studies on the D_1 receptor indicate that in most expression systems, the D_1 receptor undergoes a homologous desensitization (reviewed in SIBLEY and NEVE 1997; SIBLEY et al. 1998). This is manifested as a decrease in maximal activation (V_{max}) of cAMP accumulation and a slower decrease in receptor number upon more prolonged exposure to DA, suggesting that specific GRKs may be phosphorylating the receptor. The D_1 receptor, when co-expressed with various GRKs in a heterologous expression system, can serve as a substrate for GRK 2, GRK 3 and GRK 5 (TIBERI et al. 1996) resulting in either a rightward shift in the dose response curve for DA with little effect on V_{max} (GRK2,3), or a decrease in both potency and efficacy (GRK 5). GRK 5 belongs to a separate family of GRKs that is highly expressed in heart, lung and retina, and is characterized by being constitutively associated with the cell membrane.

The observation that D_1 receptors are endocytosed in a dynamin-dependent manner (VICKERY and VON ZASTROW 1999) and the association of internalized receptors with light vesicular membrane fractions (NG et al. 1995; TROGADIS et al. 1995) would suggest that once phosphorylated by subtype specific GRKs, D_1 receptors probably associate with β-arrestin-like molecules for subsequent recruitment into clathrin-coated pits and vesicular compartments, thereby promoting sequestration and resensitization (FERGUSON et al. 1996; ZHANG et al. 1997). The internalization of D_1 receptors can also occur in CHO cells, primary striatal cultures and rat striatal brain slices following agonist exposure (ARIANO et al. 1997a,b). In these cases concanavalin A, which blocks endocytosis, prevented sequestration without affecting stimulation of AC, indicating that sequestration and uncoupling of the receptor G-protein complex occur by biochemically distinct pathways. In one in vivo study (DUMARTIN et al. 1998), an acute and massive internalization of postsynaptic D_1 receptors in endocytic vesicles, occurring as early as 4 min. after injection, was visualized in the striatum after treatment with either a D_1-like agonist or amphetamine without modification of cAMP responsiveness. However, recruitment of D_1 receptors to the plasma membrane has also been reported following agonist stimulation (BRISMAR et al. 1998). Less is known about the desensitization pattern of the D_5/D_{1B} receptor; the human D_5 receptor can undergo homologous desensitization in Ltk cells, manifested as a decrease in maximal response (JARVIE et al. 1993) but studies on internalization and long-term desensitization are lacking.

The D_1 receptor also appears to undergo heterologous desensitization when stably expressed in Y1 mouse adreno-cortical tumour cells (OLSON and SCHIMMER 1992). A heterologous desensitization was also demonstrated in D_1-CHO mutants expressing low activities of protein kinase A, where the degree of agonist-induced desensitization was reduced (SIBLEY et al. 1998), indicating that protein kinase A phosphorylation plays a role in the D_1 acute desensitization. Site-directed mutagenesis of all potential protein kinase A phosphorylation sites in the D_{1A} receptor resulted in an attenuation of DA-induced desensitization in C-6 glioma cells (JIANG and SIBLEY 1996), which was further refined to the Thr268 residue in the 3rd CL (JIANG and SIBLEY 1997). However, a cAMP-independent mechanism for D_1 receptor desensitization has been reported in these cells (LEWIS et al. 1998), the functional magnitude of which appears not to correlate with receptor efficacy of various agonists. Whether the D_1 receptor undergoes a homologous vs heterologous desensitization will depend on the intrinsic cellular machinery, on the individual complement of GRKs within the particular cell or tissue, and perhaps on the mode of stimulation such as duration of action.

Studies on CT-tail truncations (Table 1) of the D_1 receptor indicate a decreased propensity for desensitization (JENSEN et al. 1995). This was also apparent on site-directed mutagenesis of the putative palmitoylated Cys347 residue. While the potential protein kinase A sites in the carboxyl terminus do not appear to affect desensitization or down-regulation of the D_1 receptor (KOZELL et al. 1997), perhaps loss of sites targeted by specific GRKs may mediate the attenuated desensitization response seen with these truncation mutants. However, the goldfish D_{1A} receptor which has carboxyl terminus truncated by 80 amino acids, can still undergo desensitization and down-regulation after long-term pre-treatment with agonists (FRAIL et al. 1993).

E. Distribution of D_1-Like Receptors

The localization of D_1-like receptors was first delineated by autoradiographic studies using either the tritiated- or iodinated-labelled and/or fluorescent-tagged high-affinity antagonists SCH 23390 or SCH 23892 (ARIANO et al. 1989; LIDOW et al. 1991). These studies revealed that D_1-like binding sites within the CNS are particularly enriched in the striatum (caudate-putamen, nucleus accumbens), olfactory tubercle, medial substantia nigra, and as a bilaminar distribution in the neocortex. D_1-like binding sites have also been identified in the retina, kidney and peripheral vasculature (AMENTA 1997; RICCI et al. 1997; OZONO et al. 1997). While these studies have identified D_1-like binding sites displaying the appropriate D_1 characteristics, the two D_1-like subtypes revealed by molecular cloning cannot be distinguished by autoradiography since both receptors bind with high affinity to the ligands typically used to localize D_1-like binding sites. Thus, the precise cellular or subcellular localization of the D_5/D_{1B} receptor by autoradiographic means awaits the identification of a high affinity D_5-specific ligand that displays poor affinity for the D_1 receptor.

In situ hybridization studies using specific cRNA probes have since been used to map the distribution of D_1 and D_5 receptor mRNAs in the human, primate and rat brain. These studies revealed that D_1 mRNA is highly abundant in the striatum, and is found in the olfactory tubercle, amygdala, hypothalamus, retina and neocortex (FREMEAU et al. 1991; LE MOINE et al. 1991; CHOI et al. 1995; GASPAR et al. 1995). The D_5 mRNA displays a somewhat different distribution in that low levels of this gene are expressed in the striatum, with slightly higher levels in the frontal cortex and substantia nigra and with high levels in the dentate gyrus of the hippocampus (MEADOR-WOODRUFF et al. 1994; BEISCHLAG et al. 1995; CHOI et al. 1995; LIDOW et al. 1997). D_{1B} mRNA has also been identified recently in bovine retinal pigment epithelium, and may mediate the DA-induced phagocytosis of the retinal photoreceptor outer segment (VERSAUX-BOTTERI et al. 1997).

Immunohistochemical studies with subtype-specific antibodies to D_1 or D_5 receptors have revealed intense D_1 staining in the basal ganglia, olfactory bulb and substantia nigra pars reticulata of rat brain (LEVEY et al. 1993), and for D_{1B} in frontal and parietal cortices, and CA2/CA3 region of the hippocampus and dentate gyrus. Within the prefrontal cortex, D_1 immunoreactivity is apparent in interneurons, being most prevalent in parvalbumin-containing interneurons and virtually absent in calretinin-containing interneurons (MULY et al. 1998). In the neostriatum, D_1-specific antibodies and to a lesser extent D_5-specific antibodies labelled medium spiny neurons while only D_5-specific antibodies labelled the large aspiny neurons typical of cholinergic interneurons (BERGSON et al. 1995a,b). Cells immunoreactive for D_{1A} have also been visualized in the rat mesencephalic trigeminal nucleus, specifically in the area receiving inputs from masticatory muscle spindles, and thus may process proprioceptive information from jaw-closing muscles (LAZAROV and PILGRIM 1997).

D_1 and D_5 receptors appear to be differentiated in terms of their subcellular distributions. In the caudate nucleus, D_1 labelling is localized to spines and shafts of projection neurons, whereas D_5 labelling is restricted to shafts. In the cerebral cortex and hippocampus, both D_1 and D_5 are co-expressed in pyramidal neurons and appear to be trafficked to distinct cellular compartments, spines for D_1 and shafts for D_5 (BERGSON et al. 1995b). At the ultrastructural level in the prefrontal cortex, D_1 immunoreactivity is associated with membranes of vesicles in proximal dendrites, with the plasma membrane on distal dendrites located near asymmetric synapses, and with presynaptic axon terminals giving rise to symmetric synapses. The different cellular (ARIANO et al. 1997a,b) and subcellular distributions for the D_1 and D_5 receptors suggest that these two receptors, although coupled to the same second messenger system, may subserve different functions within the brain.

F. Pharmacology of D_1-Like Receptor-Mediated Function: Behaviour and D_1-Like:D_2-Like Interactions

Relating aspects of function to D_1-like as distinct from D_2-like receptors was hindered, at least initially, by two factors: first, a paucity of pharmacological tools able to discriminate between them, and, second, resistance from the conceptual presumption that particular aspects of function would be mediated by either one receptor/receptor family or the other (WADDINGTON and O'BOYLE 1989). It is beyond the scope of this chapter to review the breadth and depth of the literature that has progressively challenged and rectified these anomalies. Hence the strategy adopted here is to briefly summarize the core processes and putative neuronal substrates, using examples from unconditioned psychomotor behaviour as they might apply to further aspects of function considered more extensively in Chaps. 20–27 (Vol. II), and to exemplify their clinical potential in terms of anti-parkinsonian therapy.

I. Core Processes at the Level of Behaviour and Their Putative Neuronal Substrates

Identification of SCH 23390 as the first selective D_1 (D_1-like) antagonist rapidly terminated the era over which functional prepotency of D_2 (D_2-like) receptors was presumed; together with selective D_1 (D_1-like) agonists, it revealed D_1 (D_1-like) receptors to play a fundamental role in multiple aspects of mammalian psychomotor behaviour. As reviewed in detail elsewhere (WADDINGTON and DALY 1993; WHITE and HU 1993; WADDINGTON et al. 1994, 1995, 1998a), the picture that has emerged is one of D_1 (D_1-like) receptors acting in concert with their D_2 (D_2-like) counterparts to regulate the great majority of DA-dependent behaviours.

These functional D_1-like:D_2-like interactions appear primarily co-operative in nature, particularly in relation to typical DA-dependent behaviours, such that tonic or phasic dopaminergic activity through D_1-like receptors "enables" or is "permissive" of phenomena initiated through D_2-like receptor activation; hence, for example, selective D_1-like antagonists attenuate typical stimulant responses (sniffing/locomotion) to selective D_2-like agonists, while selective D_1-like agonists synergize with selective D_2-like agonists to translate these stimulant responses into compulsive stereotyped behaviour. However, a very restricted range of other, atypical behaviours appears regulated by oppositional D_1-like:D_2-like interactions, such that tonic or phasic dopaminergic activity through D_1-like receptors normally inhibits such phenomena, which are then released when that inhibitory D_1-like influence is removed; hence, for example, selective D_1-like antagonists not only block typical stimulant responses to selective D_2-like agonists but also release the atypical response of myoclonic jerking.

When dopaminergic function is driven using a selective D_1-like agonist, similar principles appear to apply, though some elements of detail are less certain. Thus, while selective D_2-like antagonists attenuate typical responsivity (grooming) to selective D_1-like agonists, there is debate as to whether this reflects reciprocal co-operative/synergistic D_1-like:D_2-like interactions or a more non-specific motor depressant effect (WADDINGTON and DALY 1993; WHITE and HU 1993); selective D_2-like agonists synergize with selective D_1-like agonists to translate this response into stereotyped behaviour. However, again a restricted range of other, atypical behaviours appears regulated by oppositional D_1-like:D_2-like interactions, such that tonic or phasic dopaminergic activity through D_2-like receptors normally inhibits such phenomena, which are then released when that inhibitory D_2-like influence is removed; hence, for example, selective D_2-like antagonists not only attenuate the typical grooming response to selective D_1-like agonists but also release the atypical response of vacuous chewing.

Co-operative/synergistic D_1-like:D_2-like interactions are influenced importantly by the extent of integrity of tonic dopaminergic transmission. They are evident in the intact animal, and under conditions of acute DA depletion, but become at least in part decoupled under conditions of chronic disruption of dopaminergic tone through lesioning of DA neurons or prolonged depletion of DA; in particular, under such conditions selective antagonists act no longer in a heterologous manner but, rather, homologously block only responses to their counterpart agonists, though elements of synergism between D_1-like and D_2-like agonists endure. Oppositional D_1-like:D_2-like interactions appear to endure in the face of chronic disruption of dopaminergic tone.

As described in detail elsewhere (WHITE and HU 1993; WADDINGTON et al. 1994, 1995, 1998a), these behavioural phenomena, particularly those indicating D_1-like:D_2-like interactions, have attracted considerable efforts to identify their neuronal substrates at multiple levels of enquiry: from receptor binding site, through transduction systems and neurophysiological events, to more distal neurochemical processes. Perhaps the most enduring controversy is whether these interactions are mediated by adjacent D_1-like and D_2-like receptors that are co-localized on the same neuronal membrane, or are mediated in a more integrative manner by D_1-like and D_2-like receptors located on distinct populations of neurons whose efferents interact downstream of these locations; these issues are considered further in Chap. 11 of this volume.

II. Prototypical D_1-Like Behavioural Phenomena

The most widely accepted behavioural index of D_1-like receptor activation in rodents is the induction of grooming, particularly of "intense grooming" which refers to well-characterized, ethologically complete grooming syntax (MOLLOY and WADDINGTON 1984; WADDINGTON et al. 1995, 1998a); its origin appears to reside in a "pattern generator" within the anterior dorsolateral striatum (NEISEWANDER et al. 1995; WADDINGTON et al. 1998a). This behaviour is induced by all D_1-like agonists identified to date, is readily blocked by all known selective D_1-like antagonists, and attenuated by selective D_2-like antagonists in accordance with co-operative/synergistic D_1-like:D_2-like interactions (WADDINGTON et al. 1995, 1998a). The induction of vacuous chewing/rapid jaw movements is a more controversial index of D_1-like receptor activation that has proved less replicable between laboratories (ROSENGARTEN et al. 1993; WADDINGTON et al. 1995, 1998a); furthermore, when reported, there is contradictory evidence as to whether SK&F 38393-induced perioral movements are resistant to inactivation of 60%–90% of striatal and accumbal D_1-like receptors using peripherally-administered EEDQ (ROSENGARTEN et al. 1993), or are less robust than the grooming response but similarly sensitive to <50% inactivation of D_1-like receptors in the dorsolateral striatum (but not elsewhere in the striatum or in the nucleus accumbens) using intracerebrally administered EEDQ (NEISEWANDER et al. 1995). We have found the AC-stimulating benzazepine D_1-like agonists not to induce this behaviour when given alone but to do so following pre-treatment with a selective D_2-like antagonist in accordance with oppositional D_1-like:D_2-like interactions (MURRAY and WADDINGTON 1989). However, vacuous chewing is induced by the isochroman D_1-like agonist A 68930 (DALY and WADDINGTON 1993) in a manner that is sensitive to an isoquinoline (BW 737C) but not to benzazepine (SCH 23390), thienoazepine (LY 270411) or benzoquinoxoline (SDZ PSD 958) D_1-like antagonists (DEVENEY and WADDINGTON 1996; WADDINGTON et al. 1998b).

On classical pharmacological grounds these findings would suggest intense grooming to be mediated via a D_1-like receptor that recognizes all known chemical classes of D_1-like compounds, while vacuous chewing is mediated via a D_1-like subtype that recognizes preferentially the isochromans/isoquinolines. Furthermore, while the thienoazepine LY 270411 shares the action of all other known D_1-like antagonists to block D_1-like agonist-induced grooming, it is the only D_1-like antagonist examined to date which fails to attenuate materially the typical stimulant response of sniffing and locomotion or to release atypical myoclonic jerking to selective D_2-like agonism (DEVENEY and WADDINGTON 1996; WADDINGTON et al. 1998b); thus, the site mediating typical D_1-like agonist-induced behaviour may be dissociable pharmacologically from D_1-like sites participating in D_1-like:D_2-like interactions. However, these putative D_1-like sites are implicated in an era where molecular biology has to a considerable extent supplanted classical pharmacological criteria for receptor subtyping.

III. Paradoxes in Relation to the Defining Linkage to Adenylyl Cyclase

While linkage to the stimulation of AC endures as the defining characteristic of D_1-like receptors, an increasing body of evidence indicates that a broader perspective of D_1-like receptor second messenger/transduction is required (WADDINGTON et al. 1995, 1998a).

At one level, it is now well recognized that interpretation of D_1-like agonist efficacy in terms of AC stimulation is considerably more complex that envisaged originally (WADDINGTON and DEVENEY 1996; MAK et al. 1996): agonist efficacy appears to be dependent upon receptor reserve, such that some D_1-like agonists with high efficacy under conditions of high receptor reserve (e.g. SK&F 82958) evidence lower efficacy under conditions of low receptor reserve, while other agonists (e.g. A 68930, dihydrexidine) evidence full efficacy independent of extent of receptor reserve (WATTS et al. 1995); such phenomena could contribute to some of the differences in the behavioural profiles of D_1-like agonists between rodent and (non-human and human) primate species, which may differ in receptor reserve (GILMORE et al. 1995).

However, there is a deeper level of incongruity between stimulation of AC and functional (here, behavioural) responsivity to selective D_1-like agents. There is considerable evidence that the functional effects of D_1-like agonists are unrelated to their efficacies to stimulate AC. For example, the generally comparable maximal extent to which a broad range of structurally diverse D_1-like agonists induce grooming is in marked contrast to their widely varying efficacies to stimulate AC (MURRAY and WADDINGTON 1989; WADDINGTON et al. 1995, 1998a); these data are complemented by similar dissociations between AC efficacy and effects both at other levels of behaviour (GNANALINGHAM et al. 1995a) and, particularly, at the level of accumbal electrophysiology (JOHANSEN et al. 1991). These anomalies attain critical proportions in relation to the pharmacology of the benzazepine SK&F 83959. This agent shows high affinity and selectivity for D_1-like over D_2-like receptors, fails to stimulate AC, and inhibits the stimulation of AC induced by DA (ARNT et al. 1992; DEVENEY and WADDINGTON 1995; ANDRINGA et al. 1999); thus, it exhibits all of the defining characteristics of a D_1-like antagonist such as SCH 23390. However, SK&F 83959 readily induces grooming that is sensitive to antagonism by SCH 23390, and vacuous chewing that is insensitive to SCH 23390, together with other effects typical of D_1-like agonists; thus, its behavioural profile is essentially indistinguishable from that of the full efficacy selective D_1-like agonist A 68930 (DALY and WADDINGTON 1993). Furthermore, SCH 23390 can also induce vacuous chewing (DALY and WADDINGTON 1993; DEVENEY and WADDINGTON 1995), and can exert electrophysiological effects on rat nucleus accumbens neurons, which are very similar to those effects of D_1-like agonists that appear unrelated to their efficacies to stimulate AC (WACHTEL and WHITE 1995). SK&F 83959, like A 68930, synergizes with the D_2-like agonist quinpirole in an SCH 23390-sensitive manner to promote jaw movements in a spinal model (ADACHI et al. 1999).

Given that these effects of SK&F 83959 cannot be accounted for readily in terms of an active metabolite having a different pharmacological profile (ARNT et al. 1992; DEVENEY and WADDINGTON 1995), how else can these profound paradoxes be explained? The most radical explanation is that these (and possibly other) behavioural responses held to be mediated via D_1-like receptors involve not the AC-coupled D_{1A} (or D_{1B}) receptor; rather, they appear to involve a D_1-like receptor that is coupled to a transduction system other than/additional to AC, and which is able to respond similarly to drugs having a common affinity for this D_1-like site independent of whether they exert agonist or antagonist action at classical AC-coupled D_{1A} (or D_{1B}) receptors.

IV. D_1-Like Receptors and Anti-parkinsonian Activity

These paradoxes endure when such concepts are generalized into the area of therapeutic potential in Parkinson's disease. D_1-like agonists are active in conventional rodent models of anti-parkinsonian activity, though these acute effects can be accompanied by proconvulsant activity; furthermore, the acute "anti-parkinsonian" effects of many D_1-like agonists, particularly newer agents such as A 68930 and A 77636, show rapid desensitization on repeated administration (SHIOSAKI et al. 1998; VERMEULEN et al. 1999). Unlike effects in rodents, however, the acute efficacy of D_1-like agonists in the 1-methyl-4-phenyl-1,2,3,6-tetrahydropyridine (MPTP) non-human primate model of Parkinson's disease was considered initially to be related to their intrinsic activity, which for partial agonists (such as SK&F 38393, which exacerbated parkinsonism; but not for full agonists such as A 68930) appeared to be reduced in primates due to lower receptor reserve relative to rodents (WATTS et al. 1993; WEED et al. 1998).

The full efficacy D_1-like agonist A 86929, however, appears not to show rapid tolerance (SHIOSAKI et al. 1998); additionally, its profile of acute antiparkinsonian activity in the MPTP model, together with that of its pro-drug ABT-431 both in the MPTP model and in preliminary studies in patients (now terminated), was held to have advantages over that of L-DOPA and D_2-like agonists such as quinpirole in terms of reduced propensity to induce dyskinesias (SHIOSAKI et al. 1998; PEARCE et al. 1999). Yet the extent to which these concepts generalize to other agents remains unclear (GRONDIN et al. 1999; VERMEULEN et al. 1999). Furthermore, the clinical anti-parkinsonian activity of the full efficacy, preferential D_1-like agonist dihydrexidine appears to be accompanied by L-dopa-like dyskinesias (BLANCHET et al. 1998).

It is on this background that the anti-parkinsonian efficacy of SK&F 83959 in the MPTP non-human primate model, with induction also of oral dyskinesias, is striking (GNANALINGHAM et al. 1995a); furthermore, modest doses of SK&F 83959 synergized with quinpirole to prolong its duration, though higher doses synergized to produce hyperexcitability and seizures (GNANALINGHAM et al. 1995b). Recent studies have confirmed that SK&F 83959 has no efficacy to stimulate AC and readily inhibits the stimulation of AC induced by DA in (non-human and human) primate cells, with some affinity also for α_2 receptors (ANDRINGA et al. 1999). Thus, as in rodent studies, those in nonhuman primates contradict the involvement of AC-coupled entities in typical functional responses mediated through D_1-like receptors; additionally, they indicate these paradoxes to be of therapeutic significance.

V. Targeted Gene Deletion

While the lack of subtype-specific ligands has greatly impeded investigation of the functional roles of individual cloned D_1-like receptor subtypes in vivo, molecular biology has generated new techniques for addressing these issues. The use of antisense oligonucleotides to block the expression of specific gene-directed proteins indicates that D_1-like agonist-induced grooming is blocked by intracerebroventricular administration of a D_1-like antisense oligonucleotide; more recently, lordosis behaviour in female rats, which can be induced by D_1-like agonists and blocked by D_1-like antagonists, was blocked also by such administration of D_{1A} but not D_{1B} antisense; additionally, D_{1A} antisense blocks SK&F 38393-induced motor responses in denervated animals, while D_{1B} antisense potentiated these responses (SIBLEY 1999). These are some of the earliest indications of specific functional dissociations between D_{1A} and D_{1B} receptors at the level of behaviour.

However, parcellation of function between receptor subtypes in the absence of selective pharmacological agents has been advanced yet further by the generation of mice in which a given subtype is absent through targeted gene deletion by homologous recombination ("knockouts"; PICCIOTTO and WICKMAN 1998; SIBLEY 1999). These techniques have been applied in two laboratories to generate mutant mice having deletion of the D_{1A} receptor, though using different genetic constructs (DRAGO et al. 1994; XU et al. 1994). While both laboratories have assessed their "knockouts" for spontaneous behaviour, and find them to manifest no gross neurological impairment, some inconsistencies have emerged. Thus, DRAGO et al. (1994) have reported their D_{1A}-null mice to show reduced levels of rearing but with no reduction in locomotion or elimination of grooming, as assessed in terms of accumulation of events and line crossings by direct visual observation; conversely, XU et al. (1994a,b) reported their D_{1A}-null mice to show increased activity in terms of accumulation of photobeam interruptions and with some reduction in grooming on visual inspection. Recently, targeted gene deletion of the D_{1B} receptor has been reported (HOLLON et al. 1998); while studies are at a preliminary stage, these D_{1B}-null mice appear to develop normally without gross neurological deficit, though behaviourally they also display higher levels of some locomotor activities.

It has been suggested that D_{1A}-null mice may evidence a deficit in initiating, though not in expressing, spontaneous activity (SMITH et al. 1998), or may initiate more chains of grooming syntax though with fewer expressed to completion (CROMWELL et al. 1998). In our own studies, we have proceeded ethologically to resolve all elements of behaviour within the natural repertoire of the mouse (CLIFFORD et al. 1998): D_{1A}-null mice were characterized over initial exploration by reductions in rearing free, sifting and chewing, but significant increases in grooming, intense grooming and locomotion; sniffing and rearing to wall habituated less readily in D_{1A}-null mice such that these behaviours occurred subsequently to excess. Thus, their acute spontaneous behaviour was characterized by neither "hypoactivity" nor "hyperactivity" but, rather, by prominent topographical shifts between individual elements of behaviour; furthermore, targeted gene deletion of the D_{1A} receptor appeared to interact with the psychological processes of habituation to sculpt the changing topography of behaviour from initial exploration towards quiescence, and thus to reveal gene effects obscured over more acute time frames.

In relation to pharmacological challenge studies, a reduction both in the stimulation of photobeam interruptions by the D_1-like agonist SK&F 81297 and in the attenuation of photobeam interruptions by the D_1-like antagonist SCH 23390 has been reported in D_{1A}-null mice (Xu et al. 1994a,b). In our own ethologically based studies (CLIFFORD et al. 1999), topographies of behaviour were materially conserved in D_{1A}-null mice following challenge either with the full efficacy selective D_1-like agonist A 68930 or with the anomalous agent SK&F 83959; in particular, there was substantial conservation of intense grooming to A 68930, with a modest reduction at a single dose associated with a marked increase in locomotion and a marked reduction in rearing, while intense grooming to SK&F 83959 was unaltered; stereotyped sniffing and plodding locomotor responses to the D_2-like agonist RU 24213 were also unaltered. Interestingly, behavioural synergism between D_1-like and D_2-like agonists appears to be enhanced in mice with targeted gene deletion of the D_3 receptor, hence D_3 receptors may exert an inhibitory influence on co-operative/synergistic D_1-like:D_2-like interactions (Xu et al. 1997); recent cellular evidence suggests interplay between D_{1A}/D_5 and D_3 receptors (SCHWARTZ et al. 1998).

These findings indicate that D_{1A}-null mice, in which D_{1A} receptor binding and DA-sensitive AC are essentially absent, evidence substantial conservation of topographical responsivity of wildtypes to A 68930 and to SK&F 83959, agents that act as a full agonist and antagonist, respectively, at AC-coupled D_1-like receptors; furthermore, co-operative/synergistic and oppositional D_1-like:D_2-like interactions, through which D_{1A} receptors have been presumed to critically "enable" or inhibit the expression of, respectively, typical and atypical responses to D_2-like agonism, appear also to be conserved in D_{1A}-null mice at the level of behaviour. It would appear that, in accordance with the pharmacological analysis offered above, these behavioural events in D_{1A}-null mice involve a D_1-like receptor that is neither the D_{1A} nor indeed any other molecular biologically characterized D_1-like receptor (such as D_{1B}) having the "defining" characteristic of linkage to the stimulation of AC.

G. Are There Additional D_1-Like Receptor Subtypes?

I. D_1-Like Receptors Linked to Phosphoinositide Hydrolysis

These substantially behaviourally based notions of one or more additional D_1-like receptors having an alternative transduction system are buttressed by cellular evidence, both at biochemical (SCHOORS et al. 1991; LAITINEN 1993) and at electrophysiological (JOHANSEN et al. 1991; HARVEY and LACEY 1997) levels, for brain D_1-like receptors that are not coupled to AC. More specifically, there is a substantial body of cellular evidence for a D_1-like receptor coupled to phosphoinositide (PI) turnover (MAHAN et al. 1990; UNDIE and FRIEDMAN 1990; UNDIE et al. 1994). Because neither D_{1A} nor D_{1B} receptors appear to couple to PI turnover when expressed in several cell lines (MONSMA et al. 1990; TIBERI et al. 1991; PEDERSEN et al. 1994), these biochemical results have provided the strongest evidence for an additional D_1-like receptor subtype. Stimulation of PI turnover is induced by DA and numerous benzazepine D_1-like receptor agonists with a rank order of potency that is unrelated to that for stimulation of AC; however, stimulation of PI turnover is blocked by the D_1-like antagonist SCH 23390 but not by α or 5-HT antagonists (UNDIE and FRIEDMAN 1992; PACHECO and JOPE 1997). It would seem important to characterize as a matter of urgency the actions of SK&F 83959 on PI turnover.

As yet, there is little direct evidence for the involvement of D_1-like receptors linked to PI turnover in behavioural regulation. Since reports of the induction of vacuous chewing by the D_1-like antagonist SCH 23390 (DEVENEY and WADDINGTON 1995), SCH 23390 has been reported to be a weak partial agonist at PI-linked D_1-like receptors (UNDIE 1999). Recently, ROSENGARTEN and FRIEDHOFF (1998) have reported repetitive jaw movements induced by the D_1-like agonists SK&F 38393 and A 68930 to be conserved following pretreatment with EEDQ, with associated loss of DA-sensitive AC but preservation of DA-sensitive PI hydrolysis. We have found that synergism between A 68930, a full-efficacy D_1-like agonist, and the D_2-like agonist quinpirole in the induction of jaw movements is potentiated by low doses of the D_1-like antagonists SCH 23390 and particularly BW 737C (ADACHI et al. 1999); this would be consistent with removal by these antagonists of an inhibitory influence from AC-coupled D_1-like receptors on PI-coupled D_1-like receptors as reported by UNDIE and FRIEDMAN (1994). Most recently, our demonstration of elevated spontaneous grooming and conservation of grooming induced by A 68930 and by SK&F 83959 in D_{1A}-null mice (CLIFFORD et al. 1998, 1999) would be consistent with the essential absence of DA-sensitive AC but preservation of DA-sensitive PI hydrolysis in such mutants (FRIEDMAN et al. 1997). In the present climate, these issues are likely to be resolved only by the cloning of a novel D_1-like receptor that might be coupled to PI hydrolysis and have a pharmacological profile consistent with that evident in the above behavioural studies.

II. Expansion of the D_1-Like Receptor Family

Existence of only two mammalian D_1-like receptors was challenged by the identification of two additional D_1-like receptors from lower vertebrate species: the D_{1C} receptor, first isolated from the amphibian *Xenopus laevis* (SUGAMORI et al. 1994); and the D_{1D} receptor, cloned from the avian *Gallus domesticus* (DEMCHYSHYN et al. 1995). These receptors were isolated in addition to genes encoding corresponding orthologues to mammalian D_1/D_{1A} and D_5/D_{1B} receptors. A D_{1C}-receptor clone has been isolated from the freshwater eel species *Anguilla anguilla* (CARDINAUD et al. 1997) and from several additional vertebrate organisms (MACRAE and BRENNER 1995; LAMERS et al. 1996; HIRANO et al. 1998). The identification of a D_{1C} receptor from *Anguilla anguilla*, whose ancestors appeared earlier in evolution and is representative of ray-finned fish (actinopterygians) which diverged from flesh-finned fish (sarcopterygians) ~420 million years ago, suggests that the gene duplication events at the origin of D_1-like receptor diversity occurred prior to the separation of actinopterygians from sarcopterygians, from which the first tetrapod ancestor descended. In addition, a D_{1C}-like receptor appears to be found in *Gallus domesticus*, indicating that in the avian brain all four distinct receptor subtypes exist (CARDINAUD et al. 1998; NIZNIK et al. 1998).

The D_{1C} receptor, in accordance with its almost equivalent sequence identity to both D_1/D_{1A} and D_5/D_{1B} receptors, displays a somewhat intermediate pharmacological profile. This is evident by an intermediate affinity for the distinguishing agonists, DA and 6,7-ADTN. Overall, its affinity for antagonists resembled more closely the D_{1A} receptor than the D_{1B} receptor. One second generation benzazepine derivative, NNC 01–0012, which structurally resembles the antagonist SCH 23390, was found to show preferential selectivity for the D_{1C} receptor; it displayed a 10- to 20-fold higher affinity for this receptor subtype than for either vertebrate/mammalian D_1/D_{1A}, D_5/D_{1B} or D_{1D} receptors. Although this compound displayed full functional antagonism at the Xenopus D_{1C} receptor, NNC 01–0012 provoked an inherent partial agonist activity at both vertebrate D_{1B} and D_{1D} receptors (SUGAMORI et al. 1998).

Similar to D_5/D_{1B} receptors, the D_{1D} receptor displays ~10-fold higher affinity for the distinguishing agonists DA and 6,7-ADTN, suggesting that the D_{1D} receptor shares some pharmacological features with the D_5/D_{1B} subclass of receptor. This is consistent with the observed partial agonist response seen with NNC 01-0012 at this receptor subtype in terms of stimulation of cAMP accumulation. Unique pharmacological features of the D_{1D} receptor include ~10-fold higher affinity for the partial agonists SKF 38393, pergolide and lisuride, and observed low affinity for the antagonist haloperidol (DEMCHYSHYN et al. 1995). Recent work suggests that the series of isoquinoline D_1-like receptor antagonists typified by BW 737C are potent in displaying up to 50-fold selectivity for the D_{1D} receptor (NIZNIK et al. 1998).

H. Conclusions

While neither D_{1C} or D_{1D} receptors can stimulate PI turnover in transfected cells, the isolation of these genes provides compelling evidence for the possible existence of additional D_1-like receptors in mammals. The presence of such subtypes in mammals may give rise to the full spectrum of D_1-like receptor-mediated events that cannot be reconciled with the existence of currently recognized D_1/D_5 receptor subtypes. Although it is possible that D_{1C} and D_{1D} receptors may have been lost at the time of the divergence of the mammalian lineage, the widespread appearance of multiple D_1-like receptors throughout several vertebrate phyla would suggest some gain or loss of D_1-like receptor-mediated function in mammals. Elucidating the molecular or functional correlates of these receptor subtypes may allow for a better understanding of mammalian D_1-like receptor-mediated events. Conceivably, the use of mutant mice deficient in D_{1A} receptors (DRAGO et al. 1998), D_{1B} receptors (HOLLON et al. 1998) or both may be instrumental in delineating the contributory role of both known and putative, yet-to-be-identified subtypes (MORATALLA et al. 1996; FRIEDMAN et al. 1997; JIN et al. 1998; CLIFFORD et al. 1999). It must be emphasized that whether or not additional mammalian D_1-like receptor subtypes are found, multiple D_1-like receptors clearly do not have redundant roles in the CNS, and are essential regulators of numerous aspects of psychomotor function. While the days are long past since the D_1 receptor was considered merely to be a receptor "in search of a function", the precise and unique roles that each of these receptors encode are, however, still poorly understood.

Acknowledgements. On the recent untimely and tragic death of Hyman (Chaim) Niznik, his co-authors dedicate this chapter to the memory of a fine molecular neuroscientist and respected colleague who will be sadly missed; may he rest in peace.
   The authors' studies are supported by NIDA, the MRC of Canada, and OMHF, and by the Royal College of Surgeons in Ireland, the Higher Education Authority of Ireland, the Stanley Foundation and a Galen Followship from the Irish Brain Research Foundation.

References

Adachi K, Ikeda H, Hasegawa, Nakamura S, Waddington JL, Koshikawa N (1999) SK&F 83959 and non-cyclase-coupled dopamine D_1-like receptors in jaw movements via dopamine D_1-like/D_2-like receptor synergism, Eur J Pharmacol 367:143–149

Amenta F (1997) Light microscopic autoradiography of peripheral dopamine receptor subtypes Clin Exp Hypertension 19:27–41

Andersen PH, Jansen JA (1990) Dopamine receptor agonists: selectivity and dopamine D1 receptor efficacy Eur J Pharmacol Mol Pharmacol Sect 188:335

Andersen PH, Gronvald FC, Hohlweg R, Hansen LB, Guddal E, Braestrup C, Nielsen EB (1992) NNC-112, NNC-687 and NNC-756, new selective and highly potent dopamine D_1 receptor antagonists Eur J Pharmacol 219:45–52

Andringa G, Drukarch B, Leysen JE, Cools AR, Stoof JC (1999) The alleged dopamine D_1 receptor agonist SK&F 83959 is a dopamine D_1 receptor antagonist in primate cells and interacts with other receptors Eur J Pharmacol 364:33–41

Ariano MA, Monsma FJJr, Barton AC, Kang HC, Haugland RP, Sibley DR (1989) Direct visualisation and cellular localisation of D_1 and D_2 dopamine receptors in rat forebrain by use of fluorescent ligands Proc Natl Acad Sci USA 86:8570–8574

Ariano MA, Sortwell CE, Ray M, Altemus LL, Sibley DR, Levine MS (1997a) Agonist-induced morphologic decrease in cellular D_{1A} dopamine receptor staining Synapse 27:313–321

Ariano MA, Wang J, Noblett KL, larson ER, Sibley DR (1997b) Cellular distribution of the rat D_{1B} receptor in central nervous system using anti receptor antisera Brain Res 746:141–150

Arnt J, Hyttel J, Sanchez C. (1992) Partial and full dopamine D1 receptor agonists in mice and rats: relation between behavioural effects and stimulation of adenylate cyclase activity in vitro. Eur J Pharmacol 213:259–267

Asherson P, Mant R, Williams N, Cardno A, Jones L, Murphy K, Collier DA, Nanko S, Craddock N, Morris S, Muir W, Blackwood B, McGuffin P, Owen MJ (1998) A study of chromosome 4p markers and dopamine D5 receptor gene in schizophrenia and bipolar disorder. Mol Psychiatry 3:310–320

Baldessarini RJ (1997) Dopamine receptors and clinical medicine. In: Neve KA, Neve RL (eds) The Dopamine Receptors, Humana Press Inc., Totowa, pp 457–497

Barr CL, Wigg KG, Zovko E, Sandor P, Tsui L-C (1997) Linkage study of the dopamine D_5 receptor gene and Gilles de la Tourette syndrome. Am J Med Genet 74:58–61

Beischlag TV, Marchese A, Meador-Woodruff JH, Damask SP, O'Dowd BF, Tyndale RF, Van Tol HHM, Seeman P, Niznik HB (1995) The human dopamine D_5 receptor gene: cloning and characterization of the 5'-flanking and promoter region. Biochemistry 34:5960–5970

Beischlag TV, Nam D, Ulpian C, Seeman P, Niznik HB (1996) A polymorphic dinucleotide repeat in the human dopamine D_5 receptor gene promoter. Neurosci Lett 205:173–176

Bergson C, Levenson R, Lidow M (1996) Differential role of glycosylation in D_1 and D_5 dopamine receptor localization and function. Soc Neurosci Abstr 22:826

Bergson C, Mrzijak L, Lidow MS, Goldman-Rakic PS, Levenson R (1995a) Characterization of subtype-specific antibodies to the human D_5 dopamine receptor: studies in primate brain and transfected mammalian cells. Proc Natl Acad Sci USA 92:3468–3472

Bergson C, Mrzljak L, Smiley JF, Pappy M, Levenson R, Goldman-Rakic PS (1995b) Regional, cellular, and subcellular variations in the distribution of D_1 and D_5 dopamine receptors in primate brain. J Neurosci 15:7821–7836

Blanchet P, Bedard PJ, Britton DR, Kebabian JW (1993) Differential effect of selective D-1 and D-2 dopamine receptor agonists on levodopa-induced dyskinesia in 1-methyl-4-phenyl-1,2,3,6 tetrahydropyridine-exposed monkeys. J Pharmacol Exp Ther 267:275–278

Blanchet PJ, Fang J, Gillespie M, Sabounjian L, Locke KW (1998) Effects of the full dopamine D1 receptor agonist dihydrexidine in Parkinson's disease. Clin Neuropharmacol 21:339–343

Brett PM, Curtis D, Robertson MM, Gurling HMD (1995) The genetic susceptibility to Gilles de la Tourette Syndrome in a large multiple affected British kindred: linkage analysis excludes a role of the genes coding the dopamine D_1, D_2, D_3, D_4, D_5 receptors, dopamine beta hydroxylase, tyrosinase, and tyrosine hydroxylase. Biol Psychiatry 37:533–540

Brismar H, Asghar M, Carey RM, Greengard P, Aperia A (1998) Dopamine-induced recruitment of dopamine D_1 receptors to the plasma membrane. Proc Natl Acad Sci USA 95:5573–5578

Cardinaud B, Sugamori KS, Coudouel S, Vincent J-D, Niznik HB, Vernier P (1997) Early emergence of three dopamine D_1 receptor subtypes in vertebrates. J Biol Chem 272:2778–2787

Cardinaud B, Gilbert JM, Liu F, Sugamori KS, Vincent JD, Niznik HB, Vernier P (1998) Evolution and origin of the diversity of dopamine receptors in vertebrates. Adv in Pharmacol 42:936–940

Charpentier S, Jarvie KR, Severynse DM, Caron MG, Tiberi M (1996) Silencing of the constitutive activity of the dopamine D_{1B} receptor. J Biol Chem 271:28071–28076

Cho W, Taylor LP, Akil H (1996) Mutagenesis of residues adjacent to transmembrane prolines alters D_1 dopamine receptor binding and signal transduction. Mol Pharmacol 50:1338–1345

Choi WS, Machida CA, Ronnekleiv OK (1995) Distribution of dopamine D_1, D_2 and D_5 receptor mRNAs in the monkey brain: ribonuclease protection assay analysis. Mol Brain Res 31:86–94

Cichon S, Nothen MM, Rietschel M, Korner J, Propping P (1994) Single-strand conformation analysis (SSCA) of the dopamine D_1 receptor gene (DRD1) reveals no significant mutation in patients with schizophrenia and manic depression. Biol Psychiatry 36:850–853

Clifford JJ, Tighe O, Croke DT, Drago J, Sibley DR, Waddington JL (1998) Topographical evaluation of the phenotype of spontaneous behaviour in mice with targeted gene deletion of the D_{1A} dopamine receptor: paradoxical elevation of grooming syntax. Neuropharmacology 37:1595–1602

Clifford JJ, Tighe O, Croke DT, Kinsella A, Drago J, Sibley DR, Waddington JL (1999) Conservation of behavioural topography to dopamine D_1-like receptor agonists in mutant mice lacking the D_{1A} receptor implicates a D_1-like receptor not coupled to adenylyl cyclase. Neuroscience 93:1483–1489

Cromwell HC, Berridge KC, Drago J, Levine MS (1998) Action sequencing is impaired in D_{1A}-deficient mutant mice. Eur J Neurosci 10:2426–2432

Dahmer MK, Senogles SE (1996a) Dopaminergic inhibition of catecholamine secretion from chromaffin cells: evidence that inhibition is mediated by D_4 and D_5 dopamine receptors. J Neurochem 66:222–232

Dahmer MK, Senogles SE (1996b) Activation of D_5 dopamine receptors on chromaffin cells inhibits secretagogue-stimulated Na^+ uptake by a cAMP-independent mechanism. Soc Neurosci Abstr 22:1317

Daly G, Hawi Z, Fitzgerald M, Gill M (1999) Mapping susceptibility loci in attention deficit hyperactivity disorder: preferential transmission of parental alleles at DAT1, DBH and DRD5 to affected children. Mol Psychiatry 4:192–196

Daly SA, Waddington JL (1993) Behavioural evidence for "D-1-like" dopamine receptor subtypes in rat brain using the new isochroman agonist A 68930 and isoquinoline antagonist BW 737C. Psychopharmacology 113:45–50

Dearry AG, Gingrich J, Falardeau P, Fremeau RT, Bates MD, Caron MG (1990) Molecular cloning and expression of the gene for a human D_1 dopamine receptor. Nature 347:72–76

Demchyshyn LL, Sugamori KS, Lee FJS, Hamadanizadeh SA, Niznik HB (1995) The dopamine D_{1D} receptor: cloning and characterization of three pharmacologically distinct D_1-like receptors from Gallus domesticus. J Biol Chem 270:4005–4012

Demchyshyn LL, McConkey F, Niznik HB (1997) Metamorphosis of Dopamine D_5 receptors: Carboxyl tail substitution transforms agonist affinity and constitutive activity profile to mimic D_1 receptors. Soc Neurosci Abstr 23:1773

DeNinno MP, Schoenleber R, MacKenzie R, Britton DR, Asin KE, Briggs C, Trugman JM, Ackerman M, Artman L, Bednarz L, Bhatt R, Curzon P, Gomez E, Kang CH, Stittsworth J, Kebabian JW (1991) A 68930: a potent agonist selective for the dopamine D_1 receptor. Eur J Pharmacol 199:209–219

Deveney AM, Waddington JL (1995) Pharmacological characterization of behavioural responses to SK&F 83959 in relation to "D_1-like" dopamine receptors not linked to adenylyl cyclase. Br J Pharmacol 116:2120–2126

Deveney AM, Waddington JL (1996) Evidence for dopamine "D_1-like" receptor subtypes in the behavioural effects of two new selective antagonists, LY 270411 and BW 737C. Eur J Pharmacol 317:175–181

Dohlman HG, Thorner J, Caron MG, Lefkowitz RJ (1991) Model systems for the study of seven-transmembrane segment receptors. Annu Rev Biochem 60:653–688

Dollfus S, Campion D, Vasse T, Preterre P, Laurent C, d'Amato T, Thibaut F, Mallet J, Petit M (1996) Association study between dopamine D1, D2, D3 and D4 receptor genes and schizophrenia defined by several diagnostic systems. Biol Psychiatry 40:419–421

Drago J, Gerfen CR, Lachowicz JE, Steiner H, Hollon TR, Love PE, Ooi GT, Grinberg A, Lee EJ, Huang SP, Bartlett PF, Jose PA, Sibley DR, Westphal H (1994) Altered striatal function in a mutant mouse lacking D_{1A} dopamine receptors. Proc Natl Acad Sci USA 91:12564–12568

Drago J, Padungchaichot P, Accili D, Fuchs S (1998) Dopamine receptors and dopamine transporter in brain function and addictive behaviors: Insights from targeted mouse mutants. Dev Neurosci 20:188–203

Dumartin B, Caille I, Gonon F, Bloch B (1998) Internalization of D_1 dopamine receptor in striatal neurons in vivo as evidence of activation by dopamine agonists. J Neurosci 18:1650–1661

Dumbrille-Ross A, Niznik H, Seeman P (1985) Separation of dopamine D_1 and D_2 receptors Eur J Pharmacol 110:151–152

Eubanks JL, Altherr M, Wagner-McPherson C, McPherson JD, Wasmuth JJ, Evand GA (1992) Localization of the D_5 dopamine receptor gene to human chromosome 4p15.1–p15.3, centromeric to the Huntington's disease locus. Genomics 12:510–516

Feng J, Sobell JL, Heston LL, Cook EH, Goldman D, Sommer SS (1998) Scanning of the dopamine D_1 and D_5 receptor genes by REF in neuropsychiatric patients reveals a novel missense change at a highly conserved amino acid. Am J Med Genet 81:178–178

Ferguson SSG, Downey WE, Colapietro A-M, Barak LS, Menard L, Caron MG (1996) Role of β-arrestin in mediating agonist-promoted G protein-coupled receptor internalisation. Science 271:363–366

Frail DE, Manelli AM, Witte DG, Lin CW, Steffey ME, Mackenzie RG (1993) Cloning and characterization of a truncated dopamine D_1 receptor from goldfish retina: stimulation of cyclic AMP production and calcium mobilization. Mol Pharmacol 44:1113–1118

Fremeau RTjr, Duncan GE, Fornarretto MG, Dearry A, Gingrich JA, Breese GR, Caron MG (1991) Localisation of D_1 dopamine receptor mRNA in brain supports a role in cognitive, affective and neuroendocrine aspects of dopaminergic neurotransmission. Proc Natl Acad Sci USA 88: 3772–3776

Friedman E, Jin LQ, Cai GP, Hollon TR, Drago J, Sibley DR, Wang HY (1997) D-1-like dopaminergic activation of phosphoinositide hydrolysis is independent of D-1A dopamine receptors: evidence from D-1A knockout mice. Mol Pharmacol 51:6–11

Gaspar P, Bloch B, Le Moine C (1995) D_1 and D_2 receptor gene expression in the rat frontal cortex: cellular localization in different classes of efferent neurons. Eur J Neurosci 7:1050–1063

Gilmore JH, Watts VJ, Lawler CP, Noll EP, Nichols DE, Mailman RB (1995) "Full" Dopamine D1 agonists in human caudate: biochemical properties and therapeutic implications. Neuropharmacology 34: 481–488

Gnanalingham KK, Erol DD, Hunter AJ, Smith LA, Jenner P, Marsden CD (1995a) Differential anti-parkinsonian effects of benzazepine D1 dopamine agonists with varying efficacies in the MPTP-treated common marmoset. Psychopharmacology 117:275–286

Gnanalingham KK, Hunter AJ, Jenner P, Marsden CD (1995b) The differential behavioural effects of benzazepine D1 dopamine agonists with varying efficacies: co-administration with quinpirole in primate and rodent models of Parkinson's disease. Psychopharmacology 117:287–297

Grondin R, Doan VD, Gregoire L, Bedard PJ (1999) D1 receptor blockade improves L-Dopa-induced dyskinesia but worsens parkinsonism in MPTP monkeys. Neurology 52:771–776

Grandy DK, Allen LJ, Zhang Y, Magenis RE, Civelli O (1992) Chromosomal localization of three human D_5 dopamine receptor genes. Genomics 13:968–973

Grandy DK, Zhang Y, Bouvier C, Zhou Q-Y, Johnson RA, Allen L, Buck K, Bunzow JR, Salon J, Civelli O (1991) Multiple human D_5 dopamine receptor genes: a functional receptor and two pseudogenes. Proc Natl Acad Sci USA 88:9175–9179

Guan X-M, Amend A, Strader CD (1995) Determination of structural domains for G protein coupling and ligand binding in 3-adrenergic receptor. Mol Pharmacol 48:492–498

Harvey J, Lacey MG (1997) A postsynaptic interaction between dopamine D1 and NMDA receptors promote presynaptic inhibition in the rat nucleus accumbens via adenosine release. J Neurosci 17:5271–5280

Herve D, Levi-Strauss M, Marey-Semper I, Verney C, Tassin J-P, Glowinski J, Girault J-A (1993) Golf and Gs in rat basal ganglia: possible involvement of Golf in the coupling of dopamine D_1 receptor with adenylyl cyclase. J Neurosci 13:2237–2248

Hirano J, Archer SN, Djamgoz MB (1998) Dopamine receptor subtypes expressed in vertebrate (carp and eel) retinae: cloning, sequencing and comparison of five D_1-like and three D_2-like receptors. Rec Chan 5:387–404

Hollon TR, Gleason TC, Grinberg A, Ariano MA, Huang SP, Drago J, Drieling J, Crawley JN, Westphal H, Sibley DR (1998) Generation of D5 dopamine receptor-deficient mice by gene targeting. Soc Neurosci Abstr 24:594

Jarvie KR, Tiberi M, Silvia C, Gingrich JA, Caron MG (1993) Molecular cloning, stable expression and desensitization of the human D_{1B}/D_5 receptor. J Rec Res 13:573–591

Javitch JA, Fu D, Liapakis G, Chen J (1997) Constitutive activation of the 2 adrenergic receptor alters the orientation of its sixth membrane-spanning segment. J Biol Chem 272:18546–18549

Jensen AA, Pedersen UB, Kiemer A, Din N, Andersen PH (1995) Functional importance of the carboxyl tail cysteine residues in the human D_1 dopamine receptor. J Neurochem 65:1325–1331

Jiang D, Sibley DR (1996) Investigation of cAMP-mediated regulation of the rat D_1 dopamine receptor using site-directed mutagenesis. Soc Neurosci Abstr 22:1316

Jin H, Zastawny R, George SR, O'Dowd BF (1997) Elimination of palmitoylation sites in the human dopamine D_1 receptor does not affect receptor-G protein interaction. Eur J Pharmacol 324:109–116

Jin L-Q, Cai G, Wang H-Y, Smith C, Friedman E (1998) Characterization of the phosphoinositide-linked dopamine receptor in a mouse hippocampal-neuroblastoma hybrid cell line. J Neurochem 71:1935–1943

Johansen PA, Hu X-T, White FJ (1991) Relationship between D-1 dopamine receptors, adenylate cyclase, and the electrophysiological responses of rat nucleus accumbens neurons. J Neural Transm [Gen. Sect.] 86:97–113

Kebabian JW, Calne DB (1979) Multiple receptors for dopamine. Nature 277:93–96

Kebabian JW, Britton DR, DeNinno MP, Perner R, Smith L, Jenner P, Schoenleber R, Williams M (1992) A-77636: a potent and selective dopamine D1 receptor agonist with anti-parkinsonian activity in marmosets. Eur J Pharmacol 229:203–209

Kennedy JL, Macciardi FM (1998) Chromosome 4 workshop. Psychiatr Genet 8:67–71

Kerkman DJ, Ackerman M, Artman LD, MacKenzie RG, Johnson MC, Bednarz L, Montana W, Asin KE, Stampfli H, Kebabian JW (1989) A 69024: a non-benzazepine antagonist with selectivity for the dopamine D-1 receptor. Eur J Pharmacol 166:481–491

Kimura K, White BH, Sidhu A (1995a) Coupling of human D-1 dopamine receptors to different guanine nucleotide binding proteins. J Biol Chem 270:14672–14678

Kimura K, Sela S, Bouvier C, Grandy DK, Sidhu A (1995b) Differential coupling of D_1 and D_5 dopamine receptors to guanine nucleotide binding proteins in transfected GH4C1 rat somatomammotrophic cells. J Neurochem 64:2118–2124

Kjelsberg MA, Cotecchia S, Ostrowski J, Caron MG, Lefkowitz RJ (1992) Constitutive activation of the 1B- adrenergic receptor by all amino acid substitutions at a single site. Evidence for a region which constrains receptor activation. J Biol Chem 267:1430–1433

Konig B, Gratzel M (1994) Site of dopamine D_1 receptor binding to Gs protein mapped with synthetic peptides. Biochim Biophys Acta 1223:261–266

Kozasa T, Gilman AG (1995) Purification of recombinant G proteins from Sf9 cells by hexahistidine tagging of associated subunits. Characterization of alpha 12 and inhibition of adenylyl cyclase by alpha z. J Biol Chem 270:1734–1741

Kozell LB, Machida CA, Neve RL, Neve KA (1994) Chimeric D_1/D_2 dopamine receptors. J Biol Chem 269:30299–30306

Kozell LB, Neve KA (1997) Constitutive activity of a chimeric D_2/D_1 dopamine receptor. Mol Pharmacol 52:1137–1149

Kozell LB, Vu MN, Neve KA (1997) The role of potential phosphorylation sites in down-regulation and desensitization of the D_1 dopamine receptor. Soc Neurosci Abstr 23:1773

Krupnick JG, Benovic JL (1998) The role of receptor kinases and arrestins in G protein-coupled receptor regulation. Annu Rev Pharmacol Toxicol 38:289–319

Laduron PM (1983) Commentary: Dopamine-sensitive adenylate cyclase as a receptor site. In: Kaiser C, Kebabian JW (eds) Dopamine Receptors, American Chemical Society, Washington, D.C.

Laitinen JT (1993) Dopamine stimulates K+ efflux in the chick retina via D1 receptors independently of adenylyl cyclase activation. J Neurochem 61:1461–1469

Lamers AE, Groneveld D, de Kleijn DPV, Geeraedts FCG, Leunissen JAM, Flik G, Wendelaar Bonga SE, Marten GJM (1996) Cloning and sequence analysis of a hypothalamic cDNA encoding a D_{1C} dopamine receptor in tilapia. Biochem Biophys Acta 1308:17–22

Lazarov N, Pilgrim C (1997) Localization of D_1 and D_2 dopamine receptors in the rat mesencephalic trigeminal nucleus by immunocytochemistry and in situ hybridization. Neurosci Lett 236:83–86

Le Moine C, Normand E, Bloch B (1991) Phenotypical characterization of the rat striatal neurons expressing the D_1 dopamine receptor gene Proc Natl Acad Sci USA 88:4205–4209

Lee S-H, Minowa MT, Mouradian MM (1996) Two distinct promoters drive transcription of the human D_{1A} dopamine receptor gene. J Biol Chem 271:25292–25299

Lefkowitz RJ, Cotecchia S, Samama P, Costa T (1993) Constitutive activity of receptors coupled to guanine nucleotide regulatory proteins. Trends Pharmacol Sci 14:303–307

Lefkowitz RJ (1998) G-protein coupled receptors III. New roles for receptor kinases and b-arrestins in receptor signalling and desensitization. J Biol Chem 273:18677–18680

Levey AI, Hersch SM, Rye DB, Sunahara RK, Niznik HB, Kitt CA, Price DL, Maggio R, Brann MR, Ciliax BJ (1993) Localization of D_1 and D_2 dopamine receptors in brain with subtype-specific antibodies. Proc Natl Acad Sci USA 90:8861–8865

Lewis MM, Watss VJ, Lawler CP, Nichols DE, Mailman RB (1998) Homologous desensitization of the D_{1A} dopamine receptor: efficacy in causing desensitization dissociates from both receptor occupancy and functional potency. J Pharmacol Exp Ther 286:345–353

Lidow MS, Elsworth JD, Goldman-Rakic PS (1997) Down-regulation of the D_1 and D_5 dopamine receptors in primate prefrontal cortex by chronic treatment with antipsychotic drugs. J Pharmacol Exp Ther 281:597–603

Lidow MS, Goldman-Rakic PS, Gallager DW, Rakic P (1991) Distribution of dopaminergic receptors in the primate cerebral cortex: quantitative autoradiographic analysis using [^{3}H]raclopride, [^{3}H]spiperone and [^{3}H]SCH-23390. Neurosci 40:657–671

Liu Q, Sobell JL, Heston LL, Sommer SS (1995) Screening the dopamine D_1 receptor gene in 131 schizophrenics and eight alcoholics: identification of polymorphism but lack of functionally significant sequence changes. Am J Med Genet 60:165–171

Luttrell LM, Ostrowski J, Cotecchia S, Kendall H, Lefkowitz RJ (1993) Antagonism of catecholamine receptor signaling by expression of cytoplasmic domains of the receptors. Science 259:1453–1457

Macrae AD, Brenner S (1995) Analysis of the dopamine receptor family in the compact genome of the puffer fish Fugu rubripes. Genomics 25:436–446

Mahan LC, Burch RM, Monsma FJJ, Sibley DR (1990) Expression of striatal D1 dopamine receptors coupled to inositol phosphate production and Ca2+ mobilization in Xenopus oocytes. Proc Natl Acad Sci USA 87:2196–2200

Mak CK, Avalos M, Randall PK, Kwan SW, Abell CW, Neuymeyer JL, Whisennaud R, Wilcox RE (1996) Improved models for pharmacological null experiments: calcification of drug efficacy at recombinant D_{1A} dopamine receptors stably expressed in clonal cell lines Neuropharmacology 35:549–570

Marchese A, Beischlag T, Nguyen T, Niznik HB, Weinshank RL, George SR, O'Dowd BF (1995) Two gene duplication events in the human and primate dopamine D5 receptor gene family. Gene 154:153–158

Markstein R, Gull P, Rudeberg C, Urwyler S, Jaton AL, McAllister K, Dixon AK, Hoyer D (1996) SDZ PSD 958, a novel D_1 receptor antagonist with potential limbic selectivity. J Neural Transm 103:261–276

Meador-Woodruff JH, Grandy DK, Van Tol HHM, Damask SP, Little KY, Civelli O, Watson SJ, Jr (1994) Dopamine receptor gene expression in the human medial temporal lobe. Neuropsychopharmacology 10:239–426

Minor DL, Wyrick SD, Charifson PS, Watts VJ, Nichols DE, Mailman RB (1994) Synthesis and molecular modeling of 1-Phenyl-1,2,3,4-tetrahydroisoquinolines and related 5,6,8,9-Tetrahydro-13bH-dibenzo[a,h]quinolizines as D_1 dopamine antagonists. J Med Chem 37:4317–4328

Minowa MT, Minowa T, Mouradian MM (1993) Activator region analysis of the human D_{1A} dopamine receptor gene. J Biol Chem 268:23544–23551

Missale C, Nash SR, Robinson SW, Jaber M, Caron MG (1998) Dopamine receptors: from structure to function. Physiol Rev 78:189–225

Moffett S, Mouillac B, Bonin H, Bouvier M (1993) Altered phosphorylation and desensitization patterns of a human 2-adrenergic receptor lacking the palmitoylated Cys341. EMBO J 12:349–356

Molloy AG, Waddington JL (1984) Dopaminergic behaviour stereospecific promoted by the D1 agonist R-SK&F 38393 and selectively blocked by the D1 antagonist SCH 23390. Psychopharmacology 82:409–410

Monsma FJJr, Mahan LC, McVittie LD, Gerfen CR, Sibley DR (1990) Molecular cloning and expression of a D1 dopamine receptor linked to adenylyl cyclase activation. Proc Natl Acad Sci USA 87:6723–6727

Moratalla R, Xu M, Tonegawa S, Graybiel AM (1996) Cellular responses to psychomotor stimulant and neuroleptic drugs are abnormal in mice lacking the D_1 dopamine receptor. Proc Natl Acad Sci USA 93:14928–14933

Moro, O., Lameh, J., Hogger, P., and Sadee, W. (1993) Hydrophobic amino acid in the i2 loop plays a key role in receptor-G protein coupling. J Biol Chem 268:22273–22276

Mottola D, Brewster WK, Cook LL, Nichols DE, Mailman RB (1992) Dihydrexidine, a novel full efficacy D_1 dopamine receptor agonist. J Pharmacol Exp Ther 262:383–393

Mouradian MM, Minowa MT, Minowa T (1994) Dopamine receptor genes: Promotors and transcriptional activation. In: Niznik HB (ed) Dopamine Receptors and Transporters. Marcell Dekker, pp 205–235

Muly EC, Szigeti K, Goldman-Rakic PS (1998) D_1 receptor in interneurons of macaque prefrontal cortex: distribution and subcellular localization. J Neurosci 18:10553–10565

Murray AM, Waddington JL (1989) The induction of grooming and vacuous chewing by a series of selective D_1 dopamine receptor agonists: two directions of $D_1:D_2$ interaction. Eur J Pharmacol 160:377–384.

Murray AM, Waddington JL (1990) New putative selective agonists at the D-1 dopamine receptor: behavioural and neurochemical comparison of CY 208–243 with SK&F 101384 and SK&F 103243. Pharmacol Biochem Behav 35:105–110

Namba T, Sugimoto Y, Negishi M, Irie A, Ushikubi F, Kakizuka A, Ito S, Ichikawa A, Narumiya A (1993) Alternative splicing of C-terminal tail of prostaglandin E receptor subtype EP3 determines G-protein specificity. Nature 365:166–170

Neisewander JL, Ong A, McGonigle P (1995) Anatomical localisation of SK&F 38393-induced behaviours in rats using the irreversible monoamine receptor antagonist. EEDQ Synapse 19:134–143

Neve KA, Neve RL (1997) Molecular biology of dopamine receptors. In: Neve KA, Neve RL (eds) The Dopamine Receptors, Humana Press Inc., Totowa, pp 27–76

Ng GYK, Trogardis J, Stevens J, Bouvier M, O'Dowd BF, George SR (1995) Agonist-induced desensitization of dopamine D_1 receptor-stimulated adenylyl cyclase activity is temporally and biochemically separated from D_1 receptor internalization Proc Natl Acad Sci USA 92:10157–10161

Nisenbaum ES, Mermelstein PG, Wilson CJ, Surmeier DJ (1998) Selective blockade of a slowly inactivating potassium current in striatal neurons by (±) 6-chloro-APB hydrobromide (SK&F 82958). Synapse 29:213–224

Niznik HB (1994) Dopamine Receptors and Transporters: Pharmacology, Structure and function. Marcel Dekker, 1994

Niznik HB, Liu F, Sugamori KS, Cardinaud B, Vernier P (1998) Expansion of the dopamine D_1 receptor gene family: defining molecular, pharmacological, and functional criteria for D_{1A}, D_{1B}, D_{1C}, and D_{1D} receptors. Adv Pharmacol 42:404–408

Noda K, Saad Y, Graham RM, Karnik SS (1994) The high affinity state of the 2-adrenergic receptor requires unique interaction between conserved and nonconserved extracellular loop cysteines. J Biol Chem 269:6743–6752

O'Boyle KM, Gaitanopoulos DE, Brenner M, Waddington JL (1989) Agonist and antagonist properties of benzazepine and thienopyridine derivatives at the D-1 dopamine receptor. Neuropharmacology 28:401–405

O'Hara K, Ulpian C, Seeman P, Sunahara RK, Van Tol HHM, Niznik HB (1993) Schizophrenia: dopamine D_1 receptor sequence is normal, but has DNA polymorphisms. Neuropsychopharmacology 8:131–135

Okamoto T, Murayama T, Hayashi Y, Inagaki M, Ogata E, Nishimoto I (1991) Identification of a Gs activator region of the 2-adrenergic receptor that is autoregulated via protein kinase A-dependent phosphorylation. Cell 67:723–730

Okazawa H, Imafuku I, Minowa MT, Kanazawa I, Hamada H, Mouradian MM (1996) Regulation of striatal D_{1A} dopamine receptor gene transcription by Brn-4. Proc Natl Acad Sci USA 93:11933–11938

Okubo Y, Suhara T, Sudo Y, Toru M (1997) Possible roles of dopamine D1 receptors in schizophrenia. Mol Psychiat 2:291–292

Olson MF, Schimmer BP (1992) Heterologous desensitization of the human dopamine D_1 receptor in Y1 adrenal cells and in a desensitization-resistant Y1 mutant. Mol Endocrinol 6:1095–1102

Ozono R, O'Connell DP, Wang Z-Q, Moore AF, Sanada H, Felder RA, Carey RM (1997) Localization of the dopamine D_1 receptor protein in the human heart and kidney. Hypertension 30:725–729

Pacheco MA, Jope RS (1997) Comparison of [^{3}H]phosphatidylinositol and [^{3}H]phospha-tidylinositol 4,5-bisphosphate hydrolysis in postmortem human brain membranes and characterization of stimulation by dopamine D1 receptors. J Neurochem 69:639–644

Pauwels PJ, Wurch T (1998) Review: amino acid domains involved in constitutive activation of G- protein-coupled receptors. Mol Neurobiol 17:109–135

Pearce RKB, Jackson M, Britton DR, Shiosaki K, Jenner P, Marsden CD (1999) Actions of the D1 agonists A-77636 and A-86929 on locomotion and dyskinesia in MPTP-treated L-DOPA-primed common marmosets. Psychopharmacology 142:51–60

Pedersen UB, Norby B, Jensen AA, Schiodt M, Hansen A, Suhr-Jessen P, Scheideler M, Thastrup O, Andersen PH (1994) Characteristics of stably expressed dopamine D1a and D1b receptors: atypical behavior of the dopamine D1b receptor. Eur J Pharmacol 267:85–93

Picciotto MR, Wickman K (1998) Using knockout and transgenic mice to study neurophysiology and behaviour. Physiol Rev 78:1131–1155

Pitcher JA, Freedman NJ, Lefkowitz RJ (1998) G-protein coupled receptor kinases. Ann Rev Biochem 67:653–692

Pliszka SR, McCracken JT, Maas JW (1996) Catecholamines in attention-deficit hyperactivity disorder: current perspectives. J Am Acad Child Adolesc Psychiat 35:264–272

Pollock NJ, Manelli AM, Hutchins CW, Steffey ME, MacKenzie RG, Frail DE (1992) Serine mutations in transmembrane V of the dopamine D_1 receptor affect ligand interactions and receptor activation. J Biol Chem 267:17780–17786

Premont RT, Inglese J, Lefkowitz RJ (1995) Protein kinases that phosphorylate activated G protein-coupled receptors. FASEB J 9:175–182

Ricci A, Mariotta S, Greco S, Bisetti A (1997) Expression of dopamine receptors in immune organs and circulating immune cells. Clin Exp Hypertension 19:59–71

Riddall DR (1992) A comparison of the selectivities of SCH 23390. with BW737C89 for D1, D2 and 5-HT2 binding sites both in vitro and in vivo. Eur J Pharmacol 210:279–281

Rosengarten H, Schweitzer JW, Friedhoff AJ (1993) A subpopulation of dopamine D_1 receptors mediate repetitive jaw movements in rats. Pharmacol Biochem Behav 45:921–924

Rosengarten H, Friedhoff AJ (1998) A phosphoinositide-linked dopamine D_1 receptor mediates repetitive jaw movements in rats. Biol Psychiat 44:1178–1184

Ruskin DN, Rawji SS, Walters JR (1998) Effects of full D1 dopamine receptor agonists on firing rates in the globus pallidus and substantia nigra pars compacta In vivo: Tests for D1 receptor selectivity and comparisons to the partial agonist SK&F 38393. J Pharmacol Exp Ther 286:272–281

Sano T, Ohyama K, Yamano Y, Nakagomi Y, Nakazawa S, Kikyo M, Shirai H, Blank JS, Exton JH, Inagami TA (1997) A domain for G protein coupling in carboxyl-terminal tail of rat angiotensin II receptor type 1A. J Biol Chem 272:23631–23636

Savoye C, Laurent C, Amadeo S, Gheysen F, Leboyer M, Lejeune J, Zarifian E, Mallet J (1998) No association between dopamine D_1, D_2, and D_3 receptor genes and manic-depressive illness. Biol Psychiat 44:644–647

Scheer A, Fanelli F, Costa T, De Benedetti PG, Cotecchia S (1996) Constitutively active mutants of the α_{-1B} adrenergic. Receptor: role of highly conserved polar amino acids in receptor activation. EMBO J 15:3566–3578

Schoors DF, Vauquelin GP, De Vos H, Smets G, Velkeniers B, Vanhaelst L, Dupont AG (1991) Identification of a D_1 dopamine receptor, not linked to adenylate cyclase, on lactotroph cells. Br J Pharmacol 103:1928–1934

Schreiber RE, Prossnitz ER, Ye RD, Cochrane CG, Bokoch GM (1994) Domains of the human neutrophil N-formyl peptide receptor involved in G protein coupling. J Biol Chem 269:326–331

Schwartz J-C, Diaz J, Bordet R, Griffon N, Perachon S, Pilon C, Ridray S, Sokoloff P (1998) Functional implications of multiple dopamine receptor subtypes: the D_1/D_3 receptor coexistence. Brain Res Rev 26:236–242

Seeman P (1992) Dopamine receptor sequences. Therapeutic levels of neuroleptics occupy D_2 receptors, clozapine occupies D_4. Neuropsychopharmacology 7:261–284

Shah JH, Kline RH, Geter-Douglass B, Izenwasser S, Witkin JM, Newman AH (1996) (±)-3-[4'-(N, N-Dimethylamino)cinnamyl] benzazepine analogs: Novel dopamine D1 receptor antagonists. J Med Chem 39:3423–3428

Sherrington R, Mankoo B, Attwood J, Kalsi G, Curtis D, Buetow K, Povey S, Gurling H (1993) Cloning of the human dopamine D_5 receptor gene and identification of a highly polymorphic microsatellite for the DRD5 locus that shows tight linkage to the chromosome 4p reference marker RAF1P1. Genomics 18:423–425

Shiosaki K, Asin KE, Bedard P, Britton DR, Jenner P, Wel Lin C, Michaelides M, Williams M (1998) Efficacy of Dopamine D1 receptor agonists A-86929 and ABT-

431 in Animal Models of Parkinson's Disease. In: Jenner P, Demirdamar R (eds) Dopamine receptor subtypes: From Basic Science to Clinical Application, IOS press, Amsterdam, pp 84–98

Sibley DR, Neve KA (1997) Regulation of dopamine receptor function and expression. In: Neve KA, Neve RL (eds) The Dopamine Receptors, Humana Press Inc., Totowa, pp 328–424

Sibley DR, Ventura ALM, Jiang D, Mak C (1998) Regulation of the D_1 dopamine receptor through cAMP-mediated pathways. Adv Pharmacol 42:447–450

Sibley DR (1999) New insights into dopaminergic receptor function using antisense and genetically altered animals. Annu Rev Pharmacol Toxicol 39:313–341

Sidhu A (1997) Regulation and expression of D-1, but not D-5, dopamine receptors in human SK-N-MC neuroblastoma cells. J Rec Sig Trans Res 17:777–784

Sidhu A, Kimura K, Uh M, White BH, Patel S (1998) Multiple coupling of human D_5 dopamine receptors to guanine nucleotide binding proteins Gs and Gz J Neurochem 70:2459–2467

Smith DR, Striplin CD, Geller AM, Mailman RB, Drago J, Lawler CP, Gallagher M (1998) Behavioural assessment of mice lacking D_{1A} dopamine receptors. Neuroscience 86:135–146

Sobell JL, Lind TJ, Sigurdson DC, Zald DH, Snitz BE, Grove WM, Heston LL, Sommer SS (1995) The D_5 dopamine receptor gene in schizophrenia: identification of a nonsense change and multiple missense changes but lack of association with disease. Hum Mol Genet 4:507–514

Spano PF, Govoni S, Trabucchi M (1978) Studies on the pharmacological properties of dopamine receptor in various areas of the central nervous system. Adv Biochem Psychopharmacol 19:155–165

Sugamori KS, Demchyshyn LL, Chung M, Niznik HB (1994) D_{1A}, D_{1B}, and D_{1C} dopamine receptors from Xenopus laevis. Proc Natl Acad Sci USA 91:10536–10540

Sugamori KS, Hamadanizadeh SA, Scheideler MA, Hohlweg R, Vernier P, Niznik HB (1998) Functional differentiation of multiple dopamine D_1-like receptors by NNC 01–0012. J Neurochem 71:1685–1693

Sunahara RK, Guan H-C, O'Dowd BF, Seeman P, Laurier LG, Ng G, George SR, Torchia J, Van Tol HHM, Niznik HB (1991) Cloning of the gene for a human dopamine D_5 receptor with higher affinity for dopamine than D_1. Nature 350:614–619

Sunahara RK, Niznik HB, Weiner DM, Stormann TM, Brann MR, Kennedy JL, Gelernter JE, Rozmahel R, Yang Y, Israel Y, Seeman P, O'Dowd BF (1990) Human dopamine D_1 receptor encoded by an intronless gene on chromosome 5. Nature 347:80–83

Taylor JM, Jacob-Mosier GG, Lawton RG, Remmer AE, Neubig RR (1994) Binding of an α2-adrenergic receptor third intracellular loop peptide to G beta and the amino terminus to G alpha. J Biol Chem 269:27618–27624

Taylor JM, Jacob-Mosier GG, Lawton RG, VanDort M, Neubig RR (1996) Receptor and membrane interactions sites on G beta. A receptor-derived peptide binds to the carboxyl terminus. J Biol Chem 271:3336–3339

Thompson M, Comings DE, Feder L, George SR, O'Dowd BF (1998) Mutation screening of the dopamine D1 receptor gene in Tourette's syndrome and alcohol dependent patients. Am J Med Genet 81:241–244

Tiberi M, Caron MG (1994) High agonist-independent activity is a distinguishing feature of the dopamine D_{1B} receptor subtype. J Biol Chem 269:27925–27931

Tiberi M, Jarvie KR, Silvia C, Falardeau P, Gingrich JA, Godinot N, Bertrand L, Yang-Feng TL, Fremeau RTJ, Caron MG (1991) Cloning, molecular characterization and chromosomal assignment of a gene encoding a second D_1 receptor subtype: differential expression pattern in rat brain compared with the D_{1A} receptor. Proc Natl Acad Sci USA 88:7491–7495

Tiberi M, Nash SR, Bertrand L, Lefkowitz RJ, Caron MG (1996) Differential regulation of dopamine D_{1A} receptor responsiveness by various G protein-coupled receptor kinases. J Biol Chem 271:3771–3778

Tomic M, Seeman P, George SR, O'Dowd BF (1993) Dopamine D_1 receptor mutagenesis: role of amino acids in agonist and antagonist binding. Biochem Biophys Res Comm 191:1020–1027

Trogadis JE, Ng GYK, O'Dowd BF, George SR, Stevens JK (1995) Dopamine D_1 receptor distribution in Sf9 cells imaged by confocal microscopy: a quantitative evaluation. J Histochem Cytochem 43:497–506

Undie AS, Friedman E (1990) Stimulation of a dopamine D1 receptor enhances inositol phosphates formation in rat brain. J Pharmacol Exp Ther 253:987–922

Undie AS, Friedman E (1992) Selective dopaminergic mechanism of dopamine and SKF38393 stimulation of inositol phosphate formation in rat brain. Eur J Pharmacol 226:297–302

Undie AS, Weinstock J, Sarau HM, Friedman E (1994) Evidence for a distinct D1-like dopamine receptor that couples to activation of phosphoinositide metabolism in brain. J Neurochem 62:2045–2048

Undie AS, Friedman E (1994) Inhibition of dopamine agonist-induced phosphinositide hydrolysis by concomitant stimulation of cyclic AMP formation in brain slices. J Neurochem 63:222–230

Undie AS (1999) Relationship between dopamine agonist stimulation of inositol phosphate formation and cytidine diphosphate-diacylglycerol accumulation in brain slices. Brain Res 816:286–294

Varrault A, Nguyen DL, McClue S, Harris B, Jouin P, Bockaert J (1994) 5-HT_{1A} receptor synthetic peptides. J Biol Chem 269:16720–16725

Vermeulen RJ, Drukarch B, Wolters EC, Stoof JC (1999) Dopamine D_1 receptor agonists – the way forward for the treatment of Parkinson's disease CNS. Drugs 11:83–91

Vernier P, Cardinaud B, Valdenaire O, Phillipe H, Vincent J-D (1995) An evolutionary view of drug-receptor interaction: the bioamine receptor family. Trends Pharmacol Sci 16:375–381

Verrall S, Ishii M, Chen M, Wang L, Tram T, Coughlin SR (1997) The thrombin receptor second cytoplasmic loop confers coupling to Gq-like G proteins in chimeric receptors. J Biol Chem 272:6898–6902

Versaux-Botteri C, Gibert J-M, Nguyen-Legros J, Vernier P (1997) Molecular identification of a dopamine D_{1b} receptor in bovine retinal pigment epithelium. Neurosci Lett 237:9–12

Vickery RG, von Zastrow M (1999) Distinct dynamin-dependent and -independent mechanisms target structurally homologous dopamine receptors to different endocytic membranes. J Cell Biol 144:31–43

Wachtel SR, White FJ (1995) The dopamine D_1 receptor antagonist SCH 23390 can exert D_1 agonist-like effects on rat nucleus accumbens neurons. Neurosci Lett 199:13–16

Waddington JL, O'Boyle KM (1989) Drugs acting on brain dopamine receptors: a conceptual re-evaluation five years after the first selective D1 antagonist. Pharmacol Ther 43:1–52

Waddington JL, Daly SA (1993) Regulation of unconditioned motor behaviour by $D_1:D_2$ interactions. In: Waddington JL (ed) $D_1:D_2$ Dopamine Receptor Interactions: Neuroscience and Psychopharmacology, Academic Press, London, pp 51–78

Waddington JL, Daly SA, McCauley PG, O'Boyle KM (1994) Levels of functional interaction between D_1-like and D_2-like dopamine receptor systems. In: Niznik HB (ed) Dopamine Receptors and Transporters: Pharmacology, Structure and Function, Marcel Dekker. New York, pp 511–537

Waddington JL, Daly SA, Downes RP, Deveney AM, McCauley PG, O'Boyle KM (1995) Behavioural pharmacology of "D-1-like" dopamine receptors: further subtyping, new pharmacological probes and interactions with "D-2-like" receptors. Prog Neuro-Psychopharmacol Biol Psychiat 19:811–831

Waddington JL, Deveney AM (1996) Dopamine receptor multiplicity "D_1-like"-"D_2-like" interactions and "D_1-like" receptors not linked to adenylate cyclase. Biochem Soc Trans 24:177–182

Waddington JL, Deveney AM, Clifford JJ, Tighe O, Croke DT, Sibley DR, Drago J (1998a) D1-like dopamine receptors: regulation of psychomotor behaviour, D1-like: D2-like interactions and effects of D1A targeted gene deletion. In: Jenner P, Demirdamar R (eds) Dopamine receptor subtypes, IOS, Amsterdam, pp 45–63

Waddington JL, Deveney AM, Clifford JJ, Tighe O, Croke DT, Sibley DR, Drago J (1998a) Behavioural analysis of multiple D_1-like dopamine receptor subtypes: new agents and studies in transgenic mice with D_{1A} receptor knockout. Adv Pharmacol 42:514–517

Watts VJ, Lawler CP, Gilmore JH, Southerland SB, Nichols DE, Mailman RB (1993) Dopamine D1 receptors: efficacy of full (dihydrexidine) vs. partial (SK&F 38393) agonist in primates vs. rodents. Eur J Pharmacol 242:165–172

Watts VJ, Lawler CP, Gonzales AJ, Zhou QY, Civelli O, Nichols DE, Mailman RB (1995) Spare receptors and intrinsic activity: studies with D1 dopamine receptor agonists. Synapse 21:177–187

Weed MR, Woolverton WL, Paul IA (1998) Dopamine D_1 and D_2 receptor selectivities of phenyl-benzazepines in rhesus monkey striata. Eur J Pharmacol 361:129–142

Weinshank RL, Adham N, Macchi M, Olsen MA, Branchek TA, Hartig PR (1991) Molecular cloning and characterization of a high affinity dopamine receptor and its pseudogenes. J Biol Chem 266:22427–22435

Wess J (1997) G-protein-coupled receptors: molecular mechanisms involved in receptor activation and selectivity of G-protein recognition. FASEB J 11:346–354

White FJ, Hu X-T (1993) Electrophysiological correlates of $D_1:D_2$ interactions. In: Waddington JL (ed) $D_1:D_2$ Dopamine Receptor Interactions: Neuroscience and Psychopharmacology, Academic Press, London, pp 79–114

Wong SK, Ross EM (1994) Chimeric muscarinic cholinergic: beta-adrenergic receptors that are functionally promiscuous among G proteins. J Biol Chem 269:18968–18976

Xu M, Hu XT, Cooper DC, Moratalla R, Graybiel AM, White FJ, Tonegawa S (1994a) Elimination of cocaine-induced hyperactivity and dopamine-mediated neurophysiological effects in dopamine D_1 receptor mutant mice. Cell 79:945–955

Xu M, Koeltzow TE, Santiago GT, Moratalla R, Cooper DC, Hu X-T, White NM, Graybiel AM, White FJ, Tonegawa S (1997) Dopamine D_3 receptor mutant mice exhibit increased behavioural sensitivity to concurrent stimulation of D_1 and D_2 receptors. Neuron 19:837–848

Xu M, Moratalla R, Gold LH, Hiroi N, Koob GF, Graybiel AM, Tonegawa S (1994b) Dopamine D_1 receptor mutant mice are deficient in striatal expression of dynorphin and in dopamine-mediated behavioral responses. Cell 79:729–742

Zhang J, Barak LS, Winkler KE, Caron MG, Ferguson SSG (1997) A central role for beta-arrestins and clathrin-coated vesicle-mediated endocytosis in beta2-adrenergic receptor resensitization. Differential regulation of receptor resensitization in two distinct cell types. J Biol Chem 272:27005–27014

Zhou Q-Y, Grandy DK, Thambi L, Kushner JA, Van Tol HHM, Cone R, Pribnow D, Salon J, Bunzow JR, Civelli O (1990) Cloning and expression of human and rat D_1 dopamine receptors. Nature 347:76–80

Zhou Q-Y, Li C, Civelli O (1992) Characterization of gene organization and promoter region of the rat dopamine D_1 receptor gene. J Neurochem 59:1875–1883

CHAPTER 6
Understanding the Function of the Dopamine D$_2$ Receptor: A Knockout Animal Approach

S. TAN, B. HERMANN, C. IACCARINO, M. OMORI, A. USIELLO, and E. BORRELLI

A. Introduction

Dopamine (DA) is an important regulator of different physiological functions in the central nervous system (CNS) as well as in other organs. Known dysfunctions of the dopaminergic system lead to pituitary tumors and to diseases affecting the CNS such as Parkinson's disease (PD), Tourette's syndrome, and schizophrenia. Among these diseases however, PD is the one that has the most defined etiology. It has been established that PD arises from the degeneration of mesencephalic dopaminergic neurons with the consequent reduction of brain DA levels. However, the mechanisms leading to the degeneration of mesencephalic neurons is still awaited. In contrast, deficits in the dopaminergic system which can lead to other neuropathologies, such as schizophrenia, are still matter of debate. Schizophrenia may result from dysfunctions in several neural systems and probably is a multi-genic disease. However, efficient therapeutic protocols for the treatment of this disease utilize drugs that modulate the function of the dopaminergic system. Consequently, the principal aims of the study of the dopaminergic system are to identify and characterize the function of each of its components, and to understand how these components interact with each other and with other neural systems. Such work will not only advance our knowledge of the dopaminergic system, but also will help to develop novel therapies and strategies for the cure of neuropsychiatric diseases that are linked to the dopaminergic system.

Since the cloning of the different DA receptors information has been rapidly acquired on these proteins at the molecular level (for review see LACHOWICZ and SIBLEY 1997; MISSALE et al. 1998). Such work is quickly leading to the characterization of a very intricate series of events which control the correct function of these proteins *in vivo*. Analysis of the DA receptors extends from their transcriptional regulation to their membrane presentation and association with other receptors. Finally, it includes the generation of living animals deprived of the expression of dopaminergic components. The bulk of this research is unraveling an unexpectedly complex regulation of the dopaminergic system. Thus, alterations occurring in any of these regulatory events could lead to dysfunction of the system. Therefore, understanding each of the steps involved in the generation of functioning elements is of key importance. This review will focus on the DA D_2 receptor (D_2R), one of the major components of the dopaminergic system. We will give an overview of what has been done to date to unravel the role of this receptor in both the transduction of the dopaminergic signal as well as in its function *in vivo*. Finally, the D_2R in human diseases and psychiatric disorders will be discussed, and the possible capacity of this receptor during neuronal degeneration will be explored.

B. Transcriptional Regulation of the Dopamine D_2 Receptor

Regulation of the transcription of DA receptors is a key element in the normal function of the dopaminergic system. In this respect, it is interesting to note that the two major DA receptors, D_1 and D_2, possess promoters with housekeeping characteristics (SAMAD et al. 1997). This feature is characterized by the absence of a functional TATA box or upstream CAAT rich-regions. Thus, it appears that a specific expression pattern of DA receptors must be temporally and spatially dictated by cell-specific transcription factors.

We have analyzed the proximal promoter region of the DA D_2R, and identified and characterized a retinoic acid responsive element (RARE) in the 5′ flanking region of the DA D_2R gene at position –68. The described RARE in the D_2 promoter is composed of repeated motifs closely related to the core motif 5′-PuG(G/T)TCA-3′. The two motifs are separated by three base pairs, a spacing which has been described to favor the binding of the vitamin D_3 receptor (SAMAD et al. 1997). At present, we cannot exclude that this element might be a target of the vitamin D_3 receptor in a class of dopaminergic or target cell neurons, or in cells of the pituitary gland. In addition, a canonical Sp1 binding site is contiguous with the RARE in the D_2 promoter. It is speculated that in the D_2R promoter, an interaction between members of the retinoid receptor family and the Sp1 factor, could modulate the DNA-binding specificities of these proteins. Non-consensus responsive elements present in native promoters may result in lower binding affinities with their corresponding transcription factors, thus representing a putative regulatory mechanism to attenuate the response to specific ligands *in vivo*. It is also possible that transcription factors which recognize non-consensus motifs may be subject to cooperative interactions with other factors binding to close or adjacent sites (DAY et al. 1990; RHODES et al. 1993). The presence of these cooperative interactions could determine the strength and the cellular specificity of the transcriptional response. In the case of the D_2R promoter, the close vicinity of the D_2R RARE and the Sp1 sites might represent an example of such cooperation.

D_2R is expressed particularly by the dopaminergic cells in mesencephalic nuclei and also by post-synaptic target neurons, such as the medium spiny neurons of the striatum, cortex, hypothalamus, and other brain regions. In addition, outside of the brain the D_2R is highly expressed in the pituitary gland by two cell types, melanotrophs and lactotrophs. It is thus conceivable that the D_2R promoter might be controlled by different combinations of members of the thyroid hormone/retinoic acid and vitamin D_3R families in different cell types. However, the analysis of D_2R expression in retinoid receptor mutant mice revealed a stronger decrease of expression in double mutant animals [retinoid X receptor (RXR) and retinoid acid receptor (RAR) knockout] compared to simple mutant mice (RAR *or* RXR knockout) (KREZEL et al. 1998). This suggests that the absence of one retinoid receptor might be partially compensated by other members of the RAR/RXR family in simple mutants. The most important decrease in D_2R mRNA expression was observed in the RARβ/RXRγ double knockout mice, which strongly supports *in vitro* data (SAMAD et al. 1997) and identifies these receptors as specific transcription factors required for full expression of the D_2R gene in the striatum.

These data showed for the first time the involvement of RA in adult CNS functions. RA plays a key role in the control of D_2R transcription, and in the absence of RAR or RXR receptors there is a reduced expression of the DA D_2Rs. These data have a strong impact on the understanding of pathologies linked to D_2R expression since, as we have previously shown in mice, altered expression of this receptor results in a parkinsonian phenotype and pituitary tumors. Thus, impaired RA control of D_2R might also be involved in human diseases such as Parkinson's, schizophrenia, and endocrine tumors.

C. Dopamine D_2 Receptor Signal Transduction

I. Gi Protein Coupled Pathways and the D_2 Receptor Splice Variants

For many years the existence of multiple DA receptors was postulated to underlie the diverse behavioral and biochemical properties associated with dopaminergic neurotransmission and DA receptor activation (SEEMAN 1980). These receptors are known to belong to the family of seven transmembrane domain (7TM) G-protein-coupled receptors (GPCR). D_2Rs belong to the D_2-like subfamily of DA receptors consisting of D_2, D_3, and D_4. These receptors couple to $G\alpha_i/G\alpha_o$ members of the guanosine triphosphate (GTP)-binding proteins (G proteins), and have been mainly characterized as inhibitors of adenylyl cyclase (KEBABIAN and CALNE 1979; FISHBURN et al. 1995). However, D_2-like receptors can also activate other transduction pathways (PICETTI et al. 1997).

The D_2R exists in two isoforms, the short (D_2S, 415 aa) and the long (D_2L, 444 aa) isoform, which are generated from the same gene by alternative splicing (MONTMAYEUR et al. 1991). Interestingly, these isoforms follow the same pattern of expression, although the D_2L/D_2S ratio may change depending on the brain region (Fig. 1). Indeed, while D_2S mRNAs are highly expressed in the dopaminergic cell bodies and axons, D_2L is more prominently expressed at postsynaptic sites (MONTMAYEUR et al. 1991; KHAN et al. 1998). Moreover, the two isoforms present a very similar pharmacology when

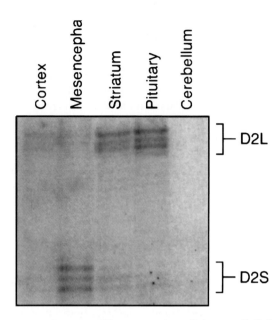

Fig. 1. S1 nuclease mapping assay. The expression of the two D_2R isoforms, D_2S and D_2L, was examined in the cortex, mesencephalon, striatum, pituitary, and cerebellum. 10 µg of total RNA was loaded per lane

expressed in established cell lines (PICETTI et al. 1997). Little is known about the regulation of the alternative splicing of D_2R. Experiments on rats show that changes in circulating sex hormone levels modulate the splicing without affecting the total amount of D_2R messenger RNA (GUIVARCH et al. 1995).

D_2L differs from D_2S only by a 29-amino acid insertion in the third intracellular loop, a well-characterized site for G-protein interaction (KOBILKA et al. 1988; LEFKOWITZ and CARON 1988; MALEK et al. 1993). G proteins are the intracellular transducers of the signals received by membrane receptors upon binding to the appropriate ligand (SIMON et al. 1991). These heterotrimeric GTP-binding proteins are composed of three subunits: α, β, and γ, whereby the α subunit is responsible for the coupling of these proteins to the receptor. To date three different inhibitory $G\alpha_i$-subtypes have been characterized ($\alpha_{i1}, \alpha_2, \alpha_3$) (JONES and REED 1987). Experiments performed *in vitro* have demonstrated that D_2S and D_2L may vary in their coupling to these subunits. D_2L interacts preferentially with the $G\alpha_{i2}$ subunit while the D_2S receptor probably binds more specifically with other $G\alpha$ subunits (MONTMAYEUR et al. 1993; PICETTI et al. 1997). Thus, differential interaction with the $G\alpha$ subunits underlies the possibility that these two receptors might have different physiological activities.

Interestingly, some of the known G proteins are localized not only to the membrane but they are also located intracellularly, raising the possibility that they might influence cellular trafficking (PIMPLIKAR and SIMONS 1993; HELMS 1995). Therefore, they may not only transduce signals from membrane receptors, but also regulate receptor presence at the cell surface. Our laboratory has demonstrated that the mRNA of the $G\alpha_{i2}$ subunit encodes two proteins, $G\alpha_{i2}$ and sG_{i2}, by an alternative splicing mechanism. sG_{i2} differs from $G\alpha_{i2}$ by the replacement of the last 24 amino acids of $G\alpha_{i2}$ with an alternative 35 amino acid segment in the C-terminal region. This novel segment targets sG_{i2} to the Golgi apparatus rather than to the plasma membrane as with $G\alpha_{i2}$ (MONTMAYEUR and BORRELLI 1994). This peptide exchange alone is necessary but not sufficient for the retention of the sG_{i2} subunit in the Golgi apparatus, strongly suggesting that this segment interacts with other regions within the sG_{i2} protein sequence (PICETTI and BORRELLI 2000). Once activated, sG_{i2} leaves the Golgi and it is found in the cytoplasm. Preliminary experiments suggest that this protein might be able to retain D_2R in the Golgi on its way to the plasma membrane (unpublished results). If this is the case *in vivo*, this might represent an alternative level of regulation of D_2R's function.

II. Kinase Pathways Involved in Signaling

In addition to the inhibition of adenylyl cyclase, other transduction pathways have been shown to be affected by D_2R activation, such as the stimulation of phospholipase A2, activation of K^+ channels, and activation or blocking of Ca^{2+} channels. D_2Rs also influence the calcium/calmodulin-dependent protein kinase's (CaMKII) function by varying the intracellular Ca^{2+} concentration (PICETTI et al. 1997).

Recently it has been demonstrated that D_2R stimulation can lead to both the phosphorylation and activation of cAMP-response element binding protein (CREB) and mitogen-activated protein kinase (MAPK) *in vitro* (YAN et al. 1999) as well as to phosphorylation of MAPK *in vivo* (CAI et al. 2000). Interestingly, CREB and MAPK seem activated by two different transduction pathways. Indeed, D_2R agonists increase intracellular Ca^{2+} and protein kinase C (PKC) activity leading to the activation of the Ras/Raf/MEK/MAPK signal transduction cascade, while intracellular Ca^{2+} and calmodulin-dependent protein kinase (CaMK) are required for the activation of CREB (YAN et al. 1999). It has been shown that activated MAPKs in the CNS are mainly cytoplasmic and localized in the cell bodies and dendrites. Conversely, activated CREB is found in the nucleus. The different subcellular localizations of these two activated components indicates that DA-induced activation of D_2Rs might result in the simultaneous stimulation of multiple targets. It has been proposed that MAPK phosphorylation, which occurs primarily in dendrites, may regulate protein synthesis, cytoskeletal dynamics, and ion channel activities at synapses. On the other hand, phosphorylated CREB may regulate gene expression (YAN et al. 1999) by acting at the nuclear level.

III. Receptor Heterodimers

A further level of regulation of D_2R-mediated signal transduction might be dependent upon intracellular interactions with membrane receptors for other neurotransmitters and neuromodulators. Indeed, a direct intramembrane interaction between D_2R and the somatostatin receptor has been recently shown (ROCHEVILLE et al. 2000). However, intra-membrane and intracellular modulation of the D_2-mediated signaling was already evoked for the adenosine A_{2A} receptor as well as for other heterologous receptors (FUXE et al. 1998). In addition, recent *in vitro* and *in vivo* studies have shown that D_1 and D_2Rs colocalize in striatal neurons (SURMEIER et al. 1996; AIZMAN et al. 2000). These observations are interesting since the described synergistic and antagonistic actions exerted by activation of D_1 and D_2 subclasses of DA receptors might well be mediated by intracellular rather than by exclusively intercellular mechanisms.

D. D_2 Receptor Function *In Vivo*

I. Generation of Knockout Mice

In order to analyze the specific function of DA D_2R *in vivo*, the D_2R gene was knocked out by homologous recombination (BAIK et al. 1995). The D_2R gene is a "split" gene composed of eight coding exons. As described previously, the D_2R transcript is spliced into two isoforms, D_2L and D_2S. Therefore, the strategy we used to generate D_2R-null mutant mice was designed to avoid formation of truncated receptors. To do so, the second coding exon of the D_2R gene was deleted, and substituted by a pGK-neomycin cassette to select recombinant embryonic stem (ES) cells (Fig. 2A). We analyzed the possibility that an alternative transcript originating in exon 1 could branch on exon 3 in the mutated gene, in the absence of exon 2. We assessed that if this happens, the transcript is not in the wild-type reading frame and if translated would create a D_2R unrelated product. Striatal membranes from D_2R mutant mice do not bind the D_2-like specific antagonist spiperone, supporting the absence of active D_2Rs. Thus, the knockout of the gene was successfully achieved, and a line of mice lacking the D_2R was generated in our laboratory (BAIK et al. 1995).

Two other groups have also recently generated D_2R-null mice by homologous recombination (KELLY et al. 1997; JUNG et al. 1999). While JUNG et al. followed a similar knockout strategy to ours, KELLY's group has used a different one. KELLY and colleagues deleted part of the C-terminal region of the receptor (Fig. 2B). Interestingly, a shorter D_2R mRNA is formed in these mice. This raises the question of whether these mice still produce a truncated receptor. Unfortunately, experiments aimed at establishing this eventuality have not been reported. However, recent work demonstrates that truncated D_2Rs can antagonize wild-type (WT) D_2Rs when they are co-expressed (LEE et al. 2000). These data suggest that D_2Rs are oligomeric and that a truncated receptor retains the ability to dimerize with the WT receptor. As described in Sect. C.III, new evidence shows that the D_2R is able to heterodimerize with other membrane receptors. Therefore, it is conceivable that a truncated D_2R could interact with other membrane receptors and lead to both pharmaco-

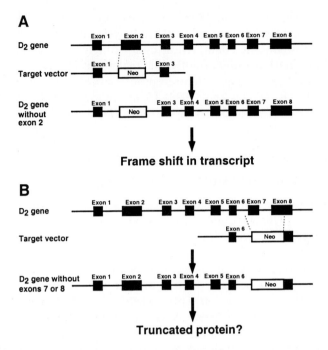

Fig. 2. Two different strategies to create the $D_2R^{-/-}$ mouse. **A** BAIK et al. (1995) removed exon 2 while JUNG et al. (1999) removed nearly all of exon 2. Both labs replaced these almost-identical regions with neo cassette, creating a frame shift in the transcript and a loss of the D_2R. **B** KELLY et al. (1997) replaced exon 7 and most of exon 8 with a neo cassette. This construct leads to a truncated transcript and possibly a truncated protein

logical and behavioral phenotypes. It is also possible that a truncated receptor could bind to DA and to D_2R agonists/antagonists, producing unpredictable results.

E. D_2 Receptor's Role as an Autoreceptor

D_2-like receptors have been shown to possess autoreceptor functions by pharmacological, neurochemical, and electrophysiological means. Indeed, both D_2R and D_3R are localized in dopaminergic neurons. This has raised the question of whether both receptors have autoreceptor functions. Therefore, D_2R-null mice represent a tool to test whether DA D_2 and D_3 receptors share the autoreceptor function *in vivo*, or if only one of them is responsible for regulating DA concentrations in the synapse.

Interestingly, the knockout of the D_2R leads to a total abrogation of the inhibitory effects of DA on both the firing of dopaminergic neurons in the substantia nigra (MERCURI et al. 1997) as well as the release of DA in the striatum (L'HIRONDEL et al. 1998; DICKINSON et al. 1999). More recently, experiments performed by *in vivo* microdialysis and voltammetry support these results (unpublished data). These data strongly indicate that D_2R is the major DA autoreceptor. Importantly, the analysis of DA $D_3^{-/-}$ mice supports this view. Indeed, in these mice autoreceptor functions are maintained in spite of lack of D_3Rs (KOELTZOW et al. 1998). However, in this study a potential role for D_3R in the control of postsynaptic short-loop feedback modulating DA release has been evoked.

F. D_2 Receptor Signaling in Physiology

D_2-like receptors are the primary sites of action of most antipsychotics (SEEMAN et al. 1975; BURT et al. 1976; KAPUR and SEEMAN 2000). Indeed, a combination of pharmacological studies with behavioral analyses on humans and laboratory animals has been central to reveal the physiological functions in which these receptors are involved. Such work has been a major driving force in the growth of fields devoted to the pharmacological characterization of the DA D_2R and its function (HORNYKIEVICZ 1973; CHASE et al. 1974; VAN KAMMEN 1979; OLSEN et al. 1980). Assays have also been used to define the effects of DA agonists and antagonists on behavior in order to identify new drugs for their therapeutic value. A major bias of these studies, however, is the lack of specificity of the available drugs for only one of the receptors. Indeed, while it is possible to discriminate D_1 from D_2-like receptors by pharmacological means, drugs directed at each member of the subfamily are still awaited. Furthermore, dopaminergic ligands may also present specificities for heterologous neurotransmitter systems.

In the last decade pharmacogenetic research has played a crucial role in providing a link between gene function in the CNS and behavior. Knockout mice are now widely utilized by behavioral neuroscientists to better understand the relevance of the molecular and cellular mechanisms underlying behavior (CRABBE et al. 1994; WYNSHAW-BORIS 1996). This section of the review will focus specifically on the DA $D_2R^{-/-}$ mice and the behavioral work that has been done to observe the role of the D_2R in motor function and drug abuse.

I. Motor Function

The basal ganglia, comprising the striatum, are a major brain system through which the cerebral cortex affects the motor system. Processing of cortical input in the striatal portion of the basal ganglia is modulated by dopaminergic input from the substantia nigra. Two major pathways involved in the control of motor function have been described in this system, the direct and the indirect. The direct pathway projects to the internal segment of the globus pallidus and substantia nigra pars reticulata, and then projects to the thalamus. The indirect pathway comprises afferents to the external segment of the globus pallidus, which then projects to the subthalamic nucleus. The subthalamic nucleus in turn projects back to both the pallidal segments and the substantia nigra. These pathways are composed of medium-sized spiny neurons, which utilize the inhibitory neurotransmitter, γ-aminobutyric acid (GABA) (ALEXANDER and CRUTCHER 1990). Neurons of the direct pathway express D_1R, dynorphin (Dyn) and substance P (SP), while those of the indirect pathway express D_2R and enkephalin (Enk).

The significance of D_1 and D_2 receptor-specific regulation of striatonigral and striatopallidal pathways is related to their opposite effects on GABAergic neurons and their regulation of the expression of Dyn, SP, and Enk. Normal movement results from a coordinated balance of cortical and thalamic excitation of the striatonigral and striatopallidal pathways, which regulate the tonic activity of substantia nigra pars reticulata neurons. DA excites the striatonigral pathway while it inhibits the striatopallidal pathway. Reduction of the dopaminergic input to the striatum results in increased expression of Enk and a decrease in SP (GERFEN et al. 1990). Disruption of this balance leads either to the production of involuntary movements or to akinesia, bradykinesia, and a shuffling gait as observed in PD patients. In this context, mice lacking DA receptors offer great opportunities to assess the specific function of each receptor in these functional interactions. Locomotor activity is reduced by DA receptor antagonists, bilateral lesions of the substantia nigra (SN) with 6-hydroxydopamine (6-OHDA), electrolytic lesions of ascending DA pathways, or by drugs such as reserpine, which deplete catecholamines. On the other hand, drugs that enhance transmission at DA synapses either increase locomotor activity or produce stereotypy, depending on the dose administered (LE MOAL and SIMON 1991).

Despite their different signal transduction mechanisms, it is believed that the D_1Rs and D_2Rs both contribute to locomotion through direct and indirect striatopallidal projections synergistically (WADDINGTON 1993). In this respect, D_2R-mutant mice have helped in the establishment of the role of this receptor in the control of locomotion. D_2R-mutant mice have impaired locomotion in contrast to D_1R-mutant mice which, under basal conditions, present normal motor functions (XU et al. 1994). The behavioral phenotype of the $D_2R^{-/-}$ mice was quantified using three different tests which are commonly employed to examine motor function: the open field, rotarod, and the ring test. In the open field, $D_2R^{-/-}$ mice present a significant reduction in both locomotion and rearing behavior when compared to the WT littermates (Fig. 3). It should be noted that the stressful conditions of the open field, dramatically exacerbate the delay in the initiation of movement in the D_2R-deficient mice versus the WT group. The rotarod apparatus instead, tests the ability of mice to coordinate movements. This test was performed in only one session to allow a direct observation of the motor function of each mouse in a natural and naïve setting. The $D_2R^{-/-}$ mice group had difficulties performing on the rotating rod and fell easily. They also spent significantly less time on the rotarod apparatus than the

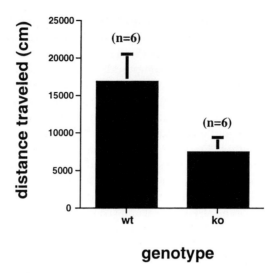

Fig. 3. Locomotion of WT and $D_2R^{-/-}$ mice in an open field test. The distance traveled (cm) was tested in 6 WT and 6 knockout mice using the open field test for 1 h. The knockout mice traveled significantly less distance than the WT mice did

WT mice (BAIK et al. 1995). Finally, we used the ring test to observe whether spontaneous movements were affected. In this test $D_2R^{-/-}$ mice showed a significant increase in time spent immobile compared with the control group. Importantly, together with the motor deficits, we documented an increase of Enk expression. This has often been used to assess the efficacy of destruction of dopaminergic neurons. These observations led us to propose that D_2R-null mice present a "parkinsonian-like phenotype," intending that lack of D_2R signaling affects movements in a similar manner to that of DA reduction in other models, although to a much lower extent.

The spontaneous behavior of D_2R-null mice was also investigated by ethological analyses. Individual elements of behavior were resolved and quantified via direct visual observations using an ethologically based rapid-time sampling behavioral checklist procedure (COLGAN 1978; CLIFFORD and WADDINGTON 1998; CLIFFORD et al. 2000). This approach allowed us to observe the mice under conditions that were relatively free of situation-dependent stress. The ethograms of spontaneous behavior showed a significant decrease in locomotion, rearing to the wall, free rearing, and grooming in the D_2R-null mice compared to the WT littermates as measured by an observer blind to the mice genotype. Nevertheless, the magnitude of the locomotor deficit in the knockout mice appeared to be less dramatic with respect to the previous investigations, which were executed in unnatural conditions.

In PD, a degeneration of dopaminergic neurons of the nigrostriatal pathway is observed. Loss of dopaminergic regulation of striatal neuron activity results in altered motor functions. Adenosine A_{2A} receptors ($A_{2A}Rs$)

and D_2Rs are colocalized in striatal medium spiny neurons. It has been proposed that adenosine binding to $A_{2A}Rs$ lowers the affinity of DA for the D_2R, thus modulating the function of DA receptors. Absence of D_2Rs in knockout mice results in impaired locomotion and coordinated movements. Recently, we explored the possibility that an $A_{2A}R$ antagonist might re-establish motor functions. Interestingly, blockade of $A_{2A}R$ rescues the behavioral parameters altered in $D_2R^{-/-}$ mice. In addition, the level of expression of enkephalin and substance, P which were altered in $D_2R^{-/-}$ mice, were re-established to normal levels after $A_{2A}R$ antagonist treatment of mutant mice. These results show that $A_{2A}R$ and D_2R have independent and antagonistic activities. Selective $A_{2A}R$ antagonists might provide a potential nondopaminergic approach to the therapeutic treatment of PD (AOYAMA et al. 2000).

In the last 2 years the two previously mentioned independent lines of D_2R-null mice have been tested for motor functions (KELLY et al.1998; JUNG et al. 1999). Using activity boxes to measure locomotion, JUNG et al. showed a clear motor impairment in their knockout mice, which was characterized by bradykinesia and postural abnormalities. These results confirm those obtained by our laboratory. In sharp contrast to our results and those of JUNG and colleagues, another $D_2R^{-/-}$ mouse line (KELLY et al. 1997, 1998) has been reported to show no evident motor abnormalities, raising the possibility that in this last line some residual D_2R activity is still present.

II. Drug Abuse

The mesolimbic dopaminergic system is believed to play a crucial role in the behavioral effects of drugs of abuse, including cocaine, amphetamine, opiates, and alcohol. It is well known that cocaine and amphetamine bind directly to the DA transporter to inhibit DA reuptake activity, thereby increasing extracellular concentrations of DA. With such a large elevation of extracellular DA following cocaine or amphetamine administration, substantial interest has turned to the role of the different DA receptors in drug abuse. Previous work reported that D_2-like receptor agonists such as quinpirole and bromocriptine induce cocaine-like behavioral effects in animals, including stimulation of locomotor activity, generation of stereotyped behavior, and positive reinforcing effects (WOOLVERTON et al. 1988). In addition, D_2-like receptor antagonists such as sulpiride can block many of the behavioral effects of cocaine and amphetamine, although there are some studies that report the opposite response to D_2 antagonists (CALLAHAN et al. 1994).

In the last few years the analysis of gene-targeted mice has provided insights into the involvement of DA receptors in the response to drug abuse. $D_1R^{-/-}$ mice revealed a complete abolition of locomotor activation in response to cocaine (XU et al. 1994). In our laboratory, the study of D_2R-null mice using the place preference paradigm has shown the absence of conditioning to morphine compared to their wild-type littermates (MALDONADO et al. 1997). This effect is specific to morphine since no difference is observed between

D_2R-deficient and WT mice in the same behavioral paradigm in response to food. Interestingly, we also demonstrated that the D_2R is not required for the development of opiate withdrawal or for the locomotor response to acute administration of morphine. Ethanol preference and sensitivity to ethanol-induced locomotor impairment are also markedly reduced in D_2R-deficient mice (PHILLIPS et al. 1998). These results demonstrate the importance of DA signaling via D_2Rs in the response to drug abuse. In contrast, D_3R mutant mice exhibit enhanced behavioral sensitivity to injections of cocaine and amphetamine. Furthermore, the mutant mice show significant place preference at low doses of amphetamine, whereas no preference is seen in the WT mice at this low dose (XU et al. 1997). D_4R-null mice are supersensitive to the stimulation of locomotor activity elicited by ethanol, cocaine, and methamphetamine, although they displayed less spontaneous locomotor activity (RUBINSTEIN et al. 1997).

What are the molecular mechanisms underlying the physiological response to drug abuse? A striking feature of the neural responses to such drugs is stimulation of cAMP which leads to the activation of the CREB (KONRADI et al. 1994) and immediate early genes in the forebrain, such as Fos/Jun (GRAYBIEL et al. 1990; HOPE et al. 1992). There is also an induction in NAC-1 mRNA in the nucleus accumbens by chronic cocaine administration (CHA et al. 1997). D_2R antagonists prevent CREB activation and the expression of c-*fos* and *zif*268 induced by acute cocaine and amphetamine treatment (YOUNG et al. 1991; DAUNAIS and McGINTY 1996). Therefore, a major aim in future studies will be the identification of the genes whose expression is differentially either stimulated or inhibited in mutant versus WT mice. This will give insights on target genes regulated by D_2R-mediated signaling.

G. Neuronal Protective Pathways via the D_2 Receptor

DA has long been associated with neurodegenerative diseases such as PD and Huntington's disease, as well as with traumatic injury and ischemia. DA can be oxidized to a semiquinone and act as a potent neurotoxic free radical, or it can be metabolized by monoamine oxidase B, leading to the production of hydrogen peroxide. By these mechanisms it is believed that DA itself can be toxic in the nervous system during pathological conditions and can contribute to neurodegeneration. However, it has also been demonstrated that DA is a potent antioxidant, with antioxidant activities equal to those of vitamin E (SAM and VERBEKE 1995; YEN and HSIEH 1997; KANG et al. 1998; SMYTHIES 1999). Recent evidence also suggests that DA D_2R pathways are protective and can lead to activation of antioxidant enzymes in the brain (SAWADA et al. 1998; IIDA et al. 1999). IIDA et al. demonstrated that activation of the D_2Rs leads to upregulation of antioxidant enzymes such as glutathione (GSH), catalase, and superoxide dismutase (SOD). SAWADA and colleagues have also shown that a D_2R-mediated pathway probably leads to upregulation of SOD in glutamate toxicity. Finally, O'NIELL et al. (1998) showed that D_2R agonists can protect from ischemia-induced hippocampal damage.

Work done in our laboratory also demonstrates that a D_2R-mediated protective pathway exists in kainate toxicity (BOZZI et al. 2000). It was shown that mice lacking the D_2R were much more sensitive to kainic acid (KA)-induced epileptic seizures. The $D_2R^{-/-}$ mice developed KA-induced seizures at doses that do not affect the WT mice. In addition, the knockout mice experience seizure-induced neurodegeneration in the dorsal CA3 region of the hippocampus. This death was shown to be apoptotic, accompanied by induction of c-*jun*, DNA fragmentation, and BAX activation in the dorsal CA3 region of the hippocampus of the $D_2R^{-/-}$ mouse. These data imply that a D_2R pathway normally acts to protect certain regions of the brain from elevated glutamate. Furthermore, these results have great implications for the treatment of neurodegenerative diseases, epilepsy, ischemia, and traumatic brain injury which all involve glutamate-induced cell death. Since DA is highly abundant in the CNS and can be both an antioxidant or contribute to oxidative stress, it is important to establish how the D_2R may regulate the function of DA under different circumstances.

H. Antiproliferative Role of Dopamine in the Pituitary

The D_2R is also strongly expressed in the anterior (AL) and intermediate lobes (IL) of the pituitary gland (BUNZOW et al. 1988; JACKSON and WESTLIND-DANIELSSON 1994; PICETTI et al. 1997), in addition to the CNS. More precisely, D_2Rs are highly expressed by the lactotroph and the melanotroph cells. Lactotrophs produce prolactin (PRL) (ELSHOLTZ et al. 1991; LEW et al. 1994; LEW and ELSHOLTZ 1995), while melanotrophs make α-melanocyte stimulating hormone (α-MSH) and β-endorphin (CHEN CL et al. 1983; COTE et al. 1986). The dopaminergic regulation on these cells has been shown to inhibit the synthesis and the release of these hormones. Both D_2L and D_2S receptors are co-expressed by these cells, and the ratio of D_2L/D_2S always favors D_2L expression. Interestingly, KUKSTAS and colleagues (KUKSTAS et al. 1991) reported the existence of two populations of lactotrophs with different D_2L/D_2S ratios that are dependent upon progesterone and testosterone regulation. This observation strongly suggests that D_2L and D_2S might have different functions *in vivo*, and that a fine regulation of the ratio of D_2L/D_2S might be critical in the control of their functions in the pituitary.

DA has been reported to control the rate of lactotroph proliferation, probably through the inhibition of the cAMP pathway (WEINER et al. 1988). In agreement with this, bromocriptine, a D_2R agonist, is successfully used to induce the regression of human lactotroph-derived pituitary tumors (prolactinomas) (BANSAL et al. 1981). This key control on lactotroph proliferation seems to be counteracted by estrogens. Indeed, estrogens positively regulate lactotroph proliferation after birth (LIEBERMAN et al. 1983; ELIAS and WEINER

1987). Estrogens may act at two levels, to stimulate the rate of transcription of the prolactin gene and to uncouple DA receptors. Thus DA and estrogens have opposing effects on lactotroph growth. Interestingly, mice lacking D_2R develop pituitary tumors of lactotroph origin (KELLY et al. 1997; SAIARDI et al. 1997). Prolactinomas in $D_2R^{-/-}$ mice were found in 100% of 1-year-old females, while they are present only in 3% of males (SAIARDI et al. 1997). This sexual dimorphism seems to be directly linked to the higher number of lactotrophs and the greater concentration of prolactin in female versus male mice. Indeed, in the absence of D_2R, female mutant mice present a robust rise in PRL levels (tenfold) in comparison with their WT controls (SAIARDI et al. 1997). The analysis of the expression of hypothalamic hormones such as thyrotropin-releasing hormone (TRH) and vasoactive intestinal peptide (VIP), known regulators of prolactin levels, showed that they were unaffected in $D_2R^{-/-}$ mice. One possible interpretation of these results is that prolactinomas in $D_2R^{-/-}$ mice are due to the lack of the inhibitory dopaminergic control on estrogen stimulation of these cells. However, we showed that $D_2R^{-/-}$ female mice are also hypoestrogenic. Estrogen levels in $D_2R^{-/-}$ females were as low as in males. These findings brought us to the conclusion that the pituitary tumors in $D_2R^{-/-}$ mice are specifically derived from the absence of dopaminergic control. One possible factor responsible for the formation of prolactinomas in $D_2R^{-/-}$ mice might be PRL. In this respect, we showed that PRL receptors are present on anterior pituitary cells. Thus, PRL may act as an autocrine modulator of lactotroph proliferation. A continuous stimulation of PRL receptors is mitogenic (MERSHON et al. 1995) and is likely to be the cause of pituitary tumors in the $D_2R^{-/-}$ mice. Therefore, the inhibitory dopaminergic tone is pivotal in the control of the rate of lactotroph proliferation by regulating PRL synthesis and release.

This hypothesis is supported by the findings that the DA transporter (DAT) knockout mice, in contrast to the $D_2R^{-/-}$ mice, show a hypoplasic pituitary. In these mice, a higher extracellular DA concentration has been reported, which may lead to an over-stimulation of D_2Rs in the pituitary. Abnormal stimulation of D_2Rs in the pituitary leads to decreased production of PRL and to a reduced number of lactotrophs. In $DAT^{-/-}$ mice, the number of somatotrophs is also reduced because the expression of the hypothalamic growth hormone-releasing hormone (GHRH) is concomitantly reduced (BOSSE et al. 1997).

The novel concept of DA as a regulator of cell growth and differentiation was also confirmed in the melanotrophs, another population of pituitary cells which express high levels of D_2Rs. $D_2R^{-/-}$ mice present a hypertrophy of the intermediate lobe, with a 40% increase in melanotroph number (SAIARDI and BORRELLI 1998). This leads to in an increase in proopiomelanocortin (POMC) transcripts (SAIARDI and BORRELLI 1998). Strikingly, the products of the cleavage of the POMC genes were altered in the $D_2R^{-/-}$ mice. POMC is expressed by two pituitary populations, the melanotroph of the intermediate lobe and the corticotrophs in the anterior lobe. The pro-hormone is normally processed by two convertases, PC1 and PC2, which are expressed in a cell-specific manner by the corticotrophs in the AL and the melanotrophs in the IL respectively. D_2R are only expressed by melanotrophs. POMC processing gives rise to adrenocorticotropin (ACTH) in corticotrophs and α-MSH and β-endorphin in melanotrophs. It was therefore surprising to observe a significant increase of the circulating levels of ACTH in $D_2R^{-/-}$ mice, despite an unaltered POMC expression in the corticotrophs. PC1 is strongly expressed in the corticotrophs, while PC2 is found in the melanotrophs (BLOOMQUIST et al. 1991; DAY et al. 1992). Interestingly, PC1 was abnormally upregulated in the melanotrophs of $D_2R^{-/-}$ mice (SAIARDI and BORRELLI 1998). PC1 upregulation results in an aberrant production of ACTH by melanotrophs in mutant mice. This result illustrates a novel function of DA in the control of pituitary cell identity. Thus, in the absence of dopaminergic control, melanotrophs produce ACTH. This leads to an aberrant control of ACTH production since POMC expression cannot be downregulated by glucocorticoids in the melanotrophs as it is in corticotrophs. ACTH overproduction causes an over-stimulation of the adrenal gland, a higher production of glucocorticoids, and a hypertrophy of the adrenal cortex, which leads to a phenotype that resembles Cushing's syndrome (SAIARDI and BORRELLI 1998). This result establishes an unprecedented link between the dopaminergic system and the etiology of Cushing's-like syndrome.

I. Genetic Association of the D_2 Receptor with Disease

D_2R antagonists are commonly used to treat schizophrenia and other diseases associated with psychosis, while D_2R agonists act as one of the major treatments for PD. The use of D_2 antagonists and agonists as treatments for disorders which exhibit a range of symptoms from psychotic episodes and personality impairments to loss of motor function and memory was an initial clue that the D_2R is a key player in coordinating many systems of the nervous system. This section of the review will focus on the major disorders involving the D_2R and its genetic association with these diseases. We will review only the work that has been done in the past few years as there have been many polymorphisms located in the D_2R gene and its promoter region, but few genetic links to any of the diseases or disorders.

I. Schizophrenia

It is clear that genetic factors contribute to the development of schizophrenia. However, it is a complex disease, which is also affected by environmental influences (KARAYIORGON and GOGOS 1997). For this reason, no genetic factors have been linked to schizophrenia to date (TERENIUS 2000). The DA hypothesis of schizophrenia speculates that DA systems in schizophrenic patients are overactive (VAN ROSSUM 1966) since all of the clinically effective antipsychotic drugs are D_2-like receptor antagonists (EMILIEN et al. 1999). Therefore, it is clear that the D_2R is important for understanding the molecular basis of the disease. Many labs have focused on the possible genetic associations between the D_2R and schizophrenia.

Both a Japanese and a Swedish group of patients were examined for the –141C Del/Ins D_2R polymorphism. In these patients the –141C Del allele frequency was significantly lower in the patients versus the controls (ARINAMI et al. 1997; JÖNSSON et al. 1999). However, contrary data were recently found in a group of British and Scottish schizophrenics who had higher levels of the –141C Del allele than the controls (BREEN et al. 1999).

Because a solid genetic link has not been established between the D_2R and schizophrenia, other genetic connections are being explored. For example, SEEMAN and colleagues looked for variations in D_2R RNA to find genetic alterations in its splice products. A unique splice site in the D_2L receptor was identified (SEEMAN et al. 2000), although no link was established between this splice variant and psychosis. Such studies could be useful in trying to understand how the D_2R may be involved in schizophrenia. Striatal D_2R DA binding is elevated in many schizophrenic patients (SEEMAN 1992; SOARS and INNIS 1999; PRINCE et al. 2000; ABI-DARGHAM et al. 2000). Exciting new work done by ABI-DARGHAM and colleagues used an innovative technique that demonstrates that schizophrenic patients have both an increase in the amount of striatal DA as well as in the quantity of striatal D_2Rs (ABI-DARGHAM et al. 2000). Perhaps a genetic factor, such as the splice variant mentioned above affects D_2R expression or its affinity for DA and increases the risk of developing schizophrenia. However, at this time there remain no clear data that genetically link the D_2R to schizophrenia.

II. Alcoholism

Two D_2R polymorphisms, *Taq*A1 and the –141C Ins, have been observed with respect to alcoholism with varying results. Some studies have shown an association while others were not able to demonstrate a link between the D_2R and alcoholism (ISHIGURO et al. 1998; LOBOS and TODD 1998; HILL et al. 1999a,b; SANDER et al. 1999). An alternate way that the D_2R may affect one's vulnerability to alcoholism is through an association with personality traits which alcoholics commonly exhibit. In one such study, HILL et al. (2000) compared personality traits between alcoholics and control subjects using the multidimensional personality questionnaire (MPQ) and found that the MPQ traits may be linked to both a D_2R polymorphism as well as to a D_4 receptor loci.

III. Parkinson's Disease

Since D_2R agonists are commonly used with levodopa to treat PD, D_2R has been examined for possible genetic linkage to the disease. One recent paper linked the disease with polymorphisms at the A1 allele (*Taq*A1) and the B1 allele (*Taq*B1), demonstrating that these two polymorphisms were more frequent in PD patients than in age-matched controls. Thus, people possessing these two polymorphisms may have an increased risk of developing PD (OLIVERI et al. 2000). Similar to the research on genetic factors in schizophrenia, many groups have failed to show any association between the D_2R and PD using various polymorphisms (COMINGS et al. 1991; NANKO et al. 1994; PASTOR et al. 1999; OLIVERI et al. 2000).

Alternatively, the relationship between D_2R polymorphisms and the likelihood of developing hallucinations following long-term treatment with levodopa and DA agonists is being explored. MAKOFF et al. (2000) demonstrated that the C allele of the *Taq*A1 polymorphism is associated with late-onset hallucinations as a result of chronic treatment with levodopa and DA agonists. Another group explored a short tandem repeat polymorphism of the D_2R gene. They showed that the 15 allele of the polymorphism is increased in PD patients and that the 13 and 14 alleles are increased in nondyskinetic PD patients versus ones that develop dyskinesia as a side effect of levodopa therapy. If a PD patient carries at least one of these two alleles, they have a reduced risk of developing dyskinesia (OLIVERI et al. 1999). While no strong genetic linkage shows that polymorphisms in the D_2R are responsible for genetic predisposition to PD, agonists to the receptor remain highly effective in treating PD. As a result, it will be imperative to continue to search for mutations associated with the D_2R's function to identify possible genetic links to PD. This type of research will ultimately provide clues to how the disease occurs. Two genes have been clearly linked to inheritable forms of PD, alphasynuclein and parkin (POLYMEROPOULOS et al. 1997; KITADA et al. 1998), yet to date no relationship has been demonstrated between these genes and the D_2R.

J. Distinct Functions of the Dopamine D_2 Receptor Isoforms

Before concluding this review on the dopamine D_2R and studies that utilize the analysis of genetically engineered animals, we would like to add a short paragraph on very recent results (USIELLO et al. 2000).

The presence of two isoforms for D_2R prompted us to generate knockout animals for the D_2L isoform. This was achieved through the deletion of exon 6 from the D_2R gene. These animals are viable and reproduce normally (USIELLO et al. 2000). The analysis of the D_2-specific binding capacities of these mice revealed a normal amount of D_2-specific binding sites. This indicated that in the absence of exon 6, D_2R gene transcription proceeds by default, generating animals that express an equal amount of D_2-binding sites which, in this case, will be formed only by D_2S. A neurochemical analysis performed by in vivo microdialysis of $D_2L^{-/-}$ mice indicated that autoreceptor functions are maintained in these animals. This finding was further supported by behavioral evidence showing that low amounts of the D_2R-specific agonist quinpirole

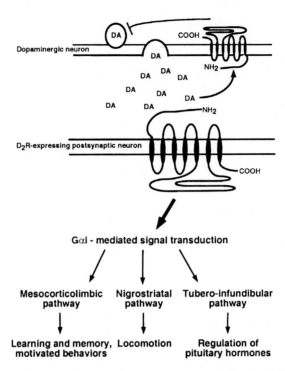

Fig. 4. Scheme of the events known to take place upon D_2R stimulation in the CNS. Dopaminergic neurons contain D_2 autoreceptors which regulate the amount of DA released from the pre-synaptic cell into the synapse. Once DA is released in the synapse, it binds to both the pre- and postsynaptic D_2Rs. In the D_2R-containing postsynaptic neuron, a $G\alpha_i$-mediated signal is induced which activates the mesocorticolimbic, nigrostriatal, and tubero-infundibular pathways. These pathways are involved in regulating learning and memory, motivated behaviors, locomotion, and the regulation of pituitary hormones respectively

induced the well-characterized sedative effect on locomotion. Furthermore, at the electrophysiological level, autoreceptor functions are also present in $D_2L^{-/-}$ mice (WANG et al. 2000). Interestingly, when D_2R-mediated postsynaptic effects of D_2-specific agonists and antagonists were tested, very reduced or absent responses were observed. In particular, the cataleptic effect of haloperidol was absent in D_2L-null mice. These results indicate that D_2L serves mainly postsynaptic functions, while D_2S presynaptic ones (USIELLO et al. 2000).

In addition, loss of D_2L revealed a functional antagonism between D_2S and D_1 receptor-mediated functions. Indeed, the behavioral response of $D_2L^{-/-}$ mice to D_1-specific agonists was highly reduced (USIELLO et al. 2000). Thus, the generation of D_2L-null mice has finally clarified that D_2L and D_2S have different functions *in vivo* and they should not be considered as functionally redundant molecules.

K. Conclusion

A general scheme of the known functions of the D_2R has been outlined in this review (Fig. 4). Great strides in the understanding of the signaling and physiological functions of D_2Rs have been made since the cloning of this receptor. The advances made to date underscore the importance of combining molecular, cellular, biochemical, pharmacological, and behavioral approaches in the study of protein function. By understanding what is occurring on a cellular level and by observing the behavioral response to these cellular and molecular events, it will be easier to untangle the many signals dictated by D_2R activation. The knockout approach has allowed, and will allow in the future, a dissection of the cellular and molecular mechanisms underlying D_2-mediated responses to DA. Molecular neuroscience is a promising and exciting area that may lead to great advances in the treatments of both psychiatric disorders such as schizophrenia as well as neurodegenerative diseases such as PD.

References

Abi-Dargham A, Rodenhiser, J, Printz D, Zea-Ponce Y, Gil R, Kegeles LS, Weiss R, Cooper TB, Mann JJ, Van Heertum RL, Gorman JM, Laruelle M (2000) Increased baseline occupancy of D_2 receptors by dopamine in schizophrenia. PNAS 97(14): 8104–8109

Aizman O, Brismar H, Uhlén P, Zettergren E, Levey AI, Forssberg H, Greengard P, Aperia A (2000) Anatomical and physiological evidence for D_1 and D_2 receptor colocalization in neostriatal neurons. Nat Neurosci 3 (3):226–230

Alexander, GE, Crutcher, MD (1990) Functional architecture of basal ganglia circuits: neural substrates of parallel processing. Trends in Neurosciences 13(7):266–271

Aoyama S, Kase H, Borrelli E (2000) Rescue of Locomotor Impairment in Dopamine D_2 Receptor-Deficient Mice by an Adenosine A2 A Receptor Antagonist. J Neurosci 20(15):5848–5852

Arinami T, Gao M, Hamaguchi H, Toru M (1997) A functional polymorphism in the promoter region of the dopamine D_2 receptor gene is associated with schizophrenia. Hum Mol Genet 6(4):577–582

Baik JA, Picetti R, Saiardi A, Thiriet G, Dierich A, Depaulis A, Le Meur M, Borrelli E (1995) Parkinsonian-like locomotor impairment in mice lacking dopamine D_2 receptor. Nature 377:424–428

Bansal S, Lee LA, Woolf PD (1981) Abnormal prolactin responsivity to dopaminergic suppression in hyperprolactinemic patients. Am J Med 71:961–970

Bloomquist BT, Eipper BA, Mains RE (1991) Prohormone-converting enzyme: regulation and evaluation of function using antisense RNA. Mol Endocrinol 5:2014–2024

Bosse R, Fumagalli F, Jaber M, Giros B, Gainetdinov RR, Wetsel WC, Missale C, Caron MG (1997) Anterior pituitary hypoplasia and dwarfism in mice lacking the dopamine transporter. Neuron 19:127–138

Bozzi Y, Vallone D, Borrelli E (2000) Neuroprotective Role of Dopamine Against Hippocampal Cell Death. J Neurosci 20:8643–8649

Breen G, Brown J, Maude S, Fox H, Collier D, Li T, Arranz M, Shaw D, St. Clair D (1999) -141C del/ins polymorphism of the dopamine receptor 2 gene is associated with schizophrenia in a British population. Am J Med Genet 88(4):407–410

Bunzow JR, Van Tol HH, Grandy DK, Albert P, Salon J, Christie M, Machida CA, Neve KA, Civelli O (1988) Cloning and expression of a rat D_2 dopamine receptor cDNA. Nature 336:783–787

Burt DR, Creese I, Snyder SH (1976) Properties of [³H]haloperidol and [³H]dopamine binding associated with dopamine receptors in calf brain membranes. Molecular Pharmacology 12(5):800–812

Cai G, Zhen X, Uryu K, Friedman E (2000). Activation of response to D_2 dopamine receptor stimulation in unilateral 6-hydroxydopamine – lesioned rats. J Neurosci 20(5):1849–1857

Callahan PM, De la Garza R 2nd, Cunningham KA (1994) Discriminative stimulus properties of cocaine: modulation by dopamine D1 receptors in the nucleus accumbens. Psychopharmacology (Berl) 115(1–2):110–114

Cha XY, Pierce RC, Kalivas PW, Mackler SA (1997) NAC-1, a rat brain mRNA, is increased in the nucleus accumbens three weeks after chronic cocaine self-administration. J Neurosci Sep 15 17(18):6864–6871

Chase TN, Woods AC, Glaubinger GA (1974) Parkinson disease treated with a suspected dopamine receptor agonist. Arch Neurol 30:383–386

Chen CL, Dionne FT, Roberts JL (1983) Regulation of the pro-opiomelanocortin mRNA levels in rat pituitary by dopaminergic compounds. Proc Natl Acad Sci USA 80:2211–2215

Clifford JJ, Waddington JL (1998) Heterogeneity of behavioural profile between three new putative selective D3 dopamine receptor antagonists using an ethologically-based approach. Psychopharmacology 136:284–290

Clifford JJ, Usiello A, Vallone D, Kinsella A, Borrelli E, Waddington JL (2000) Topographical evaluation of behavioural phenotype in a line of mice with targeted gene deletion of D_2 dopamine receptor. Neuropharmacology 39:382–390

Colgan PW (1978) Quantitative ethology. Wiley, New York

Comings DE, Comings BG, Muhleman D, Dietz G, Shahbahrami B, Tast D, Knell E, Kocsis P, Baumgarten R, Kovacs BW, Levy DL, Smith M, Borison RL, Evans D, Klein LM, MacMurray J, Tosk JM, Sverd G, Gysin R, Flanagan SD (1991) The dopamine D2 receptor locus as a modifying gene in neuropsychiatric disorders. Jama 266(13):1793–1800

Cote TE, Felder R, Kebabian JW, Sekura RD, Reisine T, Affolter HU (1986) D-2 dopamine receptor-mediated inhibition of pro-opiomelanocortin synthesis in rat intermediate lobe. Abolition by pertussis toxin or activators of adenylate cyclase. J Biol Chem 261:4555–4561

Crabbe CJ, Belknap JK, Buck KJ (1994) Genetic animal model of alcohol and drug abuse. Science 264:1715–1724

Daunais JB, McGinty JF (1996) The effects of D_1 or D_2 dopamine receptor blockade on zif/268 and preprodynorphin gene expression in rat forebrain following a short-term cocaine binge. Brain Res Mol Brain Res Jan 35(1–2):237–248

Day RN, Koike S, Sakai M, Muramatsu M, Maurer RA (1990) Both Pit-1 and the estrogen receptor are required for estrogen responsiveness of the rat prolactin gene. Molecular Endocrinology 4(12):1964–1971

Day R, Schafer MK, Watson SJ, Chretien M, Seidah NG (1992) Distribution and regulation of the prohormone convertases PC1 and PC2 in the rat pituitary. Mol Endocrinol 6:485–497

Dickinson SD, Sabeti J, Larson GA, Giardina K, Rubinstein M, Kelly MA, Grandy DK, Low MJ, Gerhardt GA, Zahniser NR (1999) Dopamine D2 receptor-deficient mice exhibit decreased dopamine transporter function but no changes in dopamine release in dorsal striatum. Journal of Neurochemistry 72(1):148–156

Elias KA, Weiner RI (1987) Inhibition of estrogen-induced anterior pituitary enlargement and arteriogenesis by bromocriptine in Fisher 344 rats. Endocrinology 120:617–621

Elsholtz HP, Lew AM, Albert PR, Sundmark VC (1991) Inhibitory control of prolactin and Pit-1 gene promoters by dopamine. Dual signaling pathways required for D_2 receptor-regulated expression of the prolactin gene. J Biol Chem 266:22919–22925

Emilien G, Maloteaux JM, Geurts M, Hoogenberg K, Cragg S (1999) Dopamine receptors-physiological understanding to therapeutic intervention potential. Pharmacol Ther 84(2):133–156

Fishburn CS, Elazar Z, Fuchs S (1995) Differential glycosylation and intracellular trafficking for the long and short isoforms of the D2 dopamine receptor. Journal of Biological Chemistry 270(50):29819–29824

Fuxe K, Ferré S, Zoli M, Agnati LF (1998) Integrated events in central dopamine transmission as analysed at multiple levels. Evidence for intramembrane adenosine A_{2A}/dopamine D_2 and adenosine A_1/dopamine D_1 receptor interactions in the basal ganglia. Brain Res Rev 26: 258–273

Gerfen CR, Engber TM, Mahan LC, Susel Z, Chase TN, Monsma FJ Jr, Sibley DR (1990) D1 and D2 dopamine receptor-regulated gene expression of striatonigral and striatopallidal neurons. Science 250(4986):1429–1432

Graybiel AM, Moratalla R, Robertson HA (1990) Amphetamine and cocaine induce drug-specific activation of the c-fos gene in striosome-matrix compartments and limbic subdivisions of the striatum. Proc Natl Acad Sci USA Sep 87(17):6912–6916

Guivarch D, Vernier P, Vincent J-D (1995) Sex steroid hormones change the differential distribution of the isoforms of the D_2 dopamine receptor messenger RNA in the rat brain. Neuroscience 69:159–166

Helms JB (1995) Role of heterotrimeric GTP binding proteins in vesicular protein transport: Indications for both classical and alternative G proteins cycles. FEBS Lett 369:84–88

Hill SY, Zezza N, Wipprecht G, Locke J, Neiswanger K (1999a) Personality traits and dopamine receptors (D_2 and D_4): linkage studies in families of alcoholics. Am J Med Genet 88(6):634–641

Hill SY, Zezza N, Wipprecht G, Xu J, Neiswanger K (1999b) Linkage studies of D_2 and D4 receptor genes and alcoholism. Am J Med Genet 88(6):676–685

Hill SY, Zezza, N, Wipprecht G, Locke J, Neiswanger K (2000) Personality traits and dopamine receptors (D2 and D4): linkage studies in families of alcoholics. Am J Med Genet 88(6):634–641

Hope B, Kosofsky B, Hyman SE, Nestler EJ (1992) Regulation of immediate early gene expression and AP-1 binding in the rat nucleus accumbens by chronic cocaine. Proc Natl Acad Sci USA Jul 1 89(13):5764–5768

Hornykievicz O (1973) Dopamine in the basal ganglia. Its role and therapeutic implications. Br Med Bull 29:172–178

Iida M, Miyazaki I, Tanaka K, Kabuto H, Iwata-Ichikawa E, Ogawa N (1999) Dopamine D_2 receptor-mediated antioxidant and neuroprotective effects of ropinirole, a dopamine agonist. Brain Res 838(1–2):51–59

Ishiguro H, Arinami T, Saito T, Akazawa S, Enomoto M, Mitushio H, Fujishiro H, Tada K, Akimoto Y, Mifune H, Shioduka S, Hamaguchi H, Toru M, Shibuya H (1998) Association study between the –141C Ins/Del and *Taq*I A polymorphisms of the dopamine D_2 receptor gene and alcoholism. Alcohol Clin Exp Res 22(4):845–848

Jackson DM, Westlind-Danielsson A (1994) Dopamine receptors: molecular biology, biochemistry and behavioral aspects. Pharmacol Ther 64:291–370

Jones DT, Reed RR (1987) Molecular cloning of five GTP-binding protein cDNA species from rat olfactory neuroepithelium. J Biol Chem 262:14241–14249

Jönsson EG, Nöthen MM, Neidt H, Forslund K, Rylander G, Mattila-Evenden M, Asberg M, Propping P, Sedvall GC (1999) Association between a promoter polymorphism in the dopamine D_2 receptor gene and schizophrenia. Schizophr Res 40(1):31–36

Jung MY, Skryabin BV, Arai M, Abbondanzo S, Fu D, Brosius J, Robakis NK, Polites HG, Pintar JE, Schmauss C (1999) Potentiation of the D_2 mutant motor phenotype in mice lacking dopamine D_2 and D_3 receptors. Neuroscience 91:911–924

Kang MY, Tsuchiya M, Packer L, Manabe M (1998) In vitro study on antioxidant potential of various drugs used in the perioperative period. Acta Anaesthesiologica Scandinavica 42(1):4–12

Kapur S, Seeman P (2000) Antipsychotic agents differ in how fast they come off the dopamine D2 receptors. Implications for atypical antipsychotic action. J Psychiatry Neurosci 25(2):161–166

Karayiorgou M, Gogos JA (1997) A turning point in schizophrenia genetics. Neuron 19(5):967–979

Kebabian JW, Calne DB (1979) Multiple receptor for dopamine Nature 277:93–96

Kelly MA, Rubinstein M, Asa SL, Zhang G, Saez C, Bunzow JR Allen RG, Hnasko R, Ben-Jonathan N, Grandy DK, Low MJ, (1997) Pituitary lactotroph hyperplasia and chronic hyperprolactinemia in D_2 receptor-deficient mice. Neuron 19:103–113

Kelly MA, Rubinstein M, Phillips TJ, Lessov CN, Burkhart-Kasch S, Zhang G, Bunzow JR, Fang Y, Gerhardt A, Grandy DK, Low MJ (1998) Locomotor activity in D_2 dopamine receptor-deficient mice is determined by gene dosage, genetic background and developmental adaptations. J Neurosci 18:3470–3479

Khan ZU, Mrzljak L, Gutierrez A, De La Calle A, and Goldman-Rakic PS (1998). Prominence of the dopamine D_2 short isoform in dopaminergic pathways. Proc Natl Acad Sci USA 95:7731–7736

Kitada T, Asakawa S, Hattori N, Matsumine H, Yamamura Y, Minoshima S, Yokochi M, Mizuno Y, Shimizu N (1998) Mutations in the parkin gene cause autosomal recessive juvenile parkinsonism. Nature 392:605–608

Kobilka BK, Kobilka TS, Daniel K, Regan JW, Caron MG, Lefkowitz RJ (1988) Chimeric alpha 2-,beta 2-adrenergic receptors: delineation of domains involved in effector coupling and ligand binding specificity. Science 240(4857):1310–1316

Koeltzow TE, Xu M, Cooper DC, Hu XT, Tonegawa S, Wolf ME, White FJ (1998) Alterations in dopamine release but not dopamine autoreceptor function in dopamine D3 receptor mutant mice. J Neurosci 18(6):2231–2238

Konradi C, Cole RL, Heckers S, Hyman SE (1994) Amphetamine regulates gene expression in rat striatum via transcription factor CREB. J Neurosci Sep;14(9): 5623–5634

Krezel W, Ghyselinck N, Samad TA, Dupé V, Kastner P, Borrelli E, Chambon P (1998) Impaired locomotion and dopamine signaling in retinoid receptor mutant mice. Science 279(5352):863–867

Kukstas LA, Domec C, Bascles L, Bonnet J, Verrier D, Israel JM Vincent JD (1991) Different expression of the two dopaminergic D_2 receptors, D_2 415 and D_2 444, in two types of lactotroph each characterized by their response to dopamine, and modification of expression by sex steroids. Endocrinology 129:1101–1103

Lachowicz JE, Sibley DR (1997) Molecular characteristics of mammalian dopamine receptors. Pharmacol Toxicol 81(3):105–113

Lee SP, O'Dowd BF, Ng GYK, Varghese G, Akil H, Mansour A, Nguyen T, George SR (2000) Inhibition of cell surface expression by mutant receptors demonstrates that D2 dopamine receptors exist as oligomers in the cell. Mol Pharmacol 58(1):120–128

Lefkowitz RJ, Caron MG (1988) Adrenergic receptors. Models for the study of receptors coupled to guanine nucleotide regulatory proteins. J Biol Chem 263(11): 4993–4996

Le Moal M, Simon H (1991) Mesocorticolimbic dopaminergic network: functional and regulatory roles. Physiol Rev 71:155–234

Lew AM, Yao H, Elsoltz HP (1994) G(i) alpha 2- and G(o) alpha-mediated signaling in the pit-1-dependent inhibition of the prolactin gene promoter. Control of transcription by dopamine D_2 receptors. J Biol Chem 269:12007–12013

Lew AM, Elsholtz HP (1995) A dopamine-responsive domain in the N-terminal sequence of Pit-1. Transcriptional inhibition in endocrine cell types. J Biol Chem 1270:7156–7160

L'hirondel M, Chéramy A, Godeheu G, Artaud F, Saiardi A, Borrelli E, Glowinski J (1998) Lack of autoreceptor-mediated inhibitory control of dopamine release in striatal synaptosomes of D2 receptor-deficient mice. Brain Res 792(2):253–262

Lieberman ME, Slabaugh MB, Rutledge JJ, Gorski J (1983) Steroids and differentiation. The role of estrogen in the differentiation of prolactin producing cells. J Steroid Biochem 19(1A):275–281

Lobos EA, Todd RD (1998) Association analysis in an evolutionary context: cladistic analysis of the DR D_2 locus to test for association with alcoholism. Am J Med Genet 81(5):411–419

Makoff AJ, Graham JM, Arranz MJ, Forsyth J, Li T, Aitchison KJ, Shaikh S, Grünewald RA (2000) Association study of dopamine receptor gene polymorphisms with drug-induced hallucinations in patients with idiopathic Parkinson's disease. Pharmacogenetics 10(1):43–48

Maldonado R, Saiardi A, Valverde O, Samad TA, Roques BP, Borrelli E (1997) Absence of opiate rewarding effects in mice lacking dopamine D_2 receptors. Nature Aug 7;388(6642):586–589

Malek D, Munch G, Palm D (1993) Two sites in the third inner loop of the dopamine D_2 receptor are involved in functional G protein-mediated coupling to adenylate cyclase. FEBS Lett 325:215–219

Mercuri NB, Saiardi A, Bonci A, Picetti R, Calabresi P, Bernardi G, Borrelli E (1997) Loss of autoreceptor function in dopaminergic neurons from dopamine D2 receptor deficient mice. Neuroscience 79(2):323–327

Mershon J, Sall W, Mitchner N, Ben-Jonathan N (1995) Prolactin is a local growth factor in rat mammary tumors. Endocrinology 136:3619–3623

Missale C, Nash SR, Robinson SW, Jaber M, Caron MG (1998) Dopamine receptors: from structure to function. Physiol Rev 78(1):189–225

Montmayeur J-P, Bausero P, Amalaiky N, Maroteaux L, Hen R, Borrelli E (1991) Differential expression of the mouse D_2 dopamine receptor isoforms. FEBS Lett 278:239–243

Montmayeur J-P, Guiramand J, Borrelli E (1993) Preferential coupling between dopamine D_2 receptors and G-proteins. Mol Endo 7(2):161–170

Montmayeur JP, Borrelli E (1994) Targeting of G alpha i2 to the Golgi by alternative spliced carboxyl-terminal region. Science 263(5143):95–98

Nanko S, Ueki A, Hattori M, Dai XY, Sasaki T, Fukuda R, Ikeda K, Kazamatsuri H (1994) No allelic association between Parkinson's disease and dopamine D2, D3, and D4 receptor gene polymorphisms. Am J Med Genet 54(4):361–364

Oliveri RL, Annesi G, Zappia M, Civitelli D, Montesanti R, Branca D, Nicoletti G, Spadafora P, Pasqua AA, Cittadella R, Andreoli V, Gambardella A, Aguglia U, Quattrone A (1999) Dopamine D_2 receptor gene polymorphism and the risk of levodopa-induced dyskinesias in PD. Neurology 53(7):1425–1430

Oliveri RL, Annesi G, Zappia M, Civitelli D, De Marco EV, Pasqua AA, Annesi F, Spadafora P, Gambardella A, Nicoletti G, Branca D, Caracciolo M, Aguglia U, Quattrone A (2000) The dopamine D_2 receptor gene is a susceptibility locus for Parkinson's disease. Mov Disord 15(1):127–131

Olsen RW, Reisine TC, Yamamura HI (1980) Neurotrasmitter receptors-biochemistry and alterations in neuropsychiatric disorders. Life Science 27:801–808

O'Neill MJ, Hicks CA, Ward MA, Cardwell GP, Reymann JM, Allain H, Bentué-Ferrer D (1998) Dopamine D_2 receptor agonists protect against ischaemia-induced hippocampal neurodegeneration in global cerebral ischaemia. Eur J Pharmacol 352(1):37–46

Pastor P, Muñoz E, Obach V, Martí MJ, Blesa R, Oliva R, Tolosa E (1999) Dopamine receptor D_2 intronic polymorphism in patients with Parkinson's disease. Neurosci Lett 273(3):151–154

Picetti R, Saiardi A, Samad TA, Bozzi Y, Baik J-H, Borrelli E (1997) Dopamine D_2 Receptors in signal transduction and behavior. Crit Rev Neurobiol 11 (2&3):121–142

Picetti R, Borrelli E (2000) A region containing a proline rich motif targets sGi2 to the Golgi apparatus. Exp Cell Res 255:258–269

Phillips TJ, Brown KJ, Burkhart-Kasch S, Wenger CD, Kelly MA, Rubinstein M, Grandy DK, Low MJ (1998) Alcohol preference and sensitivity are markedly reduced in mice lacking dopamine D_2 receptors. Nat Neurosci Nov;1(7):610–615

Pimplikar SW, Simons K (1993) Regulation of apical transport in epithelial cells by a Gs class of heterotrimeric G protein. Nature 362:456–458

Polymeropoulos M, Lavedan C, Leroy E, Ide S, Dehejia A, Dutra A, Pike B, Root H, Rubenstein J, Boyer R, Stenroos E, Chandrasekharappa S, Athanassiadou A, Papapetropoulos T, Johnson W, Lazzarini A, Duvoisin R, Di Iorio G, Golbe L, Nussbaum R (1997) Mutation in the alpha-synuclein gene identified in families with Parkinson's disese. Science 276:2045–2047

Prince JA, Harro J, Blennow K, Gottfries CG, Oreland L (2000) Putamen mitochondrial energy metabolism is highly correlated to emotional and intellectual impairment in schizophrenics. Neuropsychopharmacology 22(3):284–292

Rhodes SJ, Chen R, DiMattia GE, Scully KM, Kalla KA, Lin SC, Yu VC, Rosenfeld MG (1993) A tissue-specific enhancer confers Pit-1-dependent morphogen inducibility and autoregulation on the pit-1 gene. Gene Dev 7(6):913–932

Rocheville M, Lange DC, Kumar U, Patel SC, Patel RC, Patel YC (2000) Receptors for dopamine and somatostatin: Formation of hetero-oligomers with enhanced functional activity. Science 288:154–157

Rubinstein M, Phillips TJ, Bunzow JR, Falzone TL, Dziewczapolski G, Zhang G, Fang Y, Larson JL, McDougall JA, Chester JA, Saez C, Pugsley TA, Gershanik O, Low MJ, Grandy DK (1997) Mice lacking dopamine D4 receptors are supersensitive to ethanol, cocaine, and methamphetamine. Cell Sep 19 90(6):991–1001

Saiardi A, Bozzi Y, Baik JH, Borrelli E (1997) Antiproliferative role of Dopamine: loss of D_2 receptors causes hormonal dysfunction and pituitary hyperplasia. Neuron 19:115–126

Saiardi A, Borrelli E (1998) Absence of dopaminergic control on melanotrophs leads to Cushing's-like syndrome in mice. Mol Endocrinol 12:1133–1139

Samad TA, Krezel W, Chambon P, Borrelli E (1997) Regulation of dopaminergic pathways by retinoids: activation of the D2 receptor promoter by members of the retinoic acid receptor-retinoid X receptor family. PNAS 94(26):14349–14354

Sam EE, Verbeke N (1995) Free radical scavenging properties of apomorphine enantiomers and dopamine: possible implication in their mechanism of action in parkinsonism. Journal of Neural Transmission. Parkinsons Disease and Dementia Section 10(2–3):115–127

Sander T, Ladehoff M, Samochowiec J, Finckh U, Rommelspacher H, Schmidt LG (1999) Lack of an allelic association between polymorphisms of the dopamine D_2 receptor gene and alcohol dependence in the German population. Alcohol Clin Exp Res 23(4):578–581

Sawada H, Ibi M, Kihara T, Urushitani M, Akaike A, Kimura J, Shimohama S (1998) Dopamine D2-type agonists protect mesencephalic neurons from glutamate neurotoxicity: mechanisms of neuroprotective treatment against oxidative stress. Ann Neurol 44(1):110–119

Seeman P, Chau-Wong M, Tedesco J, Wong K (1975) Brain receptors for antipsychotic drugs and dopamine: direct binding assays. PNAS 72(11):4376–4380

Seeman P (1980) Brain dopamine receptors. Pharmacol Rev 32:229–313

Seeman P (1992) Dopamine receptor sequences. Therapeutic levels of neuroleptics occupy D_2 receptors, clozapine occupies D4. Neurosychopharmacology 7:261–284

Seeman P, Nam D, Ulpian C, Liu IS, Tallerico T (2000) New dopamine receptor, D_2 (Longer), with unique TG splice site, in human brain. Brain Res Mol Brain Res 76(1):132–141

Simon ML, Strathmann MP, Gautam N (1991) Diversity of G proteins in signal transduction. Science 252: 802–808

Smythies J (1999) Redox mechanisms at the glutamate synapse and their significance: a review. Eur J Pharmacol 370(1):1–7

Soares JC, Innis RB (1999) Neurochemical brain imaging investigations of schizophrenia. Biol Psychiatry 46(5):600–615

Surmeier, DJ; Song, WJ; Yan, Z. (1996) Coordinated expression of dopamine receptors in neostriatal medium spiny neurons. J Neurosci 16(20):6579–6591

Terenius L (2000) Schizophrenia: pathophysiological mechanisms–a synthesis. Brain Res Brain Res Rev 31(2–3):401–404

Usiello A, Baik, JH, Rouge-Pont F, Picetti R, Dierich A, LeMeur M, Piazza PV, Borrelli, E (2000) Distinct functions of the two isoforms of dopamine D2 receptors. Nature 408(6809):199–203

Van Kammen DP (1979) The dopamine hypothesis of schizophrenia revisited. Psychoneuroendocrinology 4:34–46

Van Rossum J (1966) The significance of dopamine-receptor blockade for the mechanism of action of neuroleptic drugs. Arch Int Paharmacodyn Ther 160:492–494

Waddington J (1993) $D_1:D_2$ Dopamine receptor interactions. Academic press, London San Diego New York Boston Sydney Tokyo Toronto

Wang Y, Xu R, Sasaoka T, Tonegawa S, Kung M, Sankoorikal E (2000) Dopamine D2 long receptor-deficient Mice display alterations in striatum-dependent functions. J Neurosci 20(22):8305–8314

Weiner RI, Findell PR, Kordon C (1988) Role of classic and peptide neuromediators in the neuroendocrine regulation of LH and prolactin. In: Knobil E, Neill J, 1rst ed. The Physiology of Reproduction. New York: Ravel Press, Ltd., 1235–1281

Woolverton WL, Kleven MS (1988) Multiple dopamine receptors and the behavioral effects of cocaine. NIDA Res Monogr 88:160–184

Wynshaw-Boris A (1996) Model mice and human disease. Nat Genet 13:259–260

Xu M, Hu XT, Cooper DC, Moratalla R, Graybiel AM, White FJ, Tonegawa S (1994) Elimination of cocaine-induced hyperactivity and dopamine-mediated neurophysiological effects in dopamine D1 receptor mutant mice. Cell Dec 16 79(6):945–955

Xu M, Koeltzow TE, Santiago GT, Moratalla R, Cooper DC, Hu XT, White NM, Graybiel AM, White FJ, Tonegawa S (1997) Dopamine D3 receptor mutant mice exhibit increased behavioral sensitivity to concurrent stimulation of D_1 and D_2 receptors. Neuron Oct 19(4):837–848

Yan Z, Feng J, Fienberg AA, Greengard P (1999). D_2 dopamine receptors induce mitogen-activated protein kinase and cAMP response element-binding protein phosphorylation in neurons. Proc Natl Acad Sci USA 96: 11607–11612

Yen GC, Hsieh CL (1997) Antioxidant effects of dopamine and related compounds. Bioscience, Biotechnology, and Biochemistry 61(10):1646–1649

Young ST, Porrino LJ, Iadarola MJ (1991) Cocaine induces striatal c-fos-immunoreactive proteins via dopaminergic D1 receptors. Proc Natl Acad Sci USA Feb 15 88(4):1291–1295

CHAPTER 7

The Dopamine D_3 Receptor and Its Implication in Neuropsychiatric Disorders and Their Treatments

P. SOKOLOFF and J.-C. SCHWARTZ

A. Introduction

The pleiotropic actions of dopamine, as well as of drugs used in the treatment of Parkinson's disease and schizophrenia, have long been assumed to result from interaction with only two dopamine receptors termed D_1 and D_2. In spite of previous suggestions of additional dopamine receptors, the discovery of the D_3 receptor (SOKOLOFF et al. 1990) was rather unexpected, as were those of the D_4 and D_5 receptors that followed (SUNAHARA et al. 1991; VAN TOL et al. 1991). From the beginning, attention has been attracted by the restricted distribution of the D_3 receptor in the brain, seemingly related to functions of dopamine associated with the limbic brain. Nevertheless, the initial lack of evidence of functional coupling of this receptor, as well as of selective pharmacological tools to investigate its functions, raised questions about its physiological significance.

The present chapter aims at measuring the progress accomplished a decade after its identification and reviews some anatomical, pharmacological, and genetic data currently available that now allow us to unravel functions mediated by the D_3 receptor and its possible implications in several neuropsychiatric disorders.

B. Intracellular Signaling of the D_3 Receptor

Early attempts to establish an efficient functional model has proved a difficult task in the case of the D_3 receptor. The D_3 receptor heterologously expressed in Chinese hamster ovary (CHO) cells did not initially appear to be coupled to G proteins, as indicated by the lack of regulation of dopamine binding by guanine nucleotides (SOKOLOFF et al. 1990). In the same cell line, no evidence was obtained for coupling to inhibition of adenylyl cyclase (SOKOLOFF et al. 1990, 1992), and only weak stimulation of arachidonic acid release was observed (PIOMELLI et al. 1991). Subsequent studies using transfected fibroblasts as recipient cells show either no (FREEDMAN et al. 1994; TANG et al. 1994) or weaker effects (CHIO et al. 1994; LAJINESS et al. 1995; MCALLISTER et al.

1995) on adenylyl cyclase or other classical effectors than those obtained with the close receptor homologues D_2 and D_4. Among possible explanations to this apparent weak coupling, it could be hypothesized that the endogenous G proteins and effectors present in fibroblasts were not suitable for D_3 receptor coupling. Indeed, using cell lines with a neuronal origin, likely to express a wider variety of G proteins and effectors, functional coupling could be uncovered.

When expressed in immortalized mesencephalic dopaminergic neurons (the MN9D cell line), the D_3 receptor promotes upon stimulation morphogenic changes of the cells, characterized by neuritic outgrowth (SWARZENSKI et al. 1994) and inhibits dopamine release (TANG et al. 1994). In the neuroblastoma-derived cell line NG 108-15, stably expressed D_3 receptors are present under two affinity states for dopamine, interconverting by a guanine nucleotide (PILON et al. 1994). In this cell line, D_3 receptor activation induces mitogenesis (PILON et al. 1994), an effect also observed in one CHO-transfected cell line (CHIO et al. 1994), and inhibits Ca^{2+} currents (SEABROOK et al. 1994). Stimulation of NG 108-15 cells transfected with the human D_3 receptor cDNA, also strongly inhibits cyclic adenosine monophosphate (cAMP) accumulation triggered by forskolin (GRIFFON et al. 1997). Thus, in agreement with its structural homology, the D_3 receptor seems to use similar transduction pathways as the D_2 and D_4 receptors. The differences observed among various cell lines may be related to different combinations of G protein and effector isoforms.

Rather unexpectedly, however, mitogenesis induced by D_3 receptor stimulation is markedly enhanced when cAMP was increased by forskolin. This effect actually involves cAMP, since it was reproduced by two permeable cAMP analogs and depends upon the activation of the cAMP-dependent kinase, being blocked by an inhibitor of this enzyme (GRIFFON et al. 1997). Several mechanisms probably contribute to mitogenesis, and potent mitogenic factors acting through G protein-coupled receptors increase phosphatidylinositol turnover and inhibit cAMP formation. D_3 receptor-mediated mitogenesis seems to depend upon tyrosine phosphorylations (GRIFFON et al. 1997) and activation of mitogen-activated protein kinases (CUSSAC et al. 1999). This latter response is sensitive to genistein, a tyrosine kinase inhibitor, and wortmannin, a phosphatidyl inositol 3-kinase inhibitor (SCHWARTZ et al. 1998; CUSSAC et al. 1999). In analogy with signaling mediated by other receptors, these features suggest that the mitogenesis is attributable to the βγ complex of a G_i whereas the inhibition of adenylate cyclase is presumably due to the α subunits. These data indicate that inhibition of adenylyl cyclase does not take a major part in the D_3 receptor-mediated mitogenic response, which occurs in presence of forskolin and cAMP analogs. Finally, they suggests that transduction of the D_3 receptor can involve both opposite and synergistic interactions with cAMP, which may support functional interactions between this receptor and receptors activating the cAMP cascade, such as the D_1 receptor.

C. Pre- and Postsynaptic Localizations of the D_3 Receptor in the Brain

In rat brain, in which the phenotypes of neurons expressing the D_3 receptor have been characterized, the largest receptor densities occur in granule cells of the islands of Calleja and in medium-sized spiny neurons of the rostral and ventromedial shell of nucleus accumbens which co-express the D_1 receptor, substance P, dynorphin and/or neurotensin (DIAZ et al. 1994, 1995; LE MOINE and BLOCH 1996). These output neurons from the nucleus accumbens receive their dopaminergic innervation from the ventral tegmental area and reach the entorhinal and prefrontal cortex after relaying in the ventral pallidum and mediodorsal thalamus. In turn, the shell of nucleus accumbens receives projections from the cerebral cortex (infralimbic, ventral, agranular, insular, and piriform areas), hippocampus, and amygdala and also projects to the ventral tegmental area from which originate its dopaminergic afferents (ZAHM and BROG 1992; PENNARTZ et al. 1994). These various specific connections of the shell of nucleus accumbens, a part of the "extended amygdala" (HEIMER 2000), suggest that this area is involved in a series of feedback or feed-forward loops, involving notably the prefrontal cortex and ventral tegmental area and subserving control of emotions motivation and reward.

In the human and nonhuman primate brains, the phenotype of neurons expressing the D_3 receptor are not yet identified, but several studies show their distribution to be rather similar to that in the rat (highest levels in islands of Calleja and nucleus accumbens, see Fig. 1) with, however, higher densities and larger distribution in the ventral part of the caudate putamen and the cerebral cortex (LANDWEHRMEYER et al. 1993; MEADOR-WOODRUFF et al. 1994; MURRAY et al. 1994; LAHTI et al. 1995; HALL et al. 1996; MORISSETTE et al. 1998; SUSUKI et al. 1998).

One aspect of the localization and function of the D_3 receptor which has remained highly debated is its occurrence as an autoreceptor, regulating the activity of dopamine neurons. We originally proposed the existence D_3 autoreceptors on the basis of the expression in substantia nigra and ventral tegmental area of D_3 receptor mRNA, which strongly decreases after lesion of dopamine neurons (SOKOLOFF et al. 1990). This lesion, however, also downregulates postsynaptic D_3 receptor in nucleus accumbens (LÉVESQUE et al. 1995), by deprivation of brain-derived neurotrophic factor (BDNF), an anterograde factor of dopamine neurons (see below). Hence, the lesion-induced decrease in areas of dopamine cell bodies could reflect a similar process occurring in nondopaminergic neurons. Dopamine release (TANG et al. 1994) and synthesis (O'HARA et al. 1996) are inhibited by stimulation of the D_3 receptor expressed in a transfected mesencephalic cell line, and various agonists, with limited preference for the D_3 receptor (SAUTEL et al. 1995), inhibit dopamine release, synthesis, and neuron electrical activity (see LEVANT 1997 for a review), giving support to the existence of D_3 autoreceptors. However, the selectivity of these agonists towards the D_3 receptor in vivo has been strongly questioned,

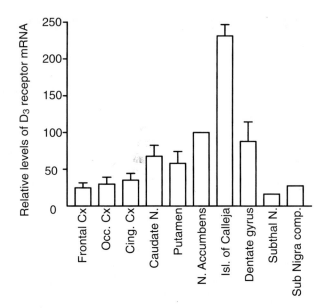

Fig. 1. D_3 receptor mRNA distribution in the human brain. In situ hybridization signals were quantified from three individuals and expressed as percentage of the level in nucleus accumbens (N. accumbens). *Frontal Cx*, frontal cortex; *Occ. Cx*, occipital cortex; *Cing. Cx*, cingulate cortex; *Caudate N.*, caudate nucleus; *Isl of Calleja*, islands of Calleja; *Subthal N.*, subthalamic nucleus; *Sub. Nigra comp.*, substantia nigra pars compacta. (Data from SUSUKI et al. 1998)

because they elicit similar inhibition of dopamine neuron activities in wild-type and D_3 receptor-deficient mice (KOELTZOW et al. 1998). In addition, dopamine autoreceptor functions are suppressed in D_2 receptor-deficient mice (MERCURI et al. 1997; L'HIRONDEL et al. 1998). Nevertheless, dopamine extra-cellular levels in the nucleus accumbens (KOELTZOW et al. 1998) and striatum (R. Gainetdinov and M.G. Caron, personal communication) are twice as high in D_3 receptor-deficient as in wild-type mice, suggesting a control of dopamine neurons activity by the D_3 receptor.

We have recently developed a selective anti-D_3 receptor antibody, the immunoreactivity of which perfectly matches D_3 receptor binding (Fig. 2) and that allowed us to confirm the presence of D_3 autoreceptors at the somato-dendritic level of all dopaminergic neurons in substantia nigra and ventral tegmental area. The function of D_3 autoreceptors remains to be established, but, together with D_2 autoreceptors (MERCURI et al. 1997), they may control the electrical activity of dopamine neurons, which would explain the elevated extracellular dopamine levels in projections areas of these neurons in D_3 receptor-deficient mice. This control could have been masked in experiments using compounds inadequately selective of the D_3 receptor (KOELTZOW et al. 1998), since the compounds used also activate the D_2 receptor. The existence of a control of dopamine release by the D_3 receptor has recently received

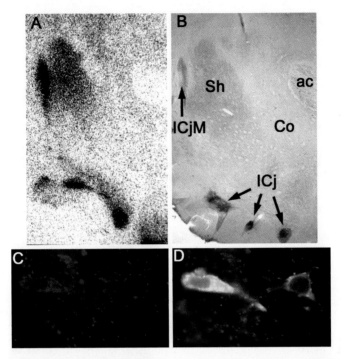

Fig. 2A–D. Immunohistochemical localization of the D_3 receptor in rat brain. Superimposable distributions of binding of $[^{125}I]trans$-7-OH-PIPAT, a D_3 receptor-selective ligand (**A**) and D_3 receptor immunoreactivity (**B**), with highest levels in the islands of Calleja (IcjM and ICj) and moderate levels in the shell of nucleus accumbens (*Sh*). *ac*, anterior commissura. Expression of D_3 receptor immunoreactivity alone (*red* in **C**) and in combination with tyrosine hydroxylase immunoreactivity (*green* in **D**). All tyrosine hydroxylase-positive neurons in the mesencephalon express the D_3 receptor. (Data from DIAZ et al. 2000)

support from the use of selective D_3 receptor antagonists (see Sect. F, this chapter). Alternatively, D_3 autoreceptors could not be operant in anesthetized animals or in vitro in brain slices used in electrophysiological studies (MERCURI et al. 1997; KOELTZOW et al. 1998), whereas dopamine extracellular levels were measured in freely moving animals. Finally, D_3 autoreceptors may mediate yet unrecognized control by dopamine of other activities of dopamine neurons, such as synthesis or release of neuropeptides co-expressed with dopamine in these neurons, e.g. neurotensin, cholecystokinin, or neurotrophins.

D. Coexisting D_1 and D_3 Receptors in Ventral Striatum Mediate Both Synergistic and Opposite Responses

The D_3 receptor is selectively distributed in the ventral striatum, a projection area of mesolimbic dopamine neurons, namely in the islands of Calleja and

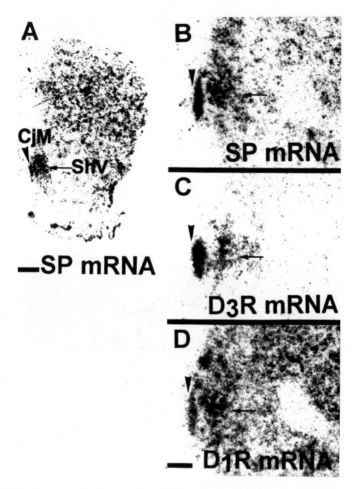

Fig. 3. A Distribution of substance P mRNA in the striatal complex. **B–D** Overlapping distributions of mRNAs of substance P and D_1 and D_3 receptors, respectively, in islands of Calleja major (*CjM*) and ventromedial shell of nucleus accumbens (*ShV*). (Data from RIDRAY et al. 1998)

the ventral part of the shell subdivision of nucleus accumbens, regions in which the D_2 receptor is not or scarcely expressed (BOUTHENET et al. 1991; DIAZ et al. 1994; DIAZ et al. 1995), but in which D_1 receptor mRNA and protein are found (FREMEAU et al. 1991; HUANG et al. 1992). We examined to which degree the D_1 and D_3 receptor mRNAs colocalize in these two areas by using in situ hybridization histochemistry on thin adjacent sections (Fig. 3, Table 1). These quantitative studies indicated that in granule cells of the islands of Calleja major, a class of substance P-containing neurons receiving dopaminergic innervation from the mesencephalon, there is a large degree of D_1/D_3 receptor coexistence. In fact, the values of Table 1 might be even underestimated inasmuch as previous studies suggested that *all* these granule cells express D_1

Table 1. Co-expression of D_1 receptor, D_3 receptor, and substance P mRNAs in ventromedial shell of nucleus accumbens (data from RIDRAY et al. 1998)

Categories of neurons	Percentage of expressing neurons[a]	
	Islands of Calleja	N. accumbens shell
$(D_1+D_3)/D_1$	82%	45%
$(D_1+D_3)/D_3$	79%	63%
D_1/SP	84%	72%
D_3/SP	65%	71%

[a] Neurons simultaneously present on paired adjacent 3-μm frontal sections and positive (≥ 4 silver grains around the nucleus) for either one or two markers were counted. Percentages based upon counting of 117–440 cells.

receptor (HUANG et al. 1992) and D_3 receptor (DIAZ et al. 1995), an underestimation presumably inherent to the method, based upon counting of cells having to be present on two adjacent sections. Taking into account this factor for the interpretation of data in the ventromedial shell of nucleus accumbens leads to the conclusion that, here also, a large majority of neurons expressing the D_3 receptor are substance P neurons also expressing the D_1 receptor. It should be noted however that the level of expression per cell and the number of expressing cells are much higher in the case of the D_1 receptor than in the case of the D_3 receptor.

The functional consequences of the D_1/D_3 receptor coexistence in these two areas were assessed by evaluating changes in gene expression triggered by costimulation of the two receptor subtypes in vivo. In the island of Calleja major, activation of the D_1 and D_3 dopamine receptor subtypes, the only ones to be expressed, leads to clearly opposite effects on the expression of the protooncogene c-*fos*. Activation of the D_1 receptor enhances whereas D_3 receptor activation reduces c-*fos* mRNAs (Fig. 4). Furthermore, the basal c-*fos* expression seems to be under tonic control by endogenous dopamine acting on both receptor subtypes since SCH 23390 and nafadotride, preferential D_1 and D_3 receptor antagonists respectively, both modified, although in opposite manner, c-*fos* mRNA levels. This opposition, demonstrated in vivo, seems in line with (1) the D_1 and D_3 receptors exerting opposite effects on cAMP generation, (2) cAMP being a potent c-*fos* gene expression activator, (3) the higher sensitivity of the D_3 than the D_1 receptor to dopamine allowing it to contribute to the response in spite of its much lower abundance.

What could be, however, the functional significance of such a paradoxical control by a single transmitter acting together on a "brake" (the D_3 receptor) and an "accelerator" (the D_1 receptor). Such a device might have a "buffering" effect, allowing it to increase the response threshold to dopamine, to maintain a low and constant level of basal gene expression, and attenuate the effects of fluctuations of phasic dopamine release. This would ensure a continuous expression of genes regulated by the transcription factor Fos.

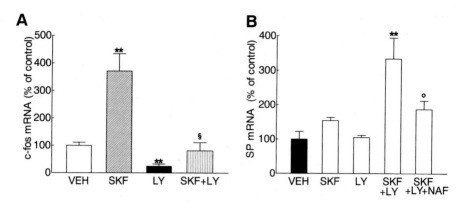

Fig. 4A,B. Opposite and synergistic interactions between D_1 and D_3 receptors. In the islands of Calleja (**A**), SKF 38393 (SKF, 10 mg/kg i.p.), a D_1 receptor agonist, increases c-*fos* mRNA level and quinpirole (*LY*, 1 mg./kg i.p.), a D_2/D_3 receptor agonists, decreases it and opposes to the effects of SKF. In the shell of nucleus accumbens (**B**), SKF increases substance P (SP) mRNA; LY has no effect alone but potentiates the effect of SKF. This potentiation is reversed by NAF. $*p < 0.05$; $**p < 0.001$ vs control; $\S p < 0.01$ vs SKF; $\mathbf{O}p < 0.01$ vs SKF+LY. (Data from Ridray et al. 1998)

In marked contrast with the above mechanism, in the shell of nucleus accumbens, D_1 and D_3 receptor activation resulted in a synergistic enhancement of substance P gene expression (Fig. 4). In view of the high degree of co-expression of the two receptor subtypes in medium-sized neurons of this area, it seems likely that the synergism occurs at this single-cell level. This may reflect the participation of the mitogen-activated protein (MAP) kinase pathway of D_3 receptor signaling that shows amplification by cAMP in transfected NG 108-15 cells.

E. D_1/D_3 Receptor Interplay in the Induction and Expression of Behavioral Sensitization: Role of Brain-Derived Neurotrophic Factor

The expression of the D_3 receptor in medium-sized neurons of the nucleus accumbens is highly dependent upon the dopaminergic innervation; ablation of the afferent neurons by unilateral 6-hydroxydopamine results in a dramatic decrease in the D_3 receptor density in ipsilateral nucleus accumbens (Lévesque et al. 1995). D_3 receptor density is also decreased in a nonhuman primate model of Parkinson's disease, i.e., in 1-methyl-4-phenyl-1,2,3,6-tetrahydropyridine (MPTP)-treated monkeys (Morissette et al. 1998) or in patients suffering from this disease (Ryoo et al. 1998). This paradoxical change (the D_2 receptor is upregulated under these circumstances) was shown to depend on the deprivation of an unidentified anterogradely transported factor from dopaminergic neurons, distinct from dopamine itself and its known peptide co-transmitters (LÉVESQUE et al. 1995). We have recently identified this factor as being BDNF, a neurotrophin synthesized by dopamine neurons (SEROOGY et al. 1994), released upon neuron depolarization (THOENEN 1995), and anterogradely transported (ALTAR et al. 1997): the D_3 receptor and Trk B, the receptor for BDNF, colocalize in the shell of nucleus accumbens and mice bearing a targeted mutation of the BDNF gene have an ablated D_3 receptor expression in this brain region (GUILLIN et al. 2001). No such changes are observed for D_1 and D_2 receptors (GUILLIN et al. 2001).

In these hemiparkinsonian rats, repeated administration of levodopa is able to induce progressively not only the recovery but even the overexpression of the receptor in the denervated nucleus accumbens (BORDET et al. 1997). Furthermore, the levodopa treatment elicits also the ectopic expression of the D_3 receptor mRNA and protein in the dorsal striatum, i.e., a brain area in which it is normally undetectable or hardly detectable. This ectopic induction takes places in substance P-containing neurons of the direct striatonigral pathway, projecting to the substantia nigra pars reticulata, in which expression of the D_3 receptor protein is also induced (BORDET et al. 2000). The process is clearly attributable to stimulation of a D_1/D_5 receptor in the denervated striatum since it is reproduced by a D_1/D_5-receptor agonist and prevented by a D_1/D_5-receptor antagonist.

Several observations indicate that the process of levodopa-induced D_3 receptor induction is causally related with the development of behavioral sensitization, which results of the progressive enhancement of sensitivity to levodopa of denervated striatal neurons (ENGBER et al. 1989; CAREY 1991): (1) the time courses of ectopic D_3 receptor appearance and disappearance were closely parallel to the changes in behavioral responsiveness to levodopa as evaluated in the Ungerstedt's rotation model (Fig. 5), (2) both processes require previous dopaminergic denervation and depend on repeated intermittent stimulation of a D_1/D_5 receptor (BORDET et al. 2000), (3) development of both processes can be partly prevented by treatment with MK-801, a NMDA receptor antagonist (BORDET et al. 2000) and (4) the expression of behavioral sensitization, i.e., the enhanced rotational response to levodopa, is blocked by co-administration of nafadotride, a preferential D_3 receptor antagonist (BORDET et al. 1997), and is mimicked by administration of BP897, a recently designed and highly selective D_3 receptor agonist, together with a D_1/D_5 receptor agonist which, alone, does not induce increased rotational response (PILLA et al. 1999). A similar process seems to occur in MPTP-treated monkeys (MORISSETTE et al. 1998). It is noteworthy that whereas overexpression of the D_2 receptor occurs as a result of dopamine denervation and may be related to disuse hypersensitivity, chronic levodopa treatment does not result in any further enhancement in this receptor abundance.

Although levodopa induces the appearance of D_3 receptor in substance P neurons, i.e., the direct striatonigral pathway that dopamine regulates via the D_1 receptor (BORDET et al. 2000), this effect does not directly originate in striatal neurons, but indirectly, via upregulation of corticostriatal BDNF: the

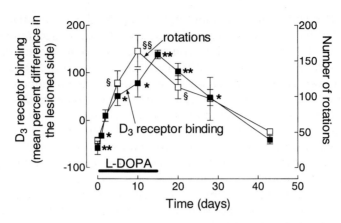

Fig. 5. Parallel changes in D_3 receptor binding and levodopa-induced rotations in hemiparkinsonian rats during and after repeated administration of levodopa. Rats with an unilateral 6-hydroxy dopamine-induced lesion of the ascending mesencephalic dopaminergic pathways placed 3 weeks earlier, received levodopa (50 mg/kg, i.p., b.i.d.) for up to 15 days and were challenged with a single same dose of levodopa. Contralateral rotations were counted 35 min afterwards, and animals killed 4 h after the challenge and D_3-receptor binding quantified on autoradiograms using [^{3}H]7-OH-DPAT as the ligand. (Data from BORDET et al. 1997)

process necessitates the integrity of cortical neurons and is blocked by a selective BDNF antagonist (GUILLIN et al. 2001). Moreover, levodopa induces BDNF mRNA in cortical pyramidal cells projecting to striatum in the lesioned side, via a D_1/D_5 receptor-mediated mechanism, and the lesion itself upregulates *Trk*B expression (NUMAN and SEROOGY 1997; GUILLIN et al. 2001). Hence, induction of D_3 receptor expression in striatum is triggered by the combination of these two processes, that are prominent in the 6-OHDA-lesioned side as compared to the control side, which accounts for the induction of D_3 receptor expression restricted to the lesioned side.

Thus, D_3 receptor overexpression causes a more pronounced disequilibrium between the two sides in responsiveness to dopamine in the direct striatonigral pathway. It appears, therefore, that stimulation of the D_3 receptor neosynthesized in striatonigral neurons enhances the D_1-receptor mediated activation of the efferent γ-aminobutyric acid (GABA)/substance P/dynorphin pathway, leading to inhibition of GABA neurons in substantia nigra pars reticulata and, consequently, disinhibition of target motor nuclei. Although the exact mechanism through which the D_3 receptor facilitates the activation of the striatonigral pathway remains conjectural, it was hypothesized that reciprocal changes in the expression and release of substance P and dynorphin, two peptide co-transmitters in this GABA pathway, may account for the sensitization (BORDET et al. 1997).

Whatever the mechanism through which the neosynthesized D_3 receptor exerts its role, induction of its expression in patients with Parkinson's disease treated with levodopa may account for both beneficial and detrimental effects of the therapy, i.e., progressive motor improvement at the beginning and abnormal involuntary movements after long-term use. It is important to underline that intermittent levodopa administration to the hemiparkinsonian rats also induces D_3 receptor overexpression in the shell of nucleus accumbens. Such a change may be responsible for the development of psychological disturbances, e.g., hallucinations, a rather common side effect of levodopa therapy.

A similar process might also be operant for induction of D_3 receptor expression during development. In the rat brain, whereas D_1 and D_2 receptors appear prenatally, the D_3 receptor appears in the ventral striatum and frontoparietal cortex during the first postnatal week (DIAZ et al. 1997). This event coincides with the time at which these areas become innervated by the dopaminergic neurons from the ventral tegmental area and it may well be that it is, in fact, triggered by BDNF.

F. D_3 Receptor-Selective Pharmacological Agents

Initial pharmacological studies with recombinant D_3 receptor showed that dopamine, as well as some of its agonists, display higher affinity at D_3 receptor than at D_2 receptor (SOKOLOFF et al. 1990, 1992). Among antagonists, antipsychotics display very similar affinities at D_2 and D_3 receptors (Table 2), but (+)AJ-76 and (+)UH 232, two aminotetralin derivatives acting as preferential dopamine autoreceptor antagonists (SVENSSON et al. 1986), show a little preference for the D_3 receptor, as compared to D_2 receptor (SOKOLOFF et al. 1990). In fact, (+)UH 232 has been shown to exhibit partial agonistic properties at the D_3 receptor (GRIFFON et al. 1995).

An important step towards identifying selective D_3 receptor ligands was the discovery that the dopamine agonist $[^3H](+)$7-OH-DPAT selectively binds in vitro to the natural D_3 receptor (LÉVESQUE et al. 1992), allowing to visualize this receptor in the rat (LÉVESQUE et al. 1992) and human (HERROELEN et al. 1994) brain slices and to confirm its pharmacological properties previously unraveled when expressed in recombinant cells. Subsequently, (+)7-OH-DPAT has been used in vivo in a flurry a animal studies aiming at assigning biological functions to stimulation of the D_3 receptor. Such a use of (+)7-OH-DPAT was particularly inappropriate, because this ligand was identified just as a selective D_3 receptor ligand in peculiar binding experimental conditions in vitro (discussed by LÉVESQUE 1996) and losses almost all its selectivity when functional tests are utilized (SAUTEL et al. 1995). Indeed, it appears now that most, if not all responses elicited by (+)7-OH-DPAT (reviewed by LEVANT 1997) are not mediated by the D_3 receptor, insofar as they are not abolished in mice bearing a targeted mutation of the D_3 receptor gene (XU et al. 1999; PERACHON et al. 2000).

Three putative D_3 receptor antagonists, nafadotride (SAUTEL et al. 1995), PNU-99194A (WATERS et al. 1993) and S14297 (MILLAN et al. 1995) have

Table 2. Compared potencies of various antipsychotic drugs at recombinant human D_2 and D_3 receptors

Drug	K_i values (nM)		$K_{i(D2)}/K_{i(D3)}$
	$D_{2(s)}$ receptor	D_3 receptor	
Amisulpride	1.3	2.4	0.53
Amperozide	322[a]	235[a]	1.37
Carpipramine	8.7	15	0.58
Chlorpromazine	2.3	5.9	0.39
Clozapine	69	479	0.14
Haloperidol	0.6	2.9	0.21
Iloperidone	3.5	4.8	0.73
Pimozide	9.8	11	0.88
Pipotiazine	0.20	0.28	0.72
Prochlorperazine	0.40	1.8	0.22
Raclopride	1.8	3.5	0.51
Remoxipride	198	2,300	0.09
Risperidone	1.0	5.4	0.19
Sertindole	0.38[a]	1.6[a]	0.24
(−)Sulpiride	10	20	0.50
Sultopride	4.5	8.1	0.55
Thioridazine	3.3	7.8	0.42

Values derived from studies of the inhibition of [^{125}I]iodosulpiride binding to h$D_{2(s)}$ and hD_3 receptors expressed by transfected CHO cells (data from SOKOLOFF et al. 1992; P. Sokoloff et al. unpublished data).
[a] Data from MALMBERG et al. 1993.

appeared later on, displaying 7–20 times higher affinity for the D_3 than the D_2 receptor. Nafadotride and PNU-99194A are pure antagonists, i.e., agents with no agonistic activity, and they increase locomotor activity at low dosage without affecting dopamine synthesis or release in rats (WATERS et al. 1993; HAADSMA-SVENSSON et al. 1995; SAUTEL et al. 1995), suggesting an inhibitory role of postsynaptic D_3 receptors. This hypothesis is consistent with the observation that D_3 receptor-mutant mice display hyperactivity in a novel environment (ACCILI et al. 1996). However, the selectivities of nafadotride and PNU-99194A have been questioned, since their stimulant effects on locomotor activity persist in D_3 receptor-mutant mice (XU et al. 1999). S14297, previously assumed as an antagonist (MILLAN et al. 1995), actually acts as a full agonist on D_3 receptor-mediated mitogenesis and inhibition of cyclic AMP accumulation in recombinant cells (PERACHON et al. 2000), which questions the nature of the effects shown with this compound. It follows that highly selective ligands were needed for assigning functional role(s) for the D_3 receptor.

Screening of a series of newly designed molecules (PILLA et al. 1999) led to the identification of BP 897 (see the chemical structure in Fig. 6). This compound displays a high affinity at the D_3 receptor (Ki = 0.92 ± 0.2nM), a 70 times lower affinity at the D_2 receptor (Ki = 61 ± 0.2nM) and much lower affinity at

D_1 and D_4 receptors, as well as at a variety of non-dopamine receptors. In NG 108-15 cells expressing the human D_3 receptor, BP 897 potently inhibits forskolin-induced cyclic AMP accumulation and is mitogenic (EC_{50} of ~1nM), however to a maximal extent (~55%) lower than that maximally elicited by dopamine or the full agonist quinpirole. In contrast, in cells expressing the D_2 receptor, BP 897 fails to either inhibit cyclic AMP accumulation or trigger mitogenesis; it reversibly antagonizes quinpirole-induced mitogenesis, however only at concentrations largely exceeding those required to stimulate the D_3 receptor. Hence, BP 897 appears as the first selective, potent, but partial (intrinsic activity ~0.6) D_3 receptor agonist, and a weak D_2 receptor antagonist in vitro.

In vivo, BP 897 at high doses (10–20mg/kg) occupies brain dopamine D_2, receptors and, as a result of blockade of these receptors, displays typical neuroleptic-like properties (induction of catalepsy, antagonism of apomorphine-induced stereotypies). This indicates that selective occupancy of D_3 receptors should be obtained at doses not exceeding 1 mg/kg. At these low doses, BP 897 does not affect spontaneous locomotor activity and body temperature. BP 897 agonist/antagonist potency in vivo were assessed on two presumably D_3 receptor-mediated responses. As mentioned in the previous section, in hemiparkinsonian rats repeatedly pretreated with levodopa, BP 897 potentiates rotations elicited by a D_1 receptor-selective agonist; this potentiation does not occur before levodopa treatment, i.e., before induction of D_3 receptor expression and is abolished by co-treatment with the nafadotride, indicating that BP 897 acts as an agonist for this response (PILLA et al. 1999). In the islands of Calleja of rats, BP 897 enhanced c-*fos* mRNA, an effect similar in direction and amplitude to that produced by nafadotride (RIDRAY et al. 1998); in addition, in contrast to D_2/D_3 receptor agonists, BP 897 potentiates the response to a D_1 receptor agonist (PILLA et al. 1999). This effect is abolished in D_3 receptor-mutant mice. Thus, in vivo, BP 897 increases D_1 receptor-mediated responses by acting as either an agonist or an antagonist, depending upon the response considered, consistent with its partial agonist properties in vitro. BP 897 acts as a receptor agonist on rotations elicited in dopamine-depleted brain and as a receptor antagonist on c-*fos* expression maintained by a dopaminergic tone. In addition, BP 897 reduces dopamine efflux measured by microdialysis (J. Costentin, personal communication), suggesting that this compound acts as an autoreceptor agonist, in agreement with the presence of D_3 receptors on dopamine neurons.

More recently, two novel and highly selective D_3 receptor antagonists have been identified. S33084 has a K_i value of 0.25nM at the D_3 receptor and a 120 times lower affinity at the D_2 receptor (MILLAN et al. 2000). SB-277011-A has a K_i value of 11nM at the D_3 receptor and a 100 times lower affinity at the D_2 receptor (REAVILL et al. 2000). Both compounds have much lower affinity at dopamine D_1 and D_4 receptors, as well as various other nondopaminergic receptors. In several functional in vitro assays, they display no agonistic activity and competitively antagonize dopamine or dopamine agonist-induced responses. Remarkably, both compounds have no effect on spontaneous locomotor activity or dopamine efflux, measured by brain microdialysis, but antagonize preferential D_3 dopamine agonist-induced inhibition of dopamine efflux. This indicates that D_3 receptors, presumably D_3 autoreceptors, exert a phasic, but not tonic control of the activity of dopamine neurons. In agreement, the two D_3 receptor antagonists do not modify spontaneous or psychostimulant-induced locomotion.

G. The D_3 Receptor and Schizophrenia

Among the novel dopamine receptor subtypes revealed by molecular approaches, the D_3 receptor was, from the beginning, considered as potentially relevant to the topic of schizophrenia, due to its localization and pharmacology (SOKOLOFF et al. 1990). A number of more recent observations have progressively strengthened this hypothesis.

Although it is clear that the therapeutic actions of antipsychotic drugs could be safely attributed to blockade of D_2 rather than D_1 receptor subtypes, the recent cloning of several D_2-like receptors, i.e., the D_2, D_3, and D_4 receptors, has raised the question of their role. The positive correlation between drug affinity at D_2-like receptor binding sites and drug plasma levels identified these receptors as a common target for all antipsychotics (SEEMAN et al. 1976; SNYDER 1976). This was apparently confirmed when D_2-receptor occupancy in striatum of patients receiving various antipsychotics at therapeutic dosage was determined by positron emission tomography and found to be ≥70%–80% in most cases (FARDE et al. 1989). It remains to be established whether it is also accompanied by significant occupancy of D_3 or D_4 receptors. Comparing the affinity of antipsychotics at recombinant D_2 and D_3 receptors (Table 2) indicates that these compounds generally show some but very limited preference for the D_2 receptor. Hence, assuming an equal access of the drugs to brain regions where each receptor subtype is mostly expressed, as seems likely (GULAT-MARNAY et al. 1985), it can be estimated that significant D_3 receptor occupancy occurs during antipsychotic treatments resulting in 70%–80% D_2 receptor occupancy. Hence, the idea that selective blockade of the D_3 receptor is sufficient to afford antipsychotic efficacy remains to be tested.

Following repeated administration of antipsychotics, supersensitivity to dopamine and, its counterpart, tolerance to dopamine antagonists develops, a process partly attributable to an enhanced expression (upregulation) of the D_2 receptor. In patients, the antipsychotic activity does not diminish upon long-term treatment. It is therefore noteworthy that repeated administration of haloperidol for 2 weeks to rats failed to trigger any significant upregulation in D_3 receptor mRNA or binding in various brain areas, whereas in the same rats D_2-receptor mRNA and binding were enhanced by up to more than 50% (FISHBURN et al. 1994; LÉVESQUE et al. 1995; TARAZI et al. 1997). In addition, the decrease in neurotensin gene transcripts in ventromedial shell of nucleus accumbens, a characteristic response to D_3-receptor blockade (DIAZ et al. 1994), does not show tolerance following repeated administration of haloperidol, whereas the reverse response in the core of nucleus accumbens, mediated by D_2 receptors, diminishes upon repeated haloperidol administration (LÉVESQUE et al. 1995). These observations tend to support the idea that the antipsychotic activity of neuroleptic drugs is related to D_3 rather than D_2 receptor blockade.

A link between the D_3 receptor and schizophrenia was suggested by the initial observation that, in rat brain, the receptor is mainly expressed in forebrain limbic areas (see Sect. C, this chapter), connected to various cortical areas, that may be directly or indirectly involved in schizophrenia. Many functional imaging and neuropsychological studies have implicated the heteromodal association neocortex, comprising several interconnected association areas (dorsal, prefrontal cortex, Broca's area, and inferior parietal cortex), in higher integrative functions (memory, speech, focused attention) and in the disorders of these functions that are observed in schizophrenia (ROSS and PEARLSON 1996). In agreement, subtle neuropathological abnormalities were also described in these brain areas of schizophrenic patients (or those to which they are connected) (ROBERTS 1991). It seems likely that dopamine plays a modulatory role on the efficiency of these heteromodal cortico-ventrostriatal circuits, in analogy with the role dopamine plays on motor cortico-dorsostriatal circuits. A dysfunction of these former circuits with functional dopamine deafferentation of the prefrontal cortex and enhanced dopamine subcortical tone was proposed to occur in schizophrenia (WEINBERGER 1987; DEUTCH 1993). It is noteworthy that dopamine seems to control these circuits not only at the level of the nucleus accumbens but also at their various relays and that the D_3 receptor is also expressed in the corresponding areas of the rat brain, e.g., in the ventral pallidum, mediodorsal thalamus, cerebral cortex, ventral tegmental area, amygdala, etc. (BOUTHENET et al. 1991; DIAZ et al. 1995).

Taken together these observations are consistent with the hypotheses that (1) schizophrenia might be associated with (caused by) a defective functioning of the circuits linking the extended amygdala and limbic cortices, and (2) dopamine regulates the gain factor of these circuits via interacting with the D_3 receptor. However, the effect of D_3 receptor activation on these circuits is still largely conjectural but was recently investigated in an indirect manner by studying the role of the D_3 receptor in disruption of prepulse inhibition of the startle reflex. This paradigm can be evaluated in schizophrenic patients in which it is altered, reflecting a deficit in sensorimotor gating, a cardinal symptom of the disease (BRAFF et al. 1992). Therefore, it is often considered as a relevant index of cognitive performance in rodents and a reliable test for antipsychotic drugs. The prepulse inhibition impairment that results from dopamine receptor stimulation by amphetamine was found unaltered in knockout mice deficient in D_3 and D_4 receptors, whereas it was suppressed in mice without D_2 receptors (RALPH et al. 1999). Although this might have reflected a lack of implication of the D_3 receptor in sensorimotor gating, this view has been challenged by the observation that the selective D_3 receptor antagonist SB-277011-A reverses disruption of prepulse inhibition in isolation-reared rats, in which deficits may reflect aspects of developmental abnormalities associated with schizophrenia (GEYER et al. 1993).

Whereas in rodents overexpression of the D_3 receptor as a mechanism of behavioral sensitization is, so far, substantiated only in the case of levodopa (see Sect. E, this chapter), a carefully performed radioligand binding study with post-mortem brain samples of schizophrenic patients, which were not taking antipsychotics for at least 1 month prior to death, shows elevated levels of D_3 receptor in the ventral striatum (GUREVICH et al. 1997). In contrast, in patients that had been treated with antipsychotics up to the time of death, D_3 receptor levels did not differ significantly from those of controls. These data were interpreted as indicating that antipsychotics actively downregulate the D_3 receptor in schizophrenic patients that, otherwise, have a higher density of this receptor in the ventral striatum. Since schizophrenic patients also have BDNF levels elevated in anterior cingulate cortex and hippocampus (TAKAHASHI et al. 2000), brain regions projecting to striatum, this neurotrophin may, as in the case of levodopa-induced sensitization, participate in D_3 receptor overexpression and the pathological manifestations of the disease.

Various lines of evidence suggest the potential contribution of sensitization as a pathophysiologic mechanism in schizophrenia (GLENTHOJ and HEMMINGSEN 1997; LIEBERMAN et al. 1997). Administration of psychostimulants such as amphetamine to healthy subjects can produce psychotic symptoms (ANGRIST et al. 1974) and exacerbate psychotic symptoms in patients with schizophrenia (JANOWSKY and DAVIS 1976; LIEBERMAN et al. 1987). The heightened inhibition in D_2-like receptor binding elicited by amphetamine, almost twice higher in drug-free schizophrenic patients than in controls, is interpreted as an enhanced dopamine release in these patients (LARUELLE et al. 1996). Although its mechanism remains conjectural, this enhanced responsiveness to amphetamine could be considered as reflecting a sensitized dopaminergic state, perhaps via a higher setting point of the feedback loops controlling dopamine release.

Two main features connect D_3 receptor ontogeny to schizophrenia pathophysiology. The first one is that the dopaminergic system appears early in brain development in higher species (VOORN et al. 1988) and a neurodevelopmental role for dopamine has been suggested (ROSENGARTEN and FRIDHOFF 1979; MILLER and FRIEDHOFF 1986). The second one is that the hypothesized role of D_3-receptor overexpression in the etiology of schizophrenia raises the question of mechanisms governing this receptor expression during development. D_3-receptor expression during embryonic and early postnatal development is characterized by an appearance of transcripts at early stages in neuroblasts or migrating neurons in the rat (DIAZ et al. 1997) or human brain (SCHWARTZ et al. 2000), in which it was already detected at week 6. However, the cortical neuroepithelium giving rise to the cerebral neocortex was heavily labeled in the human but not in the rat embryo (SCHWARTZ et al. 2000). Tentatively, the

D_3 receptor expressed in neuroepithelial cells lining the cerebral ventricles might regulate their mitotic activity since, in NG108–15 cells, activation of the recombinant receptors enhances MAP kinase activity and [^{3}H]thymidine incorporation (see Sect. B, this chapter). Putative dopaminergic neurons were seen in the human brain, adjacent to the ventricle, presumably in the germinal zone at 16 days of gestation (FREEMAN et al. 1991). Being the only dopamine receptor subtype expressed in these dividing neuroepithelial cells and owing to its high sensitivity to dopamine (allowing dopamine to act at a certain distance from its release), the D_3 receptor could have a role in the control of their proliferation elicited by dopamine and, therefore, in the number of neurons of their progeny. Among the cerebral morphometric abnormalities detected in either in vivo or post mortem studies, ventricular enlargement, reduction in neocortical and hippocampal neuronal density could result from an abnormal control of their progenitors in the neuroepithelium during early development.

A very large series of studies (nearly 50 published so far) aiming at assessing the implication of the dopamine D_3 receptor gene in schizophrenia and other psychiatric disorders have started after the identification of polymorphisms this gene (DRD_3), notably a mutation substituting a Gly^9 for a Ser^9 in the N-terminus of the receptor and creating a *Bal*I restriction site (LANNFELT et al. 1992). The first studies consisted in linkage analysis and did not favor a nonrandom segregation of either allele and schizophrenia (COON et al. 1993; WIESE et al. 1993), leading to the suggestion that the DRD_3 gene does not play a major role in the susceptibility to schizophrenia. However, it has been repeatedly proposed that, in presumably polygenic diseases such as schizophrenia, the minor contribution of a susceptibility allele to genetic predisposition might be more reliably detected in association studies. CROCQ and coworkers were the first to detect, in French and Welsh populations, evidence for an association of homozygosity of either allele 1 (Ser-Ser) or allele 2 (Gly-Gly) with schizophrenia (CROCQ et al. 1992). A flurry of other association studies has followed that has been split into confirmation and non-replications (the majority of the studies). Nevertheless, two independent meta-analyses of a large number (29–30) of association studies were recently performed in an attempt to minimize the lack of statistical power of individual studies and to control for population heterogeneity in a total population of over 2,500 patients and 2,500 controls, and convergent conclusions were obtained (DUBERTRET et al. 1998; WILLIAMS et al. 1998). Significant, although limited, excess of homozygotes for both alleles were found, clearly suggesting that having the 1–1 allele (or the 2–2 allele) of the D_3 receptor gene slightly enhances susceptibility to the disease over having the 1–2 allele.

The Ser^9-Gly^9 mutation takes place in a region of the protein (the N-terminal extracellular tail) not likely to be involved in either ligand binding or signal generation. In agreement, only minor differences were detected between the dopamine-binding affinities of the homozygous and heterozygous recombinant receptors expressed in a cell line (LUNDSTROM and TURPIN 1996).

Nevertheless, a point mutation in the same region of rhodopsin, which also belongs to the superfamily of heptahelical signaling proteins, is responsible for one of the most common forms of retinitis pigmentosa (SUNG et al. 1991). The mutation could affect the level of membrane expression of the D_3 receptor, via a linkage disequilibrium with other genetic variations in its promoter region. The association of schizophrenia was significant with the *Bal*I polymorphism but not with another polymorphism (*Msp*I) situated ~35 kb downstream in the same sample of subjects (GRIFFON et al. 1996).

In conclusion, the data obtained during the last decade concerning the dopamine D_3 receptor lead to a tentative but rather coherent picture of the receptor's role in schizophrenia. Dopamine, via the D_3 receptor, may regulate the gain of neural circuits involving the heteromodal association neocortex and controlling emotional and cognitive processes. The neural expression of the D_3 receptor is triggered during development and regulated afterwards in a complex manner, depending namely upon the activity of afferent dopamine and glutamate neurons and the release of the neurotrophic factor BDNF by these neurons. Hence, the level of D_3 receptor expression is controlled by neural loops, the activity of which is, in turn, controlled by the D_3 receptor abundance. The excessive expression of the D_3 receptor in schizophrenia, observed in one study, may result from genetic or developmental factors and account for the behavioral sensitization to psychostimulants observed in schizophrenia and at least a fraction of the psychotic symptomatology of this disease. This "D_3 receptor hypothesis" in schizophrenia has the advantage to put together a variety of recent and old observations, but has to be evaluated notably by the assessment of the clinical efficacy of selective D_3 receptor agents.

H. The D_3 Receptor and Drug Addiction

Abused drugs (alcohol, heroin, and cocaine) elicit a variety of chronically relapsing disorders by interacting with brain reward systems. Converging evidence supports the idea that the mesocorticolimbic system, which projects from the ventral tegmental area to the nucleus accumbens, frontal cortex, olfactory tubercle, amygdala, and septal area, is an important substrate for the hedonic and reinforcing effects of drugs (KOOB 1992; NESTLER 1992). The involvement of these regions in hedonistic effects is indicated by the fact that a strong electrical intracranial self-stimulation is maintained in these regions in animals (MILNER 1991). Destruction of dopaminergic neurons by 6-hydroxydopamine induces an extinction-like responding in cocaine and amphetamine self-administration, as reflected by a reduction in responding over days; this lesion also induces a decrease in the reinforcing effects of cocaine using a progressive-ratio schedule (NESTLER 1992). In addition, it is well established that all abused drugs share the property to unconditionally elevate dopamine extracellular concentrations in the nucleus accumbens (DI CHIARA and IMPERATO 1988; KOOB 1992). Dopamine receptor agonists and antagonists modulate psychostimulants self-administration in the rat, an effect interpreted as an interaction with their rewarding effect, because it is similar to the behavioral adaptation observed after changing the unit dose of self-administered drug.

One of the major projection areas of the dopaminergic mesocorticolimbic system, the shell of nucleus accumbens projects to the prefrontal cortex and receives projections from the hippocampus and amygdala (ZAHM and BROG 1992; PENNARTZ et al. 1994), which makes it a critical brain structure to control motivation and responses to drug-conditioned cues (WHITELAW et al. 1996; MEIL and SEE 1997). The D_3 receptor is highly expressed in this structure, where its expression has been found elevated in postmortem brain from cocaine addicts (STALEY and MASH 1996; SEGAL et al. 1997), suggesting either a drug-induced overexpression of the D_3 receptor or a defect in the D_3 receptor-gene regulation in vulnerable subjects. In agreement with the former hypothesis, repeated intermittent administration of drugs of abuse induces a sensitization strikingly similar to that produced by levodopa in hemiparkinsonian rats, characterized by enhanced behavioral responses (KALIVAS and STEWART 1991) and similar activation of neuropeptide synthesis and transcription factors (COLE et al. 1995). It is tempting to speculate that a D_1 receptor-mediated increase in D_3 receptor responsiveness is a common mechanism underlying both types of sensitization. As the reinforcing efficacy of drugs of abuse under repeated administration is also intensified under certain conditions (WOOLVERTON et al. 1984; LETT 1989; HORGER et al. 1990), alteration of D_3 receptor responsiveness could also be involved in drug seeking and dependence. Moreover, a weak but significant genetic association has been found between the D_3 receptor gene and heroin addiction (DUAUX et al. 1998) or drug abuse in schizophrenia (KREBS et al. 1998), suggesting that a defect in this gene, possibly in its regulatory parts, might be implicated in vulnerability to drug addiction.

Cocaine self-administration in rats is highly dependent on D_1 receptor-mediated mechanisms (CAINE and KOOB 1994). D_1 receptor-deficient mice produced by gene targeting demonstrate the essential role of the D_1 receptor in the locomotor and neurochemical effects of psychostimulants; in these mutant mice, cocaine fails to induce the typical locomotor hyperactivity (XU et al. 1998) and can no longer stimulate striatal neurons to express various immediate early genes, such as c-*fos* and *Jun*B (DRAGO et al. 1996; MORATALLA et al. 1996; XU et al. 1998). The basal and cocaine-induced expression of neuropeptides, such as dynorphin and substance P, are also altered (DRAGO et al. 1996; XU et al. 1998). Antagonists of D_1 receptor have been shown to block several behavioral effects of cocaine, including locomotor hyperactivity (CABIB et al. 1991; TELLA 1994), subjective effects measured by drug discrimination procedures (SPEALMAN et al. 1991) and rewarding effects assessed by drug self-administration paradigms (MALDONADO et al. 1993; CAINEKOOB 1994). A number of studies also suggest the involvement of the D_3 receptor in modulating cocaine reward (CAINE and KOOB 1993, 1995; SPEALMAN 1996; CAINE et al. 1997). suggesting that both D_1 and D_3 receptor stimulations contribute to mediate or enhance the reinforcing effect of cocaine. It should be stressed,

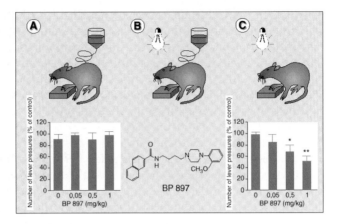

Fig. 6A–C. BP 897 inhibits cue-controlled cocaine seeking but has no reinforcing effects. **A** During repeated sessions, rats are trained to self-administer i.v. cocaine by having them press a lever. BP 897, tested when a stable pattern of self-administration is acquired, has no effect on cocaine self-administration session, i.e., no reinforcing effect. **B** Progressively, a light stimulus is associated with cocaine self-administration, which gains reinforcing properties and finally maintains drug-seeking behavior, even without drug delivery (**C**). During this last phase, the behavior of the animal, which reflects the motivation to take drug induced by the conditioned stimulus, is dose-dependently reduced by BP 897 (by ANOVA, $*p < 0.05$ and $**p < 0.01$ vs saline). (Data from PILLA et al. 1999; reprinted from SOKOLOFF and SCHWARTZ 1999)

however, that none of these studies has utilized fully D_3 receptor-selective compounds, and that the use of a fairly selective D_3 antagonist has challenged these conclusions (BAKER et al. 1999).

It was thus of high interest to assess the highly selective and partial agonist BP 897 for its ability to interact with or to substitute for cocaine in self-administration paradigms. This work was performed in B.J. Everitt's laboratory at Cambridge University (PILLA et al. 1999) and it was found that, in marked contrast with full D_2/D_3 receptor agonists (CAINE and KOOB 1993, 1995; CAINE et al. 1997), BP 897 has no effect on cocaine self-administration, i.e., does not enhance the reinforcing effects of the drug (Fig. 6). When BP 897 was substituted for cocaine, responding rapidly declined over days, indicating that the drug does not possess reinforcing properties sufficient to maintain drug-taking in animals (PILLA et al. 1999), whereas full D_2/D_3 receptor agonists do (CAINE and KOOB 1993). Thus, it appears from two sets of independent observations that BP 897 is without intrinsic reinforcing effects. The absence of effects of BP 897 might equally well be explained by either its partial agonistic property, as agonists and antagonists have opposite effects in modulating cocaine self-administration, or its D_3 receptor-selectivity, as all other compounds tested have also a high potency at the D_2 receptor.

Environmental stimuli appears to be one of the major factors that can cause relapse in abstinent drug addicts. This process is critical for psychostimulants, but also for nicotine and heroin addiction (CHILDRESS et al. 1992; O'BRIEN et al. 1992; GRECH et al. 1996; O'BRIEN and McMELLAN 1996). Moreover, in animals, such cues can induce and maintain drug-seeking behavior and also reinstate drug-seeking after extinction (DE WIT and STEWART 1981; STEWART 1983; SELF and NESTLER 1988; MEIL and SEE 1996, 1997; ARROYO et al. 1999). Reducing the motivational effects of drug-related cues might therefore provide an effective means for preventing the reinstatement of drug-seeking behavior and have potential therapeutic utility in the treatment of drug addiction (O'BRIEN and McMELLAN 1996).

A second-order schedule of reinforcement was used for evaluating the effects of BP 897 on drug cue-controlled seeking-behavior (ARROYO et al. 1999). In this paradigm, rats are first trained to self-administer intravenous cocaine and each self-administration is made contingent upon a response on a lever, paired with a light stimulus which becomes the conditioned stimulus (CS). During training, the number of lever responses to get the CS is progressively increased, as well as the number of CS required to get a cocaine infusion, so that the CS progressively gains motivational salience and, as a conditioned reinforcer, maintains and controls drug-seeking behavior (ARROYO et al. 1999). Hence, evaluating the lever-pressing of the rat under this schedule is a measurement of the motivation to self-administer cocaine following the presentation of the CS. BP 897 dose-dependently reduces cocaine-seeking behavior (Fig. 6), without disrupting the pattern of responding and without increasing the latency to initiate responding at any dose up to 1mg/kg, indicating the absence of any nonspecific, for example motor, effects that might have disrupted performance.

These results show that BP 897, although having no intrinsic reinforcing effect, is nevertheless able to reduce drug-cue-controlled cocaine-seeking behavior. This property, so far unprecedented, indicates that neural mechanisms underlying responding with conditioned reinforcement and responding for cocaine itself are pharmacologically dissociable, in agreement with neuroanatomical studies (WHITELAW et al. 1996; MEIL and SEE 1997). The partial agonistic character and D_3R selectivity of BP 897 seems essential for its dissociated actions: partial agonism of BP 897 may confer a "buffering" capacity allowing it to oppose, as an antagonist, to a chain of neural events initiated by conditioned increase of dopamine release (FONTANA et al. 1993) while maintaining, as an agonist, a moderate degree of D_3R stimulation. The idea that compounds like BP 897 could be useful in reducing relapse vulnerability, with minimal liability of dependence in humans deserves clinical verification.

I. The D_3 Receptor and Depression

Emphasis has been placed upon the putative role of nucleus accumbens dopamine systems in appetitive motivation and positive reinforcement (KOOB 1992; SALAMONE 1994). Hence, mesolimbic dopaminergic neurons projecting to the nucleus accumbens have been suggested to be involved in the neurobiology of depression and the therapeutic actions of some antidepressant drugs (KAPUR and MANN 1992; BROWN and GERSHON 1993; FIBIGER 1995; WILLNER 1997). This hypothesis postulates that decreased dopamine activity is involved in depression, while increased dopamine function contributes to mania. Accordingly, dopaminergic drugs (e.g., amphetamine or cocaine) can produce effects in humans that are remarkably similar to an idiopathic manic episode (SILVERSTONE 1985; GESSA et al. 1995) and the discontinuation of such drugs (MARKOU and KOOB 1991) or the acute administration of dopamine receptor antagonists can result in a psychopathological state similar to a depressive episode (BELMAKER and WALD 1977). Furthermore, dopaminergic drugs such as the dopamine-uptake inhibitors amineptine, nomifensine, and buprorion have been successfully used for treating major depression (ZUNG 1983; KINNEY 1985; GARATTINI 1997).

Nevertheless, in spite of the long-standing clinical use of antidepressant drugs, the neural mechanisms underlying the therapeutic effect of these drugs are still incompletely understood. Most antidepressant drugs are active after an acute administration in mouse behavioral tests suggested to predict antidepressant activity, such as the forced swimming test (PORSOLT et al. 1978) and the tail suspension test (STERU et al. 1985). However, the antidepressant effect emerges in patients after chronic but not acute treatment, enlightening the crucial importance of adaptive changes in promoting this activity. Moreover, the various antidepressant drugs have different primary pharmacological targets, which make it difficult to define a final common mechanism of action, and the situation is even less clear regarding other antidepressant treatments such as electroconvulsive therapy (ECT).

Chronic treatment with antidepressant drugs produces in rats a variety of changes in dopaminergic neurotransmission, most notably a sensitization of behavioral responses to agonists acting at dopamine D_2/D_3 receptors within the nucleus accumbens (MAJ et al. 1984; WILLNER 1997). Similar results have been obtained after chronic ECT (BARKAI et al. 1990; NOMIKOS et al. 1991; SMITH and SHARP 1997). Hence, we were wondering whether these changes in dopamine responsiveness would be attributable to changes in D_3 receptor expression (LAMMERS et al. 2000). A selective increase in D_3 receptor binding or mRNA expression in the shell of nucleus accumbens indeed accompanies all chronic antidepressant treatments including monoamine uptake inhibitors, a monoamineoxidase inhibitor, and ECT (Table 3). Some features of the response to these antidepressant treatments suggest that enhanced dopamine neurotransmission through this receptor participates in the adaptive changes leading to antidepressant activity.

First, antidepressant drugs enhanced D_3 receptor mRNA after chronic (21-day), but not acute treatment; in addition, fluoxetine and amitriptyline enhanced D_3 receptor binding after a 42-day treatment, which is in agreement with the delayed therapeutic action of these drugs in depressed patients. Several antidepressant drugs had actually an opposite effect when tested after

Table 3. Changes in D_3 receptor expression in the shell of nucleus accumbens after antidepressant treatments (data from LAMMERS et al. 2000)

Treatment	Acute D_3R mRNA (1 day)	Chronic D_3R mRNA (21 days)	D_3R binding (42 days)
Desipramine	nd	+54%[b]	nd
Imipramine	−24%[a]	+35%[b]	nd
Amitriptyline	−31%[b]	+35%[b]	+46%[a]
Fluoxetine	+24%	−30%[a]	+42%[a]
Tranylcypromine	−22%[a]	+38%[a]	nd
Electroconvulsive shocks	nd	+49%[a]	+42%[b]

nd, not determined.
[a] $p < 0.05$; [b] $p < 0.01$ vs saline-treated animals.

a single administration: imipramine, amitriptyline, and tranylcypromine significantly reduced D_3 receptor mRNA in the shell of nucleus accumbens. Interestingly, fluoxetine, which markedly differed from other antidepressant drugs in its effects on D_3 receptor mRNA, especially after short chronic treatment, also increased D_3 receptor protein in the shell of nucleus accumbens after a 42-day chronic treatment, suggesting a delayed action in rats as compared with other antidepressant drugs. A short-term exposure to fluoxetine decreases dopamine extracellular levels, whereas other antidepressant drugs have an opposite effect (ICHIKAWA and MELTZER 1995), this also exemplifies the fact that acute effects of antidepressant drugs may not be relevant to their clinical efficacy.

Secondly, whereas the various antidepressant treatments have most likely different short-term outcomes, they all increased D_3 receptor expression upon long-term repeated administration. These observations, together with those of a previous study showing that other antidepressant drugs increase D_3 receptor binding (MAJ et al. 1998), show that enhanced D_3 receptor expression is a common secondary action of antidepressant treatments (LAMMERS et al. 2000). Moreover, this common effect is restricted to the shell of nucleus accumbens, that various studies have shown to subserve the role of dopamine in appetitive motivation, positive reinforcement, pleasurable effect of reinforcing stimuli (KOOB 1992; SALAMONE 1994) and therapeutic actions of some antidepressant drugs (KAPUR and MANN 1992; BROWN and GERSHON 1993; FIBIGER 1995; WILLNER 1997). Interestingly enough, ECT produced the most rapid and robust increase in D_3 receptor protein and mRNA expression, as compared to antidepressant drugs, which is in agreement with their therapeutic efficacy in refractory depression.

Although some antidepressant drugs elicit changes in D_1 and D_2 receptor expression (DZIEDZICKA-WASYLEWSKA et al. 1997a,b), this effect is not common to all antidepressant drugs (AINSWORTH et al. 1998), or occurs in different brain regions. Several previous studies have reported that ECT treatment increases D_1 and D_2 receptor mRNA and D_2 receptor protein in the nucleus accumbens (BARKAI et al. 1990; SMITH and SHARP 1997), but these changes are limited in amplitude or short lasting. We could confirm that there were no changes in D_1 or D_2 receptor mRNAs in the nucleus accumbens 24 h after termination of chronic ECT (LAMMERS et al. 2000). In contrast, chronic ECT induced a robust and long-lasting increase in both D_3 receptor binding and mRNA in the shell of nucleus accumbens. A selective enhanced D_3 receptor expression in this brain region thus appears as a common consequence of various antidepressant treatments.

Furthermore, antidepressant treatments not only increase in D_3 receptor expression, but also elicit changes related to D_3 receptor function. In the shell of nucleus accumbens this receptor is found expressed mainly in substance P- and dynorphin-expressing neurons (LE MOINE and BLOCH 1996; RIDRAY et al. 1998) and both neuropeptide expressions are influenced by pharmacological manipulations of D_1 and D_3 receptors in various experimental circumstances (BORDET et al. 1997; RIDRAY et al. 1998; BORDET et al. 2000). In agreement with a previous observation (WALKER et al. 1991), we found that amitriptyline and fluoxetine increased substance P mRNA in the shell of nucleus accumbens, which might be the result of an increased D_3 receptor function (LAMMERS et al. 2000). Since antidepressant treatments decrease substance P concentrations (SHIRAYAMA et al. 1996), they may accelerate substance P turnover and release. Nevertheless, this hypothesis is hardly reconcilable with the antidepressant effect of a substance P receptor antagonist recently reported (KRAMER et al. 1998).

These results suggest that the increased locomotor stimulants effects of D_2/D_3 receptor agonists following repeated treatments with various antidepressant drugs (MAJ et al. 1989; SERRA et al. 1990) may result from increased D_3 receptor function, a process analogous to that elicited by levodopa in hemiparkinsonian rats (see Sect. D). Moreover, enhancement of dopamine function by repeated ECT requires concomitant activation of D_1 and D_2/D_3 receptors (SMITH and SHARP 1997), which is in line with observations showing that stimulation of the D_3 receptor potentiates D_1 receptor-mediated responses (PILLA et al. 1999). Finally, the role of D_3 receptor overexpression in one form of behavioral sensitization (BORDET et al. 1997) is consistent with the idea that a similar process participates in the behavioral changes triggered by antidepressant treatments.

Importantly, our data also show a progressive and strong downregulation of D_3 receptor binding by repeated handling and injections, even though no drug was administered. This change is not related to increasing age, which actually enhances D_3 receptor expression (GUREVICH et al. 1999) and might be interpreted as a stress-induced effect. Actually, the action of fluoxetine and amitriptyline were not strictly to enhance D_3 receptor binding in the shell of nucleus accumbens, but rather to counteract this downregulation, thus opposing the effects of stress. Accordingly, antidepressant drugs reverse chronic mild stress-induced decrease in the consumption of palatable sweet solutions

(WILLNER et al. 1992), an effect accompanied by increased D_2/D_3 receptor binding (PAPP et al. 1994) and D_2 receptor mRNA (DZIEDZICKA-WASYLEWSKA et al. 1997b). These results point to a role of the D_3 receptor in the adaptive changes induced by chronic stress, a major risk factor in depression (ANISMAN and ZACHARKO 1992; RISCH 1997; AGID et al. 1999; HARKNESS et al. 1999), that antidepressant treatments may correct by their effects on D_3 receptor expression.

One can speculate about possible mechanisms by which chronic stress and antidepressant treatments regulate the D_3 receptor gene in opposite directions. Acute stress activates the mesocortical dopaminergic system (THIERRY et al. 1976), but chronic stress triggers adaptive processes leading to decreased dopaminergic transmission in the shell of nucleus accumbens (FINLAY and ZIGMOND 1997; GAMBARANA et al. 1999). On the contrary, antidepressant drugs and ECT induce a subsensitivity of dopamine autoreceptors, leading to a persistent enhancement of dopamine neuron electrical activity (CHIODO and ANTELMAN 1980a,b; WHITE and WANG 1983). Although antidepressant drug treatments have variable effects on basal extracellular dopamine in the nucleus accumbens (AINSWORTH et al. 1998), they increase amphetamine-evoked dopamine release (BROWN et al. 1991; STEWART and RAJABI 1996), and ECT treatments enhance spontaneous dopamine release (NOMIKOS et al. 1991). There is also some evidence for a decreased dopamine metabolism and turnover in depressed patients (KAPUR and MANN 1992; D'HAENEN and BOSSUYT 1994; SHAH et al. 1997). Hence, the data in the literature are rather consistent with a decreased activity of mesolimbocortical dopamine neurons by chronic stress and in depression, which is reversed by chronic antidepressant treatments. The convergent effects of antidepressant drugs with different serotonin/norepinephrine pharmacological selectivities on dopamine neuron activity can be explained as an adaptation of these neurons to the inhibitory actions of both serotonin and norepinephrine, suggested by various anatomical and functional studies (PRISCO and ESPOSITO 1995; GRENHOFF et al. 1993). We have previously shown that D_3 receptor expression, unlike that of other dopamine receptor subtypes, depends critically upon dopamine neuron activity, through the release of an anterograde factor different from dopamine itself and its known co-transmitters (LÉVESQUE et al. 1995). Thus, the opposite variations of dopamine neuron activity in response to chronic stress or antidepressants drug treatments may determine opposite parallel variations in D_3 receptor expression in the shell of nucleus accumbens. In agreement with the contention that BDNF controls D_3 receptor expression (see Sect. E), BDNF mRNA is reduced by stress (SMITH et al. 1995) and elevated by chronic antidepressant treatments (NIBUYA et al. 1995), which suggests an important role for this neurotrophin in mediating early events elicited by chronic antidepressant treatments.

In conclusion, these results suggests that D_3 receptor expression and function are downregulated in stress and, possibly, depression, and that these changes are reversed by antidepressant treatments. As a corollary, selective D_3

receptor agonists may represent a new class of antidepressant drugs, a proposal supported by data showing that pramipexole, a preferential D_3 receptor agonist (MIERAU et al. 1995; SAUTEL et al. 1995), displays antidepressant-like effects in animals (WILLNER et al. 1994; MAJ et al. 1997) and is efficacious as antidepressant in humans (SZEGEDI et al. 1996; GOLDBERG et al. 1999).

J. Conclusions

The various data collected so far and reported above are compelling enough to assume that, a decade after its birth, the D_3 receptor has now reached its years of maturity. This age is notably marked by the development of useful experimental tools, including a selective antibody, mice bearing a targeted mutation of the D_3 receptor gene, and the first highly selective D_3 receptor ligands. The combined use of these tools has removed some of the uncertainties regarding the functions mediated by the D_3 receptor, that can be summarized below.

It is suggested that, acting as an autoreceptor, the D_3 receptor may control the phasic, but not tonic, activity of dopamine neurons. This hypothesis is consistent with the presence of D_3 receptors on all mesencephalic dopamine neurons (DIAZ et al. 2000) and with the observations that a selective D_3 receptor agonist inhibits dopamine efflux and that selective D_3 receptor antagonists reverse dopamine-agonist induced inhibition of dopamine efflux (MILLAN et al. 2000; REAVILL et al. 2000). It is also consistent with observations that selective D_3 receptor antagonists do not affect spontaneous locomotor activity nor dopamine efflux (MILLAN et al. 2000; REAVILL et al. 2000). Phasic activity of dopamine neurons may be induced by novelty, which may account for the hyperactivity displayed by D_3 receptor-deficient mice in a novel environment (ACCILI et al. 1996; XU et al. 1997), or presentation of drug-conditioned cues, which may account for the inhibition by a selective and partial D_3 receptor agonist of cocaine cue-controlled seeking behavior (PILLA et al. 1999). Further work will be necessary to directly substantiate this hypothesis.

Various converging anatomical, pharmacological, and genetic observations also suggest that the D_3 receptor might be implicated in Parkinson's disease (BORDET et al. 1997), schizophrenia (SCHWARTZ et al. 2000), drug addiction (PILLA et al. 1999), and depression (LAMMERS et al. 2000), or the treatment of these disorders. The identification of selective D_3 receptor agents and their forthcoming introduction into the clinics will be decisive for evaluating this hypothesis.

References

Accili D, Fishburn CS, Drago J, Steiner H, Lachowicz JE, Park B-H, Gauda EB, Lee EJ, Cool MH, Sibley DR, Gerfen CR, Westphal H, Fuchs S (1996) A targeted mutation of the D_3 receptor gene is associated with hyperactivity in mice. Proc Natl Acad Sci USA 93:1945–1949

Agid O, Shapira B, Zislin J, Ritsner M, Hanin B, Murad H, Troudart T, Bloch M, Heresco-Levy U, Lerer B (1999) Environment and vulnerability to major psychiatric illness: a case control study of early parental loss in major depression, bipolar disorder and schizophrenia. Mol Psychiatry 4:163–172

Ainsworth K, Smith SE, Zetterstrom TS, Pei Q, Franklin M, Sharp T (1998) Effects of antidepressant drugs on dopamine D_1 and D_2 receptor expression and dopamine release in the nucleus accumbens. Psychopharmacol 140:470–477

Altar CA, Cai N, Bliven T, Juhasz M, Conner JM, Acheson AL, Lindsay RM, Wiegand SJ (1997) Anterograde transport of brain-derived neurotrophic factor and its role in the brain. Nature 389:856–860

Angrist B, Sathananthan G, Wilk S, Gershon S (1974) Amphetamine psychosis: behavioral and biochemical aspects. J Psychiatr Res 11:13–23

Anisman H, Zacharko RM (1992) Depression as a consequence of inadequate neurochemical adaptation in response to stressors. Br J Psychiatry 15:36–43

Arroyo M, Markou A, Robbins TW, Everitt BJ (1999) Acquisition, maintenance and reinstatement of intravenous cocaine self-administration under a second-order schedule of reinforcement in rats : effects of conditioned cues and continuous access to cocaine. Psychopharmacology 140:331–344

Baker LE, Hood CA, Heidema AM (1999) Assessement of D_3 versus D_2 receptor modulation of the discriminative stimulus effects of (+)-7-OH-DPAT in rats. Behav Pharmacol 10:717–722

Barkai AI, Durkin M, Nelson HD (1990) Localized alterations of dopamine receptor binding in rat brain by repeated electroconvulsive shock: an autoradiographic study. Brain Res 529:208–213

Belmaker RH, Wald D (1977) Haloperidol in normals. Br J Psychiatry 131:222–223

Bordet R, Ridray S, Carboni S, Diaz J, Sokoloff P, Schwartz J-C (1997) Induction of dopamine D_3 receptor expression as a mechanism of behavioral sensitization to levodopa. Proc Natl Acad Sci USA 94:3363–3367

Bordet R, Ridray S, Schwartz J-C, Sokoloff P (2000) Involvement of the direct striatonigral pathway in levodopa-induced sensitization in 6-hydroxydopamine-lesioned rats. Eur J Neurosci 12: 2117–2123

Bouthenet M-L, Souil E, Martres M-P, Sokoloff P, Giros B, Schwartz J-C (1991) Localization of dopamine D_3 receptor mRNA in the rat brain using in situ hybridization histochemistry: comparison with D_2 receptor mRNA. Brain Res 564:203–219

Braff DL, Grillon C, Geyer MA (1992) Gating and habituation of the startle reflex in schizophrenic patients. Arch Gen Psychiatry 49:206–215

Brown AS, Gershon S (1993) Dopamine and depression. J Neural Transm 91:75–109

Brown EE, Nomikos GG, Wilson C, Fibiger HC (1991) Chronic desimipramine enhances the effect of locally applied amphetamine on interstitial concentrations of dopamine in the nucleus accumbens. Eur J Pharmacol 202:125–127

Cabib S, Castellano C, Cestari V, Filibeck U, Puglisi-Allegra S (1991) D1 and D2 receptor antagonists differently affect cocaine-induced locomotor hyperactivity in the mouse. Psychopharmacology (Berl) 105:335–339

Caine SB, Koob GF (1993) Modulation of cocaine self-administration in the rat through D_3 dopamine receptors. Science 260:1814–1816

Caine SB, Koob GF (1994) Effects of dopamine D-1 and D-2 antagonists on cocaine self-administration under different schedules of reinforcement in the rat. J Pharmacol Exp Ther 270:209–218

Caine SB, Koob GF (1995) Pretreatment with the dopamine agonist 7-OH-DPAT shifts the cocaine self-administration dose-effect function to the left under different schedules in the rat. Behav. Pharmacol. 6:333–347

Caine SB, Koob GF, Parsons LH, Everitt BJ, Schwartz J-C, Sokoloff P (1997) D3 receptor functional test in vitro predicts potencies of dopamine agonists to reduce cocaine self-administration. Neuroreport 8:2373–2377

Carey RJ (1991) Chronic L-DOPA treatment in the unilateral 6-OHDA rat: evidence for behavioral sensitization and biochemical tolerance. Brain Res 568:205–214

Childress E, R., Roohsenow DJ, Robbins SH, O'Brien CP (1992) Classically conditioned factors in drug dependence. In: Lowinson W, Luiz P, Millman RB and Langard JG (eds) Substance abuse: a comprehensive textbook. Williams and Wilkins, Baltimore, pp 56–69

Chio CL, Lajiness ME, Huff RM (1994) Activation of heterologously expressed D_3 dopamine receptors: comparison with D_2 dopamine receptors. Mol Pharmacol 45:51–60

Chiodo LA, Antelman SM (1980a) Electroconvulsive shock: progressive dopamine autoreceptor subsensitivity independent of repeated treatment. Science 210:799–801

Chiodo LA, Antelman SM (1980b) Repeated tricyclics induce a progressive dopamine autoreceptor subsensitivity. Nature 287:451–454

Cole RL, Konradi C, Douglass J, Hyman SE (1995) Neuronal adaptation to amphetamine and dopamine: molecular mechanisms of prodynorphin gene regulation in rat striatum. Neuron 14:813–823

Coon H, Byerley W, Holik J, Hoff M, Myles-Worsley M, Lannfelt L, Sokoloff P, Schwartz JC, Waldo M, Freedman R, Plaetke R (1993) Linkage analysis of schizophrenia with five dopamine receptor genes in nine pedigrees. Am J Hum Genet 52:327–334

Crocq MA, Mant R, Asherson P, Williams J, Hode Y, Mayerova A, Collier D, Lannfelt L, Sokoloff P, Schwartz J-C, Gill M, Macher J-P, McGuffin P, Owen MJ (1992) Association between schizophrenia and homozygosity at the dopamine D_3 receptor gene. J Med Genet 29:858–860

Cussac D, Newman-Tancredi A, Pasteau V, Millan MJ (1999) Human dopamine D_3 receptors mediate mitogen-activated protein kinase activation via phosphatidylinositol 3-kinase and an atypical protein kinase C-dependent mechanism. Mol Pharmacol 56:1025–1030

de Wit H, Stewart J (1981) Reinstatement of cocaine-reinforced responding in the rat. Psychopharmacol 75:134–143

Deutch AY (1993) Prefrontal cortical dopamine systems and the elaboration of functional corticostriatal circuits: implications for schizophrenia and Parkinson's disease. J Neural Transm [Gen Sect] 91:197–221

D'Haenen HA, Bossuyt A (1994) Dopamine D2 receptors in depression measured with single photon emission computed tomography. Biological Psychiatry 35:128–32

Di Chiara G, Imperato A (1988) Drugs abused by humans preferentially increase synaptic dopamine concentrations in the mesolimbic system of freely moving rats. Proc Natl Acad Sci USA 85:5274–5278

Diaz J, Lévesque D, Griffon N, Lammers CH, Martres M-P, Sokoloff P, Schwartz J-C (1994) Opposing roles for dopamine D_2 and D_3 receptors on neurotensin mRNA expression in nucleus accumbens. Eur J Neurosci 6:1384–1387

Diaz J, Lévesque D, Lammers CH, Griffon N, Martres M-P, Schwartz J-C, Sokoloff P (1995) Phenotypical characterization of neurons expressing the dopamine D_3 receptor. Neuroscience 65:731–745

Diaz J, Pilon C, Le Foll B, Gros C, Triller A, Schwartz J-C, Sokoloff P (2000) Dopamine D_3 receptors expressed by all mesencephalic dopamine neurons. J Neurosci 20:8677–8684

Diaz J, Ridray S, Mignon V, Griffon N, Schwartz J-C, Sokoloff P (1997) Selective expression of dopamine D_3 receptor mRNA in proliferative zones during embryonic development of the rat brain. J Neurosci 17:4282–4292.

Drago J, Gerfen CR, Westphal H, Steiner H (1996) D_1 dopamine receptor-deficient mouse: cocaine-induced regulation of immediate early gene and substance P expression in the striatum. Neuroscience 74:813–823

Duaux E, Gorwood P, Sautel F, Griffon N, Sokoloff P, Schwartz J-C, Olié J-P, Lôo H, Poirier M-F (1998) Homozygosity at the dopamine D_3 receptor gene is associated with opioid dependence. Mol Psychiatry 3:333–336

Dubertret C, Gorwod P, Ades J, Feingold J, Schwartz J-C, Sokoloff P (1998) A metaanalysis of DRD3 gene and schizophrenia: ethnic heterogeneity and significant association in Caucasians. Am J Med Genet [Neuropsychiatric Genetics] 81:318–322

Dziedzicka-Wasylewska M, Rogoz R, Klimek V, Maj J (1997) Repeated administration of antidepressant drugs affects the levels of mRNA coding for D_1 and D_2 dopamine receptors in the rat brain. J Neural Transm 104:515–524

Dziedzicka-Wasylewska M, Willner P, Papp M (1997) Changes in dopamine receptor mRNA expression following chronic mild stress and chronic antidepressant treatment. Behav Pharmacol 8:607–618

Engber TM, Susel Z, Juncos JL, Chase TN (1989) Continuous and intermittent levodopa differentially affect rotation induced by D_1 and D_2 dopamine agonists. Brain Res 552:113–118

Farde L, Wiesel FA, Nordström AL, Sedval G (1989) D1- and D2-dopamine receptor occupancy during treatment with conventional and atypical neuroleptics. Psychopharmacology 99:S28–S31

Fibiger HC (1995) Neurobiology of depression: focus on dopamine. Adv Biochem Psychopharmacol 49:1–17

Finlay JM, Zigmond MJ (1997) The effects of stress on central dopaminergic neurons: possible clinical implications. Neurochem Res 22:1387–1394

Fishburn CS, David C, Carmon S, Fuchs S (1994) The effect of haloperidol on D2 dopamine receptor subtype mRNA levels in the brain. FEBS Lett 339:63–66

Fontana DJ, Post RM, Pert A (1993) Conditioned increases in mesolimbic dopamine overflow by stimuli associated with cocaine. Brain Res 629:31–39

Freedman SB, Patel S, Marwood R, Emms F, Seabrook GR, Knowles MR, McAllister G (1994) Expression and pharmacological characterization of the human D_3 dopamine receptor. J Pharmacol Exp Ther 268:417–426

Freeman TB, Spence MS, Boss BD, Spector DH, Strecker RE, Olanow CW, Kordower JH (1991) Development of dopaminergic neurons in the human substantia nigra. Exp Neurol 113:344–353

Fremeau RT, Duncan GE, Fornaretto MG, Dearry A, Gingrich JA, Brees GR, Caron MG (1991) Localization of D_1 dopamine receptor mRNA in brain supports a role in cognitive, affective and neuroendocrine aspects of dopaminergic neurotransmission. Proc Natl Acad Sci USA 88:3772–3776

Gambarana C, Masi F, Tagliamonte A, Scheggi S, Ghiglieri O, De Montis MG (1999) A chronic stress that impairs reactivity in rats also decreases dopaminergic transmission in the nucleus accumbens: a microdialysis study. Journal of Neurochemistry 72:2039–46

Garattini S (1997) Pharmacology of amineptine, an antidepressant agent acting on the dopaminergic system: a review. Int Clin Psychopharmacol 12 Suppl. 3:S15–S19

Gessa GL, Pani L, Serra G, Fratta W (1995) Animal models of mania. Adv Biochem Psychopharmacol 49:43–66

Geyer MA, Wilkinson LS, Humby T, Robbins TW (1993) Isolation rearing of rats produces a deficit in prepulse inhibition of acoustic startle similar to that in schizophrenia. Biol Psychiatry 34:361–372

Glenthoj BY, Hemmingsen R (1997) Dopaminergic sensitization: implications for the pathogenesis of schizophrenia. Prog Neuropsychopharmacol Biol Psychiatry 21:23–46

Goldberg JF, Frye MA, Dunn RT (1999) Pramipexole in refractory bipolar depression. Am J Psychiatry 156:798

Grech DM, Spealman RD, Bergman J (1996) Self-administration of D_1 receptor agonists by squirrel monkeys. Psychopharmacology 125:97–104

Grenhoff J, Nisell M, Ferre S, Aston-jones G, Svensson TH (1993) Noradrenergic modulation of midbrain dopamine cell firing elicited by stimulation of the locus coeruleus in the rat. J Neural Transm 93:11–25

Griffon N, Pilon C, Schwartz J-C, Sokoloff P (1995) The preferential dopamine D_3 receptor ligand (+)UH 232 is a partial agonist. Eur J Pharmacol 282:R3–R4

Griffon N, Crocq MA, Pilon C, Martres M-P, Mayerova A, Uyanik G, Burgert E, Duval F, Macher J-P, Javoy-Agid F, Tamminga CA, Schwartz J-C, Sokoloff P (1996) Dopamine D_3 receptor gene: organization, transcript variants and polymorphism associated with schizophrenia. Am J Med Genet [Neuropsychiatric Genetics] 67: 58–70

Griffon N, Pilon C, Sautel F, Schwartz J-C, Sokoloff P (1997) Two intracellular pathways for the dopamine D_3 receptor : opposite and synergistic interactions with cyclic AMP. J Neurochem 67:1–9

Guillin O, Diaz J, Carroll P, Griffon N, Schwartz J-C, Sokoloff P (2001) BDNF controls dopamine D_3 receptor expression and triggers behavioural sensitization. Nature 411:86–89

Gulat-Marnay C, Lafitte A, Schwartz J-C, Protais P (1985) Effects of discriminant and non-discriminant dopamine antagonists on in vivo binding of ^{3}H-N-propylnorapomorphine in mouse striatum and tuberculum olfactorium. Naunyn Schmiedeberg's Arch Pharmacol 329:117–122

Gurevich EV, Bordelon Y, Shapiro RM, Arnold SE, Gur RE, Joyce JN (1997) Mesolimbic dopamine D_3 receptors and use of antipsychotics in patients with schizophrenia. Arch Gen Psychiatry 54:225–232

Gurevich EV, Himes JW, Joyce JN (1999) Developmental regulation of expression of the D_3 dopamine receptor in rat nucleus accumbens and islands of Calleja. J Pharmacol Exp Ther 289:587–598

Haadsma-Svensson SR, Smith MW, Svensson K, Waters N, Carlsson A (1995) The chemical structure of U99194A. J Neural Transm Gen Sect 99:1

Hall H, Halldin C, Dijkstra D, Wikstrom H, Wise LD, Pugsley TA, Sokoloff P, Pauli S, Farde L, Sedvall G (1996) Autoradiographic localisation of D_3 dopamine receptors in human brain using the selective receptor agonist (+)-[^{3}H]PD 128907. Psychopharmacology 128:240–247

Harkness KL, Monroe SM, Simons AD, Thase M (1999) The generation of life events in recurrent and non-recurrent depression. Psychological Medicine 29:134–144

Heimer L (2000) Basal forebrain in the context of schizophrenia. Brain Res Brain Res Rev 31:205–235

Herroelen L, De Backer J-P, Wilczak N, Flamez A, Vauquelin G, De Keyser J (1994) Autoradiographic distribution of D_3-type dopamine receptors in human brain using [^{3}II]7-hydroxy-N,N-di-n-propyl-2-aminotetralin. Brain Res ••:222–228

Horger BA, Shelton K, Schenk S (1990) Preexposure sensitizes rats to the rewarding effects of cocaine. Pharmacol Biochem Behav 37:707–711

Huang Q, Zhou D, Chase K, Gusella JF, Aronin N, Difiglia M (1992) Immunohistochemical localization of the D_1 dopamine receptor in rat brain reveals its axonal transport, pre- and post-synaptic localization, and prevalence in the basal ganglia, limbic system and thalamic reticular nucleus. Proc Natl Acad Sci USA 89:11988–11992

Ichikawa J, Meltzer HY (1995) Effect of antidepressants on striatal and accumbens extracellular dopamine levels. Eur J Pharmacol 281:255–261

Janowsky DS, Davis JM (1976) Methylphenidate, dextroamphetamine, and levamfetamine. Effects on schizophrenic symptoms. Arch Gen Psychiatry 33:304–308

Kalivas PW, Stewart J (1991) Dopamine transmission in the initiation and expression of drug- and stress-induced sensitization of motor activity. Brain Res Rev 16:223–244

Kapur S, Mann J (1992) Role of the dopaminergic system in depression. Biol Psychiatry 32:1–17

Kinney JL (1985) Nomifensine maleate: a new second-generation antidepressant. Clin Pharmacol:625–636

Koeltzow TE, Xu M, Cooper DC, Hu XT, Tonegawa S, Wolf ME, White FJ (1998) Alterations in dopamine release but not dopamine autoreceptor function in dopamine D_3 receptor mutant mice. J Neurosci 18:2231–2238

Koob GF (1992) Dopamine, addiction and reward. Sem Neurosci 4:139–148
Koob GF (1992) Drugs of abuse: anatomy, pharmacology and function of reward pathways. Trends Pharmacol Sci 13:177–184
Kramer MS, Cutler N, Feighner J, Shrivastava R, Carman J, Sramek JJ, Reines SA, Guanghan L, Snavely D, Wyatt-Knowles E, Hale JJ, Mills SG, MacCoss M, Swain CJ, Harrison T, Hill RG, Hefti F, Scolnick EM, Cascieri MA, Chicchi GG, Sadowski S, Williams AR, Hewson L, Smith D, Carlson EJ, Hargreaves RJ, Rupniak NMJ (1998) Distinct mechanism for antidepressant activity by blockade of central substance P receptors. Science 281:1640–1645
Krebs M-O, Sautel F, Bourdel M-C, Sokoloff P, Schwartz J-C, Olié J-P, Lôo H, Poirier M-F (1998) Dopamine D_3 receptor gene variants and substance abuse in schizophrenia. Mol Psychiatry 3:337–341
Lahti RA, Roberts RC, Tamminga CA (1995) D_2-family receptor distribution in human postmortem tissue: an autoradiographic study. Neuroreport 6:2505–2512
Lajiness ME, Chio CL, Huff RM (1995) Signaling mechanisms of D_2, D_3 and D_4 dopamine receptors determined in transfected cells. Clin Pharmacol 18:S25-S33
Lammers CH, Diaz J, Schwartz J-C, Sokoloff P (2000) Selective increase of dopamine D_3 receptor gene expression as a common effect of chronic antidepressant treatments. Mol Psychiatry 5:378–388
Landwehrmeyer B, Mengod G, Palacios JM (1993) Differential visualization of dopamine D_2 and D_3 receptor sites in rat brain: a comparative study using in situ hybridization histochemistry and ligand binding autoradiography. Eur J Neurosci 5:145–153
Lannfelt L, Sokoloff P, Martres MP, Pilon C, Giros B, Jönsson E, Sedvall G, Schwartz JC (1992) Amino acid substitution in the dopamine D_3 receptor as a useful polymorphism for investigating psychiatric disorders. Psychiatric Genetics 2:249–256
Laruelle M, Abi-Dargham A, van Dyck CH, Gil R, D'Souza CD, Erdos J, McCance E, Rosenblatt W, Fingado C, Zoghbi SS, Baldwin RM, Seibyl JP, Krystal JH, Charney DS, Innis RB (1996) Single photon emission computerized tomography imaging of amphetamine-induced dopamine release in drug-free schizophrenic subjects. Proc Natl Acad Sci U S A 93:9235–9240
Le Moine C, Bloch B (1996) Expression of the D_3 dopamine receptor in peptidergic neurons of the nucleus accumbens: comparison with the D_1 and D_2 dopamine receptors. Neuroscience 73:131–143
Lett BT (1989) Repeated exposures intensify rather than diminish the rewarding effects of amphetamine, morphine, and cocaine. Psychopharmacology 98:357–362
Levant B (1997) The D_3 dopamine receptor: neurobiology and potential clinical relevance. Pharmacol Rev 49:231–252
Lévesque D (1996) Aminotetralin drugs and D3 receptor functions. What may partially selective D3 receptor ligands tell us about dopamine D3 receptor functions? Biochem Pharmacol 52:511–518
Lévesque D, Diaz J, Pilon C, Martres M-P, Giros B, Souil E, Schott D, Morgat J-L, Schwartz J-C, Sokoloff P (1992) Identification, characterization and localization of the dopamine D_3 receptor in rat brain using 7-[^{3}H]-hydroxy-N,N di-n-propyl-2-aminotetralin. Proc Natl Acad Sci USA 89:8155–8159
Lévesque D, Martres M-P, Diaz J, Griffon N, Lammers CH, Sokoloff P, Schwartz J-C (1995) A paradoxical regulation of the dopamine D_3 receptor expression suggests the involvement of an anterograde factor from dopamine neurons. Proc Natl Acad; Sci USA 92:1719–1723
L'hirondel M, Chéramy A, Gedeheu G, Artaud F, Saiardi A, Borrelli E, Glowinski J (1998) Lack of autoreceptor-mediated inhibitory control of dopamine release in striatal synaptosomes of D_2 receptor-deficient mice. Brain Res 792:253–262
Lieberman JA, Kane JM, Alvir J (1987) Provocative tests with psychostimulant drugs in schizophrenia. Psychopharmacology 91:415–433

Lieberman JA, Sheitman BB, Kinon BJ (1997) Neurochemical sensitization in the pathophysiology of schizophrenia: deficits and dysfunction in neuronal regulation and plasticity. Neuropsychopharmacology 17:205–229

Lundstrom K, Turpin MP (1996) Proposed schizophrenia-related gene polymorphism: expression of the Ser9Gly mutant human dopamine D_3 receptor with the Semliki virus system. Bioch Biophys Res Comm 225:1068–1072

Maj J, Dziedzicka-Wasylewska M, Rogoz E, Rogoz Z (1998) Effect of antidepressant drugs administered repeatedly on the dopamine D_3 receptors in the rat brain. Eur J Pharmacol 351:31–37

Maj J, Papp M, Skuza G, Bigajska K, Zazula M (1989) The influence of repeated treatment with imipramine, (+)- and (−)-oxaprotiline on behavioural effects of dopamine D-1 and D-2 agonists. J Neural Trans 76:29–38

Maj J, Rogoz Z, Skuza G, Kolodziejczyk K (1997) Antidepressant effects of pramipexole, a novel dopamine receptor agonist. J Neural Transm 104:525–533

Maj J, Rogoz Z, Skuza G, Sowinska H (1984) Repeated treatment with antidepressant drugs potentiates the locomotor response to (+)-amphetamine. J Pharm Pharmacol 36:127–130

Maldonado R, Robledo P, Chover AJ, Caine SB, Koob GF (1993) D1 dopamine receptors in the nucleus accumbens modulate cocaine self-administration in the rat. Pharmacol Biochem Behav 45:239–242

Malmberg A, Jackson DM, Eriksson A, Mohell N (1993) Unique binding characteristics of antipsychotic agents interacting with human dopamine D_{2A}, D_{2B} and D_3 receptors. Mol Pharmacol 43:749–754

Markou A, Koob GF (1991) Postcocaine anhedonia. An animal model of cocaine withdrawal. Neuropsychopharmacology 4:17–26

McAllister G, Knowles MR, Ward-Booth SM, Sinclair HA, Patel S, Marwood R, Emms F, Smith A, Seabrook GR, Freeman SB (1995) Functional coupling of human D_2, D_3 and D_4 dopamine receptors in HEK293 cells. J of Receptor & Signal Transduction Research 15:267–281

Meador-Woodruff JH, Grandy DK, Van Tol HHM, Damask SP, Little KY, Civelli O, Watson SJ (1994) Dopamine receptor gene expression in the human medial temporal lobe. Neuropsychopharmacol 10:239–248

Meil WM, See RE (1996) Conditioned cue recovery of responding following prolonged withdrawal from self-administered cocaine in rats: an animal model of relapse. Behav Pharmacol 7:754–763

Meil WM, See RE (1997) Lesions of the basolateral amygdala abolish the ability of drug associated cues to reinstate responding during withdrawal from self-administered cocaine. Behav Brain Res 87:139–148

Mercuri NB, Daiardi A, Bonci A, Picetti R, Calabresi P, Bernardi G, Borrelli E (1997) Loss of autoreceptor function in dopaminergic neurons form dopamine D_2 receptor-deficient mice. Neuroscience 79:323–327

Mierau J, Schneider FJ, Ensinger HA, Chio CL, Lajiness ME, Huff RM (1995) Pramipexole binding and activation of cloned and expressed dopamine D_2, D_3, D_4 receptors. Eur J Pharmacol 290:29–36

Millan MJ, Gobert A, Newman-Tancredi A, Lejeune F, Cussac D, Rivet J-M, Audinot V, Dubuffet T, Lavielle G (2000) S33084, a novel potent, selective, and competitive antagonist at dopamine D_3-receptors: I receptoral, electrophysiological and neurochemical profile compared with GR218,231 and L741,626. J Pharmacol Exp Ther 293:1048–1062

Millan MJ, Peglion J-L, Vian J, Rivet J-M, Brocco M, Gobert A, A. N-T, Dacquet C, Bervoets K, Giradon S, Jacques V, Chaput C, Audinot V (1995) Functional correlates of dopamine D_3 receptor activation in the rat in vivo and their modulation by the selective antagonist, (+) S-14297 : I. activation of postsynaptic D_3 receptors mediates hypothermia, whereas blockade of D_2 receptors elicits prolactin secretion and catalepsy. J Pharmacol Exp Ther 275:885–898

Miller JC, Friedhoff AJ (1986) Prenatal neuroleptic exposure alters postnatal striatal cholinergic activity in the rat. Dev Neurosci 8:111–116

Milner PM (1991) Brain-stimulation reward: a review. Can J Psychol 45:1–36
Moratalla R, Xu M, Tonegawa S, Graybiel AM (1996) Cellular responses to psychomotor stimulant and neuroleptic drugs are abnormal in mice lacking the D1 dopamine receptor. Proc Natl Acad Sci USA 93:14928–14933
Morissette M, Goulet M, Grondin R, Blanchet P, Bedard PJ, Di Paolo T, Levesque D (1998) Associative and limbic regions of monkey striatum express high levels of dopamine D3 receptors: effects of MPTP and dopamine agonist replacement therapies. Eur J Neurosci 10:2565–2573
Murray AM, Ryoo HL, Gurevich E, Joyce JN (1994) Localization of dopamine D_3 receptors to mesolimbic and D_2 receptors to mesostriatal regions of human forebrain. Proc Natl Acad Sci USA 91:11271–11275
Nestler EJ (1992) Molecular mechanisms of drug addiction. J Neuroscience 12:2439–2450
Nibuya M, Morinobu S, Duman RS (1995) Regulation of BDNF and TrkB mRNA in rat brain by chronic electroconvulsive seizure and antidepressant drug treatments. J Neurosci 15:7539–7547
Nomikos GG, Zis AP, Damsma G, Fibiger HC (1991) Electroconvulsive shock produces large increases in interstitial concentrations of dopamine in the rat striatum: an in vivo microdialysis study. Neuropsychopharmacology 4:65–69
Numan S, Seroogy KB (1997) Increased expression of TrkB mRNA in rat caudate-putamen following 6-OHDA lesions of the nigrostriatal pathway. Eur J Neurosci 9:489–495
O'Brien CP, Childress AR, McMellan AT, Ehrman RA (1992) A learning model of addiction. Res Publ Assoc Res Nerv Ment Dis 70:157–177
O'Brien CP, McMellan AT (1996) Myths about the treatment of addiction. Lancet 347:237–240
O'Hara CM, Uhland-Smith A, O'Malley KL, Todd RD (1996) Inhibition of dopamine synthesis by dopamine D_2 and D_3 bot not D_4 receptors. J Pharmacol Exp Ther 277:186–192
Papp M, Klimek V, Willner P (1994) Parallel changes in dopamine D2 receptor binding in limbic forebrain associated with chronic mild stress-induced anhedonia and its reversal by imipramine. Psychopharmacology 115:441–446
Pennartz CM, Groenewegen HJ, Lopes da Silva FH (1994) The nucleus accumbens as a complex of functionally distinct neuronal ensembles: an integration of behavioural, electrophysiological and anatomical data. Prog Neurobiol 42:719–761
Perachon S, Betancur C, Pilon C, Rostene W, Schwartz J-C, Sokoloff P (2000) Role of dopamine D3 receptors in thermoregulation: a reappraisal. Neuroreport 11:221–225
Pilla M, Perachon S, Sautel F, Garrido F, Mann A, Wermuth CG, Schwartz J-C, Everitt BJ, Sokoloff P (1999) Selective inhibition of cocaine-seeking behaviour by a partial dopamine D_3 receptor agonist. Nature 400:371–375
Pilon C, Lévesque D, Dimitriadou V, Griffon N, Martres MP, Schwartz J-C, Sokoloff P (1994) Functional coupling of the human dopamine D_3 receptor in a transfected NG 108-15 neuroblastoma-glioma hybrid cell line. Eur J Pharmacol [Mol Pharmacol Sect] 268:129–139
Piomelli D, Pilon C, Giros B, Sokoloff P, Martres M-P, Schwartz J-C (1991) Dopamine activation of the arachidonic acid cascade via a modulatory mechanism as a basis for D_1/D_2 receptor synergism. Nature 353:164–167
Porsolt RD, Anton G, Blavet N, Jalfre M (1978) Behavioral despair in rats: a new model sensitive to antidepressant treatments. Eur J Pharmacol 47:379–391
Prisco S, Esposito E (1995) Differential effects of acute and chronic fluoxetine administration on the spontaneous activity of dopaminergic neurones in the ventral tegmental area. Br J; Pharmacol 116:1923–1931
Ralph RJ, Varty GB, Kelly MA, Wang Y-M, Caron MG, Rubinstein M, Grandy DK, Low MJ, Geyer MA (1999) The dopamine D_2, but not D_3 or D_4, receptor subtype is essential for the disruption of prepulse inhibition produced by amphetamine in mice. J Neurosci 19:4627–4633

Reavill C, Taylor SG, Wood MD, Ashmeade T, Austin NE, Avenell KY, Boyfield I, Branch CL, Cilia J, Coldwell MC, Hadley MS, Hunter AJ, Jeffrey P, Jewitt F, Johnson CN, Jones DNC, Medhurst AD, Middlemiss DN, Nash DJ, Riley GJ, Routledge C, Stemp G, Thewlis KM, Trail B, Vong AKK, Hagan JJ (2000) Pharmacological actions of a novel high-affinity, and selective human dopamine D_3 receptor antagonist, SB-277011-A. J Pharmacol Exp Ther 294:1154–1165

Ridray S, Griffon N, Souil E, Mignon V, Carboni S, Diaz J, Schwartz J-C, Sokoloff P (1998) Coexpression of dopamine D_1 and D_3 receptors in rat ventral striatum: opposite and synergistic functional interactions. Eur J Neurosci 10:1676–1686

Risch SC (1997) Recent advances in depression research: from stress to molecular biology and brain imaging. J Clin Psychiatry 58 Suppl 5:3–6

Roberts GW (1991) Schizophrenia: a neuropathological perspective. Br J Psychiatry 158:8–17

Rosengarten H, Fridhoff AJ (1979) Enduring changes in dopamine receptor cells of pups from drug administration to pregnant and nursing rats. Science 203:1133–1135

Ross CA, Pearlson GD (1996) Schizophrenia, the heteromodal association neocortex and development: potential for a neurogenetic approach. Trends Neurosci 19:717–176

Ryoo HL, Pierrotti D, Joyce JN (1998) Dopamine D_3 receptor is decreased and D_2 receptor is elevated in the striatum of Parkinson's disease. Mov Disord 13:788–797

Salamone JD (1994) The involvement of nucleus accumbens dopamine in appetitive and aversive motivation. Behav Brain Res 61:117–133

Sautel F, Griffon N, Lévesque D, Pilon C, Schwartz JC, Sokoloff P (1995) A functional test identifies dopamine agonists selective for D_3 versus D_2 receptors. Neuro Report 6:329–332

Sautel F, Griffon N, Sokoloff P, Schwartz J-C, Launay C, Simon P, Costentin J, Schoenfelder A, Garrido F, Mann A, Wermuth CG (1995) Nafadotride, a potent preferential dopamine D_3 receptor antagonist, activates locomotion in rodents. J Pharmacol Exp Ther 275:1239–1246

Schwartz JC, Diaz J, Bordet R, Griffon N, Perachon S, Pilon C, Ridray S, Sokoloff P (1998) Functional implications of multiple dopamine receptor subtypes: the D1/D3 receptor coexistence. Brain Res Rev 26:236–242

Schwartz J-C, Diaz J, Pilon C, Sokoloff P (2000) Possible implications of the D_3 receptor in schizophrenia and antipsychotic drug actions. Brain Res Rev 31:277–287

Seabrook GR, Kemp JA, Freedman SB, Patel S, Sinclair HA, McAllister G (1994) Functional expression of human D_3 receptor in differentiated neuroblastoma X glioma NG 108-15 cells. Br J Pharmacol 111:391–393

Seeman P, Lee T, Chau-Wong M, Wong K (1976) Antipsychotic drug doses and neuroleptic/dopamine receptors. Nature 261:717–719

Segal DM, Moraes CT, Mash DC (1997) Up-regulation of D_3 dopamine receptor mRNA in the nucleus accumbens of human cocaine fatalities. Mol Brain Res 45:335–339

Self DW, Nestler EJ (1988) Relapse to drug-seeking: neural and molecular mechanisms. Drug Alcohol Dep 51:49–60

Seroogy KB, Lundgren KH, Tran TMD, Guthrie KM, Isackson PJ, Gall C (1994) Dopaminergic neurons in rat ventral midbrain express brain-derived neurotrophin factor and neurotrophin-3 mRNAs. J Comp Neurol 342:321–334

Serra G, Collu M, D'Aquila PS, De Montis GM, Gessa GL (1990) Possible role of dopamine D1 receptor in the behavioural supersensitivity to dopamine agonists induced by chronic treatment with antidepressants. Brain Res 527:234–243

Shah PJ, Ogilvie AD, Goodwin GM, Ebmeier KP (1997) Clinical and psychometric correlates of dopamine D2 binding in depression. Psychological Medicine 27:1247–56

Shirayama Y, Mitsushio H, Takashima M, Ichikawa H, Takahashi K (1996) Reduction of substance P after chronic antidepressants treatment in the striatum, substantia nigra and amygdala of the rat. Brain Res 739:70–78

Silverstone T (1985) Dopamine in manic depressive illness. A pharmacological synthesis. J Affect Dis 8:225–231

Smith MA, Makino S, Kvenansky R, Post RM (1995) Stress and glucocorticoids affect the expression of brain-derived neurotrophic factor and neurotrophin-3 mRNAs in the hippocampus. J Neurosci 15:1768–1777

Smith SE, Sharp T (1997) Evidence that the enhancement of dopamine function by repeated electroconvulsive shock requires concomitant activation of D1-like and D2-like receptors. Psychopharmacology 133:77–84

Snyder SH (1976) The dopamine hypothesis of schizophrenia: focus on the dopamine receptor. Am J Psychiatry 133:197–202

Sokoloff P, Andrieux M, Besançon R, Pilon C, Martres M-P, Giros B, Schwartz J-C (1992) Pharmacology of human D_3 dopamine receptor expressed in a mammalian cell line: comparison with D_2 receptor. Eur J Pharmacol Mol Pharmacol Sect 225: 331–337

Sokoloff P, Giros B, Martres M-P, Bouthenet M-L, Schwartz J-C (1990) Molecular cloning and characterization of a novel dopamine receptor (D_3) as a target for neuroleptics. Nature 347:146–151

Sokoloff P, Schwartz J-C (1999) Une nouvelle arme pour vaincre la pharmacodépendance? Médecine/Sciences 15:1067–1069

Spealman RD (1996) Dopamine D_3 receptor agonists partially reproduce the discriminative stimulus effects of cocaine in squirrel monkeys. J Pharmacol Exp Ther 278: 1128–1137

Spealman RD, Bergman J, Madras BK, Melia KF (1991) Discriminative stimulus effects of cocaine in squirrel monkeys: involvement of dopamine receptor subtypes. J Pharmacol Exp Ther 258:945–53

Staley JK, Mash DC (1996) Adaptive increase in D3 dopamine receptors in the brain reward circuits of human cocaine fatalities. J Neuroscience 16:6100–6106

Steru L, Chermat R, Thierry B, Simon P (1985) The tail suspension test: a new method for screening antidepressants in mice. Psychopharmacology (Berl) 85:367–370

Stewart J (1983) Conditioned and unconditioned drug effects in relapse to opiate and stimulant drug-administration. Prog Neuropsychopharmacol Biol Psychiatry 7:591–597

Stewart J, Rajabi H (1996) Initial increases in extracellular dopamine in the ventral tegmental area provide a mechanism for the development of desimipramine-induced sensitization within the midbrain dopamine system. Synapse 23:258–264

Sunahara RK, Guan HC, O'Dowd BF, Seeman P, Laurier LG, Ng G, George SR, Torchia J, Van Tol HHM, Niznik HB (1991) Cloning of the gene for a human dopamine D_5 receptor with higher affinity for dopamine than D_1. Nature 350:614–619

Sung CH, Schneider BG, Agarwal N, Papermaster DS, Nathans J (1991) Functional heterogeneity of mutant rhodopsins responsible for autosomal dominant retinitis pigmentosa. Proc Natl Acad Sci USA 88:8840–8844

Susuki M, Hurd YL, Hall H, Sokoloff P, Schwartz J-C, Sedvall G (1998) D_3 dopamine receptor mRNA is widely expressed in the human brain. Brain Res 779:58–74

Svensson K, Johansson AM, Magnusson T, Carlsson A (1986) (+)-AJ 76 and (+)-UH 232: central stimulants acting as preferential dopamine autoreceptor antagonists. Naunyn-Schmiedegerg's Arch Pharmacol 334:234–245

Swarzenski BC, Tang L, Oh YJ, O'Malley KL, Todd RD (1994) Morphogenic potentials of D_2, D_3 and D_4 dopamine receptor revealed in transfected neuronal cell lines. Proc Natl Acad Sci USA 91:649–653

Szegedi A, Wetzel J, Hillert A, Kleiser E, Gaebel W, Benkert O (1996) Pramipexole, a novel selective dopamine agonist in major depression. Mov Disord 11(Suppl.1):266

Takahashi M, Shirakawa O, Toyooka K, Kitamura N, Hashimoto T, Maeda K, Koizumi S, Wakabayashi K, Takahashi H, Someya T, Nawa H (2000) Abnormal expression of brain-derived neurotrophic factor and its receptor in the corticolimbic system of schizophrenic patients. Mol Psychiatry 5:293–300

Tang L, Todd RD, Heller A, O'Malley KL (1994) Pharmacological and functional characterization of D_2, D_3 and D_4 dopamine receptors in fibroblasts and dopaminergic cell lines. J Pharmacol Exp Ther 268:495–502

Tang L, Todd RD, O'Malley KL (1994) Dopamine D_2 and D_3 receptors inhibit dopamine release. J Pharmacol Exp Ther 270:475–479

Tarazi FI, Florijn WJ, Creese I (1997) Differential regulation of dopamine receptors after chronic typical and atypical antipsychotic drug treatment. Neuroscience 78: 985–996

Tella SR (1994) Differential blockade of chronic versus acute effects of intravenous cocaine by dopamine receptor antagonists. Pharmacol Biochem Behav 48:151–9

Thierry A-M, Tassin J-P, Blanc G, Glowinski J (1976) Selective activation of mesocortical DA systems by stress. Nature 263:242–244

Thoenen H (1995) Neurotrophins and neuronal plasticity. Science 270:593–598

Van Tol HHM, Bunzow JR, Guan HC, Sunahara RK, Seeman P, Niznik HB, Civelli O (1991) Cloning of the gene for a human dopamine D_4 receptor with high affinity for the antipsychotic clozapine. Nature 350:610–614

Voorn P, Kalsbeek A, Jorritsma-Byham B, Groenewegen HJ (1988) The pre- and post-natal development of the dopaminergic cell group in the developmental mesencephalon and the dopaminergic innervation of the striatum of the rat. Neuroscience 25:857–887

Walker PD, Riley LA, Hart HP, Jonakait GM (1991) Serotonin regulation of tachykinin biosynthesis in the rat neostriatum. Brain Res 546:33–39

Waters N, Svensson K, Haadsma-Svensson SR, Smith MW, Carlsson A (1993) The dopamine D_3 receptor: a postsynaptic receptor inhibitory on rat locomotor activity. J Neural Transm [Gen Sect] 94:11–19

Weinberger DR (1987) Implications of normal brain development for the pathogenesis of schizophrenia. Arch Gen Psychiat 44:660–669

White FJ, Wang RY (1983) Differential effects of classical and atypical antipsychotic drugs on A9 and A10 dopamine neurons. Science 221:1054–1057

Whitelaw RB, Markou A, Robbins TW, Everitt BJ (1996) Excitotoxic lesions of the basolateral amygdala impair the acquisition of cocaine-seeking behaviour under a second-order schedule of reinforcement. Psychopharmacology 127:213–224

Wiese C, Lannfelt L, Kristbjarnarson H, Yang L, Zoega T, P. S, Ivorsson O, Vinogradov S, Schwartz J-C, H.W. M, Helgason T (1993) No linkage between schizophrenia and D_3 dopamine receptor gene locus in icelandic pedigrees. Psychiatry Res 46: 253–259

Williams J, Spurlock G, Holmans P, Mant R, Murphy K, Jones L, Cardno A, Asherson P, Blackwood D, Muir W, Meszaros K, Aschauer H, Mallet J, Laurent C, Pekkarinen P, Seppala J, Stefanis CN, Papadimitriou GN, Macciardi F, Verga M, Pato C, Azevedo H, Crocq M-A, Gurling H, Kalsi G, Curtis D, McGuffin P, Owen MJ (1998) A meta-analysis and transmission disequilibrium study of association between the dopamine D_3 receptor gene and schizophrenia. Mol Psychiatry 3:141–149

Willner P (1997) The mesolimbic dopamine system as a target for rapid antidepressant action. Int Clin Psychopharmacology 12 Suppl. 3:S7–S14

Willner P, Lappas S, Cheeta S, Muscat R (1994) Reversal of stress-induced anhedonia by the dopamine receptor agonist, pramipexole. Psychopharmacology (Berl) 115: 454–462

Willner P, Muscat R, Papp M (1992) Chronic mild stress-induced anhedonia: a realistic animal model of depression. Neurosci & Biobehav Rev 16:525–34

Woolverton WL, Cervo L, Johanson CE (1984) Effects of repeated methamphetamine administration on methamphetamine self-administration in rhesus monkeys. Pharmacol Biochem Behav 21:737–741

Xu M, Koeltzow TE, Cooper DC, Tonegawa S, White FJ (1999) Dopamine D_3 receptor mutant and wild-type mice exhibit identical responses to putative D_3 receptor-selective agonists and antagonists. Synapse 31:210–215

Xu M, Koeltzow TE, Tirado Santiago G, Moratella R, Cooper DC, Hu X-T, White NM, Graybiel AM, White FJ, Tonegawa S (1997) Dopamine D_3 receptor mutant mice exhibit increased behavioral sensitivity to concurrent stimulation of D_1 and D_2 receptors. Neuron 19:837–848

Xu M, Moratalla R, Gold LH, Hiroi N, Koob GF, Graybiel AM, Tonegawa S (1998) Dopamine D_1 receptor mutant mice are deficient in striatal expression of dynorphin and in dopamine-mediated behavioral responses. Neuron 79:729–742

Zahm DS, Brog JS (1992) On the significance of subterritories in the "accumbens" part of the rat ventral striatum. Neuroscience 50:751–767

Zung WW (1983) Review of placebo-controlled trials with bupropion. J Clin Psychiatry 44:104–114

CHAPTER 8
Dopamine D_4 Receptors: Molecular Biology and Pharmacology

O. CIVELLI

A. The Dopamine D_4 Receptor

I. Discovery

The discovery of the dopamine D_4 receptor stems from the same principle as the one which led to the cloning of the D_2 and the other dopamine receptors: the recognition that G protein-coupled receptors are evolutionarily related and thus may share sequence similarities (BUNZOW et al. 1988). Consequently, homology-screening approaches may lead to the identification of novel subtypes. By analyzing the mRNAs of human neuroepithelioma SK-N-MC cells with D_2 receptor cDNA probes under conditions of low stringency hybridization, a D_2-related mRNA was detected which encoded a protein of 387 residues containing seven hydrophobic domain regions. This protein exhibited an overall homology of 41% and 39% with the D_2 and D_3 receptors respectively and about 56% for both receptors within the hydrophobic regions. It was named the dopamine D_4 receptor (VAN TOL et al. 1991).

II. Pharmacological Profile

The D_4 receptor, when expressed in COS-7 cells, bound dopamine with high- and low-affinity states (VAN TOL et al. 1991). It also exhibited a saturable binding of [^{3}H]-spiperone, the prototypic D_2-like antagonist. Further pharmacological characterization revealed that the D_4 receptor bound other D_2 antagonists and agonists with similar or somewhat lower affinity than did the D_2 receptor (VAN TOL et al. 1991) (Table 1). A particularly notable exception to this profile was observed for the atypical antipsychotic clozapine, which bound to the D_4 receptor with higher affinity than to the D_2 receptor (VAN TOL et al. 1991; VAN TOL et al. 1992; LAHTI et al. 1993; MILLS et al. 1993; CHIO et al. 1994; MCHALE et al. 1994). The D_4 receptor did not show a particularly high affinity for other newer atypical antipsychotics, such as olanzapine, risperidone, and remoxipride. Interestingly, The D_4 receptor has been shown to bind and be activated by epinephrine and norepinephrine at concentrations that are in the range of those activating the adrenergic receptors (LANAU et al. 1997; NEWMAN-TANCREDI et al. 1997).

Table 1. Pharmacological profile of the D_4 receptor

	GTP-γS binding	EC50 μM
	D_4	D_2
Dopamine	0.2	9.4
Epinephrine	1.2	>100
Norepinephrine	2.7	>100
	Binding affinities	K_i nM
	D_4	D_2
Chlorpromazine	7.9	1.1
Haloperidol	1.8	0.67
Raclopride	1,650	2.3
Sulpiride	569	9
Spiperone	0.4	0.13
Olanzapine	27	11
Risperidone	3.6	1.5
Remoxipride	3,690	125
Clozapine	10	60

Affinities of different dopamine-related compounds to the D_4 receptor expressed in heterologous cell lines. Agonist affinities are measured as GTP-γS binding to membranes (LANAU et al. 1997). Antagonist affinities are measured as spiperone displacement (HARTMAN and CIVELLI 1996).

III. Biological Activities

Heterologous expression of the D_4 receptor has further shown that, depending on the cellular environment, the D_4 receptor couples to inhibition of adenylyl cyclase activity, stimulation of Na^+/H^+ exchange, and potentiation of stimulated arachidonic acid release in Chinese hamster ovary (CHO) cells (CHIO 1994; MCHALE 1994). In cultured cerebral granule cells, D_4R activation inhibits an L-type calcium current via a G_i/G_o protein, but independent of adenylyl cyclase inhibition (MEI et al. 1995). Furthermore, the D_4R repeats are capable of direct interaction with SH3-domain containing proteins (i.e., cytoskeletal proteins, serine/threonine and tyrosine kinases) for additional D_4 receptor signaling mechanisms independent of G-protein activation (OLDENHOF et al. 1998).

IV. Tissue Localization

In the CNS, the D_4 receptor is expressed in limbic and cortical areas which are involved in emotional/affective behavior and cognition. The D_4 mRNA has been detected in human frontal cortex, amygdala, thalamus, hypothalamus, cerebellum, and hippocampus (MATSUMOTO et al. 1995). In situ hybridization in human temporal lobe have detected D_4 mRNA in the dentate gyrus and the CA2 region of the hippocampus, as well as in the entorhinal cortex

(MEADOR-WOODRUFF et al. 1994). In mouse retina, D_4 mRNA has been localized in the photoreceptor cells, the inner nuclear and the ganglion cell layers (COHEN et al. 1992). D_4 mRNA has been detected in the pituitary, and high levels have been found in the rat heart (O'MALLEY et al. 1992) although this was not confirmed in human tissue (MATSUMOTO et al. 1995). Several D_4 specific antisera have been raised (ARIANO et al. 1997; DEFAGOT et al. 1997), one in particular has been reported to label γ-aminobutyric acid (GABA)ergic neurons in the cerebral cortex, hippocampus, reticular thalamic nucleus, globus pallidus, and substantia nigra (MRZLJAK et al. 1996). In the mouse brain, D_4 immunoreactivity was found in neurons located in layer II-VI of the frontal and piriform cortices, in scattered neurons in the caudate putamen and in larger neurons of the globus pallidus (MAUGER et al. 1998). Together, these data show that the D_4 receptor is expressed in the CNS but that its level of expression is low when compared to that of the D_2 receptor in the striatum.

V. Physiological Role

Thus far, little is known about the physiological role of the D_4 receptor. In retina, D_4-like activation mediates dopamine-induced contraction and elongation of cone photoreceptor cells (HILLMAN 1995) and regulates activity of serotonin N-acetyltransferase, which controls melatonin synthesis (ZAWILSKA and NOVAK 1994). In rats, D_4 receptors play an important role in the induction of behavioral sensitization to amphetamine and accompanying adaptations in pre- and postsynaptic neural systems associated with the mesolimbocortical dopamine projections (FELDPAUSCH et al. 1998) and to be active in the reversal of apomorphine-induced blockade of prepulse inhibition (MANSBACH et al. 1998). Moreover, mice devoid of D_4 receptor have been engineered. These null mutant mice were less active in open field tests but outperformed wild-type mice on the rotarod and displayed locomotor supersensitivity to ethanol, cocaine, and methamphetamine. Biochemical analyses revealed that dopamine synthesis and its conversion to DOPAC were elevated in the dorsal striatum from mutant mice suggesting that the D_4 receptor modulates normal, coordinated, and drug-stimulated motor behaviors as well as the activity of nigrostriatal dopamine neurons (RUBINSTEIN et al. 1997).

B. The Multiple Human D_4 Receptors

I. D_4 Receptors with Variable Third Cytoplasmic Loops

The D_4 receptor gene (DRD_4) is located at the tip of the short arm of the human chromosome 11p15 (GELERNTER et al. 1992) and is highly polymorphic in humans and other primates predicting great diversity in D_4 receptor proteins (Fig. 1). The most striking polymorphism involves a 48-base pair (bp) sequence which encodes an imperfect 16 amino acid repeat in the putative third cytoplasmic loop of the protein (VAN TOL et al. 1992). DRD_4 alleles differ not only in the number of these repeat units, which can vary from 2 to 10 (4

D4 GENE POLYMORPHISM

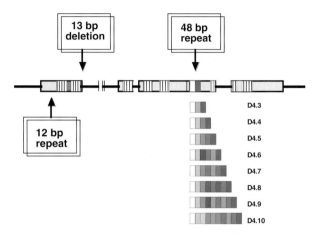

Fig. 1. Schematic representation of the human D_4 receptor gene organization and of the location of the polymorphic sites

is most common, and no alleles with 9 repeats have been found), but they also vary in nucleotide sequence and in the order in which they are arranged (LICHTER et al. 1993). In one study, 18 distinct D_4R protein sequences were identified among 178 human alleles (LICHTER et al. 1993), and although this repeat structure is not apparent in rat or mouse D_4R sequences, similar repeat units have been identified in several species of primates (LIVAK et al. 1995). The 16 amino acid repeats are proline-rich, and consist of tandem copies of consensus SH3-binding domains (OLDENHOF et al. 1998). Although the functional significance of D_4 repeat sequence variation is not clear, weak differential effects of sodium on antagonist binding have been reported depending on the number of repeat sequences (VAN TOL et al. 1992), and a somewhat reduced response to dopamine was observed in the $D_{4.7}$ variant compared to the shorter $D_{4.2}$ and $D_{4.4}$ variants (ASHGARI et al. 1995). Moreover, genetic linkage studies have found no correlation between the number of 16-amino acid repeats in the D_4R and incidence of schizophrenia (NANKO et al. 1993). On the other hand, two studies have suggested that the presence of the $D_{4.7}$ allele (containing seven 16-amino acid repeats) correlates with a higher level of "novelty seeking" behavior than that found in persons with the shorter $D_{4.4}$ allele (BENJAMIN et al. 1996; EBSTEIN et al. 1996). These results would suggest that the D_4 receptor can contribute to variation in a human psychological trait; however this linkage has been challenged by several studies showing that the dopamine D_4 receptor gene is probably not of importance to different personality dimensions (VANDENBERGH et al. 1997; JÖNSSON et al. 1997; POGUE-GEILE et al. 1998; GELERNTER et al. 1997). In another study, the $D_{4.7}$ allele has been reported to occur more frequently in children who develop attention deficit hyperactivity disorder (ADHD) (LaHoste et al. 1996) indicating that the dopamine D_4 receptor may be a susceptibility gene in ADHD (Smalley et al. 1998; Swanson et al. 1998).

II. D_4 Receptor with Different N-Termini

A second D_4DR polymorphism has been discovered involving a 12-bp repeat in the extracytoplasmic terminus bordering the first putative transmembrane domain. The more common allele (A1) contains two 12-bp repeats. The presence of the mutated allele (A2), containing only one 12-bp repeat, correlates with delusional disorder (Catalano et al. 1993) whose affected individuals respond poorly to neuroleptics, suggesting that the A2 allele may encode a D_4 receptor with altered affinities for these drugs. The two allelic receptors have, however, been shown to respond similarly to dopamine stimulation (Zenner et al. 1998). A third polymorphic site has been found in the first transmembrane domain. It leads to a nonsense mutation which has been found in humans with a frequency of about 2% in the general population but which did not correlate with schizophrenia, bipolar disorder, or Tourette's syndrome (Nothen et al. 1994). One individual homozygous for the null mutation has been identified who suffers from acoustic neurinoma, obesity, and reduced body temperature. He has four children; one suffers from Tourette's syndrome, another from a mild tic disorder, and a third from obsessive-compulsive disorder. No correlation was found between the presence of heterozygosity for the D_4 null mutation and either schizophrenia, bipolar disorder, or Tourette's syndrome (reviewed in Hartman and Civelli 1997).

C. The D_4 Receptor Involvement in Schizophrenia

The antipsychotic action of neuroleptics has been correlated with their ability to block D_2-like receptors (Creese et al. 1976; Seeman et al. 1976). Two observations pointed at the D_4 receptor as the dopamine receptor having a central role in the etiology of schizophrenia.

First was the report that claimed elevated densities of D_4R in striatal homogenates from postmortem tissue samples of schizophrenic vs control subjects (Seeman et al. 1993). Lack of a D_4R-specific radioligand led to use of an indirect method to measure D_4 receptor density by subtraction of 3H-raclopride binding sites (labels D_2R+D_3R) from 3H-emonapride sites (labels $D_2R+D_3R+D_4R$). In these tissues, D_4R levels were found to be low or undetectable in control tissues, but readily measurable in schizophrenic samples, with an estimated two- to sixfold increase over control levels (Seeman et al. 1993). The levels of D_4 receptor-like binding sites in these samples, however, is much higher than predicted from mRNA levels in these areas. In fact, since emonapride binds with high affinity to sigma and 5-HT2A sites while raclopride does not, this labeling will also reflect sigma and 5-HT2A binding sites (Tang et al. 1997; Noda-Saita et al. 1999). Confirmation of this observation awaits the use of D_4-specific ligands.

The second observation is the fact that clozapine, an effective atypical antipsychotic with a low propensity for side effects, has a higher affinity for the D_4 vs the D_2 receptor (Van Tol et al. 1991), suggesting that D_4-specific antagonists might be useful in the treatment of schizophrenia. An increasing number of D_4-specific ligands have been reported (Hidaka et al. 1996; Kula et al. 1997; Rowley et al. 1997; Unangst et al. 1997; Arlt et al. 1998; Bourrain et al. 1998; Faraci et al. 1998; Sanner et al. 1998; Löber et al. 1999), none of which show activity in animal models for antipsychotic activity. The D_4-selective antagonist, NGD94-1, did not affect locomotor activity in rats but attenuated apomorphine-induced disruption of prepulse inhibition, an effect which has been observed with other antipsychotic drugs (Tallman et al. 1997). U-101387, another D_4 antagonist, also produced no effects on locomotor activity and failed to alter dopamine neuronal firing or affect monoamine turnover, but did substantially induce c-*fos* expression in cortical areas of rat brain (Merchant et al. 1996). Two substituted 4-aminoperidines (U-99363E and U-101958) have also been reported to be high-affinity antagonists to the D_4 receptor but have poor stability and low bioavailability (Schlachter et al. 1997). Finally, L-745870 a selective high affinity D_4 antagonist did not increase dopamine metabolism or plasma prolactin levels in rodents, two responses affected by standard neuroleptics (Patel et al. 1997). Moreover, L-745870 was evaluated in phase II clinical trials of patients with acute schizophrenia and, while generally well tolerated, did not show clinical efficacy. It was thus concluded that blockade of D_4 receptors does not appear to confer antipsychotic activity (Kramer et al. 1997).

D. Conclusions

Among the new dopamine receptors, the D_4 receptor is the one that has been most heavily studied. It is a D_2-like receptor, which places it among the targets of antipsychotic drugs. It is expressed in brain areas involved in emotional/affective behavior and cognition. It exhibits two particularities that have not been found with the other dopamine receptors. One is a high degree of sequence polymorphism in human. Although variations in D_4 receptor structure have not been directly associated with prevalent psychiatric disorders, it is possible that they affect more subtle psychological traits. The other particularity of the D_4 receptor is its particular affinity for clozapine. This fact carried high hopes for discovering better atypical neuroleptics; however, the first result reported in that direction is not promising. In any case, the D_4 receptor has evolved to become a bona fide dopamine and possibly other catecholamine target and thus must participate in some aspect of the regulatory role of dopamine.

Acknowledgements. I thank D. Hartman for previous reviews which set the basis of this manuscript and L. Tamura for proofreading of the manuscript. This work was supported by an endowment of the Eric and Lila Nelson Chair in Neuropharmacology.

References

Ariano MA, Wang J, Noblett KL, Larson ER, Sibley DR (1997) Cellular distribution of the rat D4 dopamine receptor protein in the CNS using anti-receptor antisera. Brain Research, Mar 28, 752(1–2):26–34

Arlt M, Böttcher H, Riethmüller A, Schneider G, Bartoszyk GD, Greiner H, Seyfried CA (1998) SAR of novel biarylmethylamine dopamine D4 receptor ligands. Bioorganic and Medicinal Chemistry Letters, Aug 4, 8(15):2033–2038

Ashgari V, Sanyal S, Buchwaldt S, Paterson A, Jovanovic V, Van Tol, HHM (1995). Modulation of intracellular cyclic AMP levels by different human dopamine D4 receptor variants. J Neurochem 65:1157–1165

Benjamin J, Li L, Patterson C, Greenberg BD, Murphy DL, Hamer DH (1996) Population and familial association between the D4 dopamine receptor gene and measures of Novelty Seeking. Nature Genetics 12:81–84

Bourrain S, Collins I, Neduvelil JG, Rowley M, Leeson PD, Patel S, Patel S, Emms F, Marwood R, Chapman KL, Fletcher AE, Showell GA (1998) Substituted pyrazoles as novel selective ligands for the human dopamine D4 receptor. Bioorganic and Medicinal Chemistry, Oct 6(10):1731–1743

Bristow LJ, Collinson N, Cook GP, Curtis N, Freedman SB, Kulagowski JJ, Leeson PD, Patel S, Ragan CI, Ridgill M, Saywell KL, Tricklebank MD (1997) L-745,870, a subtype selective dopamine D4 receptor antagonist, does not exhibit a neuroleptic-like profile in rodent behavioral tests. Journal of Pharmacology and Experimental Therapeutics, Dec 283(3):1256–1263

Bristow LJ, Kramer MS, Kulagowski J, Patel S, Ragan CI, Seabrook GR (1997) Schizophrenia and L-745,870, a novel dopamine D4 receptor antagonist. Trends in Pharmacological Sciences, Jun 18(6):186–188

Bunzow JR, Van TH, Grandy DK, Albert P, Salon J, Christie M, Machida CA, Neve KA, Civelli O (1988) Cloning and expression of a rat D2 dopamine receptor cDNA. Nature 336:783–787

Catalano M, Nobile M, Novelli E, Nothen MM, Smeraldi E (1993) Distribution of a novel mutation in the first exon of the human dopamine D4 receptor gene in psychotic patients. Biol Psychiatry 34:459–464

Chio CL, Drong RF, Riley DT, Gill GS, Slightom JL, Huff RM (1994) D4 dopamine receptor-mediated signaling events determined in transfected Chinese hamster ovary cells. J Biol Chem 269:11813–11819

Cohen AI, Todd RD, Harmon S, O'Malley KL (1992) Photoreceptors of mouse retinas possess D4 receptors coupled to adenylate cyclase. Proc Natl Acad Sci USA 89:12093–12097

Collins I, Rowley M, Davey WB, Emms F, Marwood R, Patel S, Patel S, Fletcher A, Ragan IC, Leeson PD, Scott AL, Broten T (1998) 3-(1-piperazinyl)-4,5-dihydro-1H-benzo[g]indazoles: high affinity ligands for the human dopamine D4 receptor with improved selectivity over ion channels. Bioorganic and Medicinal Chemistry, Jun 6(6):743–753

Creese I, Burt DR, Snyder SH. (1976). Dopamine receptor binding predicts clinical and pharmacological potencies of antischizophrenic drugs. Science 192:481–483

Defagot MC, Malchiodi EL, Villar MJ, Antonelli MC (1997) Distribution of D4 dopamine receptor in rat brain with sequence-specific antibodies. Brain Research. Molecular Brain Research, Apr 45(1):1–12

Ebstein RP, Novick O, Umansky R, Priel B, Osher Y, Blaine D, Bennett ER, Nemanov L, Katz M, Belmaker RH (1996) Dopamine D4 receptor (D4DR) exon III polymorphism associated with the human personality trait of Novelty Seeking. Nature Genetics 12:78–80

Faraci WS, Zorn SH, Sanner MA, Fliri A (1998) The discovery of potent and selective dopamine D4 receptor antagonists. Current Opinion in Chemical Biology, Aug 2(4):535–540

Feldpausch DL, Needham LM, Stone MP, Althaus JS, Yamamoto BK, Svensson KA, Merchant KM (1998) The role of dopamine D4 receptor in the induction of behavioral sensitization to amphetamine and accompanying biochemical and molecular adaptations. Journal of Pharmacology and Experimental Therapeutics 286:497–508

Gelernter J, Kennedy JL, Van Tol HHM, Civelli O, Kidd KK (1992) The D4 dopamine receptor (DRD4) maps to distal 11p close to HRAS. Genomics 13:208–210

Gelernter J, Kranzler H, Coccaro E, Siever L, New A, Mulgrew CL (1997) D4 dopamine-receptor (DRD4) alleles and novelty seeking in substance-dependent, personality-disorder, and control subjects. American Journal of Human Genetics 61:1144–1152

Hartman DS, Civelli O (1996). Molecular attributes of dopamine receptors: new potential for antipsychotic drug development. Ann Med 28:211–219

Hartman DS, Civelli O (1997) Dopamine receptor diversity: molecular and pharmacological perspectives. Prog Drug Res 48:173–194

Hidaka K, Tada S, Matsumoto M, Ohmori J, Maeno K, Yamaguchi T (1996)YM-50001: a novel, potent and selective dopamine D4 receptor antagonist. Neuroreport Nov 4, 7(15–17):2543–2546

Hillman D.W, Lin D, Burnside B (1995) Evidence for D4 receptor regulation of retinomotor movement in isolated teleost cone inner-outer segments. J Neurochem 64:1326–1335

Jönsson EG, Nöthen MM, Gustavsson JP, Neidt H, Brené S, Tylec A, Propping P, Sedvall GC (1997) Lack of evidence for allelic association between personality traits and the dopamine D4 receptor gene polymorphisms. American Journal of Psychiatry, 154:697–699

Kramer M, Last B, Zimbroff D, Hafez H, Alpert M, Allan E, Rotrosen J, McElvoy J, Kane J, Kronig M, Meredith C, Silva JA, Ereshefsky L, Marder S, Wirshing W, Conley R, Getson A, Chavez-Eng C, Cheng H, Reines S (1997) The effects of a selective D4 receptor antagonist (L-745870) in acutely psychotic schizophrenic patients. Arch gen. Psychiatry 54:567–572

Kula NS, Baldessarini RJ, Kebabian JW, Bakthavachalam V, Xu L (1997) RBI-257: a highly potent dopamine D4 receptor-selective ligand. European Journal of Pharmacology Jul 23, 331(2–3):333–336

Lanau F, Zenner M-T, Civelli O, Hartman DS (1997) Epinephrine and norepinephrine act as potent agonists at the recombinant human dopamine D4 receptor. J Neurochem 68:804–812

Lahti RA, Evans DL, Stratman NC, Figur LM (1993) Dopamine D4 versus D2 receptor selectivity of dopamine receptor antagonists: possible therapeutic implications. Eur. J Pharm. 236:483–486

LaHoste GJ, Swanson JM, Wigal SB, Glabe C, Wigal T, King N, Kennedy JL (1996) Dopamine D4 receptor gene polymorphism is associated with attention deficit hyperactivity disorder. Molecular Psychiatry 1:121–124

Lichter JB, Barr CL, Kennedy JL, Van Tol H, Kidd KK, Livak KJ (1993) A hypervariable segment in the human dopamine receptor D4 (DRD4) gene. Hum Mol Genet 2:767–773

Livak KJ, Rogers J, Lichter JB (1995) Variability of dopamine D4 receptor (DRD4) gene sequence within and among nonhuman primate species. Proc Natl Acad Sci USA 92:427–431

Löber S, Hübner H, Gmeiner P (1999) Azaindole derivatives with high affinity for the dopamine D4 receptor: synthesis, ligand binding studies and comparison of molecular electrostatic potential maps. Bioorganic and Medicinal Chemistry Letters, Jan 4, 9(1):97–102

Mansbach RS, Brooks EW, Sanner MA, Zorn SH (1998) Selective dopamine D4 receptor antagonists reverse apomorphine-induced blockade of prepulse inhibition. Psychopharmacology 135:194–200

Matsumoto M, Hidaka K, Tada S, Tasaki Y, Yamaguchi T (1995a) Full-length cDNA cloning and distribution of human dopamine D4 receptor. Mol Brain Res 29:157–162

Mauger C, Sivan B, Brockhaus M, Fuchs S, Civelli O, Monsma FJ (1998) Development and characterization of antibodies directed against the mouse D4 dopamine receptor. Eur J Neurosci 10:529–537

McHale M, Coldwell MC, Herrity N, Boyfield I, Winn FM, Ball S, Cook T, Robinson J H, Gloger IS (1994). Expression and functional characterisation of a synthetic version of the human D4 dopamine receptor in a stable human cell line. Febs Lett 345:147–150

Meador-Woodruff JH, Grandy DK, Van Tol, HHM, et al (1994) Dopamine receptor gene expression in the human medial temporal lobe. Neuropsychopharm 10:239–248

Mei YA, Grifton N, Buquet C, et al (1995) Activation of dopamione D4 receptor inhibits an L-type calcium current in cerebellar granule cells. Neurosci 68:107–116

Mills A, Allet B, Bernard A, Chabert C, Brandt E, Cavegn C, Chollet A, Kawashima E (1993) Expression and characterization of human D4 dopamine receptors in baculovirus-infected insect cells. FEBS 320:130–134

Merchant KM, Gill GS, Harris DW, Huff RM, Eaton MJ, Lookingland K, Lutzke BS, Maccall RB, Piercey MF, Schreur PJKD, Sethy VH, Smith MW, Svensson KA, Tang AH, VonVoigtlander PF, Tenbrink RE (1996) Pharmacological characterization of U-101387, a dopamine D4 receptor selective antagonist. JPET 279:1392–1403

Mrzljak L, Bergson C, Pappy M, Huff R, Levenson R, Goldman-Rakic PS (1996) Localization of dopamine D4 receptors in GABAergic neurons of the primate brain. Nature 381:245–248

Nanko S, Hattori M, Ikeda K, Sasaki T, Kazamatsuri H, Kuwata S (1993) Dopamine D4 receptor polymorphism and schizophrenia. Lancet 341:689–690

Newman-Tancredi A, Audinot-Bouchez V, Gobert A, Millam MJ (1997) Noradrenaline and andrenaline are high affinity agonists at dopamine D4 receptors. Eur J Pharmacol 319:379–383

Noda-Saita K, Matsumoto M, Hidaka K, Hatanaka K, Ohmori J, Okada M, Yamaguchi T (1999) Dopamine D4-like binding sites labeled by [3H]nemonapride include substantial serotonin 5-HT2A receptors in primate cerebral cortex. Biochemical and Biophysical Research Communications 255:367–370

Nothen MM, Cichon S, Hemmer S, Hebebrand J, Remschmidt H, Lehmkuhl G, Poustka F, Schmidt M, Catalano M, Fimmers R, Korner J, Rietschel M, Propping P (1994) Human dopamine D4 receptor gene: frequent occurrence of a null allele and observation of homozygosity. Hum Mol Genet 3:2207–2212

Oldenhof J, Vickery R, Anafi M, Oak J, Ray A, Schoots O, Pawson T, von Zastrow M, Van Tol HH (1998) SH3 binding domains in the dopamine D4 receptor. Biochemistry 37:15726–15736

O'Malley KL, Harmon S, Tang L, Todd RD (1992) The rat dopamine D4 receptor: sequence, gene structure, and demonstration of expression in the cardiovascular system. New Biologist 4:137–146

Patel S, Freedman S, Chapman KL, Emms F, Fletcher AE, Knowles M Marwood, R McAllister, G Myers, J Patel, S Curtis, N Kulagowski JJ, Leeson PD, Ridgill M, Graham M, Matheson S, Rathbone D, Watt AP, Bristow LJ, Rupniak NMJ, Baskin E, Lynch JJ Ragan CI (1997) Biological profile of L-745,870, a selective antagonist with high affinity for the dopamine D4 receptor. JPET 283:636–647

Pogue-Geile M, Ferrell R, Deka R, Debski T, Manuck S (1998) Human novelty-seeking personality traits and dopamine D4 receptor polymorphisms: a twin and genetic association study. American Journal of Medical Genetics 81:44–48

Rowley M, Collins I, Broughton HB, Davey WB, Baker R, Emms F, Marwood R, Patel S, Patel S, Ragan CI, Freedman SB, Ball R, Leeson PD (1997) 4-Heterocyclylpiperidines as selective high-affinity ligands at the human dopamine D4 receptor. Journal of Medicinal Chemistry Jul 18, 40(15):2374–2385

Rubinstein M, Phillips TJ, Bunzow JR, Falzone TL, Dziewczapolski G, Zhang G, Fang Y, Larson JL, McDougall JA, Chester JA, Saez C, Pugsley TA, Gershanik O, Low MJ, Grandy DK (1997) Mice lacking dopamine D4 receptors are supersensitive to ethanol, cocaine, and methamphetamine. Cell 90:991–1001

Sanner MA, Chappie TA, Dunaiskis AR, Fliri AF, Desai KA, Zorn SH, Jackson ER, Johnson CG, Morrone JM, Seymour PA, Majchrzak MJ, Faraci WS, Collins JL, Duignan DB, Prete Di CC, Lee JS, Trozzi A (1998) Synthesis, SAR and pharmacology of CP-29,3019: a potent, selective dopamine D4 receptor antagonist. Bioorganic and Medicinal Chemistry Letters Apr 7 8(7):725–730

Schlachter SK, Poel TJ, Lawson CF, Dinh DM, Lajiness ME, Romero AG, Rees SA, Duncan JN, Smith MW (1997) Substituted 4-aminopiperidines having high in vitro affinity and selectivity for the cloned human dopamine D4 receptor. Eur J Pharm 322:283–286

Seeman P, Lee T, Chan-Wong M, Wong K (1976) Antipsychotic drug doses and neuroleptic/dopamine receptors. Nature 261:717–719

Seeman P, Guan HC, Van Tol H (1993) Dopamine D4 receptors elevated in schizophrenia. Nature 365:441–445

Showell GA, Emms F, Marwood R, O'Connor D, Patel S, Leeson PD (1998) Binding of 2,4-disubstituted morpholines at human D4 dopamine receptors. Bioorganic and Medicinal Chemistry, Jan 6(1):1–8 (UI: 98161203)

Smalley SL, Bailey JN, Palmer CG, Cantwell DP, McGough JJ, Del'Homme MA, Asarnow JR, Woodward JA, Ramsey C, Nelson SF (1998) Evidence that the dopamine D4 receptor is a susceptibility gene in attention deficit hyperactivity disorder. Molecular Psychiatry Sep 3(5):427–430

Swanson JM, Sunohara GA, Kennedy JL, Regino R, Fineberg E, Wigal T, Lerner M, Williams L, LaHoste GJ, Wigal S (1998) Association of the dopamine receptor D4 (DRD4) gene with a refined phenotype of attention deficit hyperactivity disorder (ADHD): a family-based approach. Molecular Psychiatry 3:38–41

Tang SW, Helmeste DM, Fang H, Li M, Vu R, Bunney W Jr, Potkin S, Jones EG (1997) Differential labeling of dopamine and sigma sites by [3H]nemonapride and [3H]raclopride in postmortem human brains. Brain Research 765:7–12

Tallman JF, Primus RJ, Brodbeck R, Cornfield L, Meade R, Woodruff K, Ross P, Thurkauf A, Gallager DW. (1997) NGD 94–1: identification of a novel, high-affinity antagonist at the human dopamine D4 receptor. Journal of Pharmacology and Experimental Therapeutics 282:1011–1019

Unangst PC, Capiris T, Connor DT, Doubleday R, Heffner TG, MacKenzie RG, Miller SR, Pugsley TA, Wise LD (1997) (Aryloxy)alkylamines as selective human dopamine D4 receptor antagonists: potential antipsychotic agents. Journal of Medicinal Chemistry 40(25):4026–4029

Vandenbergh DJ, Zonderman AB, Wang J, Uhl GR, Costa PT Jr (1997) No association between novelty seeking and dopamine D4 receptor (D4DR) exon III seven repeat alleles in Baltimore Longitudinal Study of Aging participants. Molecular Psychiatry 2:417–419

Van Tol HHM, Bunzow JR, Guan H-C, Sunahara RK, Seeman P, Niznik HB, Civelli O (1991) Cloning of the gene for a human dopamine D4 receptor with high affinity for the antipsychotic clozapine. Nature 350:610–614

Van Tol HHM, Wu CM, Guan H-C, Ohara K, Bunzow JR, Civelli O, Kennedy J, Seeman P, Niznik HB, Jovanovic V (1992) Multiple dopamine D4 receptor variants in the human population. Nature 358:149–152

Zawilska JB, Nowak JZ (1994) Dopamine receptor regulating serotonin N-acetyltransferase activity in chick retina represents a D4-like subtype: pharmacological characterization Neurochem. Int 24:275–280

Zenner MT, Nobile M, Henningsen R, Smeraldi E, Civelli O, Hartman DS, Catalano M (1998) Expression and characterization of a dopamine D4R variant associated with delusional disorder. Febs Letters 422:146–150

CHAPTER 9
Signal Transduction by Dopamine D_1 Receptors

J.-A. GIRAULT and P. GREENGARD

In vertebrates the D_1 family of dopamine receptors comprises at least three members, coded by different genes: D_{1a} (or D_1), D_{1b} (or D_5), and $D_{1c/d}$ (VERNIER et al. 1995). However, only D_1 and D_5 have been identified in mammals. These receptors are characterized by their high degree of sequence identity and their common ability to interact with G_s-like proteins and to stimulate adenylyl cyclase. Most of our knowledge of the signaling by these receptors has been gained by studying the D_{1a} isoform, by far the most abundant in brain, especially in striatonigral medium-size spiny neurons.

A. Historical Perspective

The D_1 receptors were the first dopamine receptors to be identified. Following the discovery that epinephrine and norepinephrine were capable of stimulating the formation of cyclic AMP (cAMP) in target cells, it was shown that dopamine had the same effect in the superior cervical ganglia and the retina of the cow (KEBABIAN and GREENGARD 1971; BROWN and MAKMAN 1972). Soon afterwards, the presence of a dopamine-sensitive adenylyl cyclase was demonstrated in rat striatum (KEBABIAN et al. 1972). Antipsychotic drugs were found to block this response and it was hypothesized that dopamine receptors might be their site of action (CLEMENT CORMIER et al. 1974; MILLER et al. 1974). Interestingly, the effects of dopamine, lost in the absence of the cytoplasmic fraction, were recovered following addition of exogenous (GTP) (CLEMENT CORMIER et al. 1975). This was the first indication that GTP was involved in the coupling of D_1 dopamine receptors to adenylyl cyclase, in agreement with the concept proposed by RODBELL (RODBELL et al. 1971). Subsequently, the distribution of dopamine-sensitive adenylyl cyclase was studied in various brain regions and non-neuronal tissues such as the parathyroid gland (BROWN et al. 1977). Although the physiological effects of stimulation of D_1 receptors in neurons were not known, in parathyroid it was demonstrated that this stimulation induced the release of parathyroid hormone. However, it was found later that some effects of dopamine did not involve stimulation of adenylyl cyclase and that drugs acting on mammotroph cells of the anterior pituitary displayed "inappropriate" activity in the adenylyl cyclase assay, leading to the discovery of D_2 receptors (for a review, see KEBABIAN and CALNE 1979). In the following years, research on dopamine D_1 receptors was pursued in three directions: the identification of the mechanism of coupling of D_1 receptors to adenylyl cyclase, the study of the biochemical and physiological effects of cAMP, and the search for targets different from the adenylyl cyclase. In spite of the considerable progress made, there are many important questions still to be answered, and these areas of research are very active.

B. GTP-Binding Proteins Associated with D_1-Family Receptors

The role of a heterotrimeric GTP-binding protein (G protein) in mediating the activation of adenylyl cyclase by dopamine was suspected as early as 1974, as indicated above, and several years later the binding of D_1 agonists was shown to be modulated by guanine nucleotides. The G protein involved was initially assumed, without experimental evidence, to be G_s. Recently, it has been shown that in striatal neurons, G_{olf} (see Fig. 1), and not G_s, couples D_1

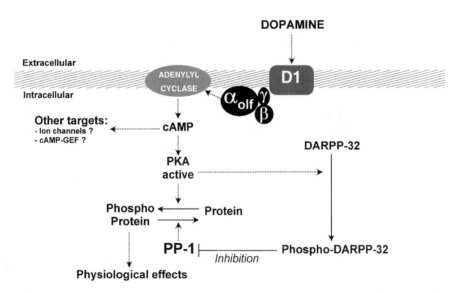

Fig. 1. Schematic representation of the main signaling pathways activated by dopamine D_1 receptors in striatonigral neurons. D_1 receptors activate adenylyl cyclase (mostly type V) through a heterotrimeric GTP-binding protein containing the α_{olf} isoform. The major target of cAMP is cAMP-dependent protein kinase (PKA), which can increase the state of phosphorylation of substrate proteins either by phosphorylating them directly, or by inhibiting protein phosphatase 1 (PP1) via the phosphorylation of DARPP-32. Identified proteins phosphorylated by PKA in striatonigral neurons in response to dopamine include ARPP-16 and ARPP-21, synapsin 1, stathmin, NMDA receptors, and CREB

receptors to adenylyl cyclase (CORVOL et al. 2001). G_{olf} is characterized by the presence of an α_{olf} subunit that is very similar to α_s (80% amino acid identity). The α_{olf} subunit was initially thought to be specific to the olfactory epithelium in which it had been identified (JONES and REED 1989). However, the α_{olf} mRNA and the corresponding protein are found in several other tissues, in particular in brain where they are highly expressed in the basal ganglia (DRINNAN et al. 1991; HERVÉ et al. 1993). In contrast, the basal ganglia contain very low levels of α_s (HERVÉ et al. 1993). The medium-sized spiny neurons, which contain high levels of D1 receptors, express α_{olf} and very little if any α_s (HERVÉ et al. 1993, 2001). In contrast, both α_s and α_{olf} transcripts are detected in cholinergic interneurons (HERVÉ et al. 2001). Stimulation of cAMP production by dopamine is virtually abolished in the striatum of α_{olf} knockout mice (CORVOL et al. 2001). However, it should be noted that α_s may couple D_1 receptors to adenylyl cyclase outside of the basal ganglia, since D_1 responses were normal in the cerebral cortex of these α_{olf} knockout mice. The phenotype of homozygous α_{olf} knockout mice is very severe, in part due to their olfactory deficit, and most die in the three postnatal weeks, making it difficult to analyze specifically the alterations in dopamine signaling (BELLUSCIO et al. 1998). Nevertheless, the cocaine-induced locomotor activity and striatal immediate early genes expression observed in wild type littermates were blocked in homozygous α_{olf} knockout mice (ZHUANG et al., 2000). Heterozygous α_{olf} +/− mice may provide a better model for studying the role of α_{olf} in dopamine actions. In the striatum of these animals the levels of α_{olf} protein are about half those in wild type, there is no compensatory change in α_s, and the activation of adenylyl cyclase by dopamine is significantly decreased (CORVOL et al. 2001). Although these heterozygous mutant mice are apparently normal, their locomotor activity in response to amphetamine injection is markedly diminished (Hervé et al. 2001). These results indicate that α_{olf} levels are a limiting parameter in D_1-mediated responses to dopamine.

In transfection experiments α_s and α_{olf} have similar properties (JONES et al. 1990) and it is unclear whether these two related gene products have different functional properties. It is possible that the selection by evolution of a two-gene system resulted from the advantage of achieving a highly specific pattern of expression. Accordingly, the regulation of the expression of the α_{olf} gene itself appears to be rather complex since at least four different transcripts are generated that are expressed in different ratios in brain and in the olfactory epithelium (HERVÉ et al. 1995). Although all these mRNAs contain the same coding sequence, they appear to be transcribed from two different promoters and to have very different abilities to be translated (HERVÉ et al. 1995).

The fact that levels of α_{olf} are limiting for the action of dopamine on adenylyl cyclase in the striatum may account for several puzzling observations. First, the levels of α_{olf} are several times higher in the somato-dendritic region of striatonigral neurons than in their nerve terminals, in the substantia nigra (HERVÉ et al. 1993). This provides an explanation for the more potent D_1 stimulation of cAMP production by dopamine in the striatum than in the substantia nigra, whereas the amounts of D_1 receptors (HERVÉ et al. 1992) and adenylyl cyclase molecules (WORLEY et al. 1986) are similar in the two regions. Second, an increased response of dopamine-stimulated cyclase has been observed in the striatum following lesion of dopamine neurons (VON VOIGTLANDER et al. 1973; MISHRA et al. 1974; KRUEGER et al. 1976; HERVE et al. 1989), in the absence of alteration of D_1 receptors or adenylyl cyclase (SAVASTA et al. 1988; HERVÉ et al. 1992). This can be explained by the levels of α_{olf}, which increase following lesion of dopamine neurons (HERVÉ et al. 1993; PENIT-SORIA et al. 1997). Interestingly, α_{olf} levels appear to be negatively regulated by the activity of dopamine D_1 receptors in the striatum since they increase following lesion of dopamine neurons (HERVÉ et al. 1993) or in D_1 receptor knockout mice, whereas they decrease in dopamine transporter knockout mice in which dopamine receptors are permanently stimulated (HERVÉ et al. 2001). In both cases there is no change in α_{olf} mRNA levels, showing that the regulation takes place at a post-transcriptional level. One possible explanation is that α_{olf} is released into the cytoplasm and degraded during its activation cycle triggered by dopamine, as has been reported for α_s following stimulation of β-adrenoreceptors (LEVIS and BOURNE 1992). Thus, the absence of dopamine or the absence of D_1 receptor would result in the accumulation of α_{olf} secondary to its lack of use. All these results indicate that the levels of α_{olf} are an important parameter of dopamine signal transduction and that changes in these levels in physiological or pathological circumstances could have important consequences.

Little is known about the other subunits which associate with α_{olf} to form a functional heterotrimeric G protein in the striatum. It has been observed that γ_7 is highly expressed in basal ganglia (WATSON et al. 1994), but the meaning of this observation is not known. Although it is possible that the G protein associated with D_1 receptor in the striatum contains α_{olf} and γ_7, it is not known whether this is also the case in other cell types, in which D_1 receptors are expressed at much lower levels. Several β subunit isoforms are expressed in the striatum (BETTY et al. 1998), but there is, as yet, no evidence for a specific association of any of them with D_1 receptors.

Coupling of D_1 receptors to other G proteins has been reported. Striatal D_1 receptors associate with α_i in reconstituted vesicles (SIDHU et al. 1991). D_1 receptors interact with G_i/G_o in transfected GH4C1 cells (KIMURA et al. 1995b) and co-immunoprecipitate with α_o (KIMURA et al. 1995a) or α_q (WANG et al. 1995). Activation of phospholipase C (PLC) in response to dopamine has been reported in kidney (FELDER et al. 1989), in several brain regions (UNDIE and FRIEDMAN 1990), and in *Xenopus* oocytes injected with D_1 receptor mRNA (MAHAN et al. 1990). However, the activation of phosphoinositide hydrolysis by dopamine was still observed in striatum from D_{1a} knockout mice, suggesting that the D_1-like dopamine receptors coupled to PLC are not D_{1a} (FRIEDMAN et al. 1997). A transmembrane protein associated with D_1 receptors has been recently identified (LEZCANO et al. 2000). This protein, termed calcyon, appears to promote Ca^{2+} release from internal stores in response to stimulation of D_1 receptors. Finally, a synergistic effect of transfected D_1 and D_2 receptors on arachidonic acid release has been reported, whereas D_1 receptors alone had no effect (PIOMELLI et al. 1991). The physiological meaning of these various observations is not yet clear, but they indicate the coupling of D_1 receptors to pathways other than the activation of adenylyl cyclase. Nevertheless, hard evidence for these possible coupling mechanisms in the physiological effects of dopamine is still missing.

C. Adenylyl Cyclases and Phosphodiesterases in the Striatum

Adenylyl cyclases (Fig. 1) are large proteins with several transmembrane domains and two cytoplasmic domains with enzymatic activity (SUNAHARA et al. 1996; HURLEY 1999). There are at least ten distinct genes coding for adenylyl cyclases in mammals. All of these isoforms are activated by α_s/GTP, but have different sensitivities to other regulators including Ca^{2+}/calmodulin and $\beta\gamma$ subunits. Type V adenylyl cyclase is the most abundant in the striatum (GLATT and SNYDER 1993). This adenylyl cyclase is negatively regulated by Ca^{2+}/calmodulin and is independent of $\beta\gamma$ subunits. Moderate levels of type IX adenylyl cyclase are also found in these neurons (ANTONI et al. 1998). This isoform is inhibited by calcium through a calcineurin-dependent mechanism (ANTONI et al. 1998). Interestingly, striatal neurons also contain high levels of the 61-kDa isoform of phosphodiesterase (PDE1B1) which is activated by the Ca^{2+}-calmodulin complex (POLLI and KINCAID 1992, 1994). These observations indicate that Ca^{2+} transients are likely to reduce the level of cAMP in striatal neurons.

D. cAMP-Dependent Protein Kinase

The major target for cAMP in eukaryotes is cAMP-dependent protein kinase A (PKA) (Fig. 1). This enzyme is a heterotetramer comprising two regulatory subunits (R) and two catalytic subunits (C) (see review by TAYLOR et al. 1988). Two molecules of cAMP bind cooperatively to the R subunits, leading to the dissociation of the holoenzyme and the diffusion of the free and fully active catalytic subunits that phosphorylate cytoplasmic proteins and have the capacity to diffuse to the nucleus (ADAMS et al. 1991). The regulatory subunits also have important targeting functions. Whereas PKA containing the RI isoforms is mostly cytosolic, that which contains RII subunits is associated with a variety of organelles. There are two RII isoforms, α and β, which interact with a number of anchoring proteins [A-kinase anchoring proteins (AKAPs); see DELL'ACQUA and SCOTT 1997]. AKAPs provide cell-specific intracellular targeting. For example in neurons, AKAP75/79 targets PKA to postsynaptic densities (CARR et al. 1992; COGHLAN et al. 1995). AKAPs are also scaffolding proteins that bring together several signal transduction enzymes at the same location in neurons. For example, AKAP79 anchors PKA, PKC and calcineurin in postsynaptic densities (KLAUCK et al. 1996). The RIIβ isoform, as well as AKAP150, are highly enriched in the striatum (GLANTZ et al. 1992; VENTRA et al. 1996). PKA activity is markedly decreased in the striatum of mice lacking a functional RIIβ gene (BRANDON et al. 1998). Although these mice have an overall normal motor behavior, they perform poorly on the rotarod, and have impaired cFos expression, and an increased locomotor sensitization in response to amphetamine (BRANDON et al. 1998).

E. cAMP-Activated Phosphorylation Pathways

PKA is a broad spectrum protein kinase and it is likely that many substrates are phosphorylated in D_1 receptor-containing neurons in response to dopamine stimulation (Fig. 1). However, only a limited number of cAMP-regulated proteins have been experimentally identified. Almost 20 years ago a systematic attempt to identify protein substrates for PKA in brain was undertaken, leading to the characterization of several proteins enriched in the dopamine-innervated brain areas (WALAAS et al. 1983a,b). They included a dopamine- and cAMP-regulated phosphoprotein with an apparent M_r of 32,000 (DARPP-32, see below), and a cAMP-regulated phosphoprotein with an apparent M_r of 21,000 (ARPP-21). ARPP-21 is highly enriched in the striatum and cerebral cortex (HEMMINGS JR. and GREENGARD 1989; HEMMINGS JR. et al. 1989; GIRAULT et al. 1990a). Its phosphorylation is increased in striatal neurons in culture by vasoactive intestinal peptide (GIRAULT et al. 1988) and in nigral slices by dopamine D_1 agonists (TSOU et al. 1993). However, to date the function of ARPP-21 is not known. Another phosphoprotein of unknown function enriched in striatal neurons is ARPP-16 (GIRAULT et al. 1990b; HORIUCHI et al. 1990). This protein is a member of a multigene family highly conserved during evolution (DULUBOVA et al. 2001). In striatonigral neurons, the phosphorylation of ARPP-16 is increased by dopamine D_1 receptors (DULUBOVA et al. 2001).

Increased phosphorylation of proteins of known function has been observed in response to stimulation of dopamine receptors. These include stathmin, a protein which regulates the stability of microtubules (CHNEIWEISS et al. 1992), and synapsin 1, a protein which regulates the availability of synaptic vesicles for release (WALAAS et al. 1989). An increase in phosphorylation of the N-methyl-D-aspartate (NMDA) receptor subunit NR1 has been demonstrated in striatal slices treated with dopamine D_1 agonists (SNYDER et al. 1998). This phosphorylation is likely to occur at Ser-897, the consensus PKA site, and to result from a direct activation of PKA as well as from an indirect inhibition of protein phosphatase 1 (PP1) by DARPP-32 (see below). This latter effect probably accounts for the concomitant increased phosphorylation of the PKC-site (Ser-890) in NR1 subunit. The physiological significance of phosphorylation of Ser-897 is not known. In contrast, phosphorylation of Ser-890 appears to increase the channel conductance, possibly by preventing its inhibition by Ca^{2+}/calmodulin (HISATSUNE et al. 1997). This may account for the facilitatory effect of D_1 receptors on NMDA currents (CEPEDA et al. 1993).

Free catalytic subunit of PKA readily accumulates in the nucleus where it can phosphorylate a number of proteins important for the regulation of transcription. One of the best-characterized nuclear substrates of PKA is the transcription factor CRE-binding protein (CREB; MONTMINY 1997). As a homodimer, CREB binds to a specific DNA sequence, the cAMP-responsive element (CRE). When phosphorylated on Ser-133, it becomes a potent activator of transcription. Dopamine agonists increase CREB phosphorylation in the striatum by stimulating D_1 receptors (KONRADI et al. 1994). Interestingly, the stimulation of NMDA receptors also appears to be necessary for this effect (KONRADI et al. 1996). This may be related to the complex regulation of CREB phosphorylation which is also a substrate for Ca^{2+}/calmodulin-dependent kinases. Alternatively, it could reflect an indirect mechanism of action. The regulation of CREB by dopamine D_1 receptors probably accounts in part for the effects of dopamine on immediate early gene expression including that of c-Fos, c-Jun, and Zif-268. Induction of these factors provides an explanation for the dopamine-induced increase in activator protein (AP)-1 activity (HUANG and WALTERS 1996). However, a pathway involving ERK (extracellular signal-regulated kinase) and the transcription factor Elk-1 is also likely to be involved in the D_1-mediated effects on immediate early genes expression (VALJENT et al. 2000). The mechanism of D_1 receptor coupling to the ERK pathway in striatal neurons is not known.

F. The Role of Protein Phosphatase 1 in the Action of D_1 Receptors

The state of phosphorylation of proteins results from an equilibrium between the activity of protein kinases and protein phosphatases. Protein phosphatase 1 (PP1) (Fig. 1) is a broad-spectrum ubiquitous protein phosphatase that plays a central role in cell signaling by catalyzing the dephosphorylation of a wide range of cellular proteins (see SHENOLIKAR and NAIRN 1991). It comprises a catalytic subunit, highly conserved among all eukaryotes, which can interact reversibly with a variety of targeting and inhibitory proteins. Several of these proteins are highly concentrated in neurons, including those which contain D_1 dopamine receptors. Two homologous PP1-binding proteins, spinophilin (ALLEN et al. 1997) and neurabin-I (neuronal actin binding protein I; NAKANISHI et al. 1997), are highly enriched in neurons. Both proteins are also capable of binding F-actin (SATOH et al. 1998). Spinophilin is concentrated in dendritic spines where it accounts for a large part of the targeting of PP1 (ALLEN et al. 1997). Neurabin-I and spinophilin have the potential to self-associate through coiled-coil interactions of their carboxy-termini. They are also likely to associate with other proteins, such as ion channels or neurotransmitter receptors, through their PDZ domain. These domains, named after the three proteins in which they were first identified (PSD-95, Dlg, ZO-1), are capable of interacting with specific C-terminal sequences. By analogy with other targeting subunits, spinophilin and neurabin I may promote activity of PP1 toward relevant substrates and inhibit activity toward other phosphoproteins (HUBBARD and COHEN 1993).

In addition to targeting subunits, PP1 is regulated by inhibitory proteins whose activity is controlled by phosphorylation. Three such inhibitors are small proteins known as inhibitor-1, inhibitor-2, and DARPP-32 (dopamine- and cAMP-regulated phosphoprotein with an apparent M_r of 32,000; see SHENOLIKAR and NAIRN 1991 for a review). Inhibitor-2 is active in its unphosphorylated state, whereas inhibitor 1 and DARPP-32 are active only when phosphorylated on a threonine residue by PKA, providing a pathway by which cAMP can regulate a protein phosphatase. Although inhibitor-1 is relatively abundant in neurons which contain D_1 receptors, including striatonigral neurons (HEMMINGS JR. et al. 1992; GUSTAFSON et al. 1991), its role in the action of D_1 receptors has not been documented. In contrast, DARPP-32, a protein related in sequence and function to inhibitor-1 and highly enriched in striatonigral neurons, has been extensively characterized and shown to play an important role in the action of dopamine (see GREENGARD et al. 1998, 1999).

G. DARPP-32

DARPP-32 (Figs. 1, 2) was originally identified as a striatal-enriched substrate for PKA, the phosphorylation of which was regulated by cAMP and dopamine in striatal slices (WALAAS et al. 1983a; WALAAS and GREENGARD 1984). When it was purified and sequenced, it was found to be similar to inhibitor-1 in its NH2-terminal region, which is important for phosphatase inhibition (HEMMINGS JR. et al. 1984b,c; WILLIAMS et al. 1986). When phosphorylated on threonine 34, DARPP-32 becomes a potent inhibitor of PP1 with an IC50<10^{-9}M (HEMMINGS JR. et al. 1984b). The interaction between PP1 and DARPP-32 is now well understood (Fig. 2) (HEMMINGS JR. et al. 1990; DESDOUITS et al. 1995b; KWON et al. 1997; HUANG et al. 1999). It involves two different binding sites: one is phosphorylation-dependent and comprises Thr-34 in DARPP-32, which is likely to interact with the phosphatase active site; the other is phosphorylation-independent and comprises the residues 7-11 (KKIQF) of DARPP-32, which interact with PP1 at a site distant from the active site. This dual interaction accounts for the mixed kinetics of inhibition of PP1 by DARPP-32 (competitive and non competitive) (HEMMINGS JR. et al. 1984b). Interestingly, the KKIQF motif in DARPP-32 (one or more basic residues followed by two hydrophobic residues separated by a variable residue) is common to other proteins which bind to the catalytic subunit of PP1, including targeting subunits (EGLOFF et al. 1997; HIRANO et al. 1997). Thus, binding of DARPP-32 and targeting subunits to PP1 is likely to be mutually exclusive.

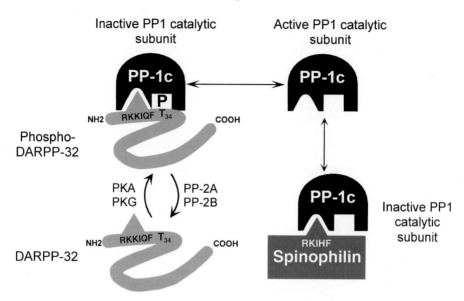

Fig. 2. Regulation of protein phosphatase 1 catalytic subunit (PP1c) in striatonigral neurons. PP1c is associated with a targeting subunit, spinophilin, which is enriched in dendritic spines where it is associated with actin. This interaction inhibits PP1c. Free PP1c is inhibited by DARPP-32 phosphorylated on Thr-34 (T_{34}). The interaction between DARPP-32 and PP1c involves phospho-Thr-34 and a sequence that is conserved in DARPP-32 and PP1c-targeting proteins (RKKIQF). Thr-34 is phosphorylated by cAMP-dependent protein kinase (PKA) and cGMP-dependent kinase (PKG). It is dephosphorylated by protein phosphatase 2A (PP-2A) and protein phosphatase 2B (PP-2B, calcineurin). DARPP-32 can also be phosphorylated on Thr-75 (T_{75}) by Cdk5, a phosphorylation which turns it into a potent inhibitor of PKA catalytic subunit (PKAc). Phospho-Thr-75 appears to be dephosphorylated by PP2-A. In addition, DARPP-32 is phosphorylated by protein casein kinase (CK)2 on Ser-45 and Ser-102 and by protein CK1 on Ser-137. Phosphorylation of Ser-137 prevents dephosphorylation of Thr-34 by calcineurin (PP2B). Phosphate groups are indicated by *P*

DARPP-32 is highly expressed in neurons which express D_1 receptors, especially the medium-sized spiny neurons in the striatum (OUIMET et al. 1984; OUIMET and GREENGARD 1990; OUIMET et al. 1992, 1998). However, its distribution is wider than that of D_1 receptors. For example, it is also found at high levels in striatopallidal neurons which contain predominantly D_2 receptors and few D_1 receptors. DARPP-32 is also expressed in other specific neuronal populations in brain and in several non-neuronal cell types including tanycytes (MEISTER et al. 1988), chromaffin cells (MEISTER et al. 1991a), parathyroid hormone producing cells (MEISTER et al. 1991b), brown adipose tissue (MEISTER et al. 1988), epithelial cells from choroid plexus (SNYDER et al. 1992), ciliary bodies (STONE et al. 1986), and kidney tubules (FRYCKSTEDT et al. 1993). Some but not all of these cells also contain D_1 receptors. This widespread distribution is in agreement with the ability of numerous extracellular signals to regulate its phosphorylation state (see GREENGARD et al. 1998, 1999).

Table 1. Signaling molecules which regulate phosphorylation of DARPP-32 on Thr-34

Tissue	Extracellular signal	Receptor	Mechanism
Signals that increase DARPP-32 phosphorylation on Thr-34			
Striatum	Dopamine	D_1	↑ cAMP
	Adenosine	A2a	↑ cAMP
	GABA	$GABA_A$	↓ Ca^{2+}
Substantia nigra	Dopamine	D_1	↑ cAMP
	NO	–	↑ cGMP
	GABA	$GABA_A$	↓ Ca^{2+}
Choroid plexus	Norepinephrine	β	↑ cAMP
	VIP	?	↑ cAMP
	Atrial natriuretic factor	ANF-R	↑ cGMP
	Serotonin	?	?
Signals that decrease DARPP-32 phosphorylation on Thr-34			
Striatum	Glutamate	NMDA-R	↑ Ca^{2+}
	Opiates	μ and δ	?
	Dopamine	D_2	?
	Cholecystokinin	CCK_B	↑ glutamate?
Substantia nigra	Depolarization	Ca^{2+} channels	↑ Ca^{2+}

The phosphorylation of DARPP-32 on Thr-34 is critical for inhibition of PP1. This threonine can be phosphorylated by cAMP- and cyclic guanosine monophosphate (cGMP)-dependent protein kinases (HEMMINGS JR. et al. 1984a). Several neurotransmitters in various cell types have been shown to act through one or the other of these protein kinases (Table 1). Dephosphorylation of Thr-34 is also catalyzed by two different protein phosphatases, protein phosphatase-2A (PP2A) and protein phosphatase-2B (PP2B or calcineurin) (HEMMINGS JR. et al. 1984b; KING et al. 1984). PP2B is highly enriched in striatonigral neurons (GOTO et al. 1986). PP2B dephosphorylates DARPP-32 in response to stimulation of NMDA receptors in the striatum (HALPAIN et al. 1990), and to depolarization in nigral nerve terminals (F. Desdouits, J.C. Siciliano, and J.-A. Girault, unpublished observations). PP2B can also be activated in striatal slices by stimulation of D_2 receptors, via a mechanism which is still to be clarified (NISHI et al. 1997). In addition to Thr-34, DARPP-32 is phosphorylated on several serine and threonine residues, which appear to have modulatory effects. Phosphorylation of Ser-102 (and perhaps Ser-45) by protein kinase CK2 (casein kinase 2) increases the ability of PKA (but not cGMP-dependent protein kinase) to phosphorylate Thr-34 (GIRAULT et al. 1989). Phosphorylation of Ser-137 by protein kinase CK1 (casein kinase 1) decreases the ability of PP2B to dephosphorylate Thr-34 (DESDOUITS et al. 1995a,c). Thus, both CK1 and CK2 enhance responses to neurotransmitters that stimulate phosphorylation of DARPP-32 at Thr-34, including dopamine. Recent results have also demonstrated that DARPP-32 is phosphorylated by cyclin-dependent protein kinases on Thr-75 (BIBB et al. 1999). In striatal neurons DARPP-32 is phosphorylated on Thr-75 by Cdk5, and this phosphorylation turns DARPP-32 into a potent inhibitor of PKA (BIBB et al. 1999). Conversely, dopamine promotes the dephosphorylation of Thr-75, providing a positive-feedback loop which amplifies dopaminergic signaling (NISHI et al. 2000). Interestingly, Cdk5 levels are regulated by ΔFosB, a transcription factor induced by chronic exposure to cocaine, and may thus contribute to the adaptive changes related to cocaine addiction (BIBB et al. 2001).

There are several important neuronal proteins for which regulation of phosphorylation by a DARPP-32-mediated inhibition of PP1 has been shown to be a critical aspect of their response to dopamine, including ion channels, receptors, enzymes, and transcription factors. The tetrodotoxin-sensitive Na^+ channel is regulated by phosphorylation by PKA and PKC (GRAY et al. 1998). It has been shown to be inhibited by dopamine D_1 receptors, through a PKA-dependent pathway in striatal (SURMEIER and KITAI 1993; SCHIFFMANN et al. 1995) and hippocampal neurons (CANTRELL et al. 1997). Interestingly, an inhibition of Na^+ currents is also achieved by injection of phospho-Thr-34-DARPP-32 or pharmacological inhibition of PP1 (SCHIFFMANN et al. 1998). The precise level of action of PP1 is unclear, however, since purified Na^+ channels are poor substrates for this phosphatase in vitro (MURPHY et al. 1993). Thus, it is possible that PP1 acts on a Na^+ channel-associated protein in intact neurons

Stimulation of dopamine D_1 receptors enhances L-type Ca^{2+} currents, and decreases N- and P-type currents (SURMEIER et al. 1995). Both effects are mediated by PKA. However, inhibition of PP1 by phospho-DARPP-32 enhances the effects of dopamine on L-type currents, whereas it attenuates its effects on N- and P-type currents (SURMEIER et al. 1995). These latter effects suggest that PKA might also increase PP1 activity in striatal neurons, by phosphorylating the targeting protein and promoting the release of the free catalytic protein, as reported in the case of glycogen-associated PP1 (DENT et al. 1990). In DARPP-32 knockout mice the modulatory effects of dopamine on Ca^{2+} currents are attenuated, indicating a physiological role for DARPP-32 in the control of Ca^{2+} influx (FIENBERG et al. 1998).

As mentioned above, DARPP-32 is also involved in the positive regulation of glutamate receptors by dopamine. DARPP-32 participates in the cascade regulating NMDA receptors reconstituted in oocytes (BLANK et al. 1997), as well as in mice (SNYDER et al. 1998). D_1 agonists prevent α-amino-3-hydroxy-5-methyl-4-isoxazolepropionic acid (AMPA) current rundown in striatal neurons (YAN et al. 1999). This stimulatory effect of dopamine on AMPA receptors appears to involve phosphorylation by PKA, inhibition by DARPP-32 of PP1-catalyzed dephosphorylation (YAN et al. 1999) and to require the precise targeting of PP1 by spinophilin (HSIEH-WILSON et al. 1999).

Dopamine acting on D_1 and D_2 receptors inhibits the activity of the electrogenic ion pump Na^+,K^+-ATPase (BERTORELLO et al. 1990; NISHI et al. 1999). This inhibition, like that induced by ouabain, is predicted to result in enhanced depolarization and greater inward current in response to low concentrations of glutamate (CALABRESI et al. 1995). Although the precise molecular mechanisms underlying the effect of dopamine receptors on Na^+, K^+-ATPase are still unclear, it is noteworthy that its inhibition by D_1 agonists is greatly attenuated in DARPP-32 knockout mice (FIENBERG et al. 1998).

In addition to the regulation of ion channels and pumps, DARPP-32 appears to be critical for other responses in striatal neurons. In DARPP-32 knockout mice, several responses to dopamine are abolished or markedly decreased. Stimulation of D_1 receptors increases GABA release in the dorsal striatum (GIRAULT et al. 1986). Although the precise mechanism of this D_1 effect is not known, it is blunted in mice lacking DARPP-32, suggesting that it involves an inhibition of PP1 by DARPP-32, among other mechanisms (FIENBERG et al. 1998). Induction of Fos by acute administration of amphetamine and induction of ΔFosB by chronic administration of cocaine are also markedly decreased in DARPP-32 knockout mice, indicating the importance of phosphatase inhibition in the effects of dopamine on gene expression (FIENBERG et al. 1998). These mutant mice also display less locomotor hyperactivity in response to cocaine. Thus, many biochemical, electrophysiological, and behavioral responses to dopamine are either blocked or diminished in the absence of DARPP-32. This is in agreement with the idea that inhibition of PP1 by DARPP-32 amplifies the responses generated by phosphorylation of specific proteins by PKA. This amplification may not be absolutely necessary in optimized breeding conditions, as indicated by the normal growth and reproduction of DARPP-32 knockout mice in a laboratory environment. However, it appears to become critical when the dopamine system is put under a higher demand.

H. Other Actions of cAMP

In recent years, new targets for cAMP have been identified, opening, at least in principle, a novel range of actions for this second messenger. A cation channel stimulated by cAMP was discovered in olfactory neurons in which it is a critical component in the detection of odorants (see ZAGOTTA and SIEGELBAUM 1996). Related cation channels, with cAMP-binding domains, have been identified in brain neurons (SANTORO et al. 1997, 1998). These channels have a pacemaker activity that is triggered by cAMP and thus could have very interesting properties in central neurons. However, nothing is known at present of their possible involvement in the effects of cAMP in response to dopamine. Another group of cAMP-activated proteins, which have been identified recently, are cAMP-GEF (cAMP-regulated guanine nucleotide exchange factor) (KAWASAKI et al. 1998), also termed Epac (exchange protein directly activated by cAMP) (DE ROOIJ et al. 1998). These proteins are capable of activating Rap1, a small GTP-binding protein, and thus, are potential activators of the MAP-kinase (mitogen-activated protein kinase) pathway in neurons. The two isoforms of cAMP-GEF are expressed, although at relatively low levels, in the striatum (KAWASAKI et al. 1998), and are candidates for being activated by cAMP in response to stimulation of D_1 receptors.

I. Concluding Remarks

Dopamine D_1 receptors play primarily a modulatory role in synaptic transmission, as discussed in other chapters of this book. In addition, they are likely to play an important role in regulating synaptic plasticity, as indicated by their role in reward systems, as well as by their ability to regulate long-term potentiation in several neuronal populations. Their discreet, although essential, role has been difficult to decipher and is, as yet, far from being fully understood. Important progress has been made by the patient dissection of the signaling pathways activated by cAMP in striatal neurons. During recent years, several of the more prominent players in neuronal electrical activity (e.g., ion channels or ion pumps) and in gene expression (e.g., transcription factors) have been shown to be regulated by dopamine and by cAMP-dependent phosphorylation. Thus, D_1 receptors are in a strategic position, capable of controlling most functions of the cells in which they are expressed and thereby modulating their physiology. Remarkably, the study of D_1 signaling has led to the discovery of a second modulatory system within the cell. DARPP-32, one of the proteins regulated by D_1 receptor-induced phosphorylation is itself a phosphatase inhibitor, capable at the same time of amplifying the effects of these receptors and of modifying those of other signaling pathways. This highly sophisticated regulatory mechanism appears to provide dopamine neurons with a capacity to tune very finely the activity of their target cells. Two major goals lie before us at present, one is to pursue the characterization of the signaling networks regulated by dopamine through D_1 receptors, the other is to determine experimentally the function of these networks in physiological and pathological situations at the level of the basal ganglia, limbic areas, and prefrontal cortex. The precise understanding of this function may be the key to understanding behavioral disorders that are known to involve these brain regions, but the mechanisms of which are still elusive.

References

Adams SR, Harootunian AT, Buechler YJ, Taylor SS, Tsien RY (1991) Fluorescence ratio imaging of cyclic AMP in single cells. Nature 349:694–697
Allen PB, Ouimet CC, Greengard P (1997) Spinophilin, a novel protein phosphatase 1 binding protein localized to dendritic spines. Proc Natl Acad Sci USA 94:9956–9961
Antoni FA, Palkovits M, Simpson J, Smith SM, Leitch AL, Rosie R, Fink G, Paterson JM (1998) Ca2+/calcineurin-inhibited adenylyl cyclase, highly abundant in forebrain regions, is important for learning and memory. J Neurosci 18:9650–9661
Belluscio L, Gold GH, Nemes A, Axel R (1998) Mice deficient in G_{olf} are anosmic. Neuron 20:69–81

Bertorello AM, Hopfield JF, Aperia A, Greengard P (1990) Inhibition by dopamine of $(Na^+ + K^+)$ATPase activity in neostriatal neurons through D_1 and D_2 dopamine receptor synergism. Nature 347:386–388

Betty M, Harnish SW, Rhodes KJ, Cockett MI (1998) Distribution of heterotrimeric G protein beta and gamma subunits in the rat brain. Neuroscience 85:475–486

Blank T, Nijholt I, Teichert U, Kügler H, Behrsing H, Fienberg A, Greengard P, Spiess J (1997) The phosphoprotein DARPP-32 mediates cAMP-dependent potentiation of striatal N-methyl-D-aspartate responses. Proc Natl Acad Sci USA 94:14859–14864

Bibb JA, Snyder GL, Nishi A, Yan Z, Meijer L, Fienberg A, Tsai LH, Kwon YT, Girault JA, Czernik AJ, Huganir R, Hemmings Jr HC, Nairn AC, Greengard P (1999) Protein kinase and protein phosphatase regulation by distinct phosphorylation sites within a single molecule. Nature 402:669–671

Bibb JA, Chen J, Taylor JR, Svenningsson P, Nishi A, Snyder GL, Yan Z, Sagawa ZK, Ouimet CC, Nairn AC, Nestler EJ, Greengard P (2001) Effects of chronic exposure to cocaine are regulated by the neuronal protein Cdk5. Nature 410:376–380

Brandon EP, Logue SF, Adams MR, Qi M, Sullivan SP, Matsumoto AM, Dorsa DM, Wehner JM, McKnight GS, Idzerda RL (1998) Defective motor behavior and neural gene expression in RIIβ protein kinase A mutant mice. J Neurosci 18:3639–3649

Brown EM, Carroll RJ, Aurbach GD (1977) Dopaminergic stimulation of cyclic AMP accumulation and parathyroid hormone release from dispersed bovine parathyroid cells. Proc Natl Acad Sci USA 74:4210–4213

Brown JH, Makman MH (1972) Stimulation by dopamine of adenylate cyclase in retinal homogenates and of adenosine-3′:5′-cyclic monophosphate formation in intact retina. Proc Natl Acad Sci USA 69:539–543

Calabresi P, De Murtas M, Pisani A, Stefani A, Sancessario G, Mercuri NB, Bernardi G (1995) Vulnerability of medium spiny striatal neurons to glutamate: Role of Na^+/K^+ ATPase. Eur J Neurosci 7:1674–1683

Cantrell AR, Smith RD, Goldin AL, Scheuer T, Catterall WA (1997) Dopaminergic modulation of sodium current in hippocampal neurons via cAMP-dependent phosphorylation of specific sites in the sodium channel α subunit. J Neurosci 17:7330–7338

Carr DW, Stofko-Hahn RE, Fraser IDC, Cone RD, Scott JD (1992) Localization of the cAMP-dependent protein kinase to the postsynaptic densities by A-Kinase Anchoring Proteins. Characterization of AKAP 79. J Biol Chem 267:16816–16823

Cepeda C, Buchwald NA, Levine MS (1993) Neuromodulatory actions of dopamine in the neostriatum are dependent upon the excitatory amino acid receptor subtypes activated. Proc Natl Acad Sci USA 90:9576–9580

Chneiweiss H, Cordier J, Sobel A (1992) Stathmin phosphorylation is regulated in striatal neurons by vasoactive intestinal peptide and monoamines via multiple intracellular pathways. J Neurochem 58:282–289

Clement Cormier YC, Kebabian JW, Petzold GL, Greengard P (1974) Dopamine-sensitive adenylate cyclase in mammalian brain: a possible site of action of antipsychotic drugs. Proc Natl Acad Sci USA 71:1113–1117

Clement Cormier YC, Parrish RG, Petzold GL, Kerabian JW, Greengard P (1975) Characterization of a dopamine-sensitive adenylate cyclase in the rat caudate nucleus. J Neurochem 25:143–149

Coghlan VM, Perrino BA, Howard M, Langeberg LK, Hicks JB, Gallatin WM, Scott JD (1995) Association of protein kinase A and protein phosphatase 2B with a common anchoring protein. Science 267:108–111

Corvol JC, Studler JM, Schonn JS, Girault JA, Herve D (2001) Galpha(olf) is necessary for coupling D1 and A2a receptors to adenylyl cyclase in the striatum. J Neurochem 76:1585–1588 de Rooij J, Zwartkruis FJ, Verheijen MH, Cool RH, Nijman SM, Wittinghofer A, Bos JL (1998) Epac is a Rap1 guanine-nucleotide-exchange factor directly activated by cyclic AMP [see comments]. Nature 396:474–477

Dell'Acqua ML, Scott JD (1997) Protein kinase A anchoring. J Biol Chem 272: 12881–12884

Dent P, Lavoinne A, Nakielny S, Caudwell FB, Watt P, Cohen P (1990) The molecular mechanism by which insulin stimulates glycogen synthesis in mammalian skeletal muscle. Nature 348:302–308

Desdouits F, Cheetham JJ, Huang H-B, Kwon Y-G, Da Cruz e Silva EF, Denefle P, Ehrlich ME, Nairn AC, Greengard P, Girault JA (1995b) Mechanism of inhibition of protein phosphatase 1 by DARPP-32: Studies with recombinant DARPP-32 and synthetic peptides. Biochem Biophys Res Commun 206:652–658

Desdouits F, Cohen D, Nairn AC, Greengard P, Girault JA (1995c) Phosphorylation of DARPP-32, a dopamine- and cAMP-regulated phosphoprotein, by casein kinase I in vitro and in vivo. J Biol Chem 270:8772–8778

Desdouits F, Siciliano JC, Greengard P, Girault JA (1995a) Dopamine- and cAMP-regulated phosphoprotein DARPP-32: Phosphorylation of Ser-137 by casein kinase I inhibits dephosphorylation of Thr-34 by calcineurin. Proc Natl Acad Sci USA 92: 2682–2685

Drinnan SL, Hope BT, Snutch TP, Vincent SR (1991) Golf in the basal ganglia. Mol Cell Neurosci 2:66–70

Dulubova I, Horiuchi A, Snyder GL, Girault JA, Czernik AJ, Shao L, Ramabhadran R, Greengard P, Nairn AC (2001) ARPP-16/ARPP-19: a highly conserved family of cAMP-regulated phosphoproteins. J Neurochem 77:229–238

Egloff MP, Johnson DF, Moorhead G, Cohen PTW, Cohen P, Barford D (1997) Structural basis for the recognition of regulatory subunits by the catalytic subunit of protein phosphatase 1. EMBO J 16:1876–1887

Felder CC, Blecher M, Jose PA (1989) Dopamine-1-mediated stimulation of phospholipase C activity in rat renal cortical membranes. J Biol Chem 264:8739–8745

Fienberg AA, Hiroi N, Mermelstein PG, Song WJ, Snyder GL, Nishi A, Cheramy A, O'Callaghan JP, Miller DB, Cole DG, Corbett R, Haile CN, Cooper DC, Onn SP, Grace AA, Ouimet CC, White FJ, Hyman SE, Surmeier DJ, Girault JA, Nestler EJ, Greengard P (1998) DARPP-32: Regulator of the efficacy of dopaminergic neurotransmission. Science 281:838–839

Friedman E, Jin LQ, Cai GP, Hollon TR, Drago J, Sibley DR, Wang HY (1997) D_1-like dopaminergic activation of phosphoinositide hydrolysis is independent of D_{1A} dopamine receptors: Evidence from D_{1A} knockout mice. Mol Pharmacol 51:6–11

Fryckstedt J, Aperia A, Snyder G, Meister B (1993) Distribution of dopamine- and cAMP-dependent phosphoprotein (DARPP-32) in the developing and mature kidney. Kidney Int 44:495–502

Girault JA, Hemmings HC, Jr, Williams KR, Nairn AC, Greengard P (1989) Phosphorylation of DARPP-32, a dopamine- and cAMP-regulated phosphoprotein, by casein kinase II. J Biol Chem 264:21748–21759

Girault JA, Horiuchi A, Gustafson EL, Rosen NL, Greengard P (1990b) Differential expression of ARPP-16 and ARPP-19, two highly related cAMP-regulated phosphoproteins, one of which is specifically associated with dopamine-innervated brain regions. J Neurosci 10:1124–1133

Girault JA, Shalaby IA, Rosen NL, Greengard P (1988) Regulation by cAMP and vasoactive intestinal peptide of phosphorylation of specific proteins in striatal cells in culture. Proc Natl Acad Sci USA 85:7790–7794

Girault JA, Spampinato U, Glowinski J, Besson MJ (1986) In vivo release of [3H]gamma-aminobutyric acid in the rat neostriatum–II. Opposing effects of D_1 and D_2 dopamine receptor stimulation in the dorsal caudate putamen. Neuroscience 19:1109–1117

Girault JA, Walaas SI, Hemmings HC, Jr, Greengard P (1990a) ARPP-21, a cAMP-regulated phosphoprotein enriched in dopamine-innervated brain regions: Tissue distribution and regulation of phosphorylation in rat brain. Neuroscience 37:317–325

Glantz SB, Amat JA, Rubin CS (1992) cAMP signaling in neurons: Patterns of neuronal expression and intracellular localization for a novel protein, AKAP 150, that anchors the regulatory subunit of cAMP-dependent protein kinase IIβ. Mol Biol Cell 3:1215–1228

Glatt CE, Snyder SH (1993) Cloning and expression of an adenylyl cyclase localized to the corpus striatum. Nature 361:536–538

Goto S, Matsukado Y, Mihara Y, Inoue N, Miyamoto E (1986) The distribution of calcineurin in rat brain by light and electron microscopic immunohistochemistry and enzyme-immunoassay. Brain Res 397:161–172

Gray PC, Scott JD, Catterall WA (1998) Regulation of ion channels by cAMP-dependent protein kinase and A-kinase anchoring proteins. Curr Opin Neurobiol 8:330–334

Greengard P, Allen PB, Nairn AC (1999) Beyond the dopamine receptor: the DARPP-32/protein phosphatase-1 cascade. Neuron in press

Greengard P, Nairn AC, Girault JA, Ouimet CC, Snyder GL, Fisone G, Allen P, Feinberg A, Nishi A (1998) The DARPP-32/protein phosphatase-1 cascade: a model for signal integration. Brain Res Rev 26:274–284

Gustafson EL, Girault JA, Hemmings HC, Jr, Nairn AC, Greengard P (1991) Immunocytochemical localization of phosphatase inhibitor-1 in rat brain. J Comp Neurol 310:170–188

Halpain S, Girault JA, Greengard P (1990) Activation of NMDA receptors induces dephosphorylation of DARPP-32 in rat striatal slices. Nature 343:369–372

Hemmings HC, Jr, Girault JA, Nairn AC, Bertuzzi G, Greengard P (1992) Distribution of protein phosphatase inhibitor-1 in brain and peripheral tissues of various species: comparison with DARPP-32. J Neurochem 59:1053–1061

Hemmings HC, Jr, Girault JA, Williams KR, LoPresti MB, Greengard P (1989) ARPP-21, a cyclic AMP-regulated phosphoprotein enriched in dopamine-innervated brain regions: amino acid sequence of the site phosphorylated by cyclic AMP in intact cells, and kinetics of its phosphorylation in vitro. J Biol Chem 264:7726–7733

Hemmings HC, Jr, Greengard P (1989) ARPP-21, a cAMP-regulated phosphoprotein Mr = 21,000 enriched in dopamine-innervated brain regions. I. Purification and characterization of the protein from bovine caudate nucleus. J Neurosci 9:851–864

Hemmings HC, Jr, Greengard P, Tung HYL, Cohen P (1984b) DARPP-32, a dopamine-regulated neuronal phosphoprotein, is a potent inhibitor of protein phosphatase-1. Nature 310:503–505

Hemmings HC, Jr, Nairn AC, Aswad DW, Greengard P (1984c) DARPP-32, a dopamine- and adenosine 3′:5′-monophosphate-regulated phosphoprotein enriched in dopamine-innervated brain regions. II Purification and characterization of the phosphoprotein from bovine caudate nucleus. J Neurosci 4:99–110

Hemmings HC, Jr, Nairn AC, Elliott JI, Greengard P (1990) Synthetic peptide analogs of DARPP-32 (M_r 32,000 dopamine- and cAMP-regulated phosphoprotein), an inhibitor of protein phosphatase-1. Phosphorylation, dephosphorylation, and inhibitory activity. J Biol Chem 265:20369–20376

Hemmings HC, Jr, Nairn AC, Greengard P (1984a) DARPP-32, a dopamine- and adenosine 3′:5′-monophosphate-regulated neuronal phosphoprotein. II. Comparison of the kinetics of phosphorylation of DARPP-32 and phosphatase inhibitor 1. J Biol Chem 259:14491–14497

Hervé D, Trovero F, Blanc G, Thierry AM, Glowinski J, Tassin JP (1989) Non-dopaminergic prefrontocortical efferent fibers modulate D_1 receptor denervation supersensitivity in specific regions of the rat striatum. J Neurosci 9:3699–3708

Hervé D, Lévi-Strauss M, Marey-Semper I, Verney C, Tassin J-P, Glowinski J, Girault JA (1993) G_{olf} and G_s in rat basal ganglia: Possible involvement of G_{olf} in the coupling of dopamine D_1 receptor with adenylyl cyclase. J Neurosci 13:2237–2248

Hervé D, Rogard M, Lévi-Strauss M (1995) Molecular analysis of the multiple G_{olf} α subunit mRNAs in the rat brain. Mol Brain Res 32:125–134

Hervé D, Trovero F, Blanc G, Glowinski J, Tassin J-P (1992) Autoradiographic identification of D_1 dopamine receptors labelled with [^{3}H]dopamine: Distribution, regulation and relationship to coupling. Neuroscience 46:687–700

Hervé D, Le Moine C, Corvol JC, Belluscio L, Ledent C, Fienberg A, Jaber M, Studler JM, Maujay J, Girault JA (2001) Galpha(olf) levels are regulated by receptor usage and control dopamine and adenosine action in the striatum. J Neurosci 21: 4390–4399

Hirano K, Phan BC, Hartshorne DJ (1997) Interactions of the subunits of smooth muscle myosin phosphatase. J Biol Chem 272:3683–3688

Hisatsune C, Umemori H, Inoue T, Michikawa T, Kohda K, Mikoshiba K, Yamamoto T (1997) Phosphorylation-dependent regulation of N-methyl-D-aspartate receptors by calmodulin. J Biol Chem 272:20805–20810

Horiuchi A, Williams KR, Kurihara T, Nairn AC, Greengard P (1990) Purification and cDNA cloning of ARPP-16, a cAMP-regulated phosphoprotein enriched in basal ganglia, and of a related phosphoprotein, ARPP-19. J Biol Chem 265:9476–9484

Hsieh-Wilson L, Allen PB, Watanabe T, Nairn AC, Greengard P (1999) Characterization of the neuronal targeting protein spinophilin and its interactions with protein phosphatase-1. Biochemistry 38:4365–4373

Huang HB, Horiuchi A, Watanabe T, Shih SR, Tsay HJ, Li HC, Greengard P, Nairn AC (1999) Characterization of the inhibition of protein phosphatase-1 by DARPP-32 and inhibitor-2. J Biol Chem 274:7870–7878

Huang KX, Walters JR (1996) Dopaminergic regulation of AP-1 transcription factor DNA binding activity in rat striatum. Neuroscience 75:757–775

Hubbard MJ, Cohen P (1993) On target with a new mechanism for the regulation of protein phosphorylation. Trends Biochem Sci 18:172–177

Hurley JH (1999) Structure, mechanism, and regulation of mammalian adenylyl cyclase. J Biol Chem 274:7599–7602

Jones DT, Masters SB, Bourne HR, Reed RR (1990) Biochemical characterization of three stimulatory GTP-binding proteins. J Biol Chem 265:2671–2676

Jones DT, Reed RR (1989) Golf: an olfactory neuron specific G-protein involved in odorant signal transduction. Science 244:790–795

Kawasaki H, Springett GM, Mochizuki N, Toki S, Nakaya M, Matsuda M, Housman DE, Graybiel AM (1998) A family of cAMP-binding proteins that directly activate Rap1. Science 282:2275–2279

Kebabian JW, Calne DB (1979) Multiple receptors for dopamine. Nature 277:93–96

Kebabian JW, Greengard P (1971) Dopamine-sensitive adenyl cyclase: possible role in synaptic transmission. Science 174:1346–1349

Kebabian JW, Petzold GL, Greengard P (1972) Dopamine-sensitive adenylate cyclase in caudate nucleus of rat brain, and its similarity to the "dopamine receptor". Proc Natl Acad Sci USA 69:2145–2149

Kimura K, Sela S, Bouvier C, Grandy DK, Sidhu A (1995b) Differential coupling of D_1 and D_5 dopamine receptors to guanine nucleotide binding proteins in transfected GH_4C_1 rat somatomammotrophic cells. J Neurochem 64:2118–2124

Kimura K, White BH, Sidhu A (1995a) Coupling of human D-1 dopamine receptors to different guanine nucleotide binding proteins. Evidence that D-1 dopamine receptors can couple to both G_s and G_o. J Biol Chem 270:14672–14678

King MM, Huang CY, Chock PB, Nairn AC, Hemmings HC, Jr, Chan KF, Greengard P (1984) Mammalian brain phosphoproteins as substrates for calcineurin. J Biol Chem 259:8080–8083

Klauck TM, Faux MC, Labudda K, Langeberg LK, Jaken S, Scott JD (1996) Coordination of three signaling enzymes by AKAP79, a mammalian scaffold protein. Science 271:1589–1592

Konradi C, Cole RL, Heckers S, Hyman SE (1994) Amphetamine regulates gene expression in rat striatum via transcription factor CREB. J Neurosci 14:5623–5634

Konradi C, Leveque JC, Hyman SE (1996) Amphetamine and dopamine-induced immediate early gene expression in striatal neurons depends on postsynaptic NMDA receptors and calcium. J Neurosci 16:4231–4239

Krueger BK, Forn J, Walters JR, Roth RH, Greengard P (1976) Stimulation by dopamine of adenosine cyclic 3′,5′-monophosphate formation in rat caudate nucleus: effect of lesions of the nigro-neostriatal pathway. Mol Pharmacol 12:639–648

Kwon YG, Huang HB, Desdouits F, Girault JA, Greengard P, Nairn AC (1997) Characterization of the interaction between DARPP-32 and protein phosphatase 1 (PP-1): DARPP-32 peptides antagonize the interaction of PP-1 with binding proteins. Proc Natl Acad Sci USA 94:3536–3541

Levis MJ, Bourne HR (1992) Activation of the alpha subunit of Gs in intact cells alters its abundance, rate of degradation, and membrane avidity. J Cell Biol 119:1297–1307

Lezcano N, Mrzljak L, Eubanks S, Levenson R, Goldman-Rakic P, Bergson C (2000) Dual signaling regulated by calcyon, a D1 dopamine receptor interacting protein, Science 287:1660–1664

Mahan LC, Burch RM, Monsma FJ, Jr, Sibley DR (1990) Expression of striatal D_1 dopamine receptors coupled to inositol phosphate production and Ca^{2+} mobilization in *Xenopus* oocytes. Proc Natl Acad Sci USA 87:2196–2200

Meister B, Askergren J, Tunevall G, Hemmings HC, Jr, Greengard P (1991b) Identification of a dopamine- and 3′,5′-cyclic adenosine monophosphate-regulated phosphoprotein of 32 kD (DARPP-32) in parathyroid hormone-producing cells of the human parathyroid gland. J Endocrinol Invest 14:655–661

Meister B, Fried G, Hokfelt T, Hemmings HC, Jr, Greengard P (1988) Immunohistochemical evidence for the existence of a dopamine- and cyclic AMP-regulated phosphoprotein (DARPP-32) in brown adipose tissue of pigs. Proc Natl Acad Sci USA 85:8713–8716

Meister B, Hokfelt T, Tsuruo Y, Hemmings HC, Jr, Ouimet CC, Greengard P, Goldstein M (1988) DARPP-32, a dopamine- and cyclic AMP-regulated phosphoprotein in tanycytes of the mediobasal hypothalamus: distribution and relation to dopamine and luteinizing hormone-releasing hormone neurons and other glial elements. Neuroscience 27:607–622

Meister B, Schultzberg M, Hemmings HC, Jr, Greengard P, Goldstein M, Hökfelt T (1991a) Dopamine- and adenosine-3′,5′-monophosphate (cAMP)-regulated phosphoprotein of 32 kDa (DARPP-32) in the adrenal gland: Immunohistochemical localization. J Auton Nerv Syst 36:75–84

Miller RJ, Horn AS, Iversen LL (1974) The action of neuroleptic drugs on dopamine-stimulated adenosine cyclic 3′,5′-monophosphate production in rat neostriatum and limbic forebrain. Mol Pharmacol 10:759–766

Mishra RK, Gardner EL, Katzman R, Makman MH (1974) Enhancement of dopamine-stimulated adenylate cyclase activity in rat caudate after lesions in substantia nigra: evidence for denervation supersensitivity. Proc Natl Acad Sci U S A 71:3883–3887

Montminy M (1997) Transcriptional regulation by cyclic AMP. Annu Rev Biochem 66:807–822

Murphy BJ, Rossie S, De Jongh KS, Catterall WA (1993) Identification of the sites of selective phosphorylation and dephosphorylation of the rat brain Na^+ channel α subunit by cAMP-dependent protein kinase and phosphoprotein phosphatases. J Biol Chem 268:27355–27362

Nakanishi H, Obaishi H, Satoh A, Wada M, Mandai K, Satoh K, Nishioka H, Matsuura Y, Mizoguchi A, Takai Y (1997) Neurabin: a novel neural tissue-specific actin filament-binding protein involved in neurite formation. J Cell Biol 139:951–961

Nishi A, Snyder GL, Greengard P (1997) Bidirectional regulation of DARPP-32 phosphorylation by dopamine. J Neurosci 17:8147–8155

Nishi A, Snyder GL, Nairn AC, Greengard P (1999) Role of calcineurin and protein phosphatase-2A in the regulation of DARPP-32 dephosphorylation in neostriatal neurons. J Neurochem 72:2015–2021

Nishi A, Bibb JA, Snyder GL, Higashi H, Nairn AC, Greengard P (2000) Amplification of dopaminergic signaling by a positive feedback loop, Proc Natl Acad Sci U S A 97:12840–12845
Ouimet CC, Greengard P (1990) Distribution of DARPP-32 in the basal ganglia: An electron microscopic study. J Neurocytol 19:39–52
Ouimet CC, LaMantia AS, Goldman-Rakic P, Rakic P, Greengard P (1992) Immunocytochemical localization of DARPP-32, a dopamine and cyclic-AMP-regulated phosphoprotein, in the primate brain. J Comp Neurol 323:209–218
Ouimet CC, Langley-Gullion KC, Greengard P (1998) Quantitative immunocytochemistry of DARPP-32-expressing neurons in the rat caudatoputamen. Brain Res 808:8–12
Ouimet CC, Miller PE, Hemmings HC, Jr, Walaas SI, Greengard P (1984) DARPP-32, a dopamine- and adenosine 3′:5′-monophosphate-regulated phosphoprotein enriched in dopamine-innervated brain regions. III Immunocytochemical localization. J Neurosci 4:111–124
Penit-Soria J, Durand C, Besson M-J, Herve D (1997) Levels of stimulatory G protein are increased in the rat striatum after neonatal lesion of dopamine neurons. Neuroreport 8:829–833
Piomelli D, Pilon C, Giros B, Sokoloff P, Martres M-P, Schwartz J-C (1991) Dopamine activation of the arachidonic acid cascade as a basis for D_1/D_2 receptor synergism. Nature 353:164–167
Polli JW, Kincaid RL (1992) Molecular cloning of DNA encoding a calmodulin-dependent phosphodiesterase enriched in striatum. Proc Natl Acad Sci USA 89: 11079–11083
Polli JW, Kincaid RL (1994) Expression of a calmodulin-dependent phosphodiesterase isoform (PDE1B1) correlates with brain regions having extensive dopaminergic innervation. J Neurosci 14:1251–1261
Rodbell M, Birnbaumer L, Pohl SL, Krans HM (1971) The glucagon-sensitive adenyl cyclase system in plasma membranes of rat liver. V. An obligatory role of guanyl-nucleotides in glucagon action. J Biol Chem 246:1877–1882
Santoro B, Grant SGN, Bartsch D, Kandel ER (1997) Interactive cloning with the SH3 domain of N-src identifies a new brain specific ion channel protein, with homology to Eag and cyclic nucleotide-gated channels. Proc Natl Acad Sci USA 94: 14815–14820
Santoro B, Liu DT, Yao H, Bartsch D, Kandel ER, Siegelbaum SA, Tibbs GR (1998) Identification of a gene encoding a hyperpolarization-activated pacemaker channel of brain. Cell 93:717–729
Satoh A, Nakanishi H, Obaishi H, Wada M, Takahashi K, Satoh K, Hirao K, Nishioka H, Hata Y, Mizoguchi A, Takai Y (1998) Neurabin-II/spinophilin. An actin filament-binding protein with one pdz domain localized at cadherin-based cell-cell adhesion sites. J Biol Chem 273:3470–3475
Savasta M, Dubois A, Benhavioles J, Scatton B (1988) Different plasticity changes in D_1 and D_2 receptors on striatal subregions following impairment of dopaminergic transmission. Neurosci Lett 85:119–124
Schiffmann SN, Desdouits F, Menu R, Greengard P, Vincent JD, Vanderhaeghen JJ, Girault JA (1998) Modulation of the voltage-gated sodium current in rat striatal neurons by DARPP-32, an inhibitor of protein phosphatase. Eur J Neurosci 10: 1312–1320
Schiffmann SN, Lledo P-M, Vincent J-D (1995) Dopamine D_1 receptor modulates the voltage-gated sodium current in rat striatal neurones through a protein kinase A. J Physiol (Lond.) 483:95–107
Shenolikar S, Nairn AC (1991) Protein phosphatases: recent progress. Adv Cyclic Nucleotide Protein Phosphorylation Res 23:1–121
Sidhu A, Sullivan M, Kohout T, Balen P, Fishman PH (1991) D_1 dopamine receptors can interact with both stimulatory and inhibitory guanine nucleotide binding proteins. J Neurochem 57:1445–1451

Snyder GL, Fienberg AA, Huganir RL, Greengard P (1998) A dopamine D_1 receptor protein kinase A dopamine- and cAMP-regulated phosphoprotein (M_r 32 kDa) protein phosphatase-1 pathway regulates dephosphorylation of the NMDA receptor. J Neurosci 18:10297–10303

Snyder GL, Girault JA, Chen JYC, Czernik AJ, Kebabian J, Nathanson JA, Greengard P (1992) Phosphorylation of DARPP-32 and protein phosphatase inhibitor-1 in rat choroid plexus: regulation by factors other than dopamine. J Neurosci 12:3071–3083

Stone RA, Laties AM, Hemmings HC, Jr, Ouimet CC, Greengard P (1986) DARPP-32 in the ciliary epithelium of the eye: a neurotransmitter-regulated phosphoprotein of brain localizes to secretory cells. J Histochem Cytochem 34: 1465–1468

Sunahara RK, Dessauer CW, Gilman AG (1996) Complexity and diversity of mammalian adenylyl cyclases. Annu Rev Pharmacol Toxicol 36:461–480

Surmeier DJ, Bargas J, Hemmings HC, Jr, Nairn AC, Greengard P (1995) Modulation of calcium currents by a D_1 dopaminergic protein kinase/phosphatase cascade in rat neostriatal neurons. Neuron 14:385–397

Surmeier DJ, Kitai ST (1993) D_1 and D_2 dopamine receptor modulation of sodium and potassium currents in rat neostriatal neurons. Prog Brain Res 99:309–324

Taylor SS, Bubis J, Toner-Webb J, Saraswat LD, First EA, Buechler JA, Knighton DR, Sowadski J (1988) CAMP-dependent protein kinase: prototype for a family of enzymes. FASEB J 2:2677–2685

Tsou K, Girault JA, Greengard P (1993) Dopamine D_1 agonist SKF 38393 increases the state of phosphorylation of ARPP-21 in substantia nigra. J Neurochem 60: 1043–1046

Undie AS, Friedman E (1990) Stimulation of a dopamine D_1 receptor enhances inositol phosphates formation in rat brain. J Pharmacol Exp Ther 253:987–992

Valjent E, Corvol J C, Pages C, Besson M J, Maldonado R, Caboche J (2000) Involvement of the extracellular signal-regulated kinase cascade for cocaine-rewarding properties, J Neurosci 20:8701–9

Ventra C, Porcellini A, Feliciello A, Gallo A, Paolillo M, Mele E, Avvedimento VE, Schettini G (1996) The differential response of protein kinase A to cyclic AMP in discrete brain areas correlates with the abundance of regulatory subunit II. J Neurochem 66:1752–1761

Vernier P, Cardinaud B, Valdenaire O, Philippe H, Vincent JD (1995) An evolutionary view of drug-receptor interaction: the bioamine receptor family. Trends Pharmacol Sci 16:375–381

Von Voigtlander PF, Boukma SJ, Johnson GA (1973) Dopaminergic denervation supersensitivity and dopamine stimulated adenyl cyclase activity. Neuropharmacology 12:1081–1086

Walaas SI, Greengard P (1984) DARPP-32, a dopamine- and adenosine 3':5'-monophosphate-regulated phosphoprotein enriched in dopamine-innervated brain regions. I Regional and cellular distribution in the rat brain. J Neurosci 4:84–98

Walaas SI, Nairn AC, Greengard P (1983b) Regional distribution of calcium- and cyclic adenosine 3':5'-monophosphate-regulated protein phosphorylation systems in mammalian brain. I. Particulate systems. J Neurosci 3:291–301

Walaas SI, Nairn AC, Greengard P (1983a) Regional distribution of calcium- and cyclic adenosine 3':5'-monophosphate-regulated protein phosphorylation systems in mammalian brain. II. Soluble systems. J Neurosci 3:302–311

Walaas SI, Sedvall G, Greengard P (1989) Dopamine-regulated phosphorylation of synaptic vesicle-associated proteins in rat neostriatum and substantia nigra. Neuroscience 29:9–19

Wang HY, Undie AS, Friedman E (1995) Evidence for the coupling of G_q protein to D_1-like dopamine sites in rat striatum: Possible role in dopamine-mediated inositol phosphate formation. Mol Pharmacol 48:988–994

Watson JB, Coulter PM, II, Margulies JE, De Lecea L, Danielson PE, Erlander MG, Sutcliffe JG (1994) G-protein gamma7 subunit is selectively expressed in medium-sized neurons and dendrites of the rat neostriatum. J Neurosci Res 39:108–116

Williams KR, Hemmings HC, Jr, LoPresti MB, Konigsberg WH, Greengard P (1986) DARPP-32, a dopamine- and cyclic AMP-regulated neuronal phosphoprotein. Primary structure and homology with protein phosphatase inhibitor-1. J Biol Chem 261:1890–1903

Worley PF, Baraban JM, Van Dop C, Neer EJ, Snyder SH (1986) Go, a guanine nucleotide-binding protein: immunohistochemical localization in rat brain resembles distribution of second messenger systems. Proc Natl Acad Sci USA 83:4561–4565

Yan Z, Hsieh-Wilson L, Feng J, Tomizawa K, Allen PB, Fienberg AA, Nairn AC, Greengard P (1999) Protein phosphatase 1 modulation of neostriatal AMPA channels: regulation by DARPP-32 and spinophilin. Nat Neurosci 2:13–17

Zagotta WN, Siegelbaum SA (1996) Structure and function of cyclic nucleotide-gated channels. Annu Rev Neurosci 19:235–263

Zhuang X, Belluscio L, Hen R (2000) Golf alpha mediates dopamine D1 receptor signaling. J Neurosci 20:RC91 (1–5)

CHAPTER 10
The Dopamine Transporter: Molecular Biology, Pharmacology and Genetics

C. PIFL and M.G. CARON

A. Introduction

More than 30 years ago, accumulation of dopamine by brain tissue was demonstrated after intraventricular injection of [^{3}H]dopamine (GLOWINSKI and IVERSEN 1966). The highest concentration was found in the striatum, the brain region with the highest innervation by dopaminergic neurons. From neuronal noradrenaline uptake, which had been discovered several years earlier (AXELROD et al. 1959; HERTTING et al. 1961) dopamine uptake was shown to be pharmacologically different: tricyclic antidepressants such as desipramine inhibited dopamine uptake with low potency in vivo (GLOWINSKI et al. 1966; CARLSSON et al. 1966; FUXE and UNGERSTEDT 1968) and in vitro (ROSS and RENYI 1967; HAMBERGER 1967; JONASON and RUTLEDGE 1968) and, in contrast to the uptake system in noradrenaline neurons, (+)- and (–)-noradrenaline were equipotent as inhibitors of [^{3}H]dopamine uptake by striatal synaptosomes (COYLE and SNYDER 1969). The potent inhibitors of dopamine uptake, nomifensine, mazindol and cocaine were radiolabelled and used as ligands to specifically label what was called the dopamine transporter (DAT) complex (JAVITCH et al. 1983; KENNEDY and HANBAUER 1983; PIMOULE et al. 1983; DUBOCOVICH and ZAHNISER 1985). Photoaffinity probes based on dopamine uptake blocking diphenylpiperazine derivatives were shown to incorporate into a glycoprotein of approximately 60 kDa apparent molecular mass (GRIGORIADIS et al. 1989; SALLEE et al. 1989), and molecular cloning finally established the DAT as a molecular entity distinct from other plasmalemmal neurotransmitter transporters (GIROS et al. 1991; KILTY et al. 1991; SHIMADA et al. 1991; USDIN et al. 1991). Interest in the function of dopamine uptake sites was especially raised by its role in the mechanism of the parkinsonism-inducing drug 1-methyl-4-phenyl-1,2,3,6-tetrahydropyridine (MPTP) (CHIBA et al. 1985; JAVITCH et al. 1985a) and in the abuse of cocaine (RITZ et al. 1987). The decisive role of the DAT not only in the pharmacology/toxicology but also in the physiology of dopaminergic neurotransmission was finally revealed by its genetic deletion in mice (GIROS et al. 1996; BOSSÉ et al. 1997; JONES et al. 1998a).

B. Molecular Biology

I. Cloning of the Dopamine Transporter

The first studies aiming at the molecular characterization of the DAT were based on the injection of poly(A)⁺RNA isolated from rat forebrain or brain stem (BLAKELY et al. 1988) or from human substantia nigra (BANNON et al. 1990) into *Xenopus laevis* oocytes which, as a result, accumulated tritiated dopamine. In a similar study *Xenopus* oocytes were injected with a transcript from a size-selected rat midbrain cDNA library and displayed uptake of dopamine in a cocaine-blockable fashion (UHL et al. 1991). However, the cDNAs encoding the rat and bovine DATs were finally obtained on the basis of their structural homology to the gamma-aminobutyric acid (GABA) and noradrenaline transporter (NET) which had been cloned previously using oligonucleotides derived from sequence data of the purified protein (GUASTELLA et al. 1990) or using a COS cell expression system (PACHOLCZYK et al. 1991), respectively. Degenerate oligonucleotide primers from conserved regions of the GABA transporter (GAT) and NET were used to amplify by polymerase chain reaction partial cDNA clones from midbrain or substantia nigra to generate probes for screening cDNA libraries. Full-length cDNAs were identified which encoded a cocaine-sensitive rat (GIROS et al. 1991; KILTY et al. 1991; SHIMADA et al. 1991) or bovine DAT (USDIN et al. 1991). The cDNA of the human DAT was cloned by screening human substantia nigra cDNA libraries with probes derived from the rat (GIROS et al. 1992; VANDENBERGH et al. 1992a; ESHLEMAN et al. 1995) or the bovine DAT (PRISTUPA et al. 1994). Functional expression and pharmacological characterization of the human transporter was reported in three of these studies (GIROS et al. 1992; PRISTUPA et al. 1994; ESHLEMAN et al. 1995). By awareness of gene homology to the GAT/NET gene family following the genome-sequencing project of *Caenorhabditis elegans*, a catecholamine transporter was cloned from this species (JAYANTHI et al. 1998). This transporter was not only inhibited by the DAT-selective drug GBR 12909 but also potently blocked by the NET-selective antagonists nisoxetine and desipramine. Although, on the basis of the higher maximal transport rate of dopamine over noradrenaline, this protein was considered to be a DAT, it may represent a non-selective catecholamine transporter in this species. Recently, an endogenous DAT in several kidney-derived cell lines was characterized and a partial cDNA clone from COS-1 cells was obtained that shared functional, pharmacological and molecular characteristics with the neuronal DAT (SUGAMORI et al. 1999).

II. Structural Features of the Cloned Dopamine Transporter

The open reading frame predicted by the respective cDNA encodes 619 amino acids for the rat, 629 for the human, 693 for the bovine and 615 for the *C. elegans* DAT with an estimated molecular mass of 68–70 kDa. The human DAT has a 92% overall identity with the rat (differing by 48 amino acids including one additional glycine at position 199 in the human protein), and is 84% identical to the bovine transporter (an additional 72 amino acids exist on the carboxy-terminal tail of the bovine transporter). The *C. elegans* catecholamine transporter exhibits ~43% identity with the known mammalian DATs. The human DAT displays 66% homology with the human NET (75% with conservative substitutions), 46% with the serotonin (RAMAMOORTHY et al. 1993) and 42% with the GABA transporter (NELSON et al. 1990). Based on amino acid sequence homology between these neurotransmitter transporters and transporters for glycine, proline, taurine, betaine and creatine, as well as transporters without identified substrates, the DAT is part of a family of Na^+/Cl^--dependent transporters as reviewed recently (NELSON 1998). As far as transport has been studied on members of this gene family, it depends on extracellular Na^+ and Cl^-. The amino acids deduced from their genes revealed a common structure of presumably 12 stretches of 20–24 hydrophobic amino acid residues that possibly represent membrane-spanning domains (transmembrane helices, TM). The lack of a signal peptide for promoting insertion into the membrane during translation, together with the 12 TMs, suggest the amino- and carboxy-terminal regions of these transporters are present in the cytoplasm. The proposed model predicts five intracellular and six extracellular loops between the TMs. The predicted position of the TMs suggests that the second extracytoplasmic loop is much larger (50–65 amino acid residues) than the other five (6–25 residues); the intracellular amino- and carboxy-terminal parts of the human DAT consist of about 68 and 44 amino acid residues, respectively.

The topology of the DAT is supported by the following experimental findings. In electron microscopic studies, immunogold labelling using an amino-terminal domain anti-peptide antiserum was associated with the cytoplasmic surface of membranes whereas an antibody directed against a second extracellular loop peptide was found extracellularly (NIRENBERG et al. 1996; HERSCH et al. 1997). In a biochemical study, the effect of membrane permeant and impermeant derivatives of methanethiosulphonate, which react with cysteines, were determined on the binding of a cocaine analogue to membranes versus intact cells heterogeneously expressing the wild-type or constructs of the DAT with mutated cysteines (FERRER and JAVITCH 1998). The rates of modulation of binding were consistent with an extracellular localization of Cys^{90} in the loop between TM1 and TM2 and Cys^{306} in the loop between TM5 and TM6 as predicted by the model. Similarly, results on mutated cysteines in positions 135 and 342 suggested the loop between TM2 and TM3 and that between TM6 and TM7 to be intracellular, again consistent with topology originally proposed by hydropathicity analysis. Like all other members of the family of Na^+/Cl^- neurotransmitter transporters, the DAT displays consensus sites for N-linked glycosylation in the large loop between TM3 and TM4 (four in the rat, three in the human and bovine, and two in the *C. elegans* transporter) which argues in favour of an extracellular localization of this loop. Controlled proteolysis of photoaffinity-labelled DAT and epitope-specific immunoprecipitation of the labelled fragments provided evidence that N-glycosylation of the transporter does occur in the region between TM3 and TM4 (VAUGHAN 1995). Two co-ordinating residues in an endogenous Zn^{2+}-binding site in the human DAT have been defined, His^{193} in the large extracellular loop and His^{375} in the loop between TM7 and TM8, which provided evidence for spatial proximity between this residues and consequently between the respective loops (NORREGAARD et al. 1998).

Putative consensus phosphorylation sites for protein kinase C were found, one in the third intracellular loop (conserved in most members of the Na^+/Cl^- neurotransmitter transporters), another near the amino-terminus. Also in the amino-terminal domain, but nearer to TM1 lie two cyclic adenosine monophosphate (cAMP)-dependent protein kinase consensus sites whereas a putative Ca^{2+}-calmodulin-dependent kinase site is found near the carboxy terminus. Phosphorylation of the transporter can modify its function (see below).

The three mammalian species homologues (human, rat, bovine) of the DAT display two "leucine zipper" motifs, that consist of at least four leucine residues periodically spaced by six amino acids in an alpha helix, in TM2 (one of the leucines replaced by an allowable methionine residue) and in TM9. This structural motif, originally thought to mediate protein–protein interactions in DNA-binding proteins (LANDSCHULZ et al. 1988) and also found in TM2 of the GAT and NET, might be involved in structural organization of the transporter. Interestingly, radiation inactivation analysis of radioligand binding to the DAT yielded target size estimates of 94–143 kDa in rat striatum (BERGER et al. 1994) and 278 kDa in canine striatum (MILNER et al. 1994), well beyond the estimate of 70 kDa derived from the deduced amino acid sequence for the cloned DAT cDNA. This suggests that the functional transporter protein might be a homo- or hetero-oligomer. Size-exclusion chromatography of solubilized membranes from primate striatum indicated binding of a cocaine analogue to a single protein with an apparent molecular weight of 170 kDa (GRACZ and MADRAS 1995). The formation of DAT oligomeric complexes was suggested to be regulated by the carboxy-terminal tail based on studies with truncation mutants (LEE et al. 1996).

III. Chimera and Mutagenesis Studies on the Cloned Dopamine Transporter

In the absence of crystallographic or high resolution nuclear magnetic resonance structural data from the DAT, analysis of functional chimeras between closely related family members and of mutant transporters represents the only way to map functional domains of the protein.

For the chimeric approach, the structural homology and functional relationship between the dopamine and NET was exploited. Uptake of the substrates dopamine, noradrenaline, and the neurotoxin 1-methyl-4-phenylpyridinium (MPP^+) and blockade of uptake by antidepressants and psychostimulants were used to determine which part of the protein is responsible for the distinct pharmacology of the transporters. Chimeric constructs based on the engineering of restriction sites unique in the cDNA of the DAT into the cDNA of the NET and expression of recombined cDNA cassettes suggested the carboxy-terminal part starting with the fourth intracellular loop and including TM8 to TM12 to be involved in substrate specificity and amphetamine stereoselectivity, the region between TM5 and TM8 to determine interaction with antidepressants, as well as cocaine, and the amino-terminal part up to TM5 to provide functions common to all members of the protein family as, for example, ionic dependence (GIROS et al. 1994). A series of chimeras constructed by a restriction site-independent method also delineated the middle part of the protein chain, TM5 to TM7, as influencing the sensitivity to NET-selective inhibitors such as desipramine, in contrast to the carboxy-terminal region including TM8–12, which did not contribute to differences between the catecholamine transporters in interacting with antidepressants (BUCK and AMARA 1995). These authors also assigned parts of the carboxy-terminal domain (TM10–11) a role in affinity for the substrate MPP^+, but, in addition, showed an effect of TM1–3 in determining apparent K_m values for transport of catecholamines and also an impact of TM1–3 on antidepressant selectivity (BUCK and AMARA 1994, 1995). A recent study attempted to explain the higher efficacy of dopamine, MPP^+ uptake and cocaine binding of the human versus the bovine DAT by the chimeric approach. Bovine TM3 in the human transporter impaired its function, but human TM3 in the bovine backbone restored only dopamine uptake (LEE et al. 1998).

The first site-directed mutations of the rat DAT (KITAYAMA et al. 1992a) focused on the aspartate residue (Asp^{79}) conserved in TM1 of the monoamine transporters but absent from the other members of the Na^+/Cl^- neurotransmitter transporters and on two serine residues in TM7 (Ser^{356} and Ser^{359} in the rat transporter). The latter are reminiscent of two closely located serines in TM5 of adrenoceptors which had been implicated in the binding of the hydroxyl group of the catechol ring (STRADER et al. 1989). Substitution of Asp^{79} to glycine or alanine resulted in loss of apparent affinity to dopamine, in a reduction of the maximal uptake rate (V_{max}) and of the affinity to the tritiated cocaine analogue (–)-2β-carbomethoxy-3β-(4-fluorphenyl)tropane (CFT); the conservative substitution to glutamate had a slightly smaller impact (KITAYAMA et al. 1992a). Mutation in TM7 serines to alanine or glycine, respectively, resulted in reduced dopamine transport and preservation of cocaine analogue binding with a minor reduction in affinity. Mutation of two other serines in TM7 of the rat transporter (Ser^{350} and Ser^{353}) had also minor effects on [^{3}H]CFT binding but enhanced the V_{max} of dopamine and, even more, of MPP^+, whereas replacement of serines and a tyrosine in TM11 (Ser^{507}, Ser^{518}, Tyr^{513}) by alanine increased substrate affinity, particularly to MPP^+ (KITAYAMA et al. 1993). Single mutations of Tyr^{533} in TM11 also affected substrate translocation more than binding of a cocaine analogue and again transport of MPP^+ in particular was changed: substitution to phenylalanine, which is present at this position in the human transporter, caused increased uptake velocity with decreased affinity (MITSUHATA et al. 1998). In this respect, observations using chimeric transporters (BUCK and AMARA 1994; GIROS et al. 1994) are in line with conclusions reached in site-directed mutagenesis in that more carboxy-terminal transmembrane domains are involved in substrate specificity.

Phenylalanines in or near the TM domains of the rat DAT were singly mutated to alanines based on the idea that catechol-π and phenyl-π interactions could allow the DAT to recognize dopamine or cocaine, respectively (LIN et al. 1999). A cluster of Phe-Ala substitutions that yielded specific reductions in transport turnover was found in TM1–2 and TM6–8 consistent with suggestions from chimeric studies on a role of the amino-terminal part in basic transport functions (GIROS et al. 1994) and the TM5–8 determining the V_{max} of substrate translocation (BUCK and AMARA 1994). Other mutants have yielded selective reduction in either dopamine or cocaine affinity. Losses of the phenyl side chains of Phe^{76}, Phe^{98}, Phe^{155}, Phe^{361} or Phe^{390} selectively reduce dopamine (Phe^{155}) or CFT affinities (Phe^{76}, Phe^{98}, Phe^{361}, Phe^{390}). A three-dimensional model of the DAT based on the primary DAT sequence and initial mutagenesis data predict a central DAT recognition pocket domain toward which these phenyl side chains are oriented (EDVARDSEN and DAHL 1994). The cocaine-selective domain around Phe^{361} in TM7, part of the structural domain conferring blocker selectivity in chimeric studies (GIROS et al. 1994; BUCK and AMARA 1995) seems to have enough space to hold a molecule of the size of cocaine which could prevent cocaine binding without affecting dopamine transport (LIN et al. 1999). Such potential cocaine antagonists are discussed as anti-addiction and anti-intoxication therapeutic agents.

Differences between the DAT and NET were the starting point for a mutagenesis study on Zn^{2+}-binding sites (NORREGAARD et al. 1998). Inhibition of uptake and potentiation of CFT-binding by micromolar Zn^{2+} in the DAT but not the NET was shown to be due to a histidine in the second extracellular loop of the DAT (His^{189}) and absent in the NET which allows co-ordination of Zn^{2+} between this site and a His^{375} conserved between DAT and NET in the fourth extracellular loop. These studies imply that substrate translocation is being modulated by the spatial proximity between the second and fourth extracellular loop.

Cysteine substitution and covalent modification can also be used to study structure-function relationships and the dynamics of protein function. The human DAT has 13 endogenous cysteines, 4 in extracellular loops (Cys^{90} in the first, Cys^{180} and Cys^{189} in the large second and Cys^{306} in the third), 2 in intracellular loops (Cys^{135} in the first, Cys^{342} in the third) and 5 in transmembrane domains (Cys^{243} in TM4, Cys^{319} in TM6 and Cys^{530} and Cys^{523} in TM11), the remaining 2 in amino- and carboxy-terminal parts (Cys^6 and Cys^{581}, respectively). Substitution of Cys^{180} and Cys^{189} by alanine in the large second but not of Cys^{90} in the first nor of Cys^{305} in the third extracellular loop of the rat transporter dramatically impaired uptake and [^{3}H]CFT binding and the reduced membrane versus perinuclear staining with an anti-DAT antiserum indicated a role in membrane insertion for the cysteines in the second extracellular loop

(WANG et al. 1995). The effect of thiol reactive methanethiosulphonates of different membrane permeability on mutants of the DAT with combinations of mutated cysteines was investigated for determination of transporter topology and interaction with cocaine (FERRER and JAVITCH 1998; see also Sect. B.II, this chapter). The reactivities of the cysteines in the intracellular loops (Cys^{135} in the first, Cys^{342} in the third) with a membrane permeant sulphhydryl-specific reagent were dramatically reduced in the presence of either cocaine or dopamine, that of the extracellular Cys^{90} in the first loop substantially potentiated by cocaine. Thus, binding of cocaine and dopamine resulted in a conformational change of the DAT protein, which suggests the substituted-cysteine accessibility method to be useful for mapping the binding site and transport pathway of the DAT.

The only truncation study of the DAT published so far removed the carboxy-terminal tail of the human transporter or substituted it by sequences of the intracellular carboxy-terminal of the D_1 or D_5 dopamine receptors which all shifted uptake of dopamine but not of noradrenaline to tenfold higher affinity but resulted in a profound loss of affinity of cocaine and other uptake inhibitors in [^{3}H]CFT binding experiments (LEE et al. 1996).

IV. Dopamine Transporter Gene

In situ chromosomal mapping revealed that the gene encoding the human DAT is localized on the most distal part of chromosome 5, at 5p15.3 (GIROS et al. 1992; VANDENBERGH et al. 1992b; locus symbol: SLC6A3, for solute carrier family 6, member 3). It spans over 64 kb and is divided into 15 exons separated by 14 introns (KAWARAI et al. 1997). Putative transmembrane domains were encoded by individual exons with the exception of TM11 and TM12. Exons encoding the TMs are interrupted by introns at codons homologous to the NET and serotonin transporter (LESCH et al. 1994; PÖRZGEN et al. 1995). The promoter sequence 5' to the transcriptional start contained neither a canonical "TATA" nor a "CAAT" box but two E-boxes (CATCTG) and eight SP-1 binding sites (GGCGGG) the relative positioning of some of them is similar to that of the human D_{1A} dopamine receptor gene and the human monoamine oxidase gene; this may be important *cis*-acting sequences for cell-specific transcription in catecholamine containing cells (KAWARAI et al. 1997). For polymorphism of the DAT gene see below (Sect. E.I.).

C. Distribution of the Dopamine Transporter

The DAT acts exclusively on the plasma membrane of dopaminergic neurons and represents the only known specific marker for this particular type of neuron. In this sense distribution data on the DAT are also relevant for the distribution of this neurotransmitter system which is implicated in a broad range of physiological functions. The function the CNS dopaminergic system may also be implicated in many neuropsychiatric disorders.

I. Distribution of Dopamine Transporter Binding

The development of radioligands that bind to the DAT was a prerequisite for determining the distribution of this protein in the brain since there are obvious limits to the spatial resolution of uptake studies and to the feasibility of transport on frozen tissue. Potent inhibitors of uptake were radiolabelled and their binding properties studied in membranes from the striatum, the brain area with the highest dopamine uptake rates. [^{3}H]Cocaine (KENNEDY and HANBAUER 1983; PIMOULE et al. 1983; SCHOEMAKER et al. 1985), [^{3}H]mazindol (JAVITCH et al. 1983, 1984), [^{3}H]nomifensine (DUBOCOVICH and ZAHNISER 1985), [^{3}H]threo-(+)methylphenidate (JANOWSKI et al. 1985), [^{3}H]GBR-12935 (BERGER et al. 1985) and [^{3}H]GBR-12783 (BONNET et al. 1986) were among the first radioligands developed to label dopamine uptake sites. The link was based on binding of these ligands being sodium-dependent similar to dopamine uptake, on the correlation of binding inhibition with uptake inhibition by different drugs and on the sensitivity of binding to destruction of dopamine nerve terminals by neurotoxins or Parkinson's disease. Highest density was found in the striatum, followed by the nucleus accumbens and the olfactory tubercle (DUBOCOVICH and ZAHNISER 1985; BONNET et al. 1986; JANOWSKY et al. 1986). Early quantitative autoradiographic studies with [^{3}H]nomifensine (SCATTON et al. 1985), [^{3}H]mazindol (JAVITCH et al. 1985b) and [^{3}H]GBR 12935 (DAWSON et al. 1986) allowed a higher spatial resolution and detected uptake sites with lower density in the lateral septum, stria terminalis, basolateral amygdaloid nucleus, lateral habenula, prefrontal cortex, and substantia nigra in agreement with dopaminergic pathways (LINDVALL and BJÖRKLUND 1983).

The drawbacks of these first ligands were low affinity ([^{3}H]cocaine, [^{3}H]threo-(+)methylphenidate), low ability to discriminate against the NET ([^{3}H]mazindol, [^{3}H]nomifensine) or binding to a transporter unrelated piperazine acceptor site ([^{3}H]GBR 12935; ANDERSEN 1987), identified as cytochrome P450IID1 (NIZNIK et al. 1990). Still, autoradiography of [^{3}H]mazindol in the presence of desipramine to occlude noradrenaline uptake sites allowed analysis of subregional differences in the striatum of the rat (MARSHALL et al. 1990) and the first detailed mapping of dopamine uptake sites in the human brain (DONNAN et al. 1991). Dopaminergic fibres of the ventral striatum, especially the nucleus accumbens, have been shown to contain a relatively low capacity for uptake. This might explain regional differences in susceptibility to MPTP-induced neuronal loss, which depends on the uptake of the toxic metabolite MPP$^+$ (MARSHALL et al. 1990; PIFL et al. 1991). Cocaine derivatives with higher affinity ([^{3}H]CFT; MADRAS et al. 1989) and higher DAT selectivity {3β-(4[^{125}iodo]phenyl)tropane-2-carboxylic acid isopropyl ester ([^{125}I]RTI-121), SCHEFFEL et al. 1992; [^{125}I]altropane, MADRAS et al. 1998a} were demonstrated to detect losses of the DAT consistent with dopamine depletion in Parkinson's diseased tissue (KISH et al. 1988; MADRAS et al. 1998b). These ligands have also been used as ligands for in vivo imaging studies (see below).

II. Distribution of the mRNA

Northern analysis with a radiolabelled hDAT cDNA-probe detect mRNAs of 4.2 kb in the human substantia nigra, slightly larger than the 3.7 kb species detected in the rat and mouse midbrain RNA samples (SHIMADA et al. 1991, 1992; LORANG et al. 1994; DONOVAN et al. 1995). By Northern analysis, no consistent hybridization was found in preparations from rat cerebral cortex, cerebellum (SHIMADA et al. 1992; LORANG et al. 1994) and human thalamus (DONOVAN et al. 1995) except low, although variable, hybridization densities in samples from rat hypothalamus (SHIMADA et al. 1992).

By a nuclease protection assay, DAT mRNA was found in human substantia nigra but not in cerebellum, cingulate cortex, hypothalamus, nucleus accumbens, putamen and caudate (BANNON et al. 1992). By the same method, DAT mRNA was detected in rat substantia nigra/ventral tegmentum and, although only at levels 5% of substantia nigra/ventral tegmentum, in hypothalamus; no signal was found in caudate/putamen, nucleus accumbens, hippocampus, olfactory bulb, locus coeruleus and prefrontal cortex (RICHTAND et al. 1995).

By in situ hybridization, the localization and distribution of DAT mRNA could be examined on a cellular level. DAT mRNA was found in perikarya of brain neurons known to be dopaminergic. The most intensely labelled DAT mRNA-containing neurons were found in the substantia nigra, pars compacta, and the ventral tegmental area of the rat (GIROS et al. 1991; SHIMADA et al. 1992; AUGOOD et al. 1993; CERRUTI et al. 1993a; MEISTER and ELDE 1993; LORANG et al. 1994). Much lower, but consistent expression, was found in cells of the dorsomedial part of the arcuate nucleus of the hypothalamus (A12 dopamine cell group), which contribute to the tuberoinfundibular pathway and in the zona incerta (A13); inconsistent were the results for the anteroventral periventricular nucleus of the hypothalamus (A14) and the rostral periaqueductal grey (A11) (CERRUTI et al. 1993; MEISTER and ELDE 1993; LORANG et al. 1994; HOFFMAN et al. 1998).

A focus of interest is whether DAT mRNA varies across subpopulations of dopaminergic neurons in the mesencephalon, since subregional differences of vulnerability with preferential loss of cells in the ventromedial substantia nigra pars compacta and relative sparing of the ventral tegmental area have been reported in MPTP-induced parkinsonism and idiopathic Parkinson's disease (SCHNEIDER et al. 1987; GIBB and LEES 1991). In rat, lower DAT mRNA expression was reported in neurons of the ventral tegmental area than in neurons of the substantia nigra pars compacta (SHIMADA et al. 1992; AUGOOD et al. 1993; BLANCHARD et al. 1994). In monkey, lower levels of hybridization for DAT mRNA were found in the dorsal tier compared to the ventral tier of midbrain dopamine neurons including the ventral tegmental area, which consists of ventral regions of the substantia nigra pars compacta (HABER et al. 1995). This pattern could explain differential effects of MPTP neurotoxicity due to differential uptake of the toxic metabolite MPP$^+$ by the DAT.

Studies on the human brain are less consistent. The ventral tegmental area displayed lower DAT expression than the substantia nigra compacta in two studies on neurologically normal individuals (HURD et al. 1994; UHL et al. 1994), but no difference was found in two other studies (BANNON and WHITTY 1997; COUNIHAN and PENNEY 1998). More consistent were findings about a significant reduction in cellular abundance of DAT mRNA in surviving nigral cells in Parkinson's disease when compared to controls; tentative explanations are that an unknown toxin entering dopamine neurons by the DAT and causing Parkinson's disease leaves behind neurons that express less transporter or that a compensatory change takes place to increase the amount of dopamine in the synaptic cleft (UHL et al. 1994; HARRINGTON et al. 1996; COUNIHAN and PENNEY 1998).

Interestingly, in midbrain of humans around the age of 20 years, a regional heterogeneity in DAT mRNA was found that disappeared during normal ageing followed by a profound loss of DAT mRNA over cells in the ventral tier of the substantia nigra (BANNON and WHITTEY 1997).

III. Distribution of Dopamine Transporter Immunoreactivity

Molecular cloning of the DAT cDNA provided peptide sequences for antibody preparation. The first anti-peptide antibodies directed at peptide sequences from the N-terminal, C-terminal and putative large second extracellular loop of the DAT were useful in immunoprecipitation and Western blot assays (VAUGHAN et al. 1993). Subsequently, antisera were produced that allowed immunohistochemical studies describing the distribution of immunoreactivity on a cellular and subcellular level in rat (for implications for the molecular topology see Sect. B.II, this chapter).

DAT immunoreactivity was concentrated in regions of dopamine cell groups and of dopaminergic innervation from neuronal somata and dendrites to axons with varicosities (CILIAX et al. 1995; FREED et al. 1995). Dense immunoreactivity was observed in the striatum, nucleus accumbens, and olfactory tubercle. Perikarya of cell groups that project to these terminal fields were immunostained with moderate density: neurons in the substantia nigra pars compacta (cell group A9) and, with less intensity, especially in medial parts, somata in the ventral tegmental area (A10) (CILIAX et al. 1995; FREED et al. 1995). In another light microscopic study, immunoreactive axons and varicosities were found in the zona incerta, the external layer of the median eminence (including fibres of the tuberoinfundibular neurons) and various parts of the amygdala; much weaker was the staining of neuronal cell bodies in the arcuate nucleus (the cell bodies of tuberoinfundibular neurons in the hypothalamus) and in the olfactory bulb (REVAY et al. 1996). In general, regional distribution of DAT immunoreactivity coincided with established dopaminergic innervation of several regions, but sporadic mismatches to the immunocytochemical distribution of tyrosine hydroxylase were apparent, for example in the cingulate cortex (CILIAX et al. 1995).

Electron microscopic immunocytochemistry in rat revealed that DAT is localized to extrasynaptic portions of the plasma membrane of striatal terminals (NIRENBERG et al. 1996; HERSCH et al. 1997), supporting the concept of volume transmission (HERKENHAM 1987). Interestingly, extrasynaptic dopamine uptake was also inferred from fast-scan cyclic voltammetry in vivo (GARRIS et al. 1994). Immunogold labelling of plasma membrane was denser in dendrites than in perikarya of neurons in substantia nigra (NIRENBERG et al. 1996; HERSCH et al. 1997) and ventral tegmental area (NIRENBERG et al. 1997a); in these cell bodies, DAT could be also visualized on subcellular membrane compartments. This could be sites of synthesis, posttranslational modification and recycling or, potentially, sites of subcellular functions of the transporter. These findings support the concept that the DAT is synthesized constitutively in the cell body and subsequently transported to and inserted into the plasma membrane. In fact, lesion of the nigrostriatal pathway by 6-hydroxydopamine produced accumulations of high-affinity dopamine uptake sites proximal to the injection 4 days later (O'DELL and MARSHALL 1988).

Immunochemical analysis of the DAT protein was also used to bring insight into the role of the transporter in vulnerability patterns of dopaminergic neurons in parkinsonism. Weaker immunolabelling of the more resistant ventral tegmental neurons in comparison to the cells in the substantia nigra pars compacta was observed (CILIAX et al. 1995; FREED et al. 1995). An ultrastructural study reported lower density of DAT gold particles per unit perimeter on plasma membranes of dopaminergic axons in the nucleus accumbens shell than in the core, which could contribute to the relative resistance of mesolimbic dopamine neurons to neurotoxic insult (NIRENBERG et al. 1997b). Interestingly, although conflicting, regional differences in the sensitivity of nucleus accumbens terminals to psychostimulants have been reported (PONTIERE et al. 1995; JONES et al. 1996). The greater extracellular diffusion distance of dopamine in prefrontal cortex might be linked to lower levels of DAT protein on dopamine axon varicosities of prelimbic prefrontal cortex than of striatum and cingulate cortex (SESACK et al. 1998).

The first DAT antibody studies in humans focused on the dramatic loss of this protein in the striatum of Parkinson's disease patients. DAT immunoreactivity, as determined by Western blotting, correlated better with dopamine loss than other neurochemical markers of dopamine terminals (WILSON et al. 1996). Immunohistochemistry revealed that the loss of DAT in the basal ganglia followed dorsal to ventral, caudal to rostral, and lateral to medial gradients, with caudolateral putamen being the most severely depleted and nucleus accumbens being least affected (MILLER et al. 1997). This pattern was first observed in a neurochemical study assessing dopamine on post-mortem idiopathic Parkinson's disease brain (KISH et al. 1988). According to a detailed analysis on the normal human brain, the distribution of DAT immunoreactivity followed established dopaminergic mesostriatal, mesolimbic, and mesocortical pathways, corroborating findings in rat brain (CILIAX et al. 1999); whereas in the rat, cortical DAT-positive axons are confined to a few regions (cingulate: CILIAX et al. 1995), DAT-immunolocalization in human cerebral cortex was widespread, albeit less dense than dopamine-immunoreactive fibres (SMILEY et al. 1992). Within the hypothalamus, dopaminergic neurons were predominantly DAT negative, but fine DAT-positive axons were scattered throughout this region (CILIAX et al. 1999). A decline in the number of DAT-containing neurons (6.1% per decade) and an age-related transition from heavy- to light-labelled neurons was seen within the human substantia nigra with age (MA et al. 1999).

D. Pharmacology

I. Uptake by the Dopamine Transporter

1. Ion Dependence of Transport

Neuronal uptake of dopamine and other biogenic amines is controlled by the transmembranal gradient of Na^+ and abolished in a medium in which Na^+ is iso-osmotically replaced with sucrose, choline or lithium (HARRIS and BALDESSARINI 1973; HOLZ and COYLE 1974). Substitution of Cl^- with any of several anions also causes a marked reduction of high-affinity uptake (KUHAR and ZARBIN 1978). A detailed evaluation of the effects of external ions on dopamine uptake shows the relationship between the initial rate of uptake and external Na^+ being sigmoidal whereas the rate of uptake versus Cl^- can be described by a rectangular hyperbola in rat striatal preparations (KRUEGER 1990; MCELVAIN and SCHENK 1992; data for Na^+: WHEELER et al. 1993), and in cells transfected with DAT cDNA (GU et al. 1994; PIFL et al. 1997; EARLES and SCHENK 1999; however hyperbolic Na^+ dependence: CHEN et al. 1999). A possible interpretation of these results is that DAT-mediated dopamine uptake requires the presence of two Na^+ ions and one Cl^- ion at the outer surface of the membrane. A multi-substrate mechanism was proposed in studies on rat brain tissue where a minimum of two Na^+ ions and a single molecule of dopamine randomly bind first to the transporter followed by the binding of a single Cl^- ion before the movement of dopamine across the membrane (MCELVAIN and SCHENK 1992; POVLOCK and SCHENK 1997). However, a fixed binding order of Na^+ binding before dopamine was observed in two independent studies on cells transfected with the human DAT (CHEN et al. 1999; EARLES and SCHENK 1999). In the presence of either 36 mM or 136 mM Na^+ neuronal dopamine uptake is optimal with 1–2 mM Mg^{2+}, K^+ or Ca^{2+}. An increase in K^+ concentrations from 0 to 10 mM, in an incubation medium containing a high Na^+ concentration, modifies the dopamine uptake according to a bell-shaped curve. Such an increase in uptake probably results from an activation of the K^+/Na^+ exchange through the Na^+/K^+-ATPase (adenosine triphosphatase) and, consequently, in a more favourable transmembrane ionic gradient (AMEJDKI-CHAB et al. 1992a; CORERA et al. 1996). The dependency of dopamine uptake on Na^+ is greatly affected by cations such as K^+ and $Tris^+$, present in many buffer systems (ZIMÁNY et al. 1989; AMEJDKI-CHAB et al. 1992a). Membrane depolarization by action of veratridine or elevated external K^+ reduces the rate of uptake (HOLZ and COYLE 1974; KRUEGER 1990). Kinetic studies provide evidence for Na^+ and Cl^- being co-transported with dopamine into the cell and are consistent with a rheogenic, that is, a net ionic current carrying process (KRUEGER 1990; AMEJDKI-CHAB et al. 1992a,b).

2. Substrates of the Dopamine Transporter

Substrates are compounds which by interacting with the DAT are translocated to the opposite side of the plasma membrane. The initial velocity of transport can be described by the Michaelis–Menten equation. The natural substrate dopamine is translocated with an apparent K_m of 0.3–1.2µM and turnover numbers of 0.3–1.5s^{-1} at 37°C depending on conditions of equilibrium exchange or zero trans entry of dopamine into rat striatal preparations (MEIERGARD and SCHENK 1994a). Turnover of the human DAT in stably transfected cells at 37°C was estimated to be 14–18s^{-1} (PIFL et al. 1996; EARLES and SCHENK 1999). The somehow higher K_m values (1–5µM) found in the majority of studies on recombinant transporters may be due to differences in post-translational modifications. However, recently, cells heterologously expressing high-affinity uptake by the human DAT have been reported (ZHANG et al. 1998; PRISTUPA et al. 1998). The DAT has considerably lower affinity for noradrenaline, K_m values 3–5 times higher than that for dopamine with V_{max} values quite similar for both catecholamines (SNYDER and COYLE 1969; PIFL et al. 1996). There is no stereoselectivity with regard to noradrenaline (SNYDER and COYLE 1969; MEIERGARD and SCHENK 1994a). The toxic metabolite MPP^+ of the parkinsonism-inducing agent MPTP is a good substrate of the DAT in rat striatal synaptosomes (CHIBA et al. 1985; JAVITCH et al. 1985a). MPP^+ obeys a kinetic similar to that of noradrenaline on the human DAT (PIFL et al. 1996). Other drugs translocated by the DAT in a Na^+- and temperature-dependent and cocaine-blockable manner are amphetamine and tyramine (PETRALI et al. 1979; ZACZEK et al. 1991; SITTE et al. 1998). The V_{max} values are less than a fourth of that measured using dopamine as a substrate (SITTE et al. 1998); an intact catechol with a primary ethylamine side chain has been shown to be necessary for optimal uptake activity (HORN 1973; MEIERGERD and SCHENK 1994a). Thermodynamic analysis suggested substrate binding to the transporter to occur with a change in entropy, by contrast the binding of uptake inhibitors with a change in enthalpy (BONNET et al. 1990). The cationic form of dopamine, perhaps including the zwitterion, is the most likely substrate of the transporter based on the pH-dependence of dopamine uptake and [^{3}H]cocaine-analog displacement (BERFIELD et al. 1999).

3. Uptake Blockers

Uptake blockers are strictly speaking drugs which by interacting with the transporter lock it in a conformational state incapable of translocation in both directions. "Pure" uptake inhibitors such as cocaine, nomifensine and benztropine are devoid of releasing effect at uptake-inhibiting concentrations in contrast to amphetamine and its derivatives (see below) (HEIKKILA et al. 1975; RAITERI et al. 1978). Drugs selectively blocking the DAT in the nanomolar range are diphenyl-substituted piperazine derivatives of the GBR series (VAN DER ZEE et al. 1980). However, this selectivity is not more than 20-fold on recombinant plasmalemmal monoamine transporters whereas drugs with a 100- to 1,000-fold selectivity for the noradrenaline or serotonin transporter are available (BUCK and AMARA 1994; ESHLEMAN et al. 1999). Competitive inhibition of uptake was reported for mazindol (KRUEGER 1990; MEIERGERD and SCHENK 1994b), nomifensine, benztropine (KRUEGER 1990; JONES et al. 1995) and amphetamine (KRUEGER 1990; CHEN et al. 1999). Different forms of inhibition of dopamine uptake by cocaine were found: competitive (KRUEGER 1990; JONES et al. 1995; CHEN et al. 1999; EARLES and SCHENK 1999), noncompetitive (MISSALE et al. 1985) and an uncompetitive type of inhibition (McELVAIN and SCHENK 1992). Allosteric interactions between cocaine and Na^+ on the DAT have been observed: its blocking action is enhanced by Na^+ (WHEELER et al. 1994; CHEN et al. 1999), and cocaine, just as mazindol, depends on a minimum of Na^+ for blockade of reverse transport (PIFL et al. 1997). Cocaine competitively inhibits the involvement of Na^+ in the uptake process in striatal preparations (McELVAIN and SCHENK 1992) but does not seem to have an effect at the Na^+ binding site of the human DAT (CHEN et al. 1999; EARLES and SCHENK 1999). On the other hand, raising Na^+ enhances the apparent affinity of substrates for the human DAT more than that of inhibitors (CHEN et al. 1999).

II. Reverse Transport by the Dopamine Transporter

Under normal conditions the DAT translocates dopamine from the extracellular compartment into the cytoplasm of dopaminergic cells. However, there are several ways to reverse the direction of transport.

First, a change of the ion gradient can make the transporter operate in a reverse mode: a decrease of extracellular NaCl, or replacement of either Cl^- or Na^+ ions by sucrose, isethionate, Li^+ or choline induces release of dopamine from striatal synaptosomes (RAITERI et al. 1979; SITGES et al. 1994) or cells transfected with the human DAT (PIFL et al. 1997). The ion gradient can also be changed by an increase of intracellular Na^+. This can be realized by blockade of the Na^+,K^+-ATPase either by metabolic inhibitors, ouabain or K^+-free medium, measures which all reverse transport and can be blocked by DAT-inhibitors (RAITERI et al. 1979; LIANG and RUTLEDGE 1982; C. Pifl and E.A. Singer, unpublished observation on recombinant DAT).

Second, substrates presented at the external face of the DAT can induce carrier-mediated release of intracellular substrates, a mechanism of exchange diffusion common to amphetamine-like drugs in striatal preparations (FISCHER and CHO 1979; RAITERI et al. 1979; PARKER and CUBEDDU 1988) and DAT-transfected cells (ESHLEMAN et al. 1994; WALL et al. 1995; PIFL et al. 1995). That amphetamine-induced release is DAT-mediated, can again be demonstrated by the blocking action of cocaine-like drugs or, especially convincing, by amphetamine being without dopamine releasing action in DAT knock-out mice (GIROS et al. 1996; JONES et al. 1998b). The importance of intracellular Na^+ also for substrate-induced release was already emphasized by the release-potentiating effect of ouabain (RUTLEDGE 1978) and gains new aspects by the channel-like features of the DAT, which allow Na^+-influx induced by substrates. The releasing action of substrates correlates with their inward-current inducing action but not with their uptake rates; furthermore, substrate-induced release is absent under conditions of low extracellular Na^+ despite ongoing uptake of the releasing drug (SITTE et al. 1998; PIFL and SINGER 1999). These findings make a point for Na^+-influx as a trigger for substrate-induced reverse transport. Reverse transport induced solely by an increase of cytoplasmic dopamine as demonstrated after intracellular injection of dopamine into Planorbis giant neurons (SULZER et al. 1995) is not supported by findings on reserpine-like compounds which are unable to induce release from striatal synaptosomes (RAITERI et al. 1979) or slices (JONES et al. 1998b) in the absence of one of the above-mentioned triggers for the DAT to transport in reverse direction.

III. The Dopamine Transporter as a Binding Site

Drugs which inhibit the dopamine carrier at nanomolar concentrations bind to the DAT protein with an affinity comparable to that of ligands for G-protein coupled receptors. Radioligands for the DAT allow mapping of this marker for dopaminergic pathways and provide tools to evaluate the molecular mechanism of dopamine transport.

1. Ligands for the Dopamine Transporter

The molecules most extensively characterized by radioligand binding techniques on striatal preparations and cells expressing the recombinant DAT of different species are phenyltropane analogs of cocaine, such as [^{3}H]CFT, also designated [^{3}H]WIN 35,428 (MADRAS et al. 1989), diphenyl-substituted piperazine derivatives, such as [^{3}H]1-[2-(diphenylmethoxy) ethyl]-4-(3-phenylpropyl)-piperazine, [^{3}H]GBR 12,935 (JANOWSKY et al. 1986) and the potent blocker of both, dopamine and noradrenaline uptake, [^{3}H]mazindol (JAVITCH et al. 1984). These ligands and [^{3}H]cocaine seem to share a common binding site (CALLIGARO and ELDEFRAWI 1988; REITH and SELMECI 1992; REFAHI-LYAMANI et al. 1995; XU and REITH 1997), that is the DAT, the putative cocaine receptor that is responsible for cocaine's stimulant and reinforcing properties (RITZ et al. 1987; CALLIGARO and ELDEFRAWI 1988).

In the absence of Tris, which has an inhibitory effect on the specific binding, Na^+ has a biphasic effect on the binding of radioligands to the DAT, stimulating at lower (<30mM) and inhibiting at higher concentrations, whereas K^+ has an inhibitory effect (BONNET et al. 1988; CALLIGARO and ELDEFRAWI 1988; ZIMÁNYI et al. 1989; REITH and COFFEY 1993). At an increased $[K^+]/[Na^+]$ ratio there is not only a decreased binding affinity of the radioligand but also a lower potency of dopamine to compete with the binding (HÉRON et al. 1996; CHEN et al. 1997). Competition of substrates for binding sites is an option to determine the binding affinity of substrates for the DAT. Whereas uptake inhibitors have about the same inhibitory potency in uptake and binding experiments, transporter substrates are typically more potent at inhibiting uptake than at inhibiting radioligand binding in striatal membranes (JAVITCH et al. 1984; SCHOEMAKER et al. 1985; DUBOCOVICH and ZAHNISER 1985; BONNET et al. 1986; JANOWSKY et al. 1986) or cells transfected with the DAT cDNA (ESHLEMAN et al. 1995, 1999) suggesting that these inhibitors may not bind to exactly the same pharmacaphores on the DAT. While there is a lack of effect of Cl^- on blocker binding (BONNET et al. 1988; REITH and COFFEY 1993) the potency of substrates to competitively inhibit blocker binding is enhanced by Cl^- (AMEJDKI-CHAB et al. 1992b; WALL et al. 1993). Thus, Cl^- participates in transport both by increasing substrate affinity and by serving as a co-transported substrate in the subsequent translocation step. A substrate inhibits uptake more potently than ligand binding because the substrate not only prevents binding of [^{3}H]dopamine, but also, by causing translocation of the [^{3}H]dopamine binding site, slows the return of the transporter to an outward-facing conformation capable of binding and transporting extracellular dopamine (ZIMÁNYI et al. 1989). The lower affinity of dopamine for the DAT at higher $[K^+]/[Na^+]$ ratios in the intracellular fluid may favour the dissociation of dopamine from the DAT on the inside which is required for translocation of the unloaded transporter, possibly the rate-limiting step of uptake or release (HÉRON et al. 1996; CHEN et al. 1997).

Photoaffinity probes developed on the basis of high-affinity ligands of the aryl-dialkylpiperazine series identified the DAT as a glycoprotein with an apparent molecular weight of about 60 kDa (GRIGORIADIS et al. 1989; SALLEE et al. 1989). The DAT seems to be glycosylated differently in striatum and nucleus accumbens (LEW et al. 1992). Comparison of the binding of photoaffinity labels based on GBR 12,935 and cocaine has shown the former becoming incorporated near TM1 and TM2 whereas the cocaine derivative incorporates closer to the carboxy-terminal near TM4–TM7, and both closely associate with transmembrane regions (VAUGHAN 1995; VAUGHAN and KUHAR 1996). A benztropine-based photoaffinity ligand was found to become incorporated in the same N-terminal region as the GBR-based one and, based on the distinct behavioural pharmacology of benztropine/GBR-compounds versus cocaine, a tempting correlation with binding properties on the DAT has been suggested (VAUGHAN et al. 1999). Differences between cocaine and other uptake inhibitors in the molecular interaction with the DAT have been studied intensively in order to find molecular substrates for differences in liability for drug abuse or potential "cocaine antagonists".

2. Imaging Techniques Based on the Dopamine Transporter

The high density of DAT in the striatum as compared to other brain regions allows in vivo assessment of transporter concentrations using ^{11}C-, ^{18}F- or ^{123}I-labelled uptake blockers by positron emission tomography (PET) or single-photon emission computed tomography (SPECT). Cerebellar uptake is routinely used as the reference value for non-specific binding since that region contains negligible DAT sites. [^{11}C]Nomifensine (AQUILONIUS et al. 1987), [^{18}F]GBR 13,119 (KILBOURN 1988), [^{11}C]cocaine (FOWLER et al. 1989), and its analogs [^{123}I]-2β-carbomethoxy-3β-(4-iodophenyl)tropane ([^{123}I]β-CIT) (INNIS et al. 1991) were the first ligands for in vivo imaging. Recently, DAT changes have been studied in various neuropsychiatric disorders with more DAT-selective, tropane-derived ligands. In early drug-naïve Parkinson's patients, DAT-binding measurements could be assessed in the subdivided striatum with marked losses in the contralateral posterior putamen (GUTTMAN et al. 1997). In multiple system atrophy and progressive supranuclear palsy, fewer marked differences were found between caudate and putamen than in Parkinson's disease (BRÜCKE et al. 1997). In non-violent alcoholics, striatal DAT densities were lower than in healthy controls (TIIHONEN et al. 1995) and increased after-alcohol withdrawal (LAINE et al. 1999). DAT density was also found reduced in caudate and putamen by 25% in abstinent methamphetamine and meth-cathinone users (MCCANN et al. 1998) and by nearly two-thirds in Lesch-Nyhan disease (WONG et al. 1996). Striatal [^{123}I]-CIT was a mean of 37% higher in subjects with Tourette's disorder than in age- and gender-matched healthy comparison subjects (MALISON et al. 1995). New radiotracers for SPECT studies of the DAT with faster kinetics than β-CIT were developed which allow analysis in a 1-day protocol (BOOJI et al. 1997; FISCHMAN et al. 1998).

IV. Electrophysiology of the Dopamine Transporter

Kinetic studies reveal that the DAT operates by translocating one molecule of dopamine with two Na$^+$ and one Cl$^-$ ions per transport cycle. This electrogenic cycle should result in an inward current that can be calculated from turnover numbers. In fact, substrates of the DAT such as dopamine, noradrenaline, amphetamine and tyramine elicit an inward current in voltage-clamped oocytes or HEK 293 cells expressing the human DAT (SONDERS et al. 1997; SITTE et al. 1998). This current is considered transport-associated since it is absent in cells without dopamine uptake, blockable by cocaine, saturates with increasing substrate concentration following the Michaelis–Menten kinetics of substrate uptake and is abolished by replacement of extracellular

Na$^+$ or Cl$^-$. However, the charge movement is higher than the dopamine concurrently accumulated by the same oocytes (Sonders et al. 1997) and is elicited in similar amounts by substrates whose maximal uptake rates differ by a factor of more than four (Sitte et al. 1998). This means, substrate-induced currents are found in addition to that arising from the stoichiometric coupling of substrates with ions. Similar extra-currents in other plasmalemmal neurotransmitter transporters have been attributed to a channel mode of these transporters (Sonders and Amara 1996).

In voltage-clamped oocytes, an increase of uptake velocity with membrane hyperpolarization can be demonstrated consistent with the transmembrane electrical potential being a thermodynamic driving force in addition to chemical gradients (Sonders et al. 1997). Substrate affinity is independent of membrane potential which supports the notion that the voltage-dependent and rate-limiting step in transport occurs subsequent to substrate binding.

Besides the transport-associated current, a leak current was also observed in oocytes and HEK 293 cells expressing the DAT (Sonders et al. 1997; Sitte et al. 1998). This current, found in the absence of any drug and inwardly directed at negative potentials, is DAT-related: (1) it is lacking in cells without DAT, (2) it is blockable by cocaine (Sonders et al. 1997; Sitte et al. 1998) and (3) its amount is related to the amplitudes of currents induced by dopamine in the same cell (Sitte et al. 1998). Blockade of the leak current by dopamine can be demonstrated if the transport-associated current is eliminated by replacement extracellular of Na$^+$ by K$^+$ or Li$^+$ (Sonders et al. 1997). From the shifts in the cocaine reversal potential in different ion-substituted buffers it can be assumed that the tonic DAT leak-conductance is Cl$^-$ impermeant but, besides Na$^+$, readily carries Li$^+$, K$^+$ and protons (Sonders et al. 1997). The physiological role of the leak current, if it occurs at all in an in vivo setting, is unclear.

The shifts of the I-V curve of transport-associated currents by dopamine and amphetamine may be predicted by an increase of ionic conductances for either Na$^+$ or Cl$^-$. Substitution experiments with acetate suggest that these currents are carried by Na$^+$ (Sitte et al. 1998). Based on its correlation with substrate-induced release and on the sensitivity of substrate-induced release to lowering of extracellular Na$^+$, transport-associated current has been suggested as a trigger for DAT-mediated release induced by substrates, including amphetamine (Sitte et al. 1998; Pifl and Singer 1999).

V. Regulation of the Dopamine Transporter

Because DAT plays such an important role in the control of the extracellular availability of dopamine, it becomes a potentially crucial target for modulation by second messenger and effector systems. Actual findings for regulation of the DAT are quite recent, and their physiological relevance is not totally appreciated. In rat hypothalamic cells in culture, increases in intracellular cAMP enhances dopamine uptake (Kadowaki et al. 1990). However, this effect could be detected in striatal preparations only by rotating disk electrode voltammetry, possibly due to the transient nature of protein kinase A stimulation (BATCHELOR and SCHENK 1998). DAT activity is also increased by stimulation of D_2 receptors: the agonist quinpirole increased, whereas the antagonist raclopride decreased, clearance rates of extracellular dopamine (MEIERGERD et al. 1993; CASS and GERHARDT 1994). This may explain why the D_2 receptor antagonist pimozide blocked up-regulation of DAT activity following repeated treatments with cocaine (PARSONS et al. 1993). The link between D_2 receptors and uptake could be through G-protein-mediated opening of K^+ channels causing hyperpolarization, which has been shown to increase transporter velocity (SONDERS et al. 1997). The link between membrane potential and DAT activity may be also the basis for the inhibitory effect of nicotinic acetylcholine-receptor stimulation, which elicits depolarization (IZENWASSER et al. 1991; YAMASHITA et al. 1995; HUANG et al. 1999).

Quite different is the mechanism of DAT modulation by protein kinase C (PKC). The decrease of V_{max} of dopamine uptake was first described on the recombinant DAT (KITAYAMA et al. 1994) and subsequently on striatal synaptosomes (COPELAND et al. 1996). PKC stimulation also diminished the transport-associated and leak current at the DAT (ZHU et al. 1997). The DAT protein was found to be phosphorylated (HUFF et al. 1997; VAUGHAN et al. 1997). Immunofluorescent confocal microscopy revealed rapid sequestration/internalization of DAT protein from the cell surface by PKC stimulation and suggests cellular trafficking as a mechanism of DAT regulation (PRISTUPA et al. 1998). This trafficking is endocytic through a clathrin-mediated mechanism; targeting to both an endosomal recycling compartment and a lysosomal/degradative pathway has been demonstrated recently – a different expression system may explain these conflicting data (DANIELS and AMARA 1999; MELIKIAN and BUCKLEY 1999). If translocation of DAT protein is really the result of transporter phosphorylation, is not yet clear; mutational analysis of the five consensus PKC/PKA sites did not abolish the effect of PKC on DAT function (C. Pifl and M.G. Caron, unpublished observation). Non-canonical sites or another PKC-sensitive protein may be involved.

The possibility that Ca^{2+}/calmodulin-dependent protein kinase II regulates the DAT was shown; whether this is mediated by phosphorylation of the transporter itself, is unclear (UCHIKAWA et al. 1995). The molecular basis of the modulation of DAT function and binding by micromolar Zn^{2+} (RICHFIELD 1993; BONNET et al. 1994) was identified as a direct interaction with the protein (NORREGAARD et al. 1998; see Sects. B.II. and B.III, this chapter).

The long-term regulation of DAT by chronic action of cocaine in vivo is due to changes in DAT expression and reviewed recently (PILOTTE 1997).

VI. Behavioural Studies Related to the Dopamine Transporter

1. Cocaine-Like Substances

This class of dopamine-uptake inhibitors (in early studies called methylphenidate-like drugs) increases locomotor activity, maintains self-administration, and increases responding maintained by electrical brain stimulation and under a fixed interval schedule (for review see FIBIGER 1977). Uptake inhibitors can be differentiated from amphetamine-like releasing drugs by reserpine. Their action depends on cell-firing induced release. Whereas amphetamine increase hypermotility in both normal and reserpinized animals, uptake-inhibitors are not only unable to stimulate motor behaviour after pre-treatment with reserpine, they also block the effect of amphetamine in vesicle-depleted animals since they block amphetamine uptake (SCHEEL-KRÜGER 1971). The orders of potency are the same for the inhibition of dopamine uptake, the central stimulatory action in normal mice, and the antagonism of the amphetamine-induced hypermotility in reserpinized mice (ROSS 1979; HEIKKILA 1981). The order of potency for effects on fixed-interval responding (operant behaviour) is equally similar to that on dopamine uptake (MCKEARNEY 1982). Finally, the potency of monoamine-uptake inhibitors in maintaining self-administration behaviour is directly related to their affinities for the DAT, more closely than to their affinities for the NET or serotonin transporter (RITZ et al. 1987). Lesions of dopamine-rich sites lead to changes in cocaine-induced locomotion and self-administration, but this is not produced by lesions of other areas (ROBERTS et al. 1975, 1977). Cocaine elevates synaptic dopamine levels in both terminal-field regions, especially nucleus accumbens, and cell-body regions, that is, the ventral tegmental area, of mesolimbic dopamine neurons (DI CHIARA and IMPERATO 1988; BRADBERRY and ROTH 1989). This action, particularly in the nucleus accumbens, is thought to underlie cocaine-induced behavioural effects (DI CHIARA and IMPERATO 1988). DAT occupancy is associated with euphorigenic effects of cocaine in humans (VOLKOW et al. 1997). However, not all dopamine uptake inhibitors are subject to abuse as is cocaine. Among inhibitors that are structurally dissimilar to cocaine, the relationship between DAT affinity and behavioural effects is distinctly different from that of analogs of cocaine (VAUGEOIS et al. 1993; IZENWASSER et al. 1994). The molecular basis for these differences is intensively investigated in binding studies (see Sect. D.III.1, this chapter); however, currently no mechanistic explanation is available.

2. Amphetamine-Like Substances

This class of substances comprises all lipophilic and therefore centrally active indirectly acting sympathomimetic amines. Profound effects on a wide range of behavioural processes have been shown for amphetamine: motor activity, attention, aggression, sexual behaviour, learning and memory, classical conditioning and operant behaviour are mostly stimulated, sleep and ingestive behaviour are suppressed, and self-administration is reinforced (for review see MOORE 1978). Amphetamine increases the concentration of both noradrenaline and dopamine in the synapse, but amphetamine-induced locomotor stereotypies and self-administration of amphetamine depend on an intact dopaminergic system (SCHECHTER and COOK 1975; CREESE and IVERSEN 1975).

That amphetamine-induced dopamine release in turn depends on an active DAT was shown by its inhibition by dopamine uptake blockers in superfused synaptosomes (Raiteri et al. 1979) and striatal microdialysis (Butcher et al. 1988; Hurd and Ungerstedt 1989). Blockade of the locomotor effect of amphetamine by selective DAT inhibitors can only be shown in reserpinized mice (Heikkila 1981). The crucial role of the DAT in the neurochemical and behavioural actions of amphetamine has recently been confirmed in animals lacking a functional DAT gene.

3. Dopamine Transporter Knock-Out

The significance of the DAT for dopaminergic neurotransmission can be assessed by the effects of uptake blockade. However, the drugs available are by no means selective and, in vivo, an anomalous increase of clearance by inhibitors has been described (Stamford et al. 1986; Ng et al. 1992; Zahniser et al. 1999).

Disruption of the DAT protein by homologous recombination techniques avoids problems of lack of pharmacological specificity. Antisense oligodeoxynucleotides or antisense plasmid of the DAT injected intranigrally reduced striatal dopamine uptake by about 30% (Silvia et al. 1997; Martres et al. 1998). After unilateral antisense-treatment, amphetamine induced contralateral rotations, but spontaneous behaviour was not affected (Silvia et al. 1997). However, behaviour was profoundly altered by complete disruption of the DAT gene by homologous recombination (Giros et al. 1996).

Spontaneous hyperlocomotion of homozygote $DAT^{-/-}$ mice was similar to that obtained with maximal doses of psychostimulants in wild-type animals. This is due to a prolonged half-life of extracellular dopamine as revealed by a 300 times slower clearance rate and a fivefold higher basal levels of extracellular dopamine as measured by cyclic voltammetry in electrically stimulated striatal slices and by microdialysis in the striatum of freely moving mice, respectively (Jones et al. 1998a). The high motor activity occurred despite marked adaptive changes, such as reductions of striatal D_1 and D_2 receptors by 50% and of tissue levels of dopamine and tyrosine hydroxylase protein by more than 95% (Giros et al. 1996; Jones et al. 1998a). A marked downregulation of D_2 receptors (−70%) was also found in pituitaries of $DAT^{-/-}$ mice with a dramatically reduced number of lactotrophs and somatotrophs and a lack of responsiveness to secretagogues (Bossé et al. 1997). The phenotype of inability to lactate and dwarfism in the DAT knock-out mice is due to an increased extracellular dopamine in the hypothalamus and the pituitary. DAT message has been found in hypothalamic neurons (Meister and Elde 1993) and DAT function in this area of the brain can now be ascertained by results from the knock-out approach (Bossé et al. 1997).

Besides other conclusions arrived at by this approach for the physiology of dopaminergic neurotransmission, such as vesicular dopamine being more dependent on recycled rather than newly synthesized dopamine which in turn seems to be preferentially degraded by monoamine oxidase (JONES et al. 1998a), corollaries for the pharmacology of psychostimulants are especially valuable. In $DAT^{-/-}$ mice, neither cocaine nor amphetamine were able to further enhance locomotor activity, even under daylight conditions when spontaneous activity is low. As expected in the mice, psychostimulants have no effect on basal overflow in striatal slices and microdialysates (GIROS et al. 1996; JONES et al. 1998b; ROCHA et al. 1998).

Amphetamine obviously depends on the plasmalemmal carrier for its dopamine-releasing action. However, amphetamine does not depend solely on the DAT for entering the cell, because it can deplete vesicles of dopamine and inhibit monoamine oxidase in $DAT^{-/-}$ mice. Using the DAT knock-out mice, it has been shown that an increase in free cytosolic dopamine by mobilizing it from vesicles is not sufficient to make the transporter act in reverse direction (JONES et al. 1998b).

Despite of its lack of effect in terms of motor excitation, cocaine still provides rewarding cues for self-administration and conditioned place preference (ROCHA et al. 1998; SORA et al. 1998). This is quite surprising considering the correlation between psychostimulant properties of drugs in tests of reward and their DAT-blocking potency (RITZ et al. 1987). Another surprising finding in DAT knock-out mice is a paradoxical behavioural effect of methylphenidate, amphetamine and cocaine: they are not only unable to further enhance locomotion, but they in fact attenuate motor activity (GAINETDINOV et al. 1999). This behaviour, reminiscent of the calming effect of these drugs in attention-deficit hyperactivity disorder (ADHD), seems to be due to their serotonin-releasing or potentiating action since it is mimicked by serotoninergic agents. Indirect evidence also suggests that the serotonin system may also mediate the initiation and maintenance of the cocaine self-administration behaviour in DAT knock-out mice (ROCHA et al. 1998). By this means, the knock-out approach shows the whole behavioural potential of a drug.

VII. The Dopamine Transporter as a Gate for Neurotoxins

Neurotoxic substrates of monoamine transporters were initially used as a tool for the topographical and functional mapping of the monoaminergic pathways in the brain. However, treatment of animals with DAT-related neurotoxins recapitulates the sequelae of dopaminergic cell loss in Parkinson's disease so closely that research on these toxins has been vigorously pursued to get insight into the aetiology of this disease. This approach was given a fresh impetus by MPP^+, a substrate of the DAT and the active metabolite of MPTP, which, as a contaminant of designer drugs, induces parkinsonism in humans (LANGSTON and IRWIN 1986).

1. 6-Hydroxydopamine

6-Hydroxydopamine competitively inhibits [^{3}H]noradrenaline uptake in brain homogenates with about twofold higher potency in the hypothalamus than in the striatum (IVERSEN 1970). Accordingly, low doses of intraventricular 6-hydroxydopamine decrease brain noradrenaline and leave dopamine unaffected (URETSKY and IVERSEN 1970). Destruction of noradrenergic or dopaminergic neurons by a higher doses can be selectively prevented by desipramine or bupropion, respectively (COOPER et al. 1980) and cultured dopaminergic neurons can be protected against the toxicity of 6-hydroxydopamine by the phencyclidine derivative N-[1-(2-benzo(b)thiopenyl)cyclohexyl]piperidine (BTCP), a selective dopamine uptake blocker (CERRUTI et al. 1993b).

2. Methamphetamine

Methamphetamine, a potent psychomotor stimulant drug that induces release of dopamine from presynaptic sites, is not a neurotoxin in the sense that neurons are lost together with their cell bodies. However, repeated high doses show large, depletions of brain dopamine and serotonin in different animals and were shown to persist for up to 4 years in monkeys (WOOLVERTON et al. 1989). Chronic users show even in abstinent state reduced DAT density in vivo (McCANN et al. 1998). The DAT was also mechanistically implicated by the protecting action of uptake inhibitors (MAREK et al. 1990). The complete resistance of $DAT^{-/-}$ mice provides direct evidence that an intact and functional DAT is required for methamphetamine neurotoxicity: striatal monoamines are unchanged and signs of astrogliosis, strongly present in wild-type animals, are lacking in $DAT^{-/-}$ mice (FUMAGALLI et al. 1998). This goes in line with extracellular striatal dopamine and free radical formation being unaffected by methamphetamine in these animals. Methamphetamine-induced increase in intracellular dopamine via disruption of vesicular storage and subsequent autooxidation of dopamine to toxic free radicals may be a critical part for toxicity of this drug as recently inferred from studies in mice deficient in the vesicular monoamine transporter (FUMAGALLI et al. 1999).

3. MPTP

MPTP was discovered as the cause of Parkinson's disease-like syndromes in individuals in the early 1980s after they had used of a special preparation of designer drugs. It is bioactivated in the brain by monoamine oxidase B to MPP^+, and this in turn is actively accumulated in dopaminergic neurons by the DAT where it mainly acts as a toxin by inhibition of complex I in the mitochondria (for review see LANGSTON and IRWIN 1986). The decisive role of the DAT for toxicity of MPTP and related agents is supported by the protective action of uptake blockade or genetic deletion (JAVITCH et al. 1985a; GAINETDINOV et al. 1997; BEZARD et al. 1999). In addition, the pattern of striatal dopamine loss corresponds to the dopamine uptake distribution (MARSHALL and NAVARRETE 1990) and the structural requirement for MPP^+ analogs with dopaminergic toxicity is to be substrate for the DAT (SAPORITO et al. 1992). Transfection of the DAT cDNA makes cells susceptible to the cytotoxicity of micromolar concentrations of MPP^+ (KITAYAMA et al. 1992b;

PIFL et al. 1993) and cell-sensitivity correlates with the expression level of the DAT (PIFL et al. 1993). The loss of other monoamine transmitters, especially noradrenaline, found in MPTP-induced parkinsonism (PIFL et al. 1991), may be explained by the even higher affinity of MPP$^+$ for the NET (PIFL et al. 1996). That noradrenergic neurons are still less susceptible than dopaminergic neurons may be due to the lower turnover rate of the NET (PIFL et al. 1996). Differences in the density of uptake sites per cell of dopaminergic and noradrenergic cells may also play a role but this is a matter of speculation due the lack of data.

4. Isoquinolines and Carbolines

Isoquinolines and carbolines are formed in the brain by condensation of biogenic amines (COLLINS et al. 1996; NAGATSU 1997). Some of these compounds induce parkinsonism with indices of dopamine toxicity (MATSURABA et al. 1998; NAGATSU and YOSHIDA 1988). They have affinity for the DAT (DRUCKER et al. 1990) and cytotoxicity appears to be correlated with their substrate affinity (McNAUGHT et al. 1996). The rather similar pattern of monoaminergic degeneration in Parkinson's disease and in parkinsonism induced by a toxic substrate of catecholamine transporters (HORNYKIEWICZ and PIFL 1994) raises the possibility of endogenous toxic monoaminergic-derived substrates as a cause for idiopathic Parkinson's disease.

E. Genetics Related to the Dopamine Transporter

I. Polymorphism

Three forms of polymorphism have been exploited in molecular genetic studies.

1. In the 3'-untranslated region the human DAT cDNA displays 3–11 copies of a 40-base repetitive element (variable number tandem repeat, VNTR) that are arrayed in a head-to-tail fashion and are absent from the rat cDNA; the ten-copy allele is the most common allele, accounting for about 70% of the alleles in Caucasians and for about 90% in Asians (VANDENBERGH et al. 1992a,b; SANO et al. 1993; DOUCETTE-STAMM et al. 1995). This 3' VNTR was used in the majority of genetic studies.
2. A restriction fragment length polymorphism (RFLP) has been published for Taq I in the 5' end of the gene; two alleles of 7 and 5.6 kb were observed with a significant racial dimorphism (VANDENBERGH et al. 1992b).
3. Finally, a highly polymorphic large-allele VNTR has been identified with 16 alleles ranging from 4.4 to 11.5 kb (BYERLEY et al. 1993a).

II. Linkage Studies

The DAT is a candidate gene for several neuropsychiatric conditions. Hyperactive dopaminergic neurotransmission has been implicated in the pathophysiology of schizophrenia and mania. The role of DAT in the termination of the activity of (extracellular) dopamine through reuptake makes it a logical candidate gene. However, neither a linkage (BYERLEY et al. 1993b; PERSICO et al. 1995; KELSOE et al. 1996; MAIER et al. 1996; KING et al. 1997) nor association (DANIELS et al. 1995; INADA et al. 1996; MAIER et al. 1996; SOUERY et al. 1996) was found in the majority of studies.

Being the direct target of cocaine action, the DAT is an obvious candidate gene for involvement in substance-abuse vulnerability. Whereas DAT gene polymorphism was not found associated with polysubstance abuse (PERSICO et al. 1993), the 9-repeat allele of the 3' VNTR occurred two times more often in cocaine-induced paranoia than in cocaine-users without this symptom (GELERNTER et al. 1994). A significant increase of the 7-repeat allele frequency was observed in Japanese alcoholics with an inactive aldehyde dehydrogenase (MURAMATSU and HIGUCHI 1995) and of the 9-repeat allele in alcoholics displaying withdrawal seizures or delirium, compared with ethnically matched non-alcoholic controls (SANDER et al. 1997).

3' VNTR polymorphism of the DAT gene was also studied in Parkinson's disease based on DAT-mediated neurotoxicity being a tentative mechanism for this disease. The rare 11-copy allele was found more common in patients of one study (LE COUTEUR et al. 1997) but allelic frequencies were not different from control in several other studies (HIGUCHI et al. 1995; LEIGHTON et al. 1997; PLANTÉ-BORDENEUVE et al. 1997; MERCIER et al. 1999).

ADHD is a common disorder of childhood that consists of inattention, excessive motor activity, impulsivity and distractibility. The most frequent treatment of choice, methylphenidate and D-amphetamine, act by inhibiting the DAT. Two independent studies examined the 3' VNTR of the DAT gene by the haplotype relative risk method and found a significant association between ADHD and the 480-bp (10-copies) allele (COOK et al. 1995; GILL et al. 1997). Analyses of parent-offspring transmission to affected children confirmed the 480-bp allele as the high-risk allele (WALDMAN et al. 1998). DAT knock-out studies, by bringing the serotonin system into play, offer an explanation for the therapeutic effect of stimulant drugs that seems paradoxical at first sight (GAINETDINOV et al. 1999). The functional consequence of the 3' VNTR is not yet known. A function of the VTNR in itself is less likely, however, by its closeness to the coding region of the DAT, the VTNR could be in linkage disequilibrium with a vulnerability-causing mutation that alters gene expression or protein structure.

Abbreviations

[^{123}I]β-CIT)	[^{123}I]-2β-carbomethoxy-3β-(4-iodophenyl)tropane
ADHD	attention deficit hyperactivity disorder
cAMP	cyclic adenosine monophosphate
BTCP	N-[1-(2-benzo(b)thiopenyl)cyclohexyl]piperidine
CFT	(–)-2β-carbomethoxy-3β-(4-fluorphenyl)tropane
DAT	dopamine transporter

GABA	gamma-aminobutyric acid
GAT	gamma-aminobutyric acid transporter
MPP$^+$	1-methyl-4-phenylpyridinium
MPTP	1-methyl-4-phenyl-1,2,3,6-tetrahydropyridine
NET	noradrenaline transporter
PET	positron emission tomography
PKC	protein kinase C
RFLP	restriction fragment length polymorphism
SPECT	single-photon emission computed tomography
TM	transmembrane helices
VNTR	variable number tandem repeat

References

Amejdki-Chab N, Benmansour S, Costentin J, Bonnet J-J (1992a) Effects of several cations on the neuronal uptake of dopamine and the specific binding of [^{3}H]GBR 12783: attempts to characterize the Na$^+$ dependence of the neuronal transport of dopamine. J Neurochem 59:1795–1804

Amejdki-Chab N, Costentin J, Bonnet J-J (1992b) Kinetic analysis of the chloride dependence of the neuronal uptake of dopamine and effect of anions on the ability of substrates to compete with the binding of the dopamine uptake inhibitor GBR 12783. J Neurochem 58:793–800

Andersen PH (1987) Biochemical and pharmacological characterization of [^{3}H]GBR 12935 binding in vitro to rat striatal membranes: labeling of the dopamine uptake complex. J Neurochem 48:1887–1896

Aquilonius S-M, Bergström K, Eckernäs S-Å, Hartvig P, Leenders KL, Lundquist H, Antoni G, Gee A, Rimland A, Uhlin J, Långström B (1987) In vivo evaluation of striatal dopamine reuptake sites using ^{11}C-nomifensine and positron emission tomography. Acta Neurol Scand 76:283–287

Augood SJ, Westmore K, McKenna PJ, Emson PC (1993) Co-expression of dopamine transporter mRNA and tyrosine hydroxylase mRNA in ventral mesencephalic neurones. Mol Brain Res 20:328–334

Axelrod J, Weil-Malherbe H, Tomchick R (1959) The physiological disposition of H^3-epinephrine and its metabolite metanephrine. J Pharmacol 127:251–256

Bannon MJ, Xue C-H, Shibata K, Dragovic LJ, Kapatos G (1990) Expression of a human cocaine-sensitive dopamine transporter in Xenopus laevis oocytes. J Neurochem 54:706–708

Bannon MJ, Poosch MS, Xia Y, Goebel DJ, Cassin B, Kapatos G (1992) Dopamine transporter mRNA content in human substantia nigra decreases precipitously with age. Proc Natl Acad Sci USA 89:7095–7099

Bannon MJ, Whitty CJ (1997) Age-related and regional differences in dopamine transporter mRNA expression in human midbrain. Neurology 48:969–977

Batchelor M, Schenk JO (1998) Protein kinase A activity may kinetically upregulate the striatal transporter for dopamine. J Neurosci 18:10304–10309

Berfield JL, Wang LC, Reith MEA (1999) Which form of dopamine is the substrate for the human dopamine transporter: the cationic or the uncharged species? J Biol Chem 274:4876–4882

Berger P, Janowsky A, Vocci F, Skolnick P, Schweri MM, Paul SM (1985) [^{3}H]GBR 12935: a specific high affinity ligand for labeling the dopamine transport complex. Eur J Pharmacol 107:289–290

Berger SP, Farrell K, Conant D, Kempner ES, Paul SM (1994) Radiation inactivation studies of the dopamine reuptake transporter protein. Mol Pharmacol 46:726–731

Bezard E, Gross CE, Fournier M-C, Dovero S, Bloch B, Jaber M (1999) Absence of MPTP-induced neuronal death in mice lacking the dopamine transporter. Exp Neurol 155:268–273

Blakely RD, Robinson MB, Amara SG (1988) Expression of neurotransmitter transport from rat brain mRNA in Xenopus laevis oocytes. Proc Natl Acad Sci USA 85:9846–9850

Blanchard V, Raisman-Vozari R, Vyas S, Michel PP, Javoy-Agid F, Uhl G, Agid Y (1994) Differential expression of tyrosine hydroxylase and membrane dopamine transporter genes in subpopulations of dopaminergic neurons of the rat mesencephalon. Mol Brain Res 22:29–40

Bonnet J-J, Protais P, Chagraoui A, Costentin J (1986) High-affinity [^{3}H]GBR 12783 binding to a specific site associated with the neuronal dopamine uptake complex in the central nervous system. Eur J Pharmacol 126:211–222

Bonnet J-J, Benmansour S, Vaugeois J-M, Costentin J (1988) Ionic requirements for the specific binding of [^{3}H]GBR 12783 to a site associated with the dopamine uptake carrier. J Neurochem 50:759–765

Bonnet J-J, Benmansour S, Costentin J, Parker EM, Cubeddu LX (1990) Thermodynamic analyses of the binding of substrates and uptake inhibitors on the neuronal carrier of dopamine labeled with [^{3}H]GBR 12783 or [^{3}H]mazindol. J Pharmacol Exp Ther 253:1206–1214

Bonnet J-J, Benmansour S, Amejdki-Chab N, Costentin J (1994) Effect of CH$_3$HgCl and several transition metals on the dopamine neuronal carrier; peculiar behaviour of Zn^{2+}. Eur J Pharmacol Mol Pharmacol Section 266:87–97

Booij J, Tissingh G, Boer GJ, Speelman JD, Stoof JC, Janssen AGM, Wolters ECH, van Royen EA (1997) [^{123}I]FP-CIT SPECT shows a pronounced decline of striatal dopamine transporter labelling in early and advanced Parkinson's disease. J Neurol Neurosurg Psychiat 62:133–140

Bossé R, Fumagalli F, Jaber M, Giros B, Gainetdinov RR, Wetsel WC, Missale C, Caron MG (1997) Anterior pituitary hypoplasia and dwarfism in mice lacking the dopamine transporter. Neuron 19:127–138

Bradberry CW, Roth RH (1989) Cocaine increases extracellular dopamine in rat nucleus accumbens and ventral tegmental area as shown by in vivo microdialysis. Neurosci Lett 103:97–102

Brücke T, Asenbaum S, Pirker W, Djamshidian S, Wenger S, Wober CH, Müller CH, Podreka I (1997) Measurement of the dopaminergic degeneration in Parkinson's disease with [^{123}I]beta-CIT and SPECT. Correlation with clinical findings and comparison with multiple system atrophy and progressive supranuclear palsy. J Neural Transm (Suppl) 50:9–24

Buck KJ, Amara SG (1994) Chimeric dopamine-norepinephrine transporters delineate structural domains influencing selectivity for catecholamines and 1-methyl-4-phenylpyridinium. Proc Natl Acad Sci USA 91:12584–12588

Buck KJ, Amara SG (1995) Structural domains of catecholamine transporter chimeras involved in selective inhibition by antidepressants and psychomotor stimulants. Mol Pharmacol 48:1030–1037

Butcher SP, Fairbrother IS, Kelly JS, Arbuthnott GW (1988) Amphetamine-induced dopamine release in the rat striatum: an in vivo microdialysis study. J Neurochem 50:346–355

Byerley W, Coon H, Hoff M, Holik J, Waldo M, Freedman R, Caron MG, Giros B (1993b) Human dopamine transporter gene not linked to schizophrenia in multigenerational pedigrees. Human Hered 43:319–322

Byerley W, Hoff M, Holik J, Caron MG, Giros B (1993a) VNTR polymorphism for the human dopamine transporter gene (DAT1). Human Mol Genet 2:335

Calligaro DO, Eldefrawi ME (1988) High affinity stereospecific binding of [^{3}H] cocaine in striatum and its relationship to the dopamine transporter. Membrane Biochem 7:87–106

Carlsson A, Fuxe K, Hamberger B, Lindqvist M (1966) Biochemical and histochemical studies on the effects of imipramine-like drugs and (+)-amphetamine on central and peripheral catecholamine neurons. Acta Physiol Scand 67:481–497

Cass WA, Gerhardt GA (1994) Direct in vivo evidence that D2 dopamine receptors can modulate dopamine uptake. Neurosci Lett 176:259–263

Cerruti C, Drian MJ, Kamenka JM, Privat A (1993b) Protection of BTCP of cultured dopaminergic neurons exposed to neurotoxins. Brain Res 617:138–142

Cerruti C, Walther DM, Kuhar MJ, Uhl GR (1993a) Dopamine transporter mRNA expression is intense in rat midbrain neurons and modest outside midbrain. Mol Brain Res 18:181–186

Chen N-H, Ding J-H, Wang Y-L, Reith MEA (1997) Modeling of the interaction of Na^+ and K^+ with the binding of the cocaine analogue 3b-(4-[^{125}I]iodophenyl)tropane-2b-carboxylic acid isopropyl ester to the dopamine transporter. J Neurochem 68:1968–1981

Chen N, Trowbridge CG, Justice JB Jr (1999) Cationic modulation of human dopamine transporter: dopamine uptake and inhibition of uptake. J Pharmacol Exp Ther 290:940–949

Chiba K, Trevor AJ, Castagnoli N (1985) Active uptake of MPP^+, a metabolite of MPTP, by brain synaptosomes. Biochem Biophys Res Commun 128:1228–1232

Ciliax BJ, Heilman C, Demchyshyn LL, Pristupa ZB, Ince E, Hersch SM, Niznik HB, Levey AI (1995) The dopamine transporter: immunochemical characterization and localization in brain. J Neurosci 15:1714–1723

Ciliax BJ, Drash GW, Staley JK, Haber S, Mobley CJ, Miller GW, Mufson EJ, Mash DC, Levey AI (1999) Immunocytochemical localization of the dopamine transporter in human brain. J Comp Neurol 409:38–56

Collins M, Neafsey EJ, Matsubara M (1996) β-Carbolines: metabolism and neurotoxicity. Biogenic Amines 12:171–180

Cook Jr EH, Stein MA, Krasowski MD, Cox NJ, Olkon DM, Kieffer JE, Leventhal BL (1995) Association of attention-deficit disorder and the dopamine transporter gene. Am J Human Genet 56:993–998

Cooper BR, Hester TJ, Maxwell RA (1980) Behavioral and biochemical effects of the antidepressant bupropion (Wellbutrin): evidence for selective blockade of dopamine uptake in vivo. J Pharmacol Exp Ther 215:127–134

Copeland BJ, Vogelsberg V, Neff NH, Hadjiconstantinou M (1996) Protein kinase C activators decrease dopamine uptake into striatal synaptosomes. J Pharmacol Exp Ther 277:1527–1532

Corera AT, Costentin J, Bonnet J-J (1996) Effect of low concentrations of K^+ and Cl^- on the Na^+-dependent neuronal uptake of [^{3}H]dopamine. Naunyn-Schmiedeberg's Arch Pharmacol 353:610–615

Counihan TJ, Penney JB Jr (1998) Regional dopamine transporter gene expression in the substantia nigra from control and Parkinson's disease brains. J Neurol Neurosurg Psychiat 65:164–169

Coyle JT, Snyder SH (1969) Catecholamine uptake by synaptosomes in homogenates of rat brain: stereospecificity in different areas. J Pharmacol Exp Ther 170:221–231

Creese I, Iversen SD (1975) The pharmacological and anatomical substrates of the amphetamine response in the rat. Brain Res 83:419–436

Daniels GM, Amara SG (1999) Regulated trafficking of the human dopamine transporter. J Biol Chem 274:35794–35801

Daniels J, Williams J, Asherson P, McGuffin P, Owen M (1995) No association between schizophrenia and polymorphisms within the genes for debrisoquine 4-hydroxylase (CYP2D6) and the dopamine transporter (DAT) Am J Med Genet (Neuropsychiat Genet) 60:85–87

Dawson TM, Gehlert DR, Wamsley JK (1986) Quantitative autoradiographic localization of the dopamine transport complex in the rat brain: use of a highly selective radioligand: [^{3}H]GBR 12935. Eur J Pharmaol 126:171–173

DiChiara G, Imperator A (1988) Drug abused by humans preferentially increase synaptic dopamine concentration in the mesolimbic system of freely moving rats. Proc Natl Acad Sci USA 85:5274–5278

Donnan GA, Kaczmarczyk SJ, Paxinos G, Chilco PJ, Kalnins RM, Woodhouse DG, Mendelsohn FAO (1991) Distribution of catecholamine uptake sites in human brain as determined by quantitative [^{3}H]mazindol autoradiography. J Comp Neurol 304:419–434

Donovan DM, Vandenbergh DJ, Perry MP, Bird GS, Ingersoll R, Nanthakumar E, Uhl GR (1995) Human and mouse dopamine transporter genes: conservation of 5′-flanking sequence elements and gene structures. Mol Brain Res 30:327–335

Doucette-Stamm LA, Blakely DJ, Tian J, Mockus S, Mao J (1995) Population genetic study of the human dopamine transporter gene (DAT1). Genet Epidemiol 12:303–308

Drucker G, Raikoff K, Neafsey EJ, Collins MA (1990) Dopamine uptake inhibitory capacities of β-carboline and 3,4-dihydro-β-carboline analogs of N-methyl-4-phenyl-1,2,3,6-tetrahydropyridine (MPTP) oxidation products. Brain Res 509:125–133

Dubocovich ML, Zahniser NR (1985) Binding characteristics of the dopamine uptake inhibitor [^{3}H]nomifensine to striatal membranes. Biochem Pharmacol 34:1137–1144

Earles C, Schenk JO (1999) Multisubstrate mechanism for the inward transport of dopamine by the human dopamine transporter expressed in HEK cells and its inhibition by cocaine. Synapse 33:230–238

Edvardsen Ë, Dahl SG (1994) A putative model of the dopamine transporter. Mol Brain Res 27:265–274

Eshleman AJ, Henningsen RA, Neve KA, Janowsky A (1994) Release of dopamine via the human transporter. Mol Pharmacol 45:312–316

Eshleman AJ, Neve RL, Janowsky A, Neve KA (1995) Characterization of a recombinant human dopamine transporter in multiple cell lines. J Pharmacol Exp Ther 274:276–283

Eshleman AJ, Carmolli M, Cumbay M, Martens CR, Neve KA, Janowsky A (1999) Characteristics of drug interactions with recombinant biogenic amine transporters expressed in the same cell type. J Pharmacol Exp Ther 289:877–885

Ferrer JV, Javitch JA (1998) Cocaine alters the accessibility of endogenous cysteines in putative extracellular and intracellular loops of the human dopamine transporter. Proc Natl Acad Sci USA 95:9238–9243

Fibiger HC (1977) On the role of the dopaminergic nigrostriatal projection in reinforcement, learning, and memory. In: Cools AR, Lohman AHM, van den Bercken JHL (eds) Psychobiology of the striatum. North Holland Publ Co, Amsterdam p 73

Fischer JF, Cho AK (1979) Chemical release of dopamine from striatal homogenates: evidence for an exchange diffusion model. J Pharmacol Exp Ther 208:203–209

Fischman AJ, Bonab AA, Babich JW, Palmer EP, Alpert NM, Elmaleh DR, Callahan RJ, Barrow SA, Graham W, Meltzer PC, Hanson RN, Madras BK (1998) Rapid detection of Parkinson's disease by SPECT with altropane: a selective ligand for dopamine transporters. Synapse 29:128–141

Fowler J, Volkow N, Wolf A, Dewey S, Schlyer D, MacGregor R, Hitzemann R, Logan J, Bendriem B, Gatley S, Christman D (1989) Mapping cocaine binding sites in human and baboon brain in vivo. Synapse 4:371–377

Freed C, Revay R, Vaughan RA, Kriek E, Grant S, Uhl GR, Kuhar MJ (1995) Dopamine transporter immunoreactivity in rat brain. J Comp Neurol 359:340–349

Fumagalli F, Gainetdinov RR, Valenzano KJ, Caron MG (1998) Role of dopamine transporter in methamphetamine-induced neurotoxicity: evidence from mice lacking the transporter. J Neurosci 18:4861–4869

Fumagalli F, Gainetdinov RR, Wang YM, Valenzano KJ, Miller GW, Caron MG (1999) Increased methamphetamine neurotoxicity in heterozygous vesicular monoamine transporter 2 knock-out mice. J Neurosci 19:2424–2431

Fuxe K, Ungerstedt U (1968) Histochemical studies on the effect of (+)-amphetamine, drugs of the imipramine group and tryptamine on central catecholamine and 5-

hydroxytryptamine neurons after intraventricular injection of catecholamines and 5-hydroxytryptamine. Eur J Pharmacol 4:135–144

Gainetdinov RR, Fumagalli F, Jones SR, Caron MG (1997) Dopamine transporter is required for in vivo MPTP neurotoxicity: evidence from mice lacking the transporter. J Neurochem 69:1322–1325

Gainetdinov RR, Wetsel WC, Jones SR, Levin ED, Jaber M, Caron MG (1999) Role of serotonin in the paradoxical calming effect of psychostimulants on hyperactivity. Science 283:397–401

Garris PA, Ciolkowski EL, Pastore P, Wightman RM (1994) Efflux of dopamine from the synaptic cleft in the nucleus accumbens of the rat brain. J Neurosci 14:6084–6093

Gelernter J, Kranzler HR, Satel SL, Rao PA (1994) Genetic association between dopamine transporter protein alleles and cocaine-induced paranoia. Neuropsychopharmacol 11:195–200

Gibb WR, Lees AJ (1991) Anatomy, pigmentation, ventral and dorsal subpopulations of the substantia nigra, and differential cell death in Parkinson's disease. J Neurol Neurosurg Psychiat 54:388–396

Gill M, Daly G, Heron S, Hawi Z, Fitzgerald M (1997) Confirmation of association between attention deficit hyperactivity disorder and a dopamine transporter polymorphism. Mol Psychiat 2:311–313

Giros B, El Mestikawy S, Bertrand L, Caron MG (1991) Cloning and functional characterization of a cocaine-sensitive dopamine transporter. FEBS Lett 295:149–154

Giros B, El Mestikawy S, Godinot N, Zheng K, Han H, Yang-Feng T, Caron MG (1992) Cloning, pharmacological characterization, and chromosome assignment of the human dopamine transporter. Mol Pharmacol 42:383–390

Giros B, Wang Y-M, Suter S, McLeskey SB, Pifl C, Caron MG (1994) Delineation of discrete domains for substrate, cocaine, and tricyclic antidepressant interactions using chimeric dopamine-norepinephrine transporters. J Biol Chem 269:15985–15988

Giros B, Jaber M, Jones SR, Wightman RM, Caron MG (1996) Hyperlocomotion and indifference to cocaine and amphetamine in mice lacking the dopamine transporter. Nature 379:606–612

Glowinski J, Iversen L (1966) Regional studies of catecholamines in the rat brain – III: Subcellular distribution of endogenous and exogenous catecholamines in various brain regions. Biochem Pharmacol 15:977–987

Glowinski J, Axelrod J, Iversen LL (1966) Regional studies of catecholamines in the rat brain. IV. Effects of drugs on the disposition and metabolism of [^{3}H]-norepinephrine and [^{3}H]-dopamine. J Pharmacol Exp Ther 153:30–41

Gracz LM, Madras BK (1995) [^{3}H]WIN 35,428 ([^{3}H]CFT) binds to multiple charge-states of the solubilized dopamine transporter in primate striatum. J Pharmacol Exp Ther 273:1224–1234

Grigoriadis DE, Wilson AA, Lew R, Sharkey JS, Kuhar MJ (1989) Dopamine transport sites selectively labeled by a novel photoaffinity probe:^{125}I-DEEP. J Neurosci 9:2664–2670

Gu H, Wall SC, Rudnick G (1994) Stable expression of biogenic amine transporters reveals differences in inhibitor sensitivity, kinetics, and ion dependence. J Biol Chem 269:7124–7130

Guastella J, Nelson N, Nelson H, Czyzyk L, Keynan S, Miedel MC, Davidson N, Lester HA, Kanner BI (1990) Cloning and expression of a rat brain GABA transporter. Science 249:1303–1306

Guttman M, Burkholder J, Kish SJ, Hussey D, Wilson A, DaSilva J, Houle S (1997) [^{11}C]RTI-32 PET studies of the dopamine transporter in early dopa-naive Parkinson's disease: Implications for the symptomatic threshold. Neurology 48:1578–1583

Haber SN, Ryoo H, Cox C, Lu W (1995) Subsets of midbrain dopaminergic neurons in monkeys are distinguished by different levels of mRNA for the dopamine transporter: comparison with the mRNA for the D2 receptor, tyrosine hydroxylase and calbindin immunoreactivity. J Comp Neurol 362:400–410

Hamberger B (1967) Reserpine-resistant uptake of catecholamines in isolated tissues of the rat. A histochemical study. Acta Physiol Scand Suppl 295:1–56

Harrington KA, Augood SJ, Kingsbury AE, Foster OJF, Emson PC (1996) Dopamine transporter (DAT) and synaptic vesicle amine transporter (VMAT2) gene expression in the substantia nigra of control and Parkinson's disease. Mol Brain Res 36:157–162

Harris JE, Baldessarini RJ (1973) The uptake of [^{3}H]dopamine by homogenates of rat corpus striatum: effects of cations. Life Sci 13:303–312

Heikkila RE (1981) Differential effects of several dopamine uptake inhibitors and releasing agents on locomotor activity in normal and in reserpinized mice. Life Sci 28:1867–1873

Heikkila RE, Orlansky H, Cohen G (1975) Studies on the distinction between uptake inhibition and release of [^{3}H]dopamine in rat brain tissue slices. Biochem Pharmacol 24:847–852

Herkenham K (1987) Mismatches between neurotransmitter and receptor localizations in brain: observations and implications. Neuroscience 23:1–38

Héron C, Billaud G, Costentin J, Bonnet J-J (1996) Complex ionic control of [^{3}H]GBR 12783 binding to the dopamine neuronal carrier. Eur J Pharmacol 301:195–202

Hersch SM, Yi H, Heilman CJ, Edwards RH, Levey AI (1997) Subcellular localization and molecular topology of the dopamine transporter in the striatum and substantia nigra. J Comp Neurol 388:211–227

Hertting G, Axelrod J, Kopin IJ, Whitby LG (1961) Lack of uptake of catecholamines after chronic denervation of sympathetic nerves. Nature 189:66

Higuchi S, Muramatsu T, Arai H, Hayashida M, Sasaki H, Trojanowski JQ (1995) Polymorphisms of dopamine receptor and transporter genes and Parkinson's disease. J Neural Transm P-D Sect 10:107–113

Hoffman BJ, Hansson SR, Mezey E, Palkovits M (1998) Localization and dynamic regulation of biogenic amine transporters in the mammalian central nervous system. Frontiers Neuroendocrinol 19:187–231

Holz RW, Coyle JT (1974) The effects of various salts, temperature, and the alkaloids veratridine and batrachotoxin on the uptake of [^{3}H]dopamine into synaptosomes from rat striatum. Mol Pharmacol 10:746–758

Horn AS (1973) Structure-activity relations for the inhibition of catecholamine uptake into synaptosomes from noradrenaline and dopaminergic neurons in rat brain homogenates. Br J Pharmacol 47:332–338

Hornykiewicz O, Pifl C (1994) The validity of the MPTP primate model for the neurochemical pathology of idiopathic Parkinson's disease. In: Briley M, Marien M (eds) Noradrenergic mechanisms in Parkinson's disease. CRC Press, Boca Raton Ann Arbor London Tokyo

Huang C-L, Chen H-C, Huang N-K, Yang D-M, Kao L-S, Chen J-C, Lai H-L, Chern Y (1999) Modulation of dopamine transporter activity by nicotinic acetylcholine receptors and membrane depolarization in rat pheochromocytoma PC12 cells. J Neurochem 72:2437–2444

Huff RA, Vaughan RA, Kuhar MJ, Uhl GR (1997) Phorbol esters increase dopamine transporter phosphorylation and decrease transport Vmax. J Neurochem 68:225–232

Hurd YL, Ungerstedt U (1989) Ca^{2+}-dependence of the amphetamine, nomifensine, and Lu 19–005 effect on in vivo dopamine transmission. Eur J Pharmacol 166:261–269

Hurd YL, Pristupa ZB, Herman MM, Niznik HB, Kleinman JE (1994) The dopamine transporter and dopamine D2 receptor messenger RNAs are differentially expressed in limbic- and motor-related subpopulations of human mesencephalic neurons. Neuroscience 63:357–362

Inada T, Sugita T, Dobashi I, Inagaki A, Kitao Y, Matsuda G, Kato S, Takano T, Yagi G, Asai M (1996) Dopamine transporter gene polymorphism and psychiatric symptoms seen in schizophrenic patients at their first episode. Am J Med Genet (Neuropsychiat Genet) 67:406–408

Innis R, Baldwin R, Sybirska E, Zea Y, Layrelle M, Al-Tikriti M, Charney D, Zoghbi S, Smith E, Wisniewksi G, Hoffer P, Wang S, Milius R, Neumeyer J (1991) Single photon emission computed tomography imaging of monoamine reuptake sites in primate brain with [^{123}I]CIT. Eur J Pharmacol 200:369–370

Iversen LL (1970) Inhibition of catecholamine uptake by 6-hydroxydopamine in rat brain. Eur J Pharmacol 10:408–410

Izenwasser S, Jacocks HM, Rosenberger JG, Cox BM (1991) Nicotine indirectly inhibits [^{3}H]dopamine uptake at concentrations that do not directly promote [^{3}H]dopamine release in rat striatum. J Neurochem 56:603–610

Izenwasser S, Terry P, Heller B, Witkin JM, Katz JL (1994) Differential relationships among dopamine transporter affinities and stimulant potencies of various uptake inhibitors. Eur J Pharmaol 263:277–283

Janowsky A, Schwer MM, Berger P, Long R, Skolnick P, Paul SM (1985) The effects of surgical and chemical lesions on striatal [^{3}H]threo(+)-methylphenidate binding: correlation with [^{3}H]dopamine uptake. Eur J Pharmacol 108:187–191

Janowsky A, Berger P, Vocci F, Labarca R, Skolnick P, Paul SM (1986) Characterization of sodium-dependent [^{3}H]GBR-12935 binding in brain: a radioligand for selective labelling of the dopamine transport complex. J Neurochem 46:1272–1276

Javitch JA, Blaustein RO, Snyder SH (1983) [^{3}H]Mazindol binding associated with neuronal dopamine uptake sites in corpus striatum membranes. Eur J Pharmacol 90:461–462

Javitch JA, Blaustein RO, Snyder SH (1984) [^{3}H]Mazindol binding associated with neuronal dopamine and norepinephrine uptake sites. Mol Pharmacol 26:35–44

Javitch JA, D'Amato RJ, Strittmater SM, Snyder SH (1985a) Parkinsonism-inducing neurotoxin, N-methyl-4-phenyl-1,2,3,6-tetrahydropyridine: uptake of the metabolite N-methyl-4-phenylpyridine by dopamine neurons explains selective toxicity. Proc Natl Acad Sci USA 82:2173–2177

Javitch JA, Strittmatter SM, Snyder SH (1985b) Differential visualization of dopamine and norepinephrine uptake sites in rat brain using [^{3}H]mazindol autoradiography. J Neurosci 5:1513–1521

Jayanthi LD, Apparsundaram S, Malone MD, Ward E, Miller DM, Eppler M, Blakely RD (1998) The *Caenorhabditis elegans* gene T23G5.5 encodes an antidepressant- and cocaine-sensitive dopamine transporter. Mol Pharmacol 54:601–609

Jonason J, Rutledge CO (1968) The effect of protriptyline on the metabolism of dopamine and noradrenaline in rabbit brain in vitro. Acta Physiol Scand 73:161–175

Jones SR, Garris PA, Wightman RM (1995) Different effects of cocaine and nomifensine on dopamine uptake in the caudate-putamen and nucleus accumbens. J Pharmacol Exp Ther 274:396–403

Jones SR, O'Dell SJ, Marshall JF, Wightman RM (1996) Functional and anatomical evidence for different dopamine dynamics in the core and shell of the nucleus accumbens in slices of rat brain. Synapse 23:224–231

Jones SR, Gainetdinov RR, Jaber M, Giros B, Wightman RM, Caron MG (1998a) Profound neuronal plasticity in response to inactivation of the dopamine transporter. Proc Natl Acad Sci USA 95:4029–4034

Jones SR, Gainetdinov RR, Wightman RM, Caron MG (1998b) Mechanisms of amphetamine action revealed in mice lacking the dopamine transporter. J Neurosci 18:1979–1986

Kadowaki K, Hirota K, Koike K, Ohmichi M, Kiyama H, Miyake A, Tonizawa O (1990) Adenosine 3',5'-cyclic monophosphate enhances dopamine accumulation in rat hypothalamic cell culture containing dopaminergic neurons. Neuroendocrinol 52:256–261

Kawarai T, Kawakami H, Yamamura Y, Nakamura S (1997) Structure and organization of the gene encoding human dopamine transporter. Gene 195:11–18

Kelsoe JR, Sadovnick AD, Kristbjarnarson H, Bergesch P, Mroczkowski-Parker Z, Drennan M, Rapaport MH, Flodman P, Spence MA, Remick RA (1996) Possible locus for bipolar disorder near the dopamine transporter on chromosome 5. Am J Med Genet 67:533–540

Kennedy LT, Hanbauer I (1983) Sodium-sensitive cocaine binding to rat striatal membrane: possible relationship to dopamine uptake sites. J Neurochem 41:172–178

Kilbourn MR (1988) In vivo binding of [^{18}F]GBR 13119 to the brain dopamine uptake system. Life Sci 42:1347–1353

Kilty JE, Lorang D, Amara SG (1991) Cloning and expression of a cocaine-sensitive rat dopamine transporter. Science 254:578–580

King N, Bassett AS, Honer WG, Masellis M, Kennedy JL (1997) Absence of linkage for schizophrenia on the short arm of chromosome 5 in multiplex Canadian families. Am J Med Genet (Neuropsychiat Genet) 74:472–474

Kish SJ, Shannak K, Hornykiewicz O (1988) Uneven pattern of dopamine loss in the striatum of patients with idiopathic Parkinson's disease. Pathophysiological and clinical implications. N Engl J Med 318:876–880

Kitayama S, Shimada S, Uhl GR (1992b) Parkinsonism-inducing neurotoxin MPP$^+$: uptake and toxicity in nonneuronal COS cells expressing dopamine transporter cDNA. Ann Neurol 32:109–111

Kitayama S, Shimada S, Xu H, Markham L, Donovan DM, Uhl GR (1992a) Dopamine transporter site-directed mutations differentially alter substrate transport and cocaine binding. Proc Natl Acad Sci USA 89:7782–7785

Kitayama S, Wang J-B, Uhl GR (1993) Dopamine transporter mutants selectively enhance MPP$^+$ transport. Synapse 15:58–62

Kitayama S, Dohi T, Uhl GR (1994) Phorbol esters alter functions of the expressed dopamine transporter. Eur J Pharmacol Mol Pharmacol Section 268:115–119

Krueger BK (1990) Kinetics and block of dopamine uptake in synaptosomes from rat caudate nucleus. J Neurochem 55:260–267

Kuhar MJ, Zarbin MA (1978) Synaptosomal transport: a chloride dependence for choline GABA, glycine and several other compounds. J Neurochem 31:251–256

Laine TPJ, Ahonen A, Torniainen P, Heikkilä J, Pyhtinen J, Räsänen P, Niemelä O, Hillbom M (1999) Dopamine transporters increase in human brain after alcohol withdrawal. Mol Psychiat 4:189–191

Landschulz WH, Johnson PF, McKnight SL (1988) The leucine zipper: a hypothetical structure common to a new class of DNA binding proteins. Science 240:1759–1764

Langston JW, Irwin I (1986) MPTP: current concepts and controversies. Clin Neuropharmacol 9:485–507

Le Couteur DG, Leighton PW, McCann SJ, Pond SM (1997) Association of a polymorphism in the dopamine-transporter gene with Parkinson's disease. Movement Disorders 12:760–763

Lee FJS, Pristupa ZB, Ciliax BJ, Levey AI, Niznik HB (1996) The dopamine transporter carboxyl-terminal tail. Truncation/substitution mutants selectively confer high affinity dopamine uptake while attenuating recognition of the ligand binding domain. J Biol Chem 271:20885–20894

Lee SH, Kang SS, Son H, Lee YS (1998) The region of dopamine transporter encompassing the 3rd transmembrane domain is crucial for function. Biochem Biophys Res Commun 246:347–352

Leighton PW, Le Couteur DG, Pang CCP, McCann SJ, Chan D, Law LK, Kay R, Pond SM, Woo J (1997) The dopamine transporter gene and Parkinson's disease in a Chinese population. Neurology 49:1577–1579

Lesch KP, Balling U, Gross J, Strauss K, Wolozin BL, Murphy DL, Riederer P (1994) Organization of the human serotonin transporter gene. J Neural Transm 95:157–162

Lew R, Patel A, Vaughan RA, Wilson A, Kuhar MJ (1992) Microheterogeneity of dopamine transporters in rat striatum and nucleus accumbens. Brain Res 584: 266–271

Liang NY, Rutledge CO (1982) Evidence for carrier-mediated efflux of dopamine from corpus striatum. Biochem Pharmacol 31:2479–2484

Lin Z, Wang W, Kopajtic T, Revay RS, Uhl GR (1999) Dopamine transporter: transmembrane phenylalanine mutations can selectively influence dopamine uptake and cocaine analog recognition. Mol Pharmacol 56:434–447

Lindvall O, Björklund A (1983) Dopamine- and norepinephrine-containing neuron systems: Their anatomy in the rat brain. In: Emson PC (ed) Chemical Neuroanatomy. Raven Press, New York, p 229

Lorang D, Amara SG, Simerly RB (1994) Cell-type-specific expression of catecholamine transporters in the rat brain. J Neurosci 14:4903–4914

Ma SY, Ciliax BJ, Stebbins G, Jaffar S, Joyce JN, Cochran EJ, Kordower JH, Mash DC, Levey AI, Mufson EJ (1999) Dopamine transporter-immunoreactive neurons decrease with age in the human substantia nigra. J Comp Neurol 409:25–37

Madras BK, Spealman RD, Fahey MA, Neumeyer JL, Saha JK, Milius RA (1989) Cocaine receptors labeled by [^{3}H]2β-carbomethoxy-3β-(4-fluorophenyl) tropane. Mol Pharmacol 36:518–524

Madras BK, Gracz LM, Fahey MA, Elmaleh D, Meltzer PC, Liang AY, Stopa EG, Babich J, Fischman AJ (1998b) Altropane, a SPECT or PET imaging probe for dopamine neurons: III. Human dopamine transporter in postmortem normal and Parkinson's diseased brain. Synapse 29:116–127

Madras BK, Meltzer PC, Liang AY, Elmaleh DR, Babich J, Fischman AJ (1998a) Altropane, a SPECT or PET imaging probe for dopamine neurons: I. Dopamine transporter binding in primate brain. Synapse 29:93–104

Maier W, Minges J, Eckstein N, Brodski C, Albus M, Lerer B, Hallmayer J, Fimmers R, Ackenheil M, Ebstein RE, Borrmann M, Lichtermann D, Wildenauer DB (1996) Genetic relationship between dopamine transporter gene and schizophrenia: linkage and association. Schizophrenia Res 20:175–180

Malison RT, McDougle CJ, van Dyck CH, Scahill L, Baldwin RM, Seibyl JP, Price LH, Leckman JF, Innis RB (1995) [^{123}I]beta-CIT SPECT imaging of striatal dopamine transporter binding in Tourette's disorder. Am J Psychiat 152:1359–1361

Marek GJ, Vosmer G, Seiden LS (1990) Dopamine uptake inhibitors block long-term neurotoxic effects of methamphetamine upon dopaminergic neurons. Brain Res 513:274–279

Marshall JF, Navarrete RJ (1990) Contrasting tissue factors predict heterogeneous striatal dopamine neurotoxicity after MPTP or methamphetamine treatment. Brain Res 534:348–351

Marshall JF, O'Dell SJ, Navarrete R, Rosenstein AJ (1990) Dopamine high-affinity transport site topography in rat brain: major differences between dorsal and ventral striatum. Neuroscience 37:11–21

Martres MP, Demeneix B, Hanoun H, Hamon M, Giros B (1998) Up- and down-expression of the dopamine transporter by plasmid DNA transfer in the rat brain. Eur J Neurosci 10:3607–3616

Matsubara K, Gonda T, Sawada H, Uezono T, Kobayashi Y, Kawamura T, Ohtaki K, Kimura K, Akaike A (1998) Endogenously occurring β-carboline induces parkinsonism in nonprimate animals: a possible causative protoxin in idiopathic Parkinson's disease. J Neurochem 70:727–735

McCann UD, Wong DF, Yokoi F, Villemagne V, Dannals RF, Ricaurte GA (1998) Reduced striatal dopamine transporter density in abstinent methamphetamine and methcathinone users: evidence from positron emission tomography studies with [^{11}C]WIN-35,428. J Neurosci 18:8417–8422

McElvain JS, Schenk JO (1992) A multisubstrate mechanism of striatal dopamine uptake and its inhibition by cocaine. Biochem Pharmacol 43:2189–2199

McKearney JW (1982) Effects of dopamine uptake inhibitors on schedule-controlled behavior in the squirrel monkey. Psychopharmacol 78:377–379

McNaught KStP, Thull U, Carrupt P-A, Altomare C, Cellamare S, Carotti A, Testa B, Jenner P, Marsden CD (1996) Toxicity of PC12 cells of isoquinoline derivatives structurally related to 1-methyl-4-phenyl-1,2,3,6-tetrahydropyridine. Neurosci Lett 206:37–40

Meiergerd SM, Patterson TA, Schenk JO (1993) D2 receptors may modulate the function of the striatal transporter for dopamine: kinetic evidence from studies in vitro and in vivo. J Neurochem 61:764–767

Meiergerd SM, Schenk JO (1994a) Striatal transporter for dopamine: catechol structure-activity studies and susceptibility to chemical modification. J Neurochem 62:998–1008

Meiergerd SM, Schenk JO (1994b) Kinetic evaluation of the commonality between the site(s) of action of cocaine and some other structurally similar and dissimilar inhibitors of the striatal transporter for dopamine. J Neurochem 63:1683–1692

Meister B, Elde R (1993) Dopamine transporter mRNA in neurons of the rat hypothalamus. Neuroendocrinol 58:388–395

Melikian HE, Buckley KM (1999) Membrane trafficking regulates the activity of the human dopamine transporter. J Neurosci 19:7699–7710

Mercier G, Turpin JC, Lucotte G (1999) Variable number tandem repeat dopamine transporter gene polymorphism and Parkinson's disease: no association found. J Neurol 246:45–47

Miller GW, Staley JK, Heilman CJ, Perez JT, Mash DC, Rye DB, Levey AI (1997) Immunochemical analysis of dopamine transporter protein in Parkinson's disease. Ann Neurol 41:530–539

Milner HE, Béliveau R, Jarvis SM (1994) The in situ size of the dopamine transporter is a tetramer as estimated by radiation inactivation. Biochim Biophys Acta 1190:185–187

Missale C, Castelletti L, Govoni S, Spano PF, Trabucchi M, Hanbauer I (1985) Dopamine uptake is differentially regulated in rat striatum and nucleus accumbens. J Neurochem 45:51–56

Mitsuhata C, Kitayama S, Morita K, Vandenbergh D, Uhl GR, Dohi T (1998) Tyrosine-533 of rat dopamine transporter: involvement in interactions with 1-methyl-4-phenylpyridinium and cocaine. Mol Brain Res 56:84–88

Moore KE (1978) Amphetamines: biochemical and behavioral actions in animals. In: Iversen LL, Iversen SD, Snyder SH (eds) Handbook of Psychopharmacology. Plenum, New York, p 41

Muramatsu T, Higuchi S (1995) Dopamine transporter gene polymorphism and alcoholism. Biochem Biophys Res Comm 211:28–32

Nagatsu T (1997) Isoquinoline neurotoxins in the brain and Parkinson's disease. Neurosci Res 29:99–111

Nagatsu T, Yoshida M (1988) An endogenous substance of the brain, tetrahydroisoquinoline, produces parkinsonism in primates with decreased dopamine, tyrosine hydroxylase and biopterin in the nigrostriatal regions. Neurosi Lett 87:178–182

Nelson N (1998) The family of Na$^+$/Cl$^-$ neurotransmitter transporters. J Neurochem 71:1785–1803

Nelson H, Mandiyan S, Nelson N (1990) Cloning of the human brain GABA transporter. FEBS Lett 269:181–184

Ng JP, Menacherry SD, Liem BJ, Anderson D, Singer M, Justice Jr JB (1992) Anomalous effect of mazindol on dopamine uptake as measured by in vivo voltammetry and microdialysis. Neurosci Lett 134:229–232

Nirenberg MJ, Vaughan RA, Uhl GR, Kuhar MJ, Pickel VM (1996) The dopamine transporter is localized to dendritic and axonal plasma membranes of nigrostriatal dopaminergic neurons. J Neurosci 16:436–447

Nirenberg MJ, Chan J, Pohorille A, Vaughan RA, Uhl GR, Kuhar MJ, Pickel VM (1997b) The dopamine transporter: comparative ultrastructure of dopaminergic axons in limbic and motor compartments of the nucleus accumbens. J Neurosci 17:6899–6907

Nirenberg MJ, Chan J, Vaughan RA, Uhl GR, Kuhar MJ, Pickel VM (1997a) Immunogold localization of the dopamine transporter: an ultrastructural study of the rat ventral tegmental area. J Neurosci 17:5255–5262

Niznik HB, Tyndale RF, Sallee FR, Gonzalez FJ, Hardwick JP, Inaba T, Kalow W (1990) The dopamine transporter and cytochrome P450IID1 (debrisoquine 4-hydroxylase) in brain: resolution and identification of two distinct [^{3}H]GBR-12935 binding proteins. Arch Biochem Biophys 276:424–432

Norregaard L, Frederiksen D, Nielsen EÈ, Gether U (1998) Delineation of an endogenous zinc-binding site in the human dopamine transporter. EMBO J 17:4266–4273

O'Dell SJ, Marshall JF (1988) Transport of [^{3}H]mazindol binding sites in mesostriatal dopamine axons. Brain Res. 460:402–406

Pacholczyk T, Blakely RD, Amara SG (1991) Expression cloning of a cocaine- and antidepressant-sensitive human noradrenaline transporter. Nature 350:350–354

Parker EM, Cubeddu LX (1988) Comparative effects of amphetamine, phenylethylamine and related drugs on dopamine efflux, dopamine uptake and mazindol binding. J Pharmacol Exp Ther 245:199–210

Parsons LH, Schad CA, Justice Jr JB (1993) Co-administration of the D2 antagonist pimozide inhibits up-regulation of dopamine release and uptake induced by repeated cocaine. J Neurochem 60:376–379

Persico AM, Vandenbergh DJ, Smith SS, Uhl GR (1993) Dopamine transporter gene polymorphisms are not associated with polysubstance abuse. Biol Psychiat 34: 265–267

Persico AM, Wang ZW, Black DW, Andreasen NC, Uhl GR, Crowe RR (1995) Exclusion of close linkage of the dopamine transporter gene with schizophrenia spectrum disorders. Am J Psychiat 152:134–136

Petrali EH, Boulton AA, Dyck LE (1979) Uptake of para-tyramine and meta-tyramine into slices of the caudate nucleus and hypothalamus of the rat. Neurochem Res 4:633–642

Pifl Ch, Schingnitz G, Hornykiewicz O (1991) Effect of 1-methyl-4-phenyl-1,2,3,6-tetrahydrpyridine on the regional distribution of brain monoamines in the rhesus monkey. Neurosience 44:591–605

Pifl C, Giros B, Caron MG (1993) Dopamine transporter expression confers cytotoxicity to low doses of the parkinsonism-inducing neurotoxin 1-methyl-4-phenylpyridinium. J Neurosci 13:4246–4253

Pifl Ch, Drobny H, Reither H, Hornykiewicz O, Singer EA (1995) Mechanism of the dopamine-releasing actions of amphetamine and cocaine: plasmalemmal dopamine transporter versus vesicular monoamine transporter. Mol Pharmacol 47:368–373

Pifl Ch, Hornykiewicz O, Giros B, Caron MG (1996) Catecholamine transporters and 1-methyl-4-phenyl-1,2,3,6-tetrahydropyridine neurotoxicity: studies comparing the cloned human noradrenaline and human dopamine transporter. J Pharmacol Exp Ther 277:1437–1443

Pifl Ch, Agneter E, Drobny H, Reither H, Singer EA (1997) Induction by low Na$^+$ and Cl$^-$ of cocaine sensitive carrier-mediated efflux of amines from cells transfected with the cloned human catecholamine transporters. Brit J Pharmacol 121:205–212

Pifl C, Singer EA (1999) Ion-dependence of carrier-mediated release in dopamine- or norepinephrine-transporter transfected cells questions the hypothesis of facilitated exchange diffusion. Mol Pharmacol 56:1047–1054

Pilotte NS (1997) Neurochemistry of cocaine withdrawal. Curr Opin Neurol 10:534–538

Pimoule C, Schoemaker H, Javoy-Agid F, Scatton B, Agid Y, Langer SZ (1983) Decrease in [^{3}H]cocaine binding to the dopamine transporter in Parkinson's disease. Eur J Pharmacol 95:145–146

Planté-Bordeneuve V, Taussig D, Thomas F, Said G, Wood NW, Marsden CD, Harding AE (1997) Evaluation of four candidate genes encoding proteins of the dopamine pathway in familial and sporadic Parkinson's disease: Evidence for association of a DRD2 allele. Neurology 48:1589–1593

Pontieri FE, Tanda G, DiChiara G (1995) Intravenous cocaine, morphine, and amphetamine preferentially increase extracellular dopamine in the "shell" as compared with the "core" of the rat nucleus accumbens. Proc Natl Acad Sci USA 92:12304–12308

Pörzgen P, Bönisch H, Brüss M (1995) Molecular cloning and organization of the coding region of the human norepinephrine transporter gene. Biochem Biophys Res Comm 215:1145–1150

Povlock SL, Schenk JO (1997) A multisubstrate kinetic mechanism of dopamine transport in the nucleus accumbens and its inhibition by cocaine. J Neurochem 69:1093-1105

Pristupa ZB, Wilson JM, Hoffman BJ, Kish SJ, Niznik HB (1994) Pharmacological heterogeneity of the cloned and native human dopamine transporter: disassociation of [^{3}H]WIN 35,428 and [^{3}H]GBR 12,935 binding. Mol Pharmacol 45:125–135

Pristuba ZB, McConkey F, Liu F, Man HY, Lee FJS, Wang YT, Niznik HB (1998) Protein kinase-mediated bidirectional trafficking and functional regulation of the human dopamine transporter. Synapse 30:79–87

Raiteri M, Cerrito F, Cervoni AM, del Carmine R, Ribera MT, Levi G (1978) Studies on dopamine uptake and release in synaptosomes. Adv Biochem Psychopharmacol 19:35–56

Raiteri M, Cerrito F, Cervoni AM, Levi G (1979) Dopamine can be released by two mechanisms differentially affected by the dopamine transport inhibitor nomifensine. J Pharmacol Exp Ther 208:195–202

Ramamoorthy S, Bauman AL, Moore KR, Han H, Yang-Feng T, Chang AS, Ganapathy V, Blakely RD (1993) Antidepressant- and cocaine-sensitive human serotonin transporter: molecular cloning, expression, and chromosomal localization. Proc Natl Acad Sci USA 90:2542–2546

Refahi-Lyamani F, Saadouni S, Costentin J, Bonnet J-J (1995) Interaction of two sulfhydryl reagents with a cation recognition site on the neuronal dopamine carrier evidences small differences between [^{3}H]GBR 12783 and [^{3}H]cocaine binding sites. Naunyn-Schmiedeberg's Arch Pharmacol 351:136–145

Reith MEA, Selmeci G (1992) Radiolabeling of dopamine uptake sites in mouse striatum: comparison of binding sites for cocaine, mazindol, and GBR 12935. Naunyn-Schmiedeberg's Arch Pharmacol 345:309–318

Reith MEA, Coffey LL (1993) Cationic and anionic requirements for the binding of 2b-carbomethoxy-3b-(4-fluorophenyl)[^{3}H]tropane to the dopamine uptake carrier. J Neurochem 61:167–177

Revay R, Vaughan R, Grant S, Kuhar MJ (1996) Dopamine transporter immunohistochemistry in median eminence, amygdala, and other areas of the rat brain. Synapse 22:93–99

Richfield EK (1993) Zinc modulation of drug binding, cocaine affinity states, and dopamine uptake on the dopamine uptake complex. Mol Pharmacol 43:100–108

Richtand NM, Kelsoe JR, Segal DS, Kuczenski R (1995) Regional quantification of dopamine transporter mRNA in rat brain using a ribonuclease protection assay. Neurosci Lett 200:73–76

Ritz MC, Lamb RJ, Goldberg SR, Kuhar MJ (1987) Cocaine receptors on dopamine transporters are related to self-administration of cocaine. Sience 237:1219–1223

Roberts DCS, Zis AP, Fibiger HC (1975) Ascending catecholaminergic pathways and amphetamine-induced locomotor activity: importance of dopamine and apparent noninvolvement of norepinephrine. Brain Res 93:441–451

Roberts DCS, Corcoran ME, Fibiger HC (1977) On the role of ascending catecholaminergic systems in intravenous self-administration of cocaine. Pharmacol Biochem Behav 6:615–620

Rocha BA, Fumagalli F, GainetdinovRR, Jones SR, Ator R, Giros B, Miller GW, Caron MG (1998) Cocaine self-administration in dopamine-transporter knockout mice. Nature Neurosci 1:132–137

Ross SB (1979) The central stimulatory action of inhibitors of the dopamine uptake. Life Sci 24:159–168

Ross SB, Renyi AL (1967) Inhibition of the uptake of tritiated catecholamines by antidepressant and related agents. Eur J Pharmacol 2:181–186

Rutledge CO (1978) Effect of metabolic inhibitors and ouabain on amphetamine- and potassium-induced release of biogenic amines from isolated brain tissue. Biochem Pharmacol 27:511–516

Sallee FR, Fogel EL, Schwartz E, Choi S-M, Curran DP, Niznik HB (1989) Photoaffinity labeling of the mammalian dopamine transporter. FEBS Lett 256:219–224

Sander T, Harms H, Podschus J, Finckh U, Nickel B, Rolfs A, Rommelspacher H, Schmidt LG (1997) Allelic association of a dopamine transporter gene polymorphism in alcohol dependence with withdrawal seizures or delirium. Biol Psychiat 41:299–304

Sano A, Kondoh K, Kakimoto Y, Kondo I (1993) A 40-nucleotide repeat polymorphism in the human dopamine transporter gene. Human Genet 91:405–406

Saporito MS, Heikkila RE, Youngster SK, Nicklas WJ, Geller HM (1992) Dopaminergic neurotoxicity of 1-methyl-4-phenylpyridinium analogs in cultured neurons: relationship to the dopamine uptake system and inhibition of mitochondrial respiration. J Pharmacol Exp Ther 260:1400–1409

Scatton B, Dubois A, Dubocovich ML, Zahniser NR, Faga D (1985) Quantitative autoradiography of ^{3}H-nomifensine binding sites in rat brain. Life Sci 36:815–822

Schechter MD, Cook PG (1975) Dopaminergic mediation of the interoceptive cue produced by d-amphetamine in rats. Psychopharmacologia 42:185–193

Scheel-Krüger J (1971) Comparative studies of various amphetamine analogues demonstrating different interactions with the metabolism of the catecholamines in the brain. Eur J Pharmacol 14:47–59

Scheffel U, Dannals RF, Wong DF, Yokoi F, Carroll FI, Kuhar MJ (1992) Dopamine transporter imaging with novel, selective cocaine analogs. Neuroreport 3:969–972

Schneider JS, Yuwiler A, Markham CH (1987) Selective loss of subpopulations of ventral mesencephalic dopaminergic neurons in the monkey following exposure to MPTP. Brain Res 411:144–150

Schoemaker H, Pimoule C, Arbilla S, Scatton B, Javoy-Agid F, Langer SZ (1985) Sodium dependent [^{3}H]cocaine binding associated with dopamine uptake sites in the rat striatum and human putamen decrease after dopaminergic denervation and in Parkinson's disease. Naunyn Schmiedeberg's Arch Pharmacol 329:227–235

Sesack SR, Hawrylak VA, Matus C, Guido MA, Levey AI (1998) Dopamine axon varicosities in the prelimbic division of the rat prefrontal cortex exhibit sparse immunoreactivity for the dopamine transporter. J Neurosci 18:2697–2708

Shimada S, Kitayama S, Lin C-L, Patel A, Nanthakumar E, Gregor P, Kuhar M, Uhl G (1991) Cloning and expression of a cocaine-sensitive dopamine transporter complementary DNA. Science 254:576–578

Shimada S, Kitayama S, Walther D, Uhl G (1992) Dopamine transporter mRNA: dense expression in ventral midbrain neurons. Mol Brain Res 13:359–362

Silvia CP, Jaber M, King GR, Ellinwood EH, Caron MG (1997) Cocaine and amphetamine elicit differential effects in rats with a unilateral injection of dopamine transporter antisense oligodeoxynucleotides. Neuroscience 76:737–747

Sitges M, Reyes A, Chiu LM (1994) Dopamine transporter mediated release of dopamine: role of chloride. J Neurosci Res 39:11–22

Sitte HH, Huck S, Reither H, Boehm S, Singer AE, Pifl C (1998) Carrier-mediated release, transport rates, and charge transfer induced by amphetamine, tyramine, and dopamine in mammalian cells transfected with the human dopamine transporter. J Neurochem 71:1289–1297

Smiley JF, Williams SM, Szigeti K, Goldman-Rakic PS (1992) Light and electron microscopic characterization of dopamine-immunoreactive axons in human cerebral cortex. J Comp Neurol 321:325–335
Snyder SH, Coyle JT (1969) Regional differences in H^3-norepinephrine and H^3-dopamine uptake into rat brain homogenates. J Pharmacol Exp Ther 165:78–86
Sonders MS, Amara SG (1996) Channels in transporters. Curr Opin Neurobiol 6:294–302
Sonders MS, Zhu S-J, Zahniser NR, Kavanaugh MP, Amara SG (1997) Multiple ionic conductances of the human dopamine transporter: the actions of dopamine and psychostimulants. J Neurosci 17:960–974
Sora I, Wichems C, Takahashi N, Li X-F, Zeng Z, Revay R, Lesch K-P, Murphy DL, Uhl GR (1998) Cocaine reward models: conditioned place preference can be established in dopamine- and in serotonin-transporter knockout mice. Proc Natl Acad Sci USA 95:7699–7704
Souery D, Lipp O, Mahieu B, Mendelbaum K, de Martelaer V, van Broeckhoven C, Mendlewicz J (1996) Association study of bipolar disorder with candidate genes involved in catecholamine neurotransmission: DRD2, DRD3, DAT1, and TH genes. Am J Med Genet (Neuropsychiat Genet) 67:551–555
Stamford JA, Kruk ZL, Millar J (1986) In vivo characterization of low affinity striatal dopamine uptake: drug inhibition profile and relation to dopaminergic innervation density. Brain Res 373:85–91
Strader CD, Sigal IS, Dixon RA (1989) Structural basis of β-adrenergic receptor function. FASEB J 3:1825–1832
Sugamori KS, Lee FJS, Pristupa ZB, Niznik HB (1999) A cognate dopamine transporter-like activity endogenously expressed in a COS-7 kidney-derived cell line. FEBS Lett 451:169–174
Sulzer D, Chen T-K, Lau YY, Kristensen H, Rayport S, Ewing A (1995) Amphetamine redistributes dopamine from synaptic vesicles to the cytosol and promotes reverse transport. J Neurosci 15:4102–4108
Tiihonen J, Kuikka J, Bergström K, Hakola P, Karhu J, Ryynänen O-P, Föhr J (1995) Altered striatal dopamine re-uptake site densities in habitually violent and nonviolent alcoholics. Nature Med 1:654–657
Uchikawa T, Kiuchi Y, Yura A, Nakachi N, Yamazaki Y, Yokomizo C, Oguchi K (1995) Ca^{2+}-dependent enhancement of [^{3}H]dopamine uptake in rat striatum: possible involvement of calmodulin-dependent kinases. J Neurochem 65:2065–2071
Uhl GR, O'Hara B, Shimada S, Zaczek R, DiGiorgianni J, Nishimori T (1991) Dopamine transporter: expression in Xenopus oocytes. Mol Brain Res 9:23–29
Uhl GR, Walther D, Mash D, Faucheux B, Javoy-Agid F (1994) Dopamine transporter messenger RNA in Parkinson's disease and control substantia nigra neurons. Ann Neurol 35:494–498
Uretsky NJ, Iversen LL (1970) Effects of 6-hydroxydopamine on catecholamine containing neurones in the rat brain. J Neurochem 17:269–278
Usdin TB, Mezey E, Chen C, Brownstein MJ, Hoffman BJ (1991) Cloning of the cocaine-sensitive bovine dopamine transporter. Proc Natl Acad Sci USA 88: 11168–11171
Van der Zee P, Koger HS, Gootjes J, Hespe W (1980) Aryl-1,4-dialk(en)ylpiperazines as selective and very potent inhibitors of dopamine uptake. Eur J Med Chem 15:363–370
Vandenbergh DJ, Persico AM, Hawkins AL, Griffin CA, Li X, Jabs EW, Uhl GR (1992b) Human dopamine transporter gene (DAT1) maps to chromosome 5p15.3 and displays a VNTR. Genomics 14:1104–1106
Vandenbergh DJ, Persico AM, Uhl GR (1992a) A human dopamine transporter cDNA predicts reduced glycosylation, displays a novel repetitive element and provides racially-dimorphic TaqI RFLPs. Mol Brain Res 15:161–166
Vaugeois JM, Bonnet J-J, Duterte-Boucher D, Costentin J (1993) In vivo occupancy of the striatal dopamine uptake complex by various inhibitors does not predict their effects on locomotion. Eur J Pharmacol 230:195–201

Vaughan RA (1995) Photoaffinity labeled ligand binding domains on dopamine transporters identified by peptide mapping. Mol Pharmacol 47:956–964

Vaughan RA, Uhl GR, Kuhar MJ (1993) Dopamine transporter recognition by antipeptide antibodies. Cell Mol Neurosci 4:209–215

Vaughan RA, Kuhar MJ (1996) Dopamine transporter ligand binding domains. Structural and functional properties revealed by limited proteolysis. J Biol Chem 271:21672–21680

Vaughan RA, Huff RA, Uhl GR, Kuhar MJ (1997) Protein kinase C-mediated phosphorylation and functional regulation of dopamine transporters in striatal synaptosomes. J Biol Chem 272:15541–15546

Vaughan RA, Agoston GE, Lever JR, Newman AH (1999) Differential binding of tropane-based photoaffinity ligands on the dopamine transporter. J Neurosci 19:630–636

Volkow ND, Wang GJ, Fischman MW, Foltin RW, Fowler JS, Abumrad NN, Vitkun S, Logand J, Gatley SJ, Pappas N, Hitzemann R, Shea CE (1997) Relationship between subjective effects of cocaine and dopamine transporter occupancy. Nature 386:827–830

Waldman ID, Rowe DC, Abramowitz A, Kozel ST, Mohr JH, Sherman SL, Cleveland HH, Sanders ML, Gard JMC, Stever C (1998) Association and linkage of the dopamine transporter gene and attention-deficit hyperactivity disorder in children: heterogeneity owing to diagnostic subtype and severity. Am J Human Genet 63:1767–1776

Wall SC, Innis RB, Rudnick G (1993) Binding of the cocaine analog 2b-carbomethoxy-3b-(4-[^{125}I]iodophenyl)tropane to serotonin and dopamine transporters: different ionic requirements for substrate and 2b-carbomethoxy-3b-(4-[^{125}I]iodophenyl)tropane binding. Mol Pharmacol 43:264–270

Wall SC, Gu H, Rudnick G (1995) Biogenic amine flux mediated by cloned transporters stably expressed in cultured cell lines: amphetamine specificity for inhibition and efflux. Mol Pharmacol 47:544–550

Wang JB, Morikawa A, Uhl GR (1995) Dopamine transporter cysteine mutants: second extracellular loop cysteines are required for transporter expression. J Neurochem 64:1416–1419

Wheeler DD, Edwards AM, Chapman BM, Ondo JG (1993) A model of the sodium dependence of dopamine uptake in rat striatal synaptosomes. Neurochem Res 18:927–936

Wheeler DD, Edwards AM, Chapman BM, Ondo JG (1994) Effects of cocaine on sodium dependent dopamine uptake in rat striatal synaptosomes. Neurochem Res 19:49–56

Wilson JM, Levey AI, Rajput A, Ang L, Guttman M, Shannak K, Niznik HB, Hornykiewicz O, Pifl C, Kish SJ (1996) Differential changes in neurochemical markers of striatal dopamine nerve terminals in idiopathic Parkinson's disease. Neurology 47:718–726

Wong DF, Harris JC, Naidu S, Yokoi F, Marenco S, Dannals RF, Ravert HT, Yaster M, Evans A, Rousset O, Bryan RN, Gjedde A, Kuhar MJ, Breese GR (1996) Dopamine transporters are markedly reduced in Lesch-Nyhan disease in vivo. Proc Natl Acad Sci USA 93:5539–5543

Woolverton WL, Ricaurte GA, Forno LS, Seiden LS (1989) Long-term effects of chronic methamphetamine administration in rhesus monkeys. Brain Res 486:73–78

Xu C, Reith MEA (1997) WIN 35,428 and mazindol are mutually exclusive in binding to the cloned human dopamine transporter. J Pharmacol Exp Ther 282:920–927

Yamashita H, Kitayama S, Zhang Y-X, Takahashi T, Dohi T, Nakamura S (1995) Effect of nicotine on dopamine uptake in COS cells possessing the rat dopamine transporter and in PC12 cells. Biochem Pharmacol 49:742–745

Zaczek R, Culp S, DeSouza EB (1991) Interactions of [^{3}H]amphetamine with rat brain synaptosomes. II. Active transport. J Pharmacol Exp Ther 257:830–835

Zahniser NR, Larson GA, Gerhardt GA (1999) In vivo dopamine clearance rate in rat striatum: regulation by extracellular dopamine concentration and dopamine transporter inhibitors. J Pharmacol Exp Ther 289:266–277

Zhang L, Elmer LW, Little KY (1998) Expression and regulation of the human dopamine transporter in a neuronal cell line. Mol Brain Res 59:66–73

Zhu S-J, Kavanaugh MP, Sonders MS, Amara SG, Zahniser NR (1997) Activation of protein kinase C inhibits uptake, currents and binding associated with the human dopamine transporter expressed in Xenopus oocytes. J Pharmacol Exp Ther 282:1358–1365

Zimányi I, Lajtha A, Reith MEA (1989) Comparison of characteristics of dopamine uptake and mazindol binding in mouse striatum. Naunyn Schmiedeberg's Arch Pharmacol 340:626–632

CHAPTER 11
Cellular Actions of Dopamine

D.J. Surmeier and P. Calabresi

A. Introduction

The neostriatum is the gateway of the basal ganglia, a group of brain nuclei that control a wide variety of psychomotor behaviors. Most of the functions of the neostriatum are dependent upon its rich dopaminergic innervation that originates in the ventral mesencephalon. For example, the loss of this dopaminergic innervation results in Parkinson's disease (Albin et al. 1989; Wooten 1990). Disordered dopaminergic signaling in the neostriatum is also thought to be a critical determinant of several other common neuropsychiatric disorders including Tourette's syndrome, schizophrenia, and drug addiction (Nemeroff and Bissette 1988; Erenberg 1992; Koob and Nestler 1997; Wise 1998).

In spite of its clinical importance, the physiological consequences of dopamine (DA) in the neostriatum have been matter of debate. Part this debate has been the outgrowth of approaching DA as a classical fast transmitter that could be thought of as excitatory or inhibitory. Recent work has shown that this is a fundamentally erroneous way of thinking about how DA modulates neuronal excitability. DA is neither excitatory nor inhibitory. DA alters the way neurons integrate synaptic input by modulating intrinsic voltage-dependent ion channels participating in the postsynaptic response to synaptic input and by modulating the synaptic process itself. Changing our conceptual framework is essential to understanding the functional role of DA in the neostriatum (and elsewhere). In this review, an attempt will be made to summarize new insights into dopaminergic modulation of the intrinsic excitability of neostriatal neurons and of synaptic transmission and plasticity.

B. DA Receptor Expression in Neostriatal Neurons

Five DA receptors have been cloned (Civelli et al. 1993; Sokoloff and Schwartz 1995; Missale et al. 1998). These can be divided into two families based upon molecular and biochemical criteria. D_1 and D_5 receptors constitute the D_1-like family of receptors. These receptors activate $G_{s/olf}$ proteins that stimulate adenylyl cyclases isoforms (Hervé et al. 1993). There is also evidence that D_1 and D_5 receptors can activate other classes of G proteins, such as G_o

and G_z proteins (SIDHU 1998). D_2, D_3, and D_4 receptors constitute the D_2-like family of receptors. These receptors activate $G_{i/o}$ proteins that can influence a variety of intracellular proteins either through α or βγ subunits (HERLITZE et al. 1996; IKEDA 1996; YAN et al. 1997).

γ-Aminobutyric acid (GABA)ergic medium spiny neurons are the principal neurons of the neostriatum, which in the rodent constitutes 90%–95% of all neostriatal neurons (CHANG et al. 1982; CHANG and KITAI 1985). In addition to having local, densely arborizing recurrent axon collaterals, medium spiny neurons innervate other basal ganglia structures and have been divided into two groups based on anatomical and biochemical grounds. Members of the "direct" pathway have axons that terminate in the substantia nigra (SN) and express high levels of substance P (SP), whereas members of the "indirect" pathway terminate in the external segment of the globus pallidus (GP) and express high levels of enkephalin (ENK) (ALEXANDER et al. 1986; GERFEN 1992).

Although this classification has had heuristic value, recent work indicates that it is oversimplified. First, neurons projecting to the SN also can send axon collaterals to the GP, arguing that "direct" and "indirect" pathways are not truly segregated (KAWAGUCHI et al. 1990). Second, although D_1 and D_2 receptors appear to be largely segregated, there is a substantial subpopulation (20%–25%) of medium spiny neurons that co-express these receptors, as well as SP and ENK (SURMEIER et al. 1996). The existence of this population was not apparent using in situ hybridization or immunocytochemical techniques and only became evident with the application of single cell reverse transcriptase polymerase chain reaction (RT-PCR) methods. These RT-PCR experiments also suggest that the less abundant DA receptors (D_3, D_4, and D_5) are expressed to varying extents, making it considerably more difficult to estimate the extent to which D_1- and D_2-class receptors are colocalized.

A key question that remains to be answered is how abundant these "other" DA receptors truly are and what functional roles they play. The limited work that has been done thus far suggests that both D_4 and D_5 are present at low levels (BERGSON et al. 1995), whereas D_3 receptors may be expressed at higher levels in a subpopulation (–40%) of neurons in the so-called direct pathways that express D_1 receptors and substance P. Interestingly, following DA depletion and the introduction of L-dopa, the expression of D_3 receptors in these neurons appears to be upregulated (BORDET et al. 1997).

In addition to medium spiny neurons, the neostriatum contains three groups of interneurons that have been defined on anatomical, histochemical, and electrophysiological grounds (KAWAGUCHI 1993). These are: (1) slow-firing, large cholinergic interneurons with prominent after-hyperpolarizations, (2) fast-spiking, parvalbumin-expressing GABAergic interneurons and (3) burst-firing, somatostatin/nitric-oxide-expressing interneurons. Each group constitutes 1%–3% of all neostriatal neurons. It has been difficult to generate an accurate picture of DA receptor expression in interneurons except for the cholinergic interneurons, which are readily identifiable by their large size. Early work using in situ hybridization suggested that cholinergic interneurons express D_2 receptors but not D_1 receptors (LeMoine et al. 1990). Single-cell RT-PCR analysis of cholinergic interneurons confirmed the presence of D_2 receptor mRNA (both short and long isoforms) but revealed that virtually all cholinergic interneurons co-express high levels of D_5 receptor mRNA (Yan et al. 1997; Yan and Surmeier 1997).

C. Dopaminergic Modulation of Intrinsic Properties of Neostriatal Neurons

Although our ultimate goal is to understand the consequences of DA receptor activation in vivo, there are substantial technical limitations to study in this setting. For mechanistic experiments, the advantages of in vitro preparations are well documented and include the ability to apply drugs at known concentrations, manipulate the extracellular ionic milieu, and obtain more stable electrophysiological recordings. In large part because of these technical advantages, studies in brain slices or acutely dissociated neostriatal cells using intracellular or whole-cell recording techniques have yielded a more consistent picture of the effects of DA than those observed in vivo. In fact, the results of these experiments can begin to explain many of the in vivo observations on the effects of DA.

In earlier studies, activation of D_1-class receptors in medium spiny neurons was reported to reduce the spike activity evoked by current injection (e.g., Uchimura et al. 1986; Akaike et al. 1987; Calabresi et al. 1987). In contrast, activation of D_2-class receptors was reported to excite medium spiny neurons (Akaike et al. 1987). More detailed, voltage-clamp studies of medium spiny neurons have put these observations on firm footing. As suggested by Calabresi et al. (1987) and later definitively shown by Surmeier et al (1992), D_1 receptor activation results in a reduction in depolarizing Na^+ currents in medium spiny neurons. This modulation is dependent upon a cyclic AMP/protein kinase A (PKA) signaling cascade that results in the phosphorylation of the pore-forming subunit of the channel (Schiffmann et al. 1995; Zhang et al. 1998).

In addition, D_1 receptor activation or application of cAMP analogs increases anomalous rectifier K^+ currents in medium spiny neurons (D.J. Surmeier and P.G. Mermelstein, unpublished observations; Pacheco-Cano et al. 1996) (see Fig. 1D). These currents are important determinants of the resting membrane potential of medium spiny neurons (see below) and are attributable to Kir2 family channels (Mermelstein et al. 1998). Interestingly, D_1 receptor/substance P-expressing medium spiny neurons have a distinct complement of Kir2 channels (Kir2.2/2.3) that do not inactivate with maintained hyperpolarization (Fig. 1A–C). The enhanced activity of these channels brought about by D_1 receptor activation should serve to suppress evoked activity from membrane potentials close to the potassium equilibrium potential.

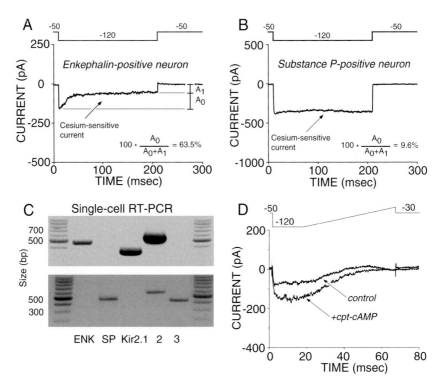

Fig. 1A–D. Inwardly rectifying K$^+$ current in medium spiny neurons are correlated with Kir2 mRNA expression and modulated by D$_1$ receptor stimulation. **A** An enkephalin-positive neuron, where a large proportion of the inwardly rectifying current inactivated during a voltage step from –50 mV to –120 mV. **B** Another neuron where the inward rectifier did not inactivate. This cell expressed substance P alone. **C** Single-cell RT-PCR profiles showing the correlation of Kir2 subunits with peptide expression. The ENK mRNA-positive neuron (*top*) expressed detectable levels of Kir2.1 and Kir2.2. On the other hand, the SP-positive neuron (*bottom*) expressed KIR2.2 and KIR2.3. KIR2.1 expression was only found in ENK-positive neurons while KIR2.3 expression was most often present in SP-positive cells. Data taken from MERMELSTEIN et al. 1998. **D** Activation of D$_1$ dopamine receptors or perfusion with cyclic adenosine monophosphate (*cAMP*) analogs increases inwardly rectifying K$^+$ currents in SP-expressing neurons

Voltage-dependent Ca^{2+} channels are also targets of the D$_1$ receptor signaling cascade. Medium spiny neurons express L-, N-, P-, Q- and R-type Ca^{2+} channels (BARGAS et al. 1994; CHURCHILL and MACVICAR 1998; MERMELSTEIN et al. 1999). Activation of D$_1$ receptors decreases N- and P/Q-type Ca^{2+} currents but increases L-type Ca^{2+} currents in medium spiny neurons (SURMEIER et al. 1995). The suppression of N- and P/Q-type Ca^{2+} channels appears to be dependent upon re-targeting of protein phosphatase 1 activity through a PKA-dependent mechanism (HUBBARD and COHEN 1993; FIENBERG et al. 1998). The modulation of L-type currents, on the other hand, is simpler. It takes place in neurons expressing the cardiac (class C) isoform of the α1 subunit of the L-type channel (SONG and SURMEIER 1996), making the modulation homologous to the well-described PKA modulation of Ca^{2+} channels in the heart (YUE et al. 1990).

The physiological consequences of activating D_2 receptors on medium spiny neurons are less well understood than the actions of D_1 receptors. Voltage-clamp studies have reported that activation of D_2-class receptors on medium spiny neurons can modulate Na^+, K^+, as well as Ca^{2+} channels, but these effects can vary from cell to cell. For example, SURMEIER et al. (1992) found that D_2 receptors could either enhance voltage-dependent Na^+ currents or suppress them. The biophysical signature of the observed enhancement was consistent with a reversal of D_1 receptor/adenylyl cyclase/PKA-mediated suppression of Na^+ currents (see above), as expected from biochemical studies (STOOF and KEBABIAN 1984). On the other hand, the suppression of currents involved a negative shift in the voltage-dependence of inactivation, without a change in peak conductance. These features resembled those reported for direct inhibition of Na^+ channels by G-protein $\beta\gamma$ subunits (MA et al. 1994). Understanding how these mechanisms might interact in shaping excitability has yet to be determined.

The effects of D_2 receptors on inwardly rectifying K^+ (IRK) channels are controversial. D_2 receptors have been reported both to activate a K^+ channel with weak inward rectification (FREEDMAN and WEIGHT 1988; FREEDMAN and WEIGHT 1989) and to suppress currents attributable to Kir2 channels (UCHIMURA and NORTH 1990). This latter observation is consistent with D_2 receptor-mediated inhibition of adenylyl cyclase and PKA enhancement of Kir2 currents (see above). D_2 receptors have also been found to enhance depolarization activated K^+ conductances in some medium spiny neurons (KITAI and SURMEIER 1992).

The origins of the apparent discrepancies in the regulation of Na^+ and K^+ channels are not clear at this point. One possibility is that there are different subtypes of medium spiny neuron with different D_2-class receptors or different ion channels. Another possibility is that the qualitative features of the modulation are dependent upon recording conditions. For example, recent work from one of our laboratories (HERNANDEZ-LOPEZ et al. 2000) suggests that D_2 receptors in medium spiny neurons are capable of mobilizing intracellular Ca^{2+} stores (VALLAR et al. 1990; TANG et al. 1994). Recordings that suppress this component of the D_2 response with intracellular Ca^{2+} chelators may observe a different pattern of effects than those that enable this component of the signaling chain. Although D_2 receptors may increase the release of Ca^{2+} from intracellular stores, transmembrane flux through voltage-dependent N-, P/Q-type Ca^{2+} channels is suppressed through a $G_{i/o}$-protein signaling pathway (HERNANDEZ-LOPEZ et al. 2000), as in heterologous systems (e.g., SEABROOK et al. 1994). Given this array of results, it is difficult to construct, at this point in time, a coherent model of how D_2 receptor activation modulates medium spiny neuron physiology.

The situation appears to be much simpler in cholinergic interneurons. DA has long been thought to directly modulate neostriatal cholinergic interneurons because of its impact on acetylcholine (ACh) release. D_2 receptor agonists have been reported to decrease ACh release (LEHMANN and LANGER 1983; STOOF et al. 1992) while D_1 receptor agonists enhance ACh release (DAMSMA et al. 1990; CONSOLO et al. 1992). Patch clamp studies have revealed that D_2 receptors negatively couple to N-type Ca^{2+} channels through a membrane-delimited signaling pathway, typical of that expected by G-protein βγ subunits (YAN et al. 1997). Given the involvement of N-type Ca^{2+} channels in transmitter release, these observations provide a cellular mechanism for the D_2 receptor mediated inhibition of ACh release. As mentioned above, cholinergic interneurons express D_5 but not D_1 DA receptors. Activation of these D_1-class receptors has been reported to enhance the responsiveness of Zn^{2+}-sensitive $GABA_A$ receptors through an adenylyl cyclase/PKA signaling pathway (YAN and SURMEIER 1997). D_5 receptor activation has also been found to enhance after-hyperpolarizing potentials, slowing discharge rates (BENNETT and WILSON 1998). Seemingly inconsistent with these two observations are reports that D_5 receptors depolarize cholinergic interneurons through a mechanism similar to substance P (AOSAKI et al. 1998), although this observation was not confirmed in a subsequent study (BENNETT and WILSON 1998). All in all, these studies argue that D_2 and D_5 receptor activation suppresses ACh release and the activity of cholinergic interneurons in a manner consistent with that seen in vivo (AOSAKI et al. 1994). They also argue that the ability of systemically applied D_1-class agonists to increase neostriatal ACh release is indirectly mediated (ABERCROMBIE and DEBOER 1997).

D. D_1 Receptor Modulation of Synaptic Transmission and Repetitive Activity in Medium Spiny Neurons

Medium spiny neurons receive excitatory synaptic inputs from several areas of the cerebral cortex (GRAYBIEL 1990). Corticostriatal afferents form synapses primarily with dendritic spines and dendritic shafts. Corticostriatal afferents form asymmetrical contacts with the heads of dendritic spines (SMITH and BOLAM 1990). Dopaminergic terminals make symmetrical contacts with necks of spines that also receive input, at their head, from corticostriatal glutamatergic terminals. The remainder of dopaminergic terminals form synapses with dendritic shafts and perikarya of spiny neurons. The close proximity of cortical glutamatergic and nigral dopaminergic terminals form an anatomical basis for possible physiological interactions at the postsynaptic level (KOETTER 1994).

To begin to understand the functional consequences of the dopaminergic modulation of ionic conductances and synaptic transmission in medium spiny neurons, it is necessary to review their electrophysiological behavior in an intact circuit. In vivo studies have shown that the stimulation of the sensorimotor cortex in the rat produces excitatory postsynaptic potentials (EPSPs) (HERRLING 1985; CALABRESI et al. 1990a). These EPSPs are mainly mediated by the activation of non-*N*-methyl-D-aspartate (NMDA) glutamate receptors (α-amino-3-hydroxy-5-methyl-4-isoxazoleprionic acid [AMPA] and/or kainate) while NMDA glutamate receptors do not seem to greatly contribute to the generation of EPSPs evoked by a single stimulus under physiological conditions in slices. In vivo, medium spiny neurons move between two membrane states, referred to as the "down-state" and the "up-state" (WILSON and KAWAGUCHI 1996). In the down-state, neurons are relatively hyperpolarized (ca. −85 mV) and do not generate action potentials. In response to temporally coherent, convergent excitatory synaptic input derived from the cortex, medium spiny neurons depolarize to the up-state during which the neurons sit at a membrane potential close to the threshold for spike generation (ca. −60 mV). Weaker, less temporally coherent synaptic input typically fails to trigger an up-state transition and the neuron falls back to the resting or down-state. The critical importance of excitatory synaptic input for the transition from the down- to the up-state is provided by the finding that decortication and disruption of thalamic efferents by transection abolished spontaneous transitions to the up-state in the neostriatum (WILSON 1993). Similarly, interruption of the hippocampal fibers that project to the nucleus accumbens is sufficient to prevent cells in this structure from entering the up-state (O'DONNELL and GRACE 1995). In vitro recordings obtained from corticostriatal slices do not show this oscillatory behavior, further supporting that an intact corticostriatal projection is required for its generation (CALABRESI et al. 1990b).

The movement between up- and down-states may be a critical determinant of the response to excitatory synaptic input (and plasticity). In vitro recordings performed in the presence of a physiological concentration of magnesium (1.2 mM) have shown that EPSPs recorded from neostriatal spiny neurons evoked by a single cortical activation are not altered by APV, a NMDA glutamate receptor antagonist, while they are blocked by CNQX, an antagonist of AMPA glutamate receptors. However, when the experiments were performed in the absence of external magnesium (to remove the voltage-dependent magnesium blockade of the NMDA receptor) the corticostriatal EPSP revealed a large NMDA-dependent component (CALABRESI et al. 1992a). These results show that NMDA receptors are present and potentially functional in medium spiny neurons. Because the magnesium block of NMDA channels is relieved by depolarization, in the up-state it will be possible for glutamatergic inputs to activate NMDA receptors. Conversely, when a similar input occurs during the hyperpolarized down-state, NMDA receptors should be largely blocked and the excitatory response will be dominated by AMPA glutamate receptors. This differential activation of the various subtypes of the ionotropic glutamate receptors during these two functional states of the spiny neurons will also have important consequences for the direction of the long-term changes of the synaptic excitability induced by repetitive cortical activation (see below).

Once an up-state transition occurs, neurons can stay in the up-state for a variable period of time, sometimes extending for a second or more (WILSON

1993). In this state, spikes can be generated in response to further depolarization. Recent work in dorsal neostriatum suggests that the state transitions of medium spiny neurons within functionally related microzones are highly correlated (STERN et al. 1998). Although medium spiny neurons are not intrinsically bistable and the transition from down- to up-state clearly is triggered by synaptic input, the probability of making a successful transition to the up-state, its kinetics, duration, and membrane potential envelope all are influenced by intrinsic membrane conductances. In the down-state, inwardly rectifying Kir2 channels dominate the membrane conductance, tending to hold the cell close to the K^+ equilibrium potential (WILSON and KAWAGUCHI 1996; MERMELSTEIN et al. 1998). In response to synaptic depolarization, voltage-dependent Na^+, K^+, and Ca^{2+} conductances come into play, determining the trajectory of the membrane potential (WILSON 1993; NISENBAUM et al. 1994). The enhancement of L-type Ca^{2+} channels by D_1 receptor activation may be important in promoting this bistability between up- and down-states (HERNANDEZ-LOPEZ et al. 1997).

Based upon the studies discussed, two predictions can be made about how DA will modulate the evoked activity of D_1 receptor expressing neurons. First, from negative membrane potentials that are governed by inwardly rectifying Kir2 channels, D_1 receptor activation will make it more difficult to evoke spikes by excitatory input. This is a consequence not only of the increased ability of Kir2 channels to clamp the membrane potential near the K^+ equilibrium potential but also the diminished capacity of voltage dependent Na^+ channels to boost sub-threshold depolarization to spike threshold. In current clamp experiments in slices, this is in fact what is seen (CALABRESI et al. 1987; HERNANDEZ-LOPEZ et al. 1997) (Fig. 2A). However, if medium spiny neurons succeed in making an up-state transition, Kir2 channels cease to be a factor because of their inward rectification and L-type Ca^{2+} channels begin to open. Thus, a second prediction is that when cells are in the up-state, D_1 receptor activation will depolarize the cell further by enhancing L-type Ca^{2+} currents and thus promote more vigorous spiking. Current clamp experiments where the up-state has been simulated by intracellular current injection have provided support for this prediction by showing that the response to depolarizing input and spike generation is in fact enhanced by D_1 receptor activation and augmentation of L-type Ca^{2+} currents (HERNANDEZ-LOPEZ et al. 1997). D_1 receptor mediated suppression of Na^+/K^+ adenosine triphosphatase (ATPase) should support this enhancement of evoked activity at depolarized membrane potentials (BERTORELLO et al. 1990).

Are these predictions born out in studies examining excitatory synaptic input to medium spiny neurons? This issue has yet to be fully resolved. First, in agreement with this model, D_1 receptor stimulation has been shown to enhance AMPA and NMDA receptor mediated EPSPs at depolarized membrane potentials through modulation of L-type Ca^{2+} channels (GALARRAGA et al. 1997; CEPEDA et al. 1998). On the other hand, D_1 receptor activation has been reported to decrease AMPA EPSPs at depolarized membrane potentials

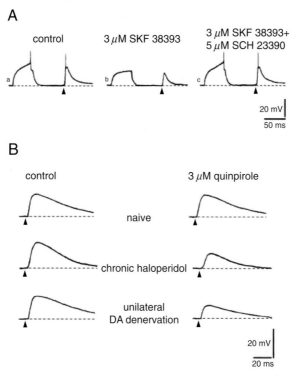

Fig. 2A,B. Effects of D_1 and D_2 DA receptor activation on synaptic transmission in the striatum. **A** The injection of depolarizing current pulse (0.9 nA) and the synaptic activation of corticostriatal fibers produced action potentials in a striatal neuron recorded in vitro. The application of the D_1 DA receptor agonist SKF 38393 blocked the firing discharge evoked by both current injection and synaptic stimulation (*middle*). This inhibitory action of SKF 38393 was antagonized by the co-administration of the D_1 DA receptor antagonist SCH 23390 (*right*). **B** In striatal neurons intracellularly recorded from naïve animals, the application of the D_2 DA receptor agonist quinpirole did not significantly affect the amplitude of corticostriatal EPSPs (*upper traces*). This pharmacological treatment, conversely, produced an inhibition of the EPSP amplitude in striatal neurons recorded after chronic haloperidol treatment (*middle traces*) or unilateral nigral lesion (*lower traces*). In this figure and in the following one, the *black triangle* indicates when the synaptic stimulation is delivered

(CALABRESI et al. 1987, 1988). This suppression could be mediated by D_1 receptor suppression of Na^+ and N/P/Q-type Ca^{2+} channels (SURMEIER et al. 1992, 1995). The biological bases for these apparently discrepant observations have yet to be worked out, but it may indicate that there are functional microdomains within the dendritic tree of medium spiny neurons that respond to D_1 receptor stimulation in different ways depending upon the relative density of Na^+ and Ca^{2+} channels.

Taken together, the experimental findings thus far suggest that D_1 receptor activation can suppress the response to weak excitatory synaptic input, making it more difficult to make the transition from down- to up-states. However, once the up-state transition has been achieved, D_1 receptor stimulation indirectly enhances NMDAR-mediated EPSPs by augmenting L-type Ca^{2+} currents activated by the synaptic input. The impact of D_1 receptors on AMPA receptor EPSPs remains to be fully worked out. Nevertheless, acting in this way, D_1 receptor stimulation may focus activity in only those neostriatal ensembles that receive highly convergent excitatory input, effectively decreasing system level noise. These results also explain the apparent paradox in the literature reporting that activation of D_1 receptors can either be excitatory or inhibitory. That is, when neurons are in the up-state, activation of D_1 receptors enhances the response to excitatory input. But in the down-state, activation of D_1 receptors suppresses the response to excitatory input. This view is largely consistent with a diverse in vivo literature showing a variety of effects on spontaneous and driven neostriatal activity (e.g., KIYATKIN and REBEC 1996). It should also be noted that changes in the strength of the excitatory input to medium spiny neurons due to the generation of long-term potentiation (LTP) or long-term depression (LTD) would likely also greatly influence the spatial and temporal dynamics of up- and down-state transitions.

E. Modulation of Synaptic Transmission by D_2-Like Receptors

D_2-class receptor agonists are currently used to attenuate parkinsonian symptoms. One major target of these ligands appears to be synaptic transmission in the neostriatum. However, there are fundamental disagreements about the conditions under which such a modulation is seen and its mechanisms. In the normosensitive neostriatum, D_2-class agonists have been reported to have a variety of effects. For example, D_2 receptor activation has been reported to depress excitatory synaptic responses mediated by AMPA receptors (Hsu et al. 1995; LEVINE et al. 1996b; UMEMIYA and RAYMOND 1997). This modulation was attributed to postsynaptic mechanisms because it was not accompanied by a change in paired-pulse facilitation and was mimicked during iontophoresis of glutamate (or AMPA) (LEVINE et al. 1996b; CEPEDA et al. 1992). Others have suggested a presynaptic site of action (Hsu et al. 1995). In other studies, activation of D_2-class receptors failed to cause significant changes in either AMPA-mediated synaptic potentials (CALABRESI et al. 1988, 1992b, 1993) or of exogenous applied AMPA and glutamate (CALABRESI et al. 1995). Recently, MALENKA's group (NICOLA and MALENKA 1998) has reported similar findings. In this study, D_2-class agonists did not modulate synaptic transmission in the dorsal neostriatum (where the previous studies were performed) but did observe a presynaptic inhibition of excitatory, glutamatergic synaptic transmission in the nucleus accumbens.

Following neostriatal DA depletion, modulation of synaptic transmission by D_2-class receptors is easier to observe (CALABRESI et al. 1988, 1992c, 1993). For example, in slices obtained from rats in which the neostriatal content of DA was decreased by reserpine treatment, D_2-class receptor agonists (bromocriptine and lisuride) and DA induced a significant depression of excitatory synaptic transmission. A similar inhibitory effect was observed in slices taken from rats chronically treatment with haloperidol (CALABRESI et al. 1992c) (Fig. 2B). In these animals, a presynaptic site of action was inferred from the inability of DA or D_2-class agonists to alter the response to exogenous glutamate. Similar results were obtained in rats unilaterally depleted of DA by nigral injection of 6-hydroxydopamine. In the DA-depleted neostriatum, D_2-class receptor agonists exerted a dose-dependent inhibition on corticostriatal EPSPs (CALABRESI et al. 1993) (Fig. 2B). There also was an increase in spontaneous excitatory depolarizing potentials. These potentials were reversibly inhibited by the D_2-class agonist quinpirole, suggesting that DA denervation increases neostriatal excitability in part by increasing glutamate release. The inference that presynaptic D_2 receptors are involved in this effect is consistent with studies showing that the frequency of 4-aminopyridine-induced spontaneous EPSPs is reduced by D_2-class agonists (FLORES-HERNÀNDEZ et al. 1997). By analogy with miniature EPSPs, it is assumed that alterations in the frequency of these synaptic events reflect a presynaptic site of action. The mechanisms underlying the change in intrastriatal D_2-class receptor function remain to be clearly determined. The modest changes in receptor number that occur following 6-hydroxydopamine (6-OHDA) lesioning or D_2 agonist treatment are unlikely to explain super-sensitivity (LAHOSTE and MARSHALL 1992). A redirection of D_2 receptor signaling is one possibility (THOMAS et al. 1992) as are alterations in the expression of regulatory RGS proteins that affect D_2 receptor signaling (INGI et al. 1998; BURCHETT et al. 1999; RAHMAN et al. 1999). The studies done to date have not effectively isolated the contribution of D_2 and D_4 receptors, both of which are expressed by cortical neurons projecting to the neostriatum (GASPAR et al. 1995; TARAZI et al. 1998). Studies in DA receptor knockout animals or the development of more selective pharmacological tools should help clarify the situation.

There is also growing appreciation that dendrites are not passive conductors of synaptic events but actively shape somatically conducted potentials (COLBERT et al. 1997; MAGEE et al. 1998; COOK and JOHNSTON 1999). This insight potentially has fundamental implications for our understanding of synaptic transmission and its modulation by DA. Given the overwhelming predominance of dendritic postsynaptic DA receptors in the neostriatum (e.g., DELLE DONNE et al. 1997), it will be interesting to see how our views of dopaminergic modulation change as we gain access to the dendrites of medium spiny neurons.

F. Dopaminergic Regulation of Corticostriatal Synaptic Plasticity

Long-term changes in synaptic transmission have served as a hypothetical substrate of memory storage and learning for over 30 years. Both LTP and LTD in synaptic strength have been described (JOHNSTON and WU 1995). Although experimental analysis of this type of synaptic plasticity has focused on hippocampal and cortical regions, it is now clear that similar changes take place in a variety of other brain structures – including the neostriatum.

One of the most important subcortical regions implicated in learning and memory is the basal ganglia. For example, positron emission tomography in humans (SEITZ et al. 1990) has suggested that storage of motor memory involves the basal ganglia. The corticostriatal projection also is thought to play a major part in learning (KIMURA and GRAYBIEL 1995). In humans, non-humans primates, and rodents, the neostriatum appears to participate in neural circuitry that is important for procedural learning (GRAYBIEL et al. 1994). LTP and LTD are potential substrates for this learning. Both LTP and LTD have been described at corticostriatal synapses (CALABRESI et al. 1992a,b, 1997a, 1998; LOVINGER and CHOI 1995; CHARPIER and DENIAU 1997; CHOI and LOVINGER 1997; CHARPIER et al. 1999). It has been hypothesized that these two forms of synaptic plasticity might represent the cellular mechanisms for the long-term changes in basal ganglia activity and motor learning (CALABRESI et al. 1996a).

DA appears to be a critical regulator of neostriatal LTD. In fact, endogenous DA appears to be necessary for the expression of neostriatal LTD– neostriatal slices obtained from 6-OHDA-lesioned rats do not express LTD (CALABRESI et al. 1992b). Interestingly, D_1- and D_2-class receptors may play a cooperative role in the expression of this form of synaptic plasticity. This hypothesis is supported by two kinds of experimental evidence (CALABRESI et al. 1992b). First, the induction of LTD is blocked either by D_1- or D_2-class receptor antagonists. Second, in DA-denervated slices, LTD can be restored by co-administration of D_1- and D_2-class receptor agonists, but not by the application of a single class of receptor agonist alone (Fig. 3). Co-activation of D_1 and D_2 receptors on the same medium spiny neuron (SURMEIER et al. 1992, 1996) may result in the activation of phospholipase A_2 (PIOMELLI et al. 1991) and subsequent activation of protein kinase C – enabling LTD. In agreement with this view, inhibition of phospholipase A_2 alters LTD (CALABRESI et al. 1992b).

More recently, DA receptor transgenic mice have been used to investigate the role of DA receptors in neostriatal synaptic plasticity. Mice in which D_2 receptors have been "knocked out" display partial motor impairments, as expected from previous studies (BAIK et al. 1995). In these animals, tetanic stimulation of corticostriatal fibers produced LTP in medium spiny neurons rather than the LTD seen in wild-type mice (CALABRESI et al. 1997b). This LTP was observed in the presence of physiological concentrations of magnesium, whereas this form of synaptic plasticity is inducible in slices wild-type rodents only in the absence of magnesium (CALABRESI et al. 1992a). The LTP observed in the D_2 receptor knockout mice was not coupled with major changes in the response to intracellular somatic current injection or resting membrane potential. Nor were there obvious changes in the response to exogenous NMDA or

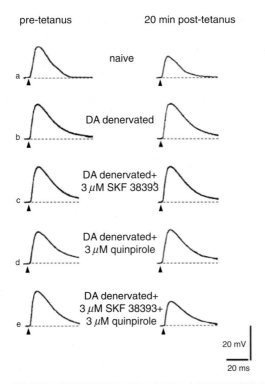

Fig. 3a–e. Effects of unilateral DA denervation on corticostriatal long-term depression (LTD). **a** In naïve animals, high-frequency stimulation (3 trains, 100 Hz, 3 s duration, 20 s interval) of corticostriatal fibers produced a LTD of excitatory transmission in the neostriatum. **b** LTD was absent in slices obtained from DA-denervated rats. **c** In these animals, the administration of the D_1 receptor agonist SKF 38393 did not re-establish LTD. **d** Application of the D_2 receptor agonist quinpirole was also ineffective. **e** This form of synaptic plasticity was present in DA-lesioned rats following the administration of both these DA agonists

the pharmacological properties of the corticostriatal EPSP evoked by a single stimulus. These latter observations seem to suggest that in D_2 receptor-deficient animals there is a synaptic abnormality that is revealed only when the corticostriatal projection is activated in a repetitive manner. As we have previously reported, in magnesium-free medium, LTP was seen in wild-type mice and found to be enhanced by L-sulpiride, whereas it was reversed into LTD by quinpirole, a D_2 receptor agonist. In D_2 receptor-null mice, this modulation was lost. These findings suggest that D_2 receptors play a key role in mechanisms underlying the direction of long-term changes in synaptic efficacy in the neostriatum.

In the 6-OHDA-lesioned neostriatum, tetanic stimulation in a magnesium-free medium fails to induce LTP (CALABRESI et al. 1997b). In light of the inference from the work with D_2 receptor knockout mice, that these receptors suppress the induction of neostriatal LTP, these results are rather surprising. However, DA denervation obtained by unilateral nigral lesion undoubtedly produces other changes that may alter the properties of LTP. In fact, morphological studies have shown that after 6-OHDA denervation, dendritic spines of neostriatal neurons are numerically reduced and abnormal in size and shape (INGHAM et al. 1989; NITSCH and RIESENBERG 1995). Similar data were found in brains of parkinsonian patients (MCNEILL et al. 1988). Dendritic spines are a principal site of long-term synaptic modifications in other CNS areas (DESMOND and LEVY 1990) and may serve a similar function in medium spiny neurons (CALABRESI et al. 1996b, 1997b). Thus, the remodeling of these neuronal structures might cause an impairment of corticostriatal LTP.

G. Summary and Conclusions

In recent years, dramatic progress has been made in unraveling the physiological impact of DA in the neostriatum. The application of voltage-clamp techniques and single-cell RT-PCR has begun to provide a much clearer understanding of the intrinsic mechanisms by which DA controls excitability. To be sure, controversies still exist but these appear to be dissipating. Nowhere is this more evident than in studies of D_1 receptor effects on the principal neurons of the neostriatum – the medium spiny neuron. In these neurons, D_1 receptors exquisitely orchestrate the modulation of a variety of ion channels. Rather than leading to simply inhibition or excitation, this pattern of modulation leads to suppression in responsiveness in the quiescent down-state but an enhancement of responsiveness in the active up-state. The consequences of this state-dependent change have yet to be fully worked out but they suggest that one of the principal actions of DA at D_1 receptors is to focus neuronal activity in those ensembles that receive the most temporally and spatially convergent excitatory input.

In spite of the advances made in the last few years, much remains to be done in the cellular and molecular analysis of dopaminergic modulation of excitability. For example, relatively little is known about the linkages of D_2 DA receptors in medium spiny neurons. Studies from one of our labs show that these receptors utilize a novel linkage to intracellular Ca^{2+} metabolism (HERNANDEZ-LOPEZ 2000). Obviously, intracellular Ca^{2+} is a key determinant not only of short-term regulation of ion channels but of longer-term changes in synaptic plasticity and gene expression. Tracing the cellular consequences of this linkage will undoubtedly lead to as rich and interesting a story as has begun to unfold for D_1 receptors in medium spiny neurons.

The advances in understanding the intrinsic integrative mechanisms is paralleled by our rapidly growing understanding of DA's role in regulating synaptic transmission and plasticity. Although considerably more complicated, it is clear that both D_1 and D_2 DA receptors are key determinants of corticostriatal transmission. Moreover, activation of both receptors is necessary for the induction of LTD in the neostriatum. Work in D_2 knockout mice has shown that D_2 receptors also regulate the induction of LTP. What remains to be determined is precisely how DA receptors interact in promoting these two forms of synaptic plasticity.

Acknowledgements. This work was supported by USPHS NIH grants NS 34696 and DA 12958 to DJS and BIOMED grant (BMH4-97-2215) to PC.

References

Abercrombie ED, DeBoer P (1997) Substantianigra D1 receptors and stimulation of striatal cholinergic interneurons by dopamine: A proposed circuit mechanism. J Neurosci 17:8498–505

Akaike A, Ohno Y, Sasa M, Takaori S (1987) Excitatory and inhibitory effects of dopamine on neuronal activity of the caudate neurons in vitro. Brain Res 418:262–272

Albin RL, Young AB, Penney JB (1989) The functional anatomy of basal ganglia disorders. TINS 12:366–375

Alexander GE, DeLong MR, Strick PL (1986) Parallel organization of functionally segregated circuits linking basal ganglia and cortex. Ann Rev Neurosci 9:357–381

Aosaki T, Kiuchi K, Kawaguchi Y (1998) Dopamine D1-like receptor activation excites rat striatal large aspiny neurons in vitro. J Neurosci 18:5180–5190

Aosaki T, Tsubokawa H, Ishida A, Watanabe K, Graybiel AM, Kimura M (1994) Responses of tonically active neurons in the primate's striatum undergo systematic changes during behavioral sensorimotor conditioning. J Neurosci 14:3969–3984

Baik J-H, Picetti R, Saiardi A et al (1995) Parkinsonian-like locomotor impairment in mice lacking dopamine D2 receptors. Nature 377:424–428

Bargas J, Howe A, Eberwine J, Cao Y, Surmeier DJ (1994) Cellular and molecular characterization of Ca2+ currents in acutely-isolated, adult rat neostriatal neurons. J Neurosci 14:6667–6686

Bennett BD, Wilson CJ (1998) Synaptic regulation of action potential timing in neostriatal cholinergic interneurons. J Neurosci 18:8539–8549

Bergson C, Mrzljak L, Smiley JF, Pappy M, Levenson R, Goldman-Rakic PS (1995) Regional, cellular and subcellular variation in the distribution of D1 and D5 dopamine receptors in primate brain. J Neurosci 15:7821–7836

Berke JD, Paletzki RF, Aronson GJ, Hyman SE, Gerfen CR (1998) A complex program of striatal gene expression induced by dopaminergic stimulation. J Neurosci 18: 5301–5310

Bernardi G, Marciani MG, Morocutti C, Pavone F, Stanzione P (1978) The action of dopamine on rat caudate neurons intracellularly recorded. Neurosci Lett 8:235–240

Bertorello AM, Hopfield JF, Aperia A, Greengard P (1990) Inhibition by dopamine of (Na-K) ATPase activity in neostriatal neurons through D1 and D2 dopamine receptor synergism. Nature 347:386–388

Bliss TVP, Lømo T (1973) Long-lasting potentiation of synaptic transmission in the dentate area of anaesthetized rabbit following stimulation of the perforant path. J Physiol 232:331–356

Bordet R, Ridray S, Carboni S, Diaz J, Sokoloff P, Schwartz JC (1997) Induction of dopamine D3 receptor expression as a mechanism of behavioral sensitization to levodopa. Proc Natl Acad Sci USA 94:3363–3367

Burchett SA, Bannon MJ, Granneman JG (1999) RGS mRNA expression in rat striatum: modulation by dopamine receptors and effects of repeated amphetamine administration. J Neurochem 72:1529–1533

Calabresi P, Benedetti M, Mercuri NB, Bernardi G (1988) Endogenous dopamine and dopaminergic agonists modulate synaptic excitation in neostriatum: intracellular studies from naive and catecholamine-depleted rats. Neurosci 27:145–157

Calabresi P, Centonze D, Gubellini P, Pisani A, Bernardi G (1998) Blockade of M2-like muscarinic receptors enhances long-term potentiation at corticostriatal synapses [In Process Citation]. Eur J Neurosci 10:3020–3023

Calabresi P, De Murtas M, Bernardi G (1997a) The neostriatum beyond the motor function: experimental and clinical evidence. Neuroscience 78:39–60

Calabresi P, De Murtas M, Mercuri NB, Bernardi G (1992d) Chronic neuroleptic treatment: D2 dopamine receptor supersensitivity and striatal glutamatergic transmission. Ann Neurol 31:366–373

Calabresi P, De Murtas M, Pisani A, Stefani A, Sancesario G, Mercuri NB, Bernardi G (1995) Vulnerability of medium spiny striatal neurons to glutamate: role of Na/K ATPase. Eur J Neurosci 7:1674–1683

Calabresi P, Maj R, Pisani A, Mercuri NB, Bernardi G (1992b) Long-term synaptic depression in the striatum: physiological and pharmacological characterization. J Neurosci 12:4224–4233

Calabresi P, Maj R, Mercuri NB, Bernardi G (1992a) Coactivation of D1 and D2 dopamine receptors is required for long-term synaptic depression in the striatum. Neurosci Lett 142:95–99

Calabresi P, Mercuri NB, Bernardi G (1990a) Synaptic and intrinsic control of membrane excitability of neostriatal neurons. I. An in vivo analysis. J Neurophysiol 63:651–662

Calabresi P, Mercuri NB, Bernardi G (1990b) Synaptic and intrinsic control of membrane excitability of neostriatal neurons. II. An in vitro analysis. J Neurophysiol 63:663–675

Calabresi P, Mercuri N, Stanzione P, Stefani A, Bernardi G (1987) Intracellular studies on the dopamine-induced firing inhibition of neostriatal neurons in vitro: Evidence for D1 receptor involvement. Neurosci 20:757–771

Calabresi P, Pisani A, Centonze D, Bernardi G (1996b) Role of Ca^{2+} in striatal LTD and LTP. Semin Neurosci 8:321–328

Calabresi P, Pisani A, Mercuri NB, Bernardi G (1992c) Long-term potentiation in the striatum is unmasked by removing the voltage-dependent blockade of NMDA receptor channel. Eur J Neurosci 4:929–935

Calabresi P, Pisani A, Mercuri NB, Bernardi G (1996a) The corticostriatal projection: from synaptic plasticity to basal ganglia disorders. Trends Neurosci 19:19–24

Calabresi P, Saiardi A, Pisani A, Baik JH, Centonze D, Mercuri NB, Bernardi G, Borrelli E (1997b) Abnormal synaptic plasticity in the striatum of mice lacking dopamine D2 receptors. J Neurosci 17:4536–4544

Cepeda C, Buchwald NA, Levine MS (1992) Neuromodulatory actions of dopamine in the neostriatum are dependent upon the excitatory amino acid receptor subtypes activated. Proc Natl Acad Sci USA 90:9576–9580

Cepeda C, Colwell CS, Itri JN, Chandler SH, Levine MS (1998) Dopaminergic modulation of NMDA-induced whole cell currents in neostriatal neurons in slices: contribution of calcium conductances. J Neurophysiol 79:82–94

Chang HT, Kitai ST (1985) Projection neurons of the nucleus accumbens: an intracellular labeling study. Brain Res 347:112–116

Chang HT, Wilson CJ, Kitai ST (1982) A Golgi study of rat neostriatal neurons: Light microscopic analysis. J Comp Neurol 208:107–126

Charpier S, Deniau JM (1997) In vivo activity-dependent plasticity at cortico-striatal connections: evidence for physiological long-term potentiation. Proc Natl Acad Sci U S A 94:7036–7040

Charpier S, Mahon S, Deniau JM (1999) In vivo induction of striatal long-term potentiation by low-frequency stimulation of the cerebral cortex [In Process Citation]. Neuroscience 91:1209–1222

Choi S, Lovinger DM (1997) Decreased frequency but not amplitude of quantal synaptic responses associated with expression of corticostriatal long-term depression. J Neurosci 17:8613–8620

Churchill D, MacVicar BA (1998) Biophysical and pharmacological characterization of voltage-dependent Ca2+ channels in neurons isolated from rat nucleus accumbens. J Neurophysiol 79:635–647

Colbert CM, Magee JC, Hoffman DA, Johnston D (1997) Slow recovery from inactivation of Na+ channels underlies the activity- dependent attenuation of dendritic action potentials in hippocampal CA1 pyramidal neurons. J Neurosci 17:6512–6521

Consolo S, Girotti P, Russi G, Di Chiara G (1992) Endogenous dopamine facilitates striatal in vivo acetylcholine release by acting on D1 receptors localized in the striatum. J Neurochem 59:1555–1557 issn:0022–3042

Cook EP, Johnston D (1999) Voltage-dependent properties of dendrites that eliminate location-dependent variability of synaptic input. J Neurophysiol 81:535–543

Damsma G, Tham CS, Robertson GS, Fibiger HC (1990) Dopamine D1 receptor stimulation increases striatal acetylcholine release in the rat. Eur J Pharmacol 186:335–338

Delle Donne KT, Sesack SR, Pickel VM (1997) Ultrastructural immunocytochemical localization of the dopamine D2 receptor within GABAergic neurons of the rat striatum. Brain Res 746:239–255

Desmond NL, Levy WB (1990) Morphological correlates of long-term potentiation imply the modification of existing synapses, not synaptogenesis, in the hippocampal dentate gyrus. Synapse 5:139–143

Erenberg G (1992) Treatment of Tourette syndrome with neuroleptic drugs. Adv Neurol 58:241–243

Fienberg AA, Hiroi N, Mermelstein PG, Song W, Snyder GL, Nishi A, Cheramy A, O'Callaghan JP, Miller DB, Cole DG, Corbett R, Haile CN, Cooper DC, Onn SP, Grace AA, Ouimet CC, White FJ, Hyman SE, Surmeier DJ, Girault J, Nestler EJ, Greengard P (1998) DARPP-32: regulator of the efficacy of dopaminergic neurotransmission. Science 281:838–842

Flores-Hernandez J, Galarraga E, Bargas J (1997) Dopamine selects glutamatergic inputs to neostriatal neurons. Synapse 25:185–195

Freedman JE, Weight FF (1988) Single K+ channels activated by D2 dopamine receptors in acutely dissociated neurons from rat corpus striatum. Proc Natl Acad Sci (USA) 85:3618–3622

Freedman JE, Weight FF (1989) Quinine potently blocks single K+ channels activated by dopamine D-2 receptors in rat corpus striatum neurons. Eur J Pharmacol 164:341–346

Freund TF, Powell JF, Smith AD (1984) Tyrosine hydroxylase-immunoreactive boutons in synaptic contact with identified striatonigral neurons, with particular reference to dendritic spines. Neuroscience 13:1189–1215

Galarraga E, Hernandez-Lopez S, Reyes A, Barral J, Bargas J (1997) Dopamine facilitates striatal EPSPs through an L-type Ca2+ conductance. Neuroreport 8:2183–2186

Gaspar P, Bloch B, Le Moine C (1995) D1 and D2 receptor gene expression in the rat frontal cortex: cellular localization in different classes of efferent neurons. Eur J Neurosci 7:1050–1063

Gerfen CR (1992) The neostriatal mosaic: multiple levels of compartmental organization in the basal ganglia. Ann Rev Neurosci 15:285–320

Graybiel AM (1990) Neurotransmitters and neuromodulators in the basal ganglia. TINS 13:244–254

Graybiel AM, Aosaki T, Flaherty AW, Kimura M (1994) The basal ganglia and adaptive motor control. Science 265:1826–1831

Groves PM (1983) A theory of the functional organization of the neostriatum and the neostriatal control of voluntary movement. Brain Res 286:109–132

Harvey J, Lacey MG (1996) Endogenous and exogenous dopamine depress EPSCs in rat nucleus accumbens in vitro via D1 receptors activation. J Physiol (Lond) 492:143–154

Harvey J, Lacey MG (1997) A postsynaptic interaction between dopamine D1 and NMDA receptors promotes presynaptic inhibition in the rat nucleus accumbens via adenosine release. J Neurosci 17:5271–580

Herlitze S, Garcia DE, Mackie K, Hille B, Scheuer T, Catterall WA (1996) Modulation of Ca2+ channels by G-protein beta gamma subunits. Nature 380:258–262

Hernandez-Lopez S, Bargas J, Surmeier DJ, Reyes A, Galarraga E (1997) D1 receptor activation enhances evoked discharge in neostriatal medium spiny neurons by modulating an l-type Ca2+ conductance. J Neurosci 17:3334–3342

Hernandez-Lopez S, Tkatch T, Perez-Garci E, Galarraga E, Bargas J, Hamm H, and Surmeier DJ (2000) D_2 dopamine receptors in striatal medium spiny neurons reduce L-type Ca^{2+} currents and excitability via a novel PLCbeta1-IP3-calcineurin-signaling cascade. J Neurosci 20:8987–8995

Herrling PL (1985) Pharmacology of the corticocaudate excitatory postsynaptic potential in the cat: evidence for its mediation by quisqualate- or kainate- receptors. Neuroscience 14:417–426

Hervé D, Levi-Strauss M, Marey-Semper I, Verney C, Tassin J-P, Glowinski J, Girault J-A (1993) Golf and Gs in rat basal ganglia: Possible involvement of Golf in the coupling of dopamine D1 receptor with adenylyl cyclase. J Neurosci 13:2237–2248

Hsu KS, Huang CC, Yang CH, Gean PW (1995) Presynaptic D2 dopaminergic receptors mediate inhibition of excitatory synaptic transmission in rat neostriatum. Brain Res 690:264–268

Hubbard MJ, Cohen P (1993) On target with a new mechanism for the regulation of protein phosphorylation. TIBS 18:172–176

Ikeda SR (1996) Voltage-dependent modulation of N-type calcium channels by G-prote $\beta\gamma$ subunits. Nature 380:255–258

Ingham CA, Hood SH, Arbuthnott GW (1989) Spine density on neostriatal neurones changes with 6-hydroxydopamine lesions and with age. Brain Res 503:334–338

Ingi T, Krumins AM, Chidiac P, Brothers GM, Chung S, Snow BE, Barnes CA, Lanahan AA, Siderovski DP, Ross EM, Gilman AG, Worley PF (1998) Dynamic regulation of RGS2 suggests a novel mechanism in G-protein signaling and neuronal plasticity. J Neurosci 18:7178–7188

Ito M (1989) Long-term depression. Annu Rev Neurosci 12:85–102

Johnston D, Wu SM-S (1995) Foundations of Cellular Neurophysiology. Bradford, Cambridge, MA

Kawaguchi Y (1993) Physiological, morphological, and histochemical characterization of three classes of interneurons in rat neostriatum. J Neurosci 13:4908–4923

Kawaguchi Y, Wilson CJ, Emson PC (1990) Projection subtypes of rat neostriatal matrix cells revealed by intracellular injection of biocytin. J Neurosci 10:3421–3438

Kimura M, Graybiel AM (1995) Role of basal ganglia in sensory motor association learning. In: Kimura M, Graybiel AM (eds) Functions of the cortico-basal ganglia loop. Springer, Tokyo, pp 2–17

Kitai ST, Sugimori M, Kocsis JD (1976) Excitatory nature of dopamine in the nigro-caudate pathway. Exp Brain Res 24:351–363

Kitai ST, Surmeier DJ (1992) Cholinergic and dopaminergic modulation of potassium conductances in neostriatal neurons. In: Advances in Neurology (Narabayashi H, Nagatsu T, Yanagisawa N, Mizuno Y, eds.), Raven, New York, pp 40–52

Kiyatkin EA, Rebec GV (1996) Dopaminergic modulation of glutamate-induced excitations of neurons in the neostriatum and nucleus accumbens of awake, unrestrained rats. J Neurophysiol 75:142–153

Koetter R (1994) Postsynaptic integration of glutamatergic and dopaminergic signals in the striatum. Prog Neurobiol 44:163–196

Koob GF, Nestler EJ (1997) The neurobiology of drug addiction. J Neuropsychiatry Clin Neurosci 9:482–497

LaHoste GJ, Marshall JF (1992) Dopamine supersensitivity and D1/D2 synergism are unrelated to changes in striatal receptor density. Synapse 12:14–26

Lehmann J, Langer SZ (1983) The striatal cholinergic interneuron: Synaptic target of dopaminergic terminals? Neurosci 10:1105–1120

LeMoine C, Tison F, Bloch B (1990) D2 dopamine receptor gene expression by cholinergic neurons in the rat striatum. Neurosci Lett 117:248–252

Levine MS, Altemus KL, Cepeda C, Cromwell HC, Crawford C, Ariano MA, Drago J, Sibley DR, Westphal H (1996a) Modulatory actions of dopamine on NMDA receptor-mediated responses are reduced in D1A-deficient mutant mice. J Neurosci 16:5870–5882

Levine MS, Li Z, Cepeda C, Cromwell HC, Altemus KL (1996b) Neuromodulatory actions of dopamine on synaptically-evoked neostriatal responses in slices. Synapse 24:65–78

Lovinger DM, Choi S (1995) Activation of adenosine A1 receptors initiates short-term synaptic depression in rat striatum. Neurosci Lett 199:9–12

Ma JY, Li M, Catterall WA, Scheuer T (1994) Modulation of brain Na+ channels by a G-protein-coupled pathway. Proc Natl Acad Sci U S A 91:12351–12355

Magee J, Hoffman D, Colbert C, Johnston D (1998) Electrical and calcium signaling in dendrites of hippocampal pyramidal neurons. Annu Rev Physiol 60:327–346

Maura G, Giardi A, Raiteri M (1988) Release-regulating D2 dopamine receptors are located on striatal glutamatergic nerve terminals. J Pharmacol Exp Ther 247: 680–684

McNeill TH, Brown SA, Rafols JA, Shoulson I (1988) Atrophy of medium spiny I striatal dendrites in advanced Parkinson's disease. Brain Res 445:148–152

Mermelstein PG, Song WJ, Tkatch T, Yan Z, Surmeier DJ (1998) Inwardly rectifying potassium (IRK) currents are correlated with IRK subunit expression in rat nucleus accumbens medium spiny neurons. J Neurosci 18:6650–6661

Mermelstein PG, Tkatch T, Song W-J, Foehring RC, Surmeier DJ (1999) Properties of Q-type calcium channels in striatal and cortical neurons are correlated with b subunit expression. J Neurosci (in press)

Nemeroff CB, Bissette G (1988) Neuropeptides, dopamine and schizophrenia. Ann N Y Acad Sci 537:273–291

Nicola SM, Kombian SB, Malenka RC (1996) Psychostimulants depress excitatory synaptic transmission in the nucleus accumbens via presynaptic D1-like dopamine receptors. J Neurosci 16:1591–1604

Nicola SM, Malenka RC (1997) Dopamine depresses excitatory and inhibitory synaptic transmission by distinct mechanisms in the nucleus accumbens. J Neurosci 17:5697–5710

Nicola SM, Malenka RC (1998) Modulation of synaptic transmission by dopamine and norepinephrine in ventral but not dorsal striatum. J Neurophysiol 79:1768–1776

Nisenbaum ES, Xu ZC, Wilson CJ (1994) Contribution of a slowly inactivating potassium current to the transition to firing of neostriatal spiny projection neurons. J Neurophysiol 71:1174–1189

Nitsch C, Riesenberg R (1995) Synaptic reorganisation in the rat striatum after dopaminergic deafferentation: an ultrastructural study using glutamate decarboxylase immunocytochemistry. Synapse 19:247–263

Pacheco-Cano MT, Bargas J, Hernandez-Lopez S, Tapia D, Galarraga E (1996) Inhibitory action of dopamine involves a subthreshold Cs(+)-sensitive conductance in neostriatal neurons Exp Brain Res 110:205–211

Pennartz CM, Ameerun RF, Groenewegen HJ, Lopes da Silva FH (1993) Synaptic plasticity in an in vitro slice preparation of the rat nucleus accumbens. Eur J Neurosci 5:107–117

Piomelli D, Pilon C, Giros B, Sokoloff P, Martres MP, Schwartz JC (1991) Dopamine activation of the arachidonic acid cascade as a basis for D1/D2 receptor synergism Nature 353:164–167

Rahman Z, Gold SJ, Potenza MN, Cowan CW, Ni YG, He W, Wensel TG, Nestler EJ (1999) Cloning and characterization of RGS9-2: a striatal-enriched alternatively spliced product of the RGS9 gene. J Neurosci 19:2016–2026

Schiffmann SN, Lledo PM, Vincent JD (1995) Dopamine D1 receptor modulates the voltage-gated sodium current in rat striatal neurones through a protein kinase A. J Physiol (Lond) 483:95–107

Seabrook GR, McAllister G, Knowles MR, Myers J, Sinclair H, Patel S, Freedman SB, Kemp JA (1994) Depression of high, threshold calcium currents by activation of human D2 (short) dopamine receptors expressed in differentiated NG108 (15 cells. Brit J Pharmacol 111:1061–1066

Seitz RJ, Roland E, Bohm C, Greitz T, Stone-Elander S (1990) Motor learning in man: a positron emission tomographic study. Neuroreport 1:57–60

Sidhu A (1998) Coupling of D1 and D5 dopamine receptors to multiple G proteins: Implications for understanding the diversity in receptor-G protein coupling [In Process Citation]. Mol Neurobiol 16:125–134

Smith AD, Bolam JP (1990) The neural network of the basal ganglia as revealed by the study of synaptic connections of identified neurones. Trends Neurosci 13:259–265

Song W-J, Surmeier DJ (1996) Voltage-dependent facilitation of calcium currents in rat neostriatal neurons. J Neurophysiol 76:2290–2306

Stanton PK, Sejnowski TJ (1989) Associative long-term depression in the hippocampus induced by hebbian covariance. Nature 339:215–218

Stern EA, Jaeger D, Wilson CJ (1998) Membrane potential synchrony of simultaneously recorded striatal spiny neurons in vivo. Nature 394:475–478

Stoof JC, Drukarch B, de Boer P, Westerink BH, Groenewegen HJ (1992) Regulation of the activity of striatal cholinergic neurons by dopamine. Neuroscience 47:755–770

Stoof JC, Kebabian JW (1984) Two dopamine receptors: Biochemistry, physiology and pharmacology. Life Sci 35:2281–2296

Surmeier DJ, Bargas J, Hemmings HC, Jr, Nairn AC, Greengard P (1995) Modulation of calcium currents by a D1 dopaminergic protein kinase/phosphatase cascade in rat neostriatal neurons. Neuron 14:385–397

Surmeier DJ, Eberwine J, Wilson CJ, Stefani A, Kitai ST (1992) Dopamine receptor subtypes co-localize in rat striatonigral neurons. Proc Natl Acad Sci (USA) 89: 10178–10182

Surmeier DJ, Song WJ, Yan Z (1996) Coordinated expression of dopamine receptors in neostriatal medium spiny neurons. J Neurosci 16:6579–6591

Tang L, Todd RD, Heller A, O'Malley KL (1994) Pharmacological and functional characterization of D2, D3 and D4 dopamine receptors in fibroblast and dopaminergic cell lines [published erratum appears in J Pharmacol Exp Ther 1994 Sep;270(3):1397]. J Pharmacol Exp Ther 268:495–502

Tarazi FI, Campbell A, Yeghiayan SK, Baldessarini RJ (1998) Localization of dopamine receptor subtypes in corpus striatum and nucleus accumbens septi of rat brain: comparison of D1-, D2-, and D4- like receptors. Neuroscience 83:169–176

Teyler TJ, Discenna P (1984) Long-term potentiation as a candidate mnemonic device. Brain Res 319:15–28

Thomas KL, Rose S, Jenner P, Marsden CD (1992) Dissociation of the striatal D-2 dopamine receptor from adenylyl cyclase following 6-hydroxydopamine-induced denervation. Biochem Pharmacol 44:73–82

Uchimura N, Higashi H, Nishi S (1986) Hyperpolarizing and depolarizing actions of dopamine via D-1 and D-2 receptors on nucleus accumbens neurons. Brain Res 375:368–372

Uchimura N, North RA (1990) Actions of cocaine on rat nucleus accumbens neurones in vitro. Br J Pharmacol 99:736–740

Vallar L, Muca C, Magni M, Albert P, Bunzow J, Meldolesi J, Civelli O (1990) Differential coupling of dopaminergic D2 receptors expressed in different cell types. Stimulation of phosphatidylinositol 4, 5-bisphosphate hydrolysis in LtK- fibroblasts, hyperpolarization, and cytosolic-free Ca2+ concentration decrease in GH4C1 cells J Biol Chem 265:10320–10326

Verma A, Moghaddam B (1998) Regulation of striatal dopamine release by metabotropic glutamate receptors. Synapse 28:220–226

Westenbroek RE, Hell JW, Warner C, Dubel SJ, Snutch TP, Catterall WA (1992) Biochemical properties and subcellular distribution of an N-type calcium channel a1 subunit. Neuron 9:1099–1115

Wilson CJ (1993) The generation of natural firing patterns in neostriatal neurons. In: Arbuthnott GW, Emson PC (eds) Chemical Signaling in the Basal Ganglia, Elsevier, Amsterdam, pp 277–297

Wilson CJ, Kawaguchi Y (1996) The origins of two-state spontaneous membrane potential fluctuations of neostriatal spiny neurons. J Neurosci 16:2397–2410

Wise RA (1998) Drug-activation of brain reward pathways. Drug Alcohol Depend 51:13–22

Wooten GF (1990) Parkinsonism. In: Pearlman AL, Collins RC (eds) Neurobiology of Disease, Oxford U. Press, New York, pp 454–468

Yan Z, Song WJ, Surmeier J (1997) D2 dopamine receptors reduce N-type Ca2+ currents in rat neostriatal cholinergic interneurons through a membrane-delimited, protein-kinase-C- insensitive pathway. J Neurophysiol 77:1003–1015

Yan Z, Surmeier DJ (1997) D5 dopamine receptors enhance Zn2+-sensitive GABA(A) currents in striatal cholinergic interneurons through a PKA/PP1 cascade. Neuron 19:1115–1126

Yue DT, Herzig S, Marban E (1990) Beta-adrenergic stimulation of calcium channels occurs by potentiation of high-activity gating modes. Proc Natl Acad Sci (USA) 87:753–757

Zhang XF, Hu XT, White FJ (1998) Whole-cell plasticity in cocaine withdrawal: reduced sodium currents in nucleus accumbens neurons. J Neurosci 18:488–498

CHAPTER 12
Dopamine and Gene Expression

E.J. NESTLER

A. Introduction

When dopamine is released from dopaminergic neurons at the synapse, it quickly binds to its receptors that are located on dendrites or nerve terminals of target neurons as well as on dopaminergic neurons themselves. Interactions between dopamine and its receptors then leads rapidly (from a few hundred milliseconds to several seconds) to electrophysiological changes in those target neurons. These electrophysiological changes mediate the acute effects of dopaminergic transmission on the functioning of neural circuits and thereby on behavior.

The electrophysiological changes elicited by activation of dopamine receptors are mediated by cascades of postreceptor, intracellular messenger pathways. Perturbation of these cascades by dopamine, in addition to leading to very rapid changes in the electrical excitability of neurons, also mediates large numbers of adaptive responses in the neurons. These adaptive responses can also occur relatively rapidly, but can persist for longer periods of time than the dopamine signal outside of the neuron. Among the many types of adaptive responses that are elicited by dopamine is the regulation of gene expression. Such changes in gene expression would be expected to occur over hours or days, and likely reflect the net level of dopaminergic transmission over relatively prolonged periods of time. The changes also would be expected to be particularly stable and would therefore appear to mediate longer-lasting adaptive responses to the dopamine signal. In some sense, then, dopamine's effects on gene expression can be viewed as the ultimate endpoint of dopamine's signal transduction pathways.

B. Dopamine Signaling to the Nucleus
I. Overview of Dopamine's Signal Transduction Pathways

Dopamine produces its biological effects via the activation of two main classes of dopamine receptors, termed D_1-like and D_2-like. D_1-like receptors include the D_1 and D_5 receptors; D_2-like receptors include the D_2 (which exists in two alternatively spliced forms), D_3, and D_4 receptors (MISSALE et al. 1998; SIBLEY 1999).

D_1-like receptors produce their biological effects predominantly via the activation of the cAMP (cyclic adenosine monophosphate) pathway. The receptors, via coupling with the G protein, G_s, activate adenylyl cyclase, the enzyme that catalyzes the synthesis of cAMP; cAMP then activates protein kinase A, which produces a myriad of biological effects via the phosphorylation of diverse types of neuronal proteins (NESTLER and GREENGARD 1999). For example, phosphorylation of ion channels is thought to mediate the rapid electrophysiological effects of D_1 receptor activation on target neurons (SURMEIER et al. 1995).

In contrast, D_2-like receptors produce their biological effects via coupling to the $G_{i/o}$ family of G proteins. This coupling leads to the activation of inwardly rectifying K^+ channels and inhibition of voltage-gated Ca^{2+} channels via the G protein $\beta\gamma$ subunits (NESTLER and DUMAN 1999). The coupling also leads to inhibition of adenylyl cyclase and, as a result, reduced activation of protein kinase A, which could further affect the activity of specific ion channels.

These initial effects of dopamine receptor activation on G proteins, the cAMP pathway, and ion channels would set in stage many additional effects on the target neurons. Altered activity of protein kinase A would, via the phosphorylation of many other proteins, modify many aspects of the neurons' functioning. Similarly, altered fluxes of Ca^{2+} into the neurons, via direct effects on Ca^{2+} channels or indirect effects on neuronal firing, would alter the activity of many types of Ca^{2+}-dependent protein kinases in the neurons, which would further alter many aspects of neuronal function. Two major classes of Ca^{2+}-dependent kinases would be affected: Ca^{2+}/calmodulin-dependent or CaM-kinases and protein kinase C (NESTLER and GREENGARD 1999).

Activation of dopamine receptors is reported to exert many additional effects on intracellular messenger cascades. There is evidence that D_1-like and D_2-like receptors can regulate the activity of the phosphatidylinositol pathway, which would alter cellular Ca^{2+} levels and protein kinase C activity. It remains unclear whether this effect is mediated indirectly via the receptors' actions on the cAMP pathway or directly via distinct effects of the receptors (see MISSALE et al. 1998; SIBLEY 1999; LEZCANO et al. 2000). Also, activation of D_2-like receptors leads to activation of MAP (mitogen-activated protein)-kinase cascades, in particular, the ERK (extracellular signal regulated kinase) pathway (LUO et al. 1998; WELSH et al. 1998). This is an action shared by many $G_{i/o}$-coupled receptors, and is mediated by the G protein $\beta\gamma$ subunits (see NESTLER and DUMAN 1999). Activation of MAP-kinase cascades would elicit still additional effects of dopamine receptor activation on target neurons.

II. Regulation of Gene Expression

Perturbation of these various intracellular signal transduction cascades would be expected to produce two major types of responses. First, as mentioned above, it would alter the phosphorylation of many proteins and lead to adaptive responses in the neurons. For example, phosphorylation of receptors would alter the responsiveness of the neurons to many types of neurotransmitters, and phosphorylation of synaptic vesicle proteins would alter the amount of neurotransmitter released by the target neurons. Second, perturbation of signal transduction cascades would alter the total amounts of certain proteins present within the neurons, which would also contribute to adaptive responses. For example, changes in the total amount of receptors or synaptic vesicle proteins in the neurons would modulate the efficacy of synaptic transmission.

The amount of a protein present within a neuron, a process termed gene expression, is controlled at every conceivable level between the gene and the final protein product (see HYMAN and NESTLER 1999). In order for a gene to be actively transcribed, its region on the chromosome must be in a suitable conformational state; DNA tightly bound in chromatin's nucleosomal structures may not be expressed effectively. The expression of genes that are available for transcription is controlled by the process of transcription initiation, whereby the macromolecular RNA polymerase II complex binds to the transcription initiation site present within the gene. Basal rates of transcription are relatively low and must be increased by regulatory elements located at other regions in the gene; some are close to the transcription initiation site, whereas others can be thousands of nucleotides away. The ability of response elements to promote transcription is mediated by a series of proteins, termed transcription factors, that bind to these response elements and then interact with proteins in the RNA polymerase II complex. Transcription factors can also inhibit transcription via analogous mechanisms. A typical gene contains response elements for numerous transcription factors, which means that the rate of transcription of most genes is determined by the summated effects of many transcription factors.

Once a gene is transcribed, the primary RNA transcript must be modified by splicing events and the addition of a poly-A tail, after which it is transported to the cytoplasm as a mature mRNA. These steps are highly regulated in neurons, although the underlying mechanisms remain poorly described at the molecular level. The mRNA is translated into the protein product on ribosomes. One prominent mechanism that regulates protein translation is the stability of the mRNA. Many mRNAs contain specific sequences in their untranslated 3' (and possibly 5') regions that bind specific proteins, which control the stability of the mRNA and, therefore, the amount of protein that can be translated (see Ross 1996).

A protein product of a gene undergoes many levels of processing. It may be cleaved from an immature product to its mature form, as is the case for many peptide hormones and neurotransmitters. The protein must also be transported form the ribosome to its correct location within the cell. Finally, the life cycle of the protein will be influenced by its degradation by proteases. The molecular basis by which each of these steps is controlled is becoming increasingly well-known at the molecular level (SEIDAH and CHRETIEN 1997).

Dopamine, via perturbation of intracellular signaling pathways, could conceivably regulate the level of a protein product in a target neuron at any

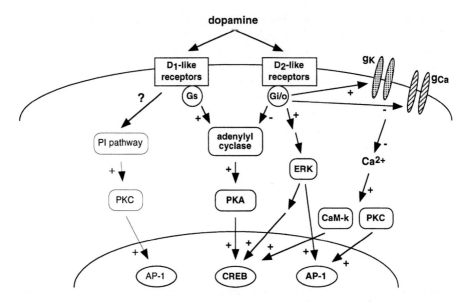

Fig. 1. Scheme showing the mechanisms by which dopamine can regulate gene expression. Activation of D_1-like receptors would activate the cAMP pathway, leading to the activation of protein kinase A (*PKA*) and to the phosphorylation and activation of cAMP-response element binding protein (*CREB*). Activation of D_2-like receptors would inhibit cell firing (via activation of K^+ channels). This action, plus inhibition of voltage-gated Ca^{2+} channels, could alter Ca^{2+} flux into neurons and alter the activity of CaM-kinases and protein kinase C. These receptors would also be expected to activate the ERK pathway. Perturbation of these various kinases would be expected to regulate CREB and Fos/Jun proteins. In addition, activation of dopamine receptors is reported to perturb the phosphatidylinositol pathway, which would further recruit Ca^{2+}-dependent mechanisms, although the precise mechanisms involved remain incompletely understood

of these aforementioned steps (Fig. 1). This would occur via the activation or inhibition of specific protein kinases or protein phosphatases, which, by altering the phosphorylation state of various nuclear, ribosomal, and protease-related proteins, would regulate the transcription, translation, and post-translational processing of a protein. While dopamine probably does regulate gene expression via such diverse actions, the only level of regulation that has been well studied at present is the regulation of transcription via alterations in transcription factors. The remainder of this chapter reviews what is known about dopamine regulation of transcription factors, with an emphasis on the types of regulation that have been documented in the brain in vivo.

C. Acute Effects of Dopamine on Gene Expression

Current knowledge of dopamine's effects on gene expression in vivo is based largely on studies of drugs that influence the functioning of dopamine receptors. This includes psychostimulant drugs (e.g., cocaine, amphetamine), which act as indirect dopamine receptor agonists by decreasing dopamine reuptake or increasing dopamine release, and typical antipsychotic drugs, which act as antagonists of D_2 and perhaps other D_2-like receptors. These studies have documented that acute perturbation of dopamine systems causes potent changes in specific transcription factors in target brain regions. However, as will be seen below, the target genes whose expression is altered as a consequence of these changes in transcription factors have been difficult to identify with certainty.

I. Regulation of CREB Family Proteins

CREB (cAMP-response element binding protein), a member of the leucine zipper family of transcription factors, was first discovered as a protein that mediates many of the effects of the cAMP pathway on gene expression (see GOLDMAN et al. 1996; FINKBEINER and GREENBERG 1998; DE CESARE et al. 1999). In the basal state, CREB dimers (bound to one another via their leucine zipper domains) bind to a specific sequence in the regulatory region of certain genes, called cAMP response elements (CREs), which show the consensus sequence TCACGTCA. To a first approximation, the binding of CREB dimers to CREs does not exert potent transcriptional effects, although more subtle consequences have been suggested (XING and QUINN 1993). Activation of the cAMP pathway leads to the phosphorylation of CREB on ser133 by protein kinase A. This enables CREB to bind to CREB-binding protein (CBP), which is a component of the RNA polymerase II complex, and to thereby regulate transcription. More-recent studies have shown that ser133 of CREB can also be catalyzed by CaM-kinases (particularly CaM-kinase IV) and various kinases activated by the ERK pathway (e.g., RSK family of kinases) (FINKBEINER and GREENBERG 1998). Based on this knowledge, activation of several types of dopamine receptors would be expected to alter the phosphorylation and activational state of CREB.

It is not surprising, then, that psychostimulants, which increase dopaminergic transmission in the brain, stimulate CREB phosphorylation in specific brain regions (KONRADI et al. 1994; KANO et al. 1995; TURGEON et al. 1997; SELF et al. 1998). This occurs most prominently in striatal regions (caudate-putamen and nucleus accumbens), which are the major targets of midbrain dopamine neurons. These effects can be blocked by administration of D_1-like receptor antagonists, which is consistent with the scheme (illustrated in Fig. 1) that dopamine produces this effect via activation of D_1-like receptors and of the cAMP pathway.

The target genes whose expression is then regulated in striatal neurons as a consequence of dopamine-elicited perturbations in CREB have remained largely obscure. The transcription factor, c-Fos, is one potential target. The ability of a psychostimulant to acutely induce c-Fos in striatum is reported to be antagonized by the local infusion of antisense oligonucleotides to CREB into this brain region (KONDRADI et al. 1994). This possibility is discussed in greater detail below. Another potential target is the neuropeptide dynorphin. The prodynorphin gene contains CRE sites, which have been shown to potently regulate the expression of the gene in vitro (COLE et al. 1995). Psychostimulants, which stimulate CREB phosphorylation, also stimulate dynorphin expression in striatum (see KREEK 1996; GERFEN et al. 1998; McGINTY and WANG 1998). Moreover, increased expression of CREB in striatal neurons, achieved via viral-mediated gene transfer, increases dynorphin expression in these neurons, while increased expression of a dominant negative mutant of CREB (which functions as a CREB antagonist) produces the opposite effect (CARLEZON et al. 1998). (The mutant form of CREB – termed mutCREB – used in this study lacks ser133 and therefore cannot be activated. However, it can still dimerize with wildtype CREB, and CREB-mutCREB heterodimers lack transcriptional activity since interactions with CBP require that both members of the dimer be phosphorylated. As a result, if mutCREB is expressed at high enough levels in the cell, it successfully antagonizes the activity of endogenous CREB.)

There are presumably many additional genes whose expression is altered by dopamine through CREB-dependent mechanisms. Many neural genes contain CREs, and have been shown to be regulated by CREB in vitro (see GOLDMAN et al. 1996; FINKBEINER and GREENBERG 1998; DE CESARE et al. 1999), but it has been difficult to obtain causative information with respect to dopamine action in vivo.

Another area that requires analysis is the potential regulation by dopamine of many CREB-like proteins in dopamine-responsiveness neurons. ATF1 (activating transcription factor-1) and various isoforms of CREM (CRE modulator protein) also bind to CRE sites in target genes and regulate gene expression (see DE CESARE et al. 1999). Interestingly, some CREM isoforms activate transcription and thereby function like CREB, whereas others (e.g., ICER, inducible CRE repressor) inhibit transcription via a mechanism similar to that described above for mutCREB. Little information is available, however, with respect to the regulation of these CREB-like transcription factors by dopamine.

II. Regulation of Fos Family Proteins

c-Fos and other Fos-like proteins, which include FosB, Fra-1 (Fos-related antigen-1), and Fra-2, are products of immediate early genes (MORGAN and CURRAN 1995). This means that the proteins can be induced very rapidly within cells, and that this induction is mediated via regulation of transcription factors (like CREB), which are present in cells at relatively high levels under basal conditions. In cultured cells, c-*fos* mRNA can be detected within 10 min of cellular activation, which indicates the rapidity of this transcriptional response.

Fos family proteins dimerize with a Jun family protein (see below) to form an active transcription factor complex called AP-1 (activator protein-1). Fos and Jun proteins, like CREB, contain leucine zippers, which mediate this dimerization. In vitro, there is some evidence of cross-dimerization between CREB and Fos–Jun family members, although the in vivo significance of this phenomenon remains obscure. AP-1 complexes then bind to AP-1 sites present within the regulatory regions of certain genes. AP-1 sites show the consensus sequence TCAG(or C)TCA. Note that the consensus AP-1 site is different from the consensus CRE site by only a single nucleotide, yet this difference is sufficient to direct regulation by Fos–Jun versus CREB.

Psychostimulants cause the rapid, but transient, induction of several Fos family proteins (c-Fos, FosB, and Fra-1) in striatal regions, and to a lesser extent in other dopamine target regions in brain, such as prefrontal cortex (e.g., GRAYBIEL et al. 1990; YOUNG et al. 1991; HOPE et al. 1992; GERFEN et al. 1998; McGINTY and WANG 1998). Induction of c-Fos is maximal within 2h after acute administration of cocaine or amphetamine and reverts to normal within 4–6h (Fig. 2). Induction of FosB and Fra-1 is somewhat more delayed, becoming maximal at 4–6h and reverting to normal within 12–18h. The transient nature of the induction of these proteins is due to the fact that the mRNAs for the proteins, and the proteins themselves, are highly unstable. Induction of Fos family proteins by psychostimulants in striatum is mediated by activation of D_1-like receptors. Moreover, the induction appears to be selective for the γ-aminobutyric acid (GABA)ergic medium spiny neurons that comprise the striatonigral (direct) pathway: those cells that project directly back to the midbrain and coexpress the neuropeptides, dynorphin and substance P (BERRETTA et al. 1993; KOSOFSKY et al. 1995; GERFEN et al. 1998).

A similar response of Fos-like proteins is caused by acute administration of typical antipsychotic drugs (e.g., D_2-like receptor antagonists) (e.g., NGUYEN et al. 1992; DEUTCH 1994; FIBIGER 1994; ROBERTSON et al. 1995). Second generation antipsychotic drugs such as clozapine (which show lower relative affinity toward D_2-like receptors and greater affinity for certain non-dopamine receptors) produce much less induction of Fos family proteins in dorsal striatum, but equivalent or even stronger induction in ventral striatum and prefrontal cortex. Induction of Fos family proteins by D_2-like receptor antagonists in striatum appears to be selective for the GABAergic medium spiny neurons that comprise the striatopallidal (indirect) pathway, which project to the globus pallidus-ventral pallidum and coexpress the neuropeptide, enkephalin (HIROI and GRAYBIEL 1996; GERFEN et al. 1998).

The c-fos gene has been characterized in detail, and many response elements located within the 5' promoter region of the gene have been shown to regulate its transcription in vitro (JANKNECHT 1995). However, little is known about the mechanism by which dopamine induces c-Fos in vivo. As stated above, there is evidence that CREB mediates c-Fos induction in striatum seen in response to psychostimulants. On the other hand, there are lingering concerns over the limitations of antisense oligonucleotide technology, particularly in vivo, which means that other approaches are needed to verify CREB's role in this phenomenon. Even less is known about how D_2-like antagonists induce c-Fos. One possibility is that this induction is mediated indirectly via gluta-

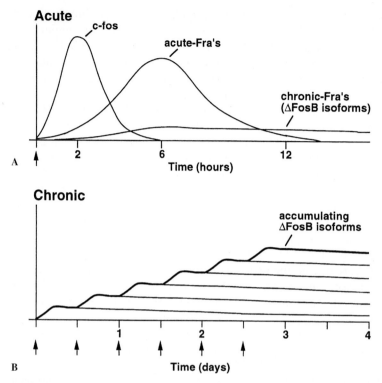

Fig. 2A,B. Scheme showing the gradual accumulation of ΔFosB versus the rapid and transient induction of acute Fra's in brain. **A** Several waves of Fra's are induced in neurons by many acute stimuli, including psychostimulants and antipsychotic drugs. c-Fos is induced rapidly and degraded within several hours of the acute stimulus, whereas other "acute Fra's" (e.g., FosB, ΔFosB, Fra-1, in some cases perhaps Fra-2) are induced somewhat later and persist somewhat longer than c-Fos. "Chronic Fra's" are biochemically modified forms of ΔFosB; they, too, are induced (although at low levels) following a single acute stimulus but persist in brain for long periods due to their extraordinary stability. In a complex with Jun-like proteins, these waves of Fra's form AP-1 binding complexes with shifting composition over time. **B** With repeated (e.g., twice daily) stimulation, each acute stimulus induces a low level of ΔFosB. This is indicated by the lower set of overlapping lines, which indicate ΔFosB induced by each acute stimulus. The result is a gradual increase in the total levels of ΔFosB with repeated stimuli during a course of chronic treatment. This is indicated by the *increasing stepped line* in the graph. The increasing levels of ΔFosB with repeated stimulation result in the gradual induction of significant levels of a long-lasting AP-1 complex, which is proposed to underlie persisting forms of neural plasticity in the brain. (Modified from HOPE et al. 1994)

mate: according to this scheme, blockade of D_2-like receptors on glutamatergic nerve terminals in striatum leads to increased glutamate release and increased activation of certain GABAergic medium spiny neurons. Another theoretical possibility is that c-Fos induction by D_2-like antagonists is mediated via activation of ERK pathways, although this has not yet been investigated directly.

Similar to the situation for CREB, it has been very difficult to identify target genes for AP-1 in the brain. The genes that have received the most attention to date are those that encode various neuropeptides, for example, dynorphin and neurotensin. The dynorphin gene contains an AP-1 site, and increases in dynorphin expression are reported following acute administration of psychostimulants (see above). It is possible that induction of AP-1 mediates this effect, although direct evidence is lacking. The neurotensin gene also contains AP-1 sites, and increases in neurotensin expression occur in response to D_2-like antagonist administration (MERCHANT et al. 1994; ROBERSON et al. 1995). While it is plausible that AP-1 may mediate this effect, direct evidence has not yet been obtained. Many other neurally expressed genes are known to contain AP-1 sites (see HIROI et al. 1998). This includes the genes for certain dopamine and glutamate receptors, neurofilament subunits, and protein kinase subunits. The goal of current research is to use tools of viral-mediated gene transfer (described above) or genetic mutations in mice (described below) to provide direct evidence for a role of AP-1 in mediating the ability of dopaminergic transmission to regulate the expression of these and other putative target genes in vivo.

III. Regulation of Jun Family Proteins

Jun family members include c-Jun, JunB, and JunD. The genes for c-Jun and JunB, like those encoding Fos family members, are immediate early genes and can be induced rapidly and transiently in brain in response to many types of acute stimuli. Indeed, there are reports that acute administration of psychostimulants can induce these Jun family members in striatum and other dopamine target brain regions (e.g., HOPE et al. 1992; MORATALLA et al. 1996). In contrast, the gene for JunD is not as highly regulated and may be expressed constitutively.

There is growing evidence that the activity of Jun family members is regulated by certain SAP kinases (stress-activated protein kinases) that form one component of MAP-kinase cascades (see NESTLER and GREENGARD 1999). SAP kinases, also called JNK (or Jun N-terminal kinases), are activated in cultured cells in response to noxious stimuli, such as osmotic stress, UV irradiation, and certain toxins. Activation of the kinases leads to the phosphorylation of Jun proteins; this activates their transcriptional activity and is thought to mediate the responses of challenged cells to these noxious stimuli. The specifics, like which genes are altered as a consequence of Jun phosphorylation and the net effect of such targets on cell function, are only now being revealed in cultured cells.

While the role of the JNK pathway in mediating the effects of dopamine on gene expression in vivo remains unclear, recent work has demonstrated that the pathway can be stimulated in primary cultures of striatal neurons

(SCHWARZSCHILD et al, 1997) This activation seems to be mediated via NMDA (*N*-methyl-D-aspartate) glutamate receptors, which provides insight into the types of in vivo situations in which such regulation might occur.

The notion that dopamine may under certain circumstances regulate gene expression via activation of JNK pathways raises some interesting questions with respect to the specificity of AP-1 regulation of transcription. This is because activation of Fos-like proteins, by dimerizing with recently induced or preexisting Jun proteins, may form one type of active AP-1 complex, whereas phosphorylation of preexisting Jun proteins (dimerizing with themselves in the absence of Fos proteins) may form another type of active AP-1 complex. A major challenge for the future is to better understand the potential functional differences between these two types of AP-1 complexes and ultimately to identify the target genes they regulate.

IV. Regulation of Egr Family Proteins

The Egr family of transcription factors (also called Zif268 or NGFI-A) are classified as zinc finger proteins and are structurally very different from the CREB and Fos-Jun families. The Egr family consists of three proteins, termed Egr-1, -2, and -3. The gene for each of these proteins is regulated as an immediate early gene and can be induced in specific brain regions by a wide array of acute stimuli. For example, acute administration of psychostimulants induces Egr proteins in striatal regions with a time course similar to that seen for c-Fos (O'DONOVAN et al. 1999). Interestingly, Egr3 appears to be somewhat more stable than the other Egr proteins or c-Fos and, as a result, persists in brain for up to 24 h after psychostimulant administration. The mechanism by which psychostimulants induce Egr proteins is poorly understood. Egr proteins bind to GC (guanine-cytosine)-rich regions of DNA and are believed to thereby regulate gene expression. However, it has been very difficult to identify physiological targets for Egr proteins both in vitro and in vivo.

D. Chronic Effects of Dopamine on Gene Expression

By analogy with the acute situation, much of what we know about dopamine's longer-lasting changes in gene expression comes from studies of the chronic effects of psychostimulants and antipsychotic drugs. Knowledge of dopamine's influence on gene expression over the long term also can be inferred from the effects of dopaminergic lesions (e.g., by 6-hydroxydopamine) or genetic mutations in mice that alter dopaminergic transmission (e.g., dopamine transporter knockout; GAINETDINOV et al. 1998). The levels of many proteins are known to be altered under these various circumstances. However, with a few notable exceptions, the transcriptional or post-transcriptional mechanisms involved remain unknown. However, some of these chronic conditions have been associated with some interesting effects on CREB and Fos family transcription factors; these actions provide some insight into the types of mechanisms by which dopaminergic transmission may produce relatively long-lived changes in gene expression in target neurons.

I. Regulation of CREB

Chronic administration of psychostimulants leads to a sustained increase in CREB phosphorylation in striatal regions (COLE et al. 1995; TURGEON et al. 1997). This may occur as a result of increased levels of protein kinase A that are known to be induced in these regions by chronic psychostimulant exposure (see NESTLER and AGHAJANIAN 1997), although direct evidence for this possibility is lacking.

A recent study suggests a possible functional consequence of this alteration in CREB. Overexpression of CREB in ventral striatum, achieved by viral-mediated gene transfer, was found to decrease an animal's sensitivity to the rewarding effects of cocaine (CARLEZON et al. 1998). In fact, low doses of cocaine were made aversive upon overexpressing CREB. Conversely, overexpression of mutCREB (the dominant negative mutant mentioned above) in this brain region increases an animal's sensitivity to the rewarding effects of cocaine. These findings suggest that the sustained phosphorylation of CREB caused by chronic psychostimulant exposure may represent a negative feedback mechanism (i.e., a mechanism for tolerance) that serves to dampen the animal's sensitivity to subsequent drug exposure.

This effect of CREB may be mediated via regulation of dynorphin expression. Recall from earlier in the chapter that overexpression of CREB increases dynorphin expression whereas overexpression of mutCREB decreases dynorphin expression in ventral striatum. Dynorphin, in ventral striatal circuits, is known to exert an aversive effect on an animal and oppose drug reward by activating κ opioid receptors, which then inhibit the dopamine neurons that mediate acute reward (SHIPPENBERG and REA 1997). Direct evidence for this scheme is that the aversive effects of low doses of cocaine, seen in animals that receive viral vectors encoding CREB in ventral striatum, can be blocked by administration of a κ opioid antagonist (CARLEZON et al. 1998). While dynorphin is likely just one of many target genes for CREB in these neurons, these findings do provide one scheme by which dopamine-induced changes in gene transcription underlie an important form of behavioral plasticity (Fig. 3).

II. Regulation of Fos Family Proteins: Induction of ΔFosB

Repeated administration of psychostimulants produces very different effects on Fos proteins as compared to acute exposure. Repeated exposure is associated with desensitization of the induction of c-Fos and the other Fos family members that are induced acutely in striatal regions (HOPE et al. 1994; NYE et al. 1995). Instead, such treatments cause the gradual accumulation of novel isoforms of ΔFosB in these regions. ΔFosB is a truncated splice variant of the

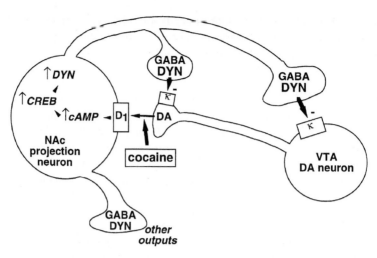

Fig. 3. Scheme showing the role of CREB and dynorphin in modulating psychostimulant-induced neural plasticity. Repeated exposure to psychostimulants causes the sustained phosphorylation and activation of CREB in ventral striatal neurons, which may be mediated by upregulation of the cAMP pathway that is known to occur under these conditions (NESTLER and AGHAJANIAN 1997). The sustained activation of CREB leads to the induction of dynorphin in these neurons. Dynorphin then inhibits dopaminergic neurons via the activation of κ opioid receptors located on dopaminergic nerve terminals within the ventral striatum as well as on dopaminergic cell bodies in the midbrain. Inhibition of dopaminergic transmission reduces the effects of subsequent exposure to psychostimulants, which are dependent on dopamine release. In this way, the rewarding properties of the drug are dampened via the sustained activation of CREB and induction of dynorphin

*fos*B gene. Unlike all other Fos family members, certain isoforms of ΔFosB are highly stable proteins. Whereas ΔFosB is induced to only a slight extent after an acute stimulus, it accumulates in brain after repeated stimulation due to this extraordinary stability (Fig. 2). The mechanism responsible for the enhanced stability of these chronic ΔFosB isoforms remains unknown, but could involve their phosphorylation. Psychostimulants induce ΔFosB selectively within the GABAergic medium spiny neurons of the striatonigral pathway (see above), the same neurons where c-Fos is induced by acute drug exposure (NYE et al. 1995; MORATALLA et al. 1996).

Induction of ΔFosB thereby provides one mechanism by which a chronic treatment can cause changes in gene expression that outlive the treatment itself. Thus, whereas the sustained increase in CREB phosphorylation seen after chronic psychostimulant administration reverts to normal within days, the ΔFosB induced by the same treatment persists for weeks or longer and can thereby mediate still longer-lasting consequences of drug exposure. In fact, ΔFosB is the longest-lasting molecular adaptation yet documented under these conditions. On the other hand, ΔFosB per se cannot mediate the near-permanent changes in striatal function that have been documented after chronic psychostimulant administration (e.g., ROBINSON and KOLB 1997). Either ΔFosB could initiate such near-permanent changes or still longer-lasting molecular signals, not yet discovered, are involved.

Recent evidence suggests that induction of ΔFosB in striatonigral neurons increases an animal's sensitivity to the rewarding and locomotor-activating effects of cocaine. This comes from transgenic mice in which ΔFosB can be induced, selectively within this subset of striatal neuron, of adult animals (KELZ et al. 1999). This inducible, tissue-specific system thereby avoids many of the complications of traditional mutagenesis systems in which the mutation in question occurs at the earliest stages of development and is expressed throughout the brain and peripheral tissues.

The AMPA (α-amino-3-hydroxy-5-methyl-4-isoxazolepropionic acid) glutamate receptor subunit, GluR2, may be one target through which ΔFosB accumulation increases an animal's sensitivity to cocaine. Induction of ΔFosB in the inducible transgenic mice mentioned above leads to increased levels of GluR2 immunoreactivity in the ventral striatum (KELZ et al. 1999). This could reflect a direct effect of ΔFosB on the GluR2 gene, since the promoter for this gene contains multiple AP-1 sites. However, indirect effects cannot be excluded: for example, it is possible that ΔFosB regulates the expression of GluR2 via intervening proteins (e.g., regulation of another transcription factor). Induction of GluR2 by ΔFosB is specific to ventral striatum and is not seen in the dorsal striatum. Moreover, other glutamate receptor subunits are not altered in these regions upon ΔFosB accumulation. Overexpression of GluR2 within ventral striatum, achieved by viral-mediated gene transfer, increases an animal's sensitivity to cocaine, which indicates that GluR2 induction could account for the behavioral changes seen upon ΔFosB accumulation (KELZ et al. 1999).

GluR2 has a major effect on the conductance properties of AMPA receptor channels. AMPA receptors that contain GluR2 show lower overall currents and no Ca^{2+} permeability compared to receptors that lack GluR2 (SEEBURG et al. 1998). Therefore, ΔFosB-mediated induction of GluR2 in ventral striatum would be expected to cause the formation of a greater number of GluR2-containing AMPA receptors in these neurons and, consequently, to decrease AMPA-mediated currents. Such an electrophysiological effect has been demonstrated directly in these neurons after chronic cocaine administration (WHITE and KALIVAS 1998). It is not known why a reduction in activity of ventral striatal neurons would sensitize an animal to cocaine's effects; such an understanding will require a great deal more research at the neural systems level. Nevertheless, these results establish a mechanism by which repeated exposure to cocaine leads to the accumulation of a long-lived transcription factor, which in turn alters the expression of a specific target gene to produce features of neural and behavioral plasticity that are associated with drug addiction (Fig. 4) (KELZ et al. 1999).

ΔFosB also is induced within striatal regions by chronic exposure to antipsychotic drugs (DOUCET et al. 1996; HIROIAND GRAYBIEL 1996; ATKINS et

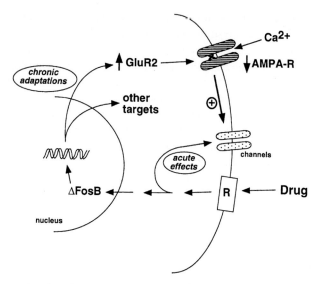

Fig. 4. Scheme showing the role of ΔFosB and GluR2 in modulating psychostimulant-induced neural plasticity. Acute psychostimulant exposure rapidly inhibits the activity of ventral striatal neurons, which mediates – in part– the acute rewarding effects of the drug. Repeated psychostimulant exposure leads to the accumulation of ΔFosB, which induces the expression of GluR2. The resulting increase in GluR2-containing AMPA receptors leads to reduced overall AMPA-mediated current and Ca^{2+} fluxes. This change in the excitability of the neurons potentiates the acute inhibitory effects of a subsequent drug exposure. In this way, the rewarding properties of the drug are enhanced via induction of ΔFosB and of GluR2

al. 1999). As seen for the induction of c-Fos by acute exposure to antipsychotic drugs, the induction of ΔFosB by chronic drug exposure occurs in the striatopallidal (enkephalin-containing) subset of GABAergic medium spiny neuron (HIROI et al. 1996). Also like the acute situation, robust induction of ΔFosB in dorsal striatum is seen with typical antipsychotic drugs, with much lower levels of induction seen with several second generation antipsychotic drugs (e.g., risperidone and olanzapine) and little if any induction seen with clozapine. Thus, induction of ΔFosB in dorsal striatum, which appears to be mediated by prolonged antagonism of D_2-like dopamine receptors, reflects the extrapyramidal liability of antipsychotic drugs. Further work is needed to determine whether ΔFosB plays a role in the generation of these side effects (e.g., tardive dyskinesia). In contrast to the acute situation, however, little induction of ΔFosB is seen in ventral striatum or prefrontal cortical regions after chronic exposure to either typical or second-generation antipsychotic drugs (ATKINS et al. 1999). ΔFosB is also induced in striatum upon 6-hydroxydopamine-induced lesions of midbrain dopamine neurons, although few data are yet available about the cellular specificity and functional consequences of this response (DOUCET et al. 1996).

The induction of ΔFosB in striatal regions by both psychostimulants and antipsychotic drugs highlights an important consideration: that the functional effects of a molecular event depend on the cell type in which it occurs. Such a conclusion may seem trivial, but is often overlooked. As stated above, induction of ΔFosB in striatonigral neurons of transgenic mice increases their sensitivity to cocaine. In a distinct line of mouse, where ΔFosB is induced largely in striatopallidal neurons, no such change in cocaine sensitivity is observed (M.B. Kelz and E.J. Nestler, unpublished observations). Thus, although psychostimulants and antipsychotic drugs both induce the same transcription factor in the same brain region, the functional consequences of the induction differs substantially since it occurs in distinct subsets of neurons.

Acknowledgements. Preparation of this review was supported by the National Institute on Drug Abuse and National Institute of Mental Health and by the Abraham Ribicoff Research Facilities of the Connecticut Mental Health Center, State of Connecticut Department of Mental Health and Addiction Services.

References

Atkins J, Carlezon WA, Chlan J, Ny HE, Nestler EJ (1999) Region-specific induction of ΔFosB by repeated administration of typical versus atypical antipsychotic drugs. Synapse 33:118–128

Berretta S, Robertson HA, Graybiel AM (1993) Neurochemically specialized projection neurons of the striatum respond differentially to psychomotor stimulants. Prog Brain Res 99:201–205

Carlezon WA Jr, Thome J, Olson VG, Lane-Ladd SB, Brodkin ES, Hiroi N, Duman RS, Neve RL, Nestler EJ (1998) Regulation of cocaine reward by CREB. Science 282:2272–2275

Cole RL, Konradi C, Douglass J, Hyman SE (1995) Neuronal adaptation to amphetamine and dopamine: molecular mechanisms of prodynorphin gene regulation in rat striatum. Neuron 14:813–823

De Cesare D, Fimia GM, Sassone-Corsi P (1999) Signaling routes to CREM and CREB: plasticity in transcriptional activation. Trends Biochem Sci 24:281–285

Deutch AY (1994) Identification of the neural systems subserving the actions of clozapine: clues from immediate-early gene expression. J Clin Psychiatry 55 Suppl B:37–42

Doucet J-P, Nakabeppu Y, Bedard PJ, Hope BT, Nestler EJ, Jasmin B, Chen JS, Iadarola MJ, St-Jean M, Wigle N, Blanchet P, Grondin R, Robertson GS (1996) Chronic alterations in dopaminergic neurotransmission produce a persistent elevation of striatal ΔFosB expression. Eur J Neurosci 8:365–381

Fibiger HC (1994) Neuroanatomical targets of neuroleptic drugs as revealed by Fos immunochemistry. J Clin Psychiatry 55 Suppl B:33–36

Finkbeiner S, Greenberg ME (1998) Ca^{2+} channel-regulated neuronal gene expression. J Neurobiol 37:171–189

Gainetdinov RR, Jones SR, Fumagalli F, Wightman RM, Caron MG (1998) Re-evaluation of the role of the dopamine transporter in dopamine system homeostasis. Brain Res Rev 26:148–153

Gerfen CR, Keefe KA, Steiner H (1998) Dopamine-mediated gene regulation in the striatum. Adv Pharmacol 42:670–673

Goldman PS, Tran VK, Goodman RH (1996) The multifunctional role of the co-activator CBP in transcriptional regulation. Rec Prog Hormone Res 52:103–119

Graybiel AM, Moratalla R, Robertson HA (1990) Amphetamine and cocaine induce drug-specific activation of the c-fos gene in striosome-matrix compartments and limbic subdivisions of the striatum. Proc Natl Acad Sci USA 87:6912–6916

Hiroi N, Graybiel AM (1996) Atypical and typical neuroleptic treatments induce distinct programs of transcription factor expression in the striatum. J Comp Neurol 374:70–83

Hiroi N, Brown J, Ye H, Saudou F, Vaidya VA, Duman RS, Greenberg ME, Nestler EJ (1998) Essential role of the fosB gene in molecular, cellular, and behavioral actions of electroconvulsive seizures. J Neurosci 18:6952–6962

Hope B, Kosofsky B, Hyman SE, Nestler EJ (1992) Regulation of IEG expression and AP-1 binding by chronic cocaine in the rat nucleus accumbens. Proc Natl Acad Sci USA 89:5764–5768

Hope BT, Nye HE, Kelz MB, Self DW, Iadarola MJ, Nakabeppu Y, Duman RS, Nestler EJ (1994) Induction of a long-lasting AP-1 complex composed of altered Fos-like proteins in brain by chronic cocaine and other chronic treatments. Neuron 13:1235–1244

Hyman SE, Nestler EJ (1999) Principles of molecular biology. In: Charney DS, Nestler EJ, Bunney BS (eds) Neurobiological Foundations of Psychiatry. Oxford University Press, pp 73–85

Janknecht R (1995) Regulation of the c-fos promoter. Immunobiology 193:137–142

Kano T, Suzuki Y, Shibuya M, Kiuchi K, Hagiwara M (1995) Cocaine-induced CREB phosphorylation and c-Fos expression are suppressed in Parkinsonism model mice. Neuroreport 6:2197–200

Kelz MB, Chen JS, Carlezon WA, Whisler K, Gilden L, Beckmann AM, Steffen C, Zhang Y-J, Marotti L, Self SW, Tkatch R, Baranauskas G, Surmeier DJ, Neve RL, Duman RS, Picciotto MR, Nestler EJ (1999) Expression of the transcription factor ΔFosB in the brain controls sensitivity to cocaine. Nature 401:272–276

Konradi C, Cole RL, Heckers S, Hyman SE (1994) Amphetamine regulates gene expression in rat striatum via transcription factor CREB. J Neurosci 14:5623–5634

Kosofsky BE, Genova LM, Hyman SE (1995) Substance P phenotype defines specificity of c-fos induction by cocaine in developing rat striatum. J Comp Neurol 351:41–50

Kreek MJ (1996) Cocaine, dopamine and the endogenous opioid system. J Addictive Dis 15:73–96

Lezcano N, Mrzljak L, Eubanks S, Levenson R, Goldman-Rakic P, Bergson C (2000) Dual signaling regulated by calcyon, a D1 dopamine receptor interacting protein. Science 287:1660–1664

Luo Y, Kokkonen GC, Wang X, Neve KA, Roth GS (1998) D2 dopamine receptors stimulate mitogenesis through pertussis toxin-sensitive G proteins and Ras-involved ERK and SAP/JNK pathways in rat C6-D2L glioma cells. J Neurochem 71:980–990

McGinty JF, Wang JQ (1998) Drugs of abuse and striatal gene expression. Adv Pharmacol 42:1017–1019

Merchant KM, Dobie DJ, Filloux FM, Totzke M, Aravagiri M, Dorsa DM (1994) Effects of chronic haloperidol and clozapine treatment on neurotensin and c-fos mRNA in rat neostriatal subregions. J Pharmacol Exp Ther 271:460-471

Missale C, Nash SR, Robinson SW, Jaber M, Caron MG (1998) Dopamine receptors: from structure to function. Physiol Rev 78:189–225

Morgan JI, Curran T (1995) Immediate-early genes: ten years on. Trends Neurosci 18:66–67

Moratalla R, Elibol B, Vallejo M, Graybiel AM (1996) Network-level changes in expression of inducible Fos-Jun proteins in the striatum during chronic cocaine treatment and withdrawal. Neuron 17:147–156

Nestler EJ, Aghajanian GK (1997) Molecular and cellular basis of addiction. Science 278:58–63

Nestler EJ, Duman RS (1999) G proteins, In: Siegel GJ, Agranoff BW, Alberts RW, Fisher SK, Uhler MD (eds) Basic Neurochemistry, 6th ed., Lippincott-Raven Publishers, pp 401–414

Nestler EJ, Greengard P (1999) Serine and threonine phosphorylation, In: Basic Neurochemistry, 6th ed., ed. by GJ Siegel, BW Agranoff, RW Alberts, SK Fisher, MD Uhler, Lippincott-Raven Publishers, pp 471–496

Nguyen TV, Kosofsky BE, Birnbaum R, Cohen BM, Hyman SE (1992) Differential expression of c-fos and zif268 in rat striatum after haloperidol, clozapine, and amphetamine. Proc Natl Acad Sci USA 89:4270–4274

Nye H, Hope BT, Kelz M, Iadarol, M, Nestler EJ (1995) Pharmacological studies of the regulation by cocaine of chronic Fra (Fos-related antigen) induction in the striatum and nucleus accumbens. J Pharmacol Exp Ther 275:1671–1680

O'Donovan KJ, Tourtellotte WG, Millbrandt J, Baraban JM (1999) The EGR family of transcription-regulatory factors: progress at the interface of molecular and systems neuroscience. Trends Neurosci 22:167–173

Robertson GS, Tetzlaff W, Bedard A, St-Jean M, Wigle N (1995) C-fos mediates antipsychotic-induced neurotensin gene expression in the rodent striatum. Neuroscience 67:325–344

Robinson TE, Kolb B (1997) Persistent structural modifications in nucleus accumbens and prefrontal cortex neurons produced by previous experience with amphetamine. J Neurosci 17:8491–8497

Ross J (1996) Control of messenger RNA stability in higher eukaryotes. Trends Genetics 12:171–175

Schwarzschild MA, Cole RL, Hyman SE (1997) Glutamate, but not dopamine, stimulates stress-activated protein kinase and AP-1-mediated transcription in striatal neurons. J Neurosci 17:3455–3466

Seeburg PH, Higuchi M, Sprengel R (1998) RNA editing of brain glutamate receptor channels: mechanism and physiology. Brain Res Rev 26:217–229

Seidah NG, Chretien M (1997) Eukaryotic protein processing: endoproteolysis of precursor proteins. Curr Op Biotechnol 8:602–607

Self DW, Genova LM, Hope BT, Barnhart WJ, Spencer JJ, Nestler EJ (1998) Involvement of cAMP-dependent protein kinase in the nucleus accumbens in cocaine self-administration and relapse of cocaine-seeking behavior. J Neurosci 18:1848-1859

Shippenberg TS, Rea W (1997) Sensitization to the behavioral effects of cocaine: modulation by dynorphin and kappa-opioid receptor agonists. Pharmacol Biochem Behav 57:449–455

Sibley DR (1999) New insights into dopaminergic receptor function using antisense and genetically altered animals. Annu Rev Pharmacol Toxicol 39:313–341

Surmeier DJ, Bargas J, Hemmings HC Jr, Nairn AC, Greengard P (1995) Modulation of calcium currents by a D1 dopaminergic protein kinase/phosphatase cascade in rat neostriatal neurons. Neuron 14:385–397

Turgeon SM, Pollack AE, Fink JS (1997) Enhanced CREB phosphorylation and changes in c-Fos and FRA expression in striatum accompany amphetamine sensitization. Brain Res 749:120–126

Welsh GI, Hall DA, Warnes A, Strange PG, Proud CG (1998) Activation of microtubule-associated protein kinase (Erk) and p70 S6 kinase by D2 dopamine receptors. J Neurochem 70:2139–2146

White FJ, Kalivas PW (1998) Neuroadaptations involved in amphetamine and cocaine addiction. Drug Alcohol Dependence 51:141–153

Xing L, Quinn PG (1993) Involvement of 3′,5′-cyclic adenosine monophosphate regulatory element binding protein (CREB) in both basal and hormone-mediated expression of the phosphoenolpyruvate carboxykinase (PEPCK) gene. Mol Endocrinol 7:1484–1494

Young ST, Porrino LJ, Iadarola MJ (1991) Cocaine induces striatal c-fos-immunoreactive proteins via dopaminergic D1 receptors. Proc Natl Acad Sci USA 88:1291–1295

Subject Index

A 68930 130
A 86929 130
acetylcholine (ACh) 9
acoustic neurinoma 227
ACTH, see Adrenocortcotropic hormone 173
activating transcription factor 1 (ATF 1) 326
adaptive reponses 321
adenosine A_{2A} receptor 164
adenosine triphosphatase 269, 306
adenylyl cyclase 11, 121, 163, 185, 236
ADHD, see attention deficit hyperactivity disorder
adrenal gland 173
adrenal medullary hormones 25
adrenocorticotropic hormone (ACTH) 173
aggression 276
akinesia 167
alcohol 33, 169, 203
alpha $(\alpha)_{olf}$ protein 237
alpha (α)receptor adrenergic 10
alpha's protein 237
alpha-(α) melanocyte stimulating hormone (α-MSH) 171
alpha-(α) methyltyrosine 33
amacrine cells 92
amineptine 206
aminotetralin 195
amisulpride 196
amitriptyline 206
AMPA receptor channels 333
amperozide 196
amphetamine 12, 33, 135, 169, 270, 276, 325
amphetamine stereoselectivity 262
amygdala 55, 81, 137
antidepressant drugs 206
antipsychotic treatment 199
antisense oligonucleotides 143
antisense-treatment 277
apomorphine 12, 35

appetitive Conditions 81
arachidonic acid 185, 223
arcuate nucleus 89, 97
aripiprazole 36
asynaptic release 100
ATF1, see Activating transcription factor 1
attention 276
attention deficit hyperactivity disorder (ADHD) 278
autism 128
autoreceptor 34
aversive conditions 81

basal ganglia 31, 166, 310
bed nucleus 55
benzazepine 130, 140
benzonaphtazepine 131
benztropine 270
beta (β) adrenoreceptor 10, 238
beta (β)-arrestin 135
beta (β)-endorphin 173
biogenic amines 29
bipolar cell 92
bipolar disorder 127, 227
bradykinesia 167
brain development 201
brain-derived neurotrophic factor (BDNF) 187
bromocriptine 12, 171, 309
brown adipose tissue 243
buproprion 206
butaclamol 125, 131

calbindin (CaBP) 44, 87
calcineurin 244
calcium channels 302
calcium chelators 303
calcium-binding protein, see calbindin 87
calmodulin 239
calretinin 87
cAMP-dependent kinase 186

cAMP-response element binding protein (CREB) 163, 241
carpipramine 196
Catalase 170
caudate neuron activity 9
caudate nucleus 5, 68
central (amygdala) nucleus 55
cerebellar peduncle, superior 44
c-fos 191, 325
chlorpromazine 24, 30, 196, 223
cholera toxin 133
cholezystokinin 189
choline 268
choline acetyltransferase 73
choroid plexus 243
chromaffin cells 243
clathrin 275
clozapine 33, 196, 223, 228, 327
cocaine 30, 169, 197, 203, 257, 325
cocaine antagonists 262, 273
cognition 43
concanavalin A 135
cortex, temporal 85
cortical afferents 78
cortical areas, sensory 84
cortico-ventrotriatal circuits 199
COS-7 cells 223
CREB, see cAMP-response element binding protein
Cushing's syndrome 173
cyclic adenosine monophosphate (cAMP) 132, 322

D_1 (Dopamine$_1$) receptor 12, 74, 121, 321
D_1 (dopamine$_1$) receptor
– labelling 75
– distribution 88
– molecular cloning 121
– point mutations 124
– truncation 125
D_{1C} (dopamine$_{1C}$) receptor 145
D_2 (dopamine$_2$) receptor 12, 35, 74
– agonists 308
– knockout 165
D_3 (dopamine$_3$) receptor 78
– coupling 186
– overexpression 194
D_4 (dopamine$_4$) receptor
– gene 225
– labelling 89
D_5 (dopamine$_5$) receptor 74
DARPP-32 242
DARPP-32 phosphorylation 244
densocellular area 52
densocellular group 47

dentate gyrus 52
desipramine 207, 257, 264
dextran amine, biotinylated 65
diagonal band nuclei 79
dihydrexidine 130, 141
beta (ß)-3,4-dihydroxyphenylethylamine 2
3,4-dihydroxyphenylserine (DOPS) 4
diphenylpiperazine 257
dopa decarboxylase 2, 23
dopa decarboxylase inhibitors 5
dopamine axons 67
– transmitting elements 96
dopamine denervation 312
dopamine neurons   ultrastructure 63
dopamine pathway, nigrostriatal 6
dopamine projection gradient 46
dopamine receptor activation 301
dopamine receptor agonist 34
dopamine receptor antagonist 34
dopamine receptor concept 11
dopamine signal transduction 238
dopamine transmission 64, 103
dopamine transporter (DAT) 66, 76
– topology 259
– gene 263
– immunoreactivity 266
– polymorphism 280
– expression 89
dopamine uptake inhibitors 206, 272, 275
dopamine varicosities 67
dopamine,
– vasopressor action
– regional distribution 26
dopamine-beta(ß)-hydroxylase (DBH) 66
dorsal tier 46
drug addiction 299
dynamine 135
dynorphin 187, 204, 326

electroconvulsive shocks 207
emonapride 227
endocrine tumors 161
enkephalin 72, 169, 302, 327
epileptic seizures 171
epinephrine 223
ergot alkaloids 12
ethanol, see alcohol
excitatory postsynaptic potential (EPSP) 304
excitatory synaptic input 305
exocytosis 30, 100
extrapyramidal side effects 34

Subject Index

fiber degeneration 54
fibroblasts 186
fluoxetine 206
forebrain 43
forskolin 186

G protein 101, 322
G protein activation 134
G protein coupled receptors (GPCR) 122, 161
gamma-(γ)aminobutyric acid (GABA) 11, 68, 167, 300
gamma-(γ)aminobutyric acid (GABA) A receptors 304
gamma-(γ)aminobutyric acid (GABA) dendrites 86
gene expression 99
glia 87
globus pallidus 79, 166
glutamate 9, 54
glutamate receptor 245
glutamate receptor subunit 2 (Glu R2) 333
glutamate terminals 72
glutamate transmission 82
glutaraldehyde 64
glutathioine 170
golgi impregnation subtypes 74
grooming 139
growth hormone-releasing hormone 172
GTP-binding protein 162, 236

hallucinations 175
haloperidol 32, 196, 223, 307
heart 225
heroin 203
hippocampus, dentate gyrus 137
histofluorescence method 6
homovanillic acid 10
Huntington's chorea 26, 36, 127, 170
hydrogen peroxide 170
5-hydroxydopamine 64
6-hydroxydopamine 7, 65, 167, 192, 203, 278
D,L-5-hydroxytryptophan 3
5-hydroxytryptophan 26
3-hydroxytyramine 2
hyperlocomotion 277
hypermotility 276
hypodopaminergia 36
hypothalamus 265

iloperidone 196
imipramine 30, 207
immunogold labeling method 75, 267

incerto-hypothalamic system 89
indolophenanthridine 130
inducible CRE repressor (ICER) 326
intercalated cell groups 81
interneurons, cortical 87
intervaricose segments 70
intrinsic membrane conductance 306
inwardly rectifying K^+ (IRK) channels 303
ionic conductance 304
ischemia 170
islands of Calleja 187
isochroman 130
isoquinoline 131, 140

Jun N-terminal kinase (JNK) 329

K^+ channels 163
kainic acid 171
Kir2 channels 301

lactotroph cells 161
L-Dopa 5, 27, 142, 174, 193
learning 43, 276, 309
levodopa, see L-Dopa
light-dark adaptation 92, 99
limbic areas, forebrain 6
limbic brain 54, 185
limbic cortex 6
lisuride 309
lithium 268
local circuit neurons 102
locomotion 140
locus ceruleus 91
lordosis behaviour 143
Ltk cells 136
luteinizing hormone-releasing hormone (LH-RH) 90
LY 270411 131
lysergic acid diethylamide (LSD) 24

magnesium block 305
mania 206, 281
mazindol 257, 270
median eminence 90, 97
medium spiny neurons 300
melanotroph cells 161
membrane presentation 159
memory 276, 309
mesencephalon 265
mesocortical projection 83
mesocorticolimbic pathway 176
metamphetamine 170, 279
methadone 14
3-methoxytyramine 32

1-methyl-4-phenyl-1,2,3,6-tetrahydropyridine (MPTP) 7, 36, 142, 192, 257
1-methyl-4-phenylpyridinium (MPP+) 260

Michaelis-Menten-equation 269
midbrain tegmentum 6
mitochondria 100
mitogen-activated protein 192
mitogen-activated protein kinase (MAPK) 163, 246, 322
mitogenesis 186
monoamine axons 100
monoamine cortical systems 84
monoamine oxidase (MAO) inhibitor 30
monoamine oxidase B 8, 170, 279
morphine 14
motivation 43
motor activity 276
motor control 43
motor function 98, 166
motor memory 310
motor regions 53
MPP⁺ 8
multiple system atrophy 273
myoclonic jerking 139

nafadotride 193
neocortex 137
neostriatum 299, 308
neurabin 1 241
neuromelanin 44
neuropeptide Y 73
neuropil 87
neurosurgery, functional 9
neurotensin 78, 82, 187, 329
neurotensin gene transcripts 199
nicotinamide adenine dinucleotide phosphate (NADPH) 73
nicotine 205
nigrostriatal pathway 6, 176
N-methyl-D-aspartate 240
nociception 97
nomifensine 206, 257, 270
noradrenaline (NA) 8, 209, 223
noradrenaline transporter (NET) 258
norepinephrine (NE) fibers 64
norepinephrine (NE), see Noradrenaline
normethanephrine 32
northern analysis 265
nuclease protection assay 265
nucleus accumbens 6, 50, 65, 68, 79, 136, 187, 264
– shell 79

nucleus basalis of Meynert 79
nucleus ruber 44

olanzapine 223, 334
olfactory neurons 246
olfactory tubercle 6, 77, 137, 264
open field test 168
opiates 33, 169
opioid peptides 97
osmium tetroxide 64
oxidative stress 171
oxytocin 91

pain 98
pallidum, ventral 79
parabrachial pigmented nuclei 43
paranigral nucleus 43
parathyroid hormone 235
paraventricular nucleus 91
Parkinsonism, reserpin induced 4, 26
Parkinsonism, postencephalitic 31
Parkinson's disease 104, 142, 159, 170, 267, 281, 299
– unilateral 10
parvalbumin 87
pathways, nigrostriatal 44
pathways, striatonigral 44
pertussis toxin 133
phenoxybenzamine 32
phenylethanolamnie N-methyltransferase (PNMT) 81
phosphatidylinositol 322
phosphoinositide (PI) 145
phospholipase A2 163, 310
phospholipase C (PLC) 238
photoreceptor 95
pimozide 196, 275
pipotiazine 196
pituitary (gland) 90, 171, 225
pituitary hormones 98
pituitary tumors 159, 172
pituitary, anterior 6
plexiform layer 93
PNMT antibodies 82
POU transcription factor 122
3-PPP 34
pramipexole 210
premotor regions 53
prochlorperazine 196
progesterone 171
progressive supranuclear palsy 273
projection system, striato-nigro-striatal 50
projection, central striatum 49
projection, striatonigral 48
projections, corticostriatal 52

Subject Index

projections, corticothalamic 52
prolactin 171
prolactinoma 172
proopiomelanocortin (POMC) 172
protein kinase A 136, 239, 322
protein phosphatase 1 (PP1) 241
projection, dorsolateral striatal 48
pseudo-rabies-virus 86
psychosis 173
psychostimulants 201
psychotic reaction 13
puncta adherentia 102
putamen 5, 68
pyramidal neurons 85

quinpirole 142, 275

raclopride 196, 223, 227, 275
receptor surface, intracellular 132
receptors, presynaptic 34
red nucleus 44
remoxipride 196, 223
reserpine 24, 167, 309
reserpine syndrom 4
retina 137, 225
retinitis pigmentosa 202
retinoic acid responsive element 160
retinoid X receptor 161
retrorubral cell groups 43
reward 33, 43
rhinal cortex 83
ring test 168
risperidone 196, 223, 334
RNA polymerase II complex 323
rotarod apparatus 167

SCH 23390 130, 136
SCH 23892 136
schizophrenia 14, 35, 127, 159, 173, 227, 281
semiquinone 170
sensorimotor gating 200
septal nucleus 55
serine 223
serotonin 209
serotonin N-acetyltransferase 225
sertindole 196
seven transmembrane domain (7TM) 161
sexual behaviour 276
signal detection 88
signal transduction cascades 323
SK & F 89145 130
SK & F 89615 130
SKF 38393 129
sniffing 140

somatostatin 82
somatostatin receptor 164
spectrophotofluorimeter 24
spinal cord 98
spinophilin 241
spiny dendrites 72
spiperone 223
stabilizers, dopaminergic 36
stathmin 240
stochastic resonance 103
stress 209
stress-activated protein (SAP) kinase 329
stria terminalis 55
stria terminalis
– bed nucleus 79
– septal complex 80
striatal region
– cognitive 52
– limbic 52
– motor 52
striatonigral fibers, ventromedial 49
striatum 44, 102, 264
striatum, dorsolateral 50
striatum, functional regions 45
striatum, ventral 34
striatum, ventromedial shell 45
striatum, ventromedial 50
subiculum 52
substance P 72, 169, 187, 204, 302
substantia innominata 79
substantia nigra, pars compacta (SNC) 43
– dorsal group 44
– densocellular region 44
– cell columns 44
subthalamic nucleus 167
sucrose 268
sulpiride 196, 223
sultopride 196
superoxide dismutase 170
suprarhinal area 83
synapses, axo-axonic 85
synapses, axodendritic 70, 81
synapses, axosomatic 81, 85
synapses, axospinous 68, 70
synapses, symmetric 68
synaptic incidence 84
synaptic plasticity 310
synaptic transmission 304, 307
synaptic vesicle proteins 323
synaptic vesicles 29
synaptology, specific differences 77

tanycytes 90, 243
tardive dyskinesia 334

testosterone 171
tetrodotoxin 245
thalamus 72, 166
thienoazepine 140
thienopyridine 130
thiophenanthrene 130
thioridazine 196
threonine 223
thyroid hormone 161
thyrotropin releasing hormone (TRH) 172
Tourette's syndrome 127, 159, 227, 273, 299
transcription factor 323
transcription initiation site 323
transcriptional regulation 159
transmission, cholinergic 73
tranylcypromine 207
tuberoinfundibular pathway 89, 176, 265

tyramine 269
tyrosine hydroxylase (TH) 65, 267
tyrosine kinase 223

Ungerstedt's rotation model 193

vacuous chewing 140
vasoactive intestinal peptide (VIP) 172, 240
ventral periventricular nucleus 89
ventral striatum 76
ventral tegmental area (VTA) 43, 67, 266
ventral tier 46
vitamin D3 receptor 160
vitamin E 170
voltage-clamp studies 301, 312
voltage-clamp studies 312

zona incerta 265

Printing (Computer to Film): Saladruck Berlin
Binding: Stürtz AG, Würzburg